Saline Water and Gases in Crystalline Rocks

Edited by

P. Fritz
Department of Earth Sciences, University of Waterloo,
Waterloo, Ontario
Canada N2L 3G1

S.K. Frape
Department of Earth Sciences, University of Waterloo,
Waterloo, Ontario
Canada N2L 3G1

Geological Association of Canada Special Paper 33

1987

Main entry under title:
Saline waters and gases in crystalline rocks

(Geological Association of Canada special paper: 33)
Includes index.
Bibliography: p.
ISBN 0-919216-32-3

1. Rocks, Crystalline. I. Fritz, P. (Peter), 1937- II. Frape, Shaun K., 1949- III. Geological Association of Canada IV. Series.

QE461.S24 1987 552′.4 C87-090080-3

Geological Association of Canada
Department of Earth Sciences,
Memorial University of Newfoundland,
St. John's, Newfoundland
Canada A1B 3X5

The GEOLOGICAL ASSOCIATION OF CANADA is Canada's national society for the geosciences. It was established in 1947 to advance geology and its understanding among both professionals and the general public. The GAC membership of 3000 includes representatives of all geological disciplines from across Canada and many parts of the world employed in government, industry and academia. There are specialist divisions for environmental earth sciences, geophysics, mineral deposits, paleontology, Precambrian, sedimentology, tectonics, and volcanology. Regional sections of GAC have been set up in Victoria, Vancouver, Edmonton, Winnipeg, and St. John's, and there are affiliated groups in Toronto and the Maritimes.

GAC activities include the organization and sponsorship of conferences, seminars, short courses, field trips, lecture tours, and student and professional awards and grants. The Association publishes the quarterly journal *Geoscience Canada* and the quarterly newsletter *Geolog*, a *Special Paper* series, Short Course Notes, and several other continuing series. GAC also maintains liaison with other earth science societies and provides advice to government and the public on geological issues. The Association was incorporated under the Canada Corporations Act in January, 1984.

For information contact: Geological Association of Canada, Department of Earth Sciences, Memorial University of Newfoundland, St. John's, Newfoundland A1B 3X5, Canada.

L'ASSOCIATION GÉOLOGIQUE DU CANADA est la société nationale canadienne pour les sciences de la Terre. Créée en 1947, elle a comme double objectif de faire progresser la géologie et de sensibiliser les spécialistes et les membres du grand public aux sciences de la Terre. Ses trois mille membres représentent toutes les disciplines géologiques; ils viennent de toutes les régions du Canada et de nombreux autres pays; ils oeuvrent dans le secteur public, dans le secteur industriel et dans le monde universitaire. L'Association comprend des divisions de spécialistes en géophysique, en gisements minéreaux, en paléontologie, en Précambrien, en sédimentologie, en tectonique, en volcanologie et en sciences de la Terre touchant à l'environnement. Des sections régionales existent à Victoria, Vancouver, Edmonton, Winnipeg et St. John's, et des groupes affiliés se sont constitués à Toronto et dans les provinces maritimes.

Les activités de l'Association géologique du Canada comprennent l'organisation et le parrainage de conférences, de colloques, de cours de formation de courte durée, de visites sur le terrain et de tournées de conférences. Elle décerne des octrois et des bourses aux étudiants et aux personnes qui travaillent dans le domaine des sciences de la Terre. L'Association publie un journal trimestriel, *Geoscience Canada*, et un bulletin trimestriel d'information, *Geolog*, une série de mémoires, des notes de cours et diverses autres séries de publications. Elle assure en outre la liaison avec d'autres sociétés en sciences de la Terre et fournit des conseils au gouvernement et au grand public sur des questions géologiques. L'Association a été constituée en corporation en janvier 1984, en vertu de la Loi sur les corporations canadiennes.

Pour renseignements: Association géologique du Canada, Department of Earth Sciences, Memorial University of Newfoundland, St. John's (Terre-Neuve), A1B 3X5, Canada.

Design: M.J. Kiel
Typesetting: Graph Comp Design Ltd., Ottawa
Printing: The Runge Press Limited, Ottawa

TABLE OF CONTENTS

Preface

The permanent removal of toxic materials from the biosphere has become a major task for industrial societies. It was against this background and with the financial backing of governments interested in the storage or permanent disposal of waste in crystalline rocks that most studies reported upon in this volume were initiated.

Crystalline rocks are often perceived as being impervious and non-reactive, i.e. suitable for permanent containment of toxic or radioactive waste. However, hydrogeological and geochemical studies in granitic rocks in Sweden, in the crystalline Canadian Shield and, more recently in the Soviet Union, have clearly revealed that such *a priori* assumptions were not justified. Instead, many deep boreholes drilled as part of mineral exploration projects yielded substantial amounts of fluids and gases. It was also somewhat unexpected that many of the discharging fluids were hypersaline brines with salt contents typically above 200 grams per litre. Furthermore, in many situations active rock-water interactions with moving fluids is very probable: crystalline rock does not present itself as a non-reactive, silent partner in disposal schemes.

A further surprise was the widespread occurrence of gas that discharges from dry holes or in association with the brines. Almost invariably methane is a dominant component. As gas migration from waste repositories is of potential importance, a closer investigation of the nature and origin of these gases was warranted. However, these studies may also be relevant in the context of the origin of economically significant hydrocarbon deposits where mantle or crustal contributions are sometimes thought to be essential.

This volume is an attempt to summarize the research efforts which deal with brines and gases in crystalline rocks. It is the result of a symposium held during the 1985 Annual Meeting of the Geological Association of Canada in Fredericton, N.B. under the sponsorship of Atomic Energy of Canada Ltd., the International Association of Hydrologists and the International Association of Hydrological Sciences, and with financial support from the National Engineering and Sciences Research Council of Canada.

The Geological Association of Canada acknowledges the financial support for publication of this volume received from the Natural Sciences and Engineering Research Council, the Canadian Geological Foundation and Atomic Energy of Canada Ltd. This support is greatly appreciated and permits continued production of the Special Paper series. The index was compiled by Patricia Sheahan and the abstracts were translated into French by Pascale Côté.

All contributions in this volume were reviewed, as this was an essential element in the preparation. Specifically we wish to acknowledge with thanks the critical comments and contributions by: T.E.W. Appleyard (Canada), H. Baadsgaard (Canada), J.F. Barker (Canada), D. Bottomley (Canada), O. Brotzen (Sweden), J.A. Cherry (Canada), S.N. Davis (USA), W.M. Edmunds (UK), H.P. Eugster (USA), E. Galimov (USSR), M. Gascoyne (Canada), W. Giggenbach (NZ), B. Hitchon (Canada), W. Kelly (USA), Y. Kharaka (USA), H.H. Loosli (Switzerland), G. Matthess (Germany), R.H. McNutt (Canada), C. Palmer (USA), F.M. Phillips (USA), E.J. Reardon (Canada), E. Roedder (USA), M. Schoell (USA), D.M. Shaw (Canada), Y. Tardy (France), T.T. Vandergraff (Canada), R.O. VanEverdingen (Canada), J. Veizer (Canada), J. Welhan (Canada) and A. Zuber (Poland).

P. FRITZ
S.K. FRAPE

The Fluid Inventory of the Crust and its Influences on Crustal Dynamics

W.S. Fyfe
Department of Geology, University of Western Ontario, London, Canada N6A 5B7

Fyfe, W.S., The fluid inventory of the crust and its influences on crustal dynamics; in Saline water and gases in crystalline rocks, Editors: Fritz, P., and Frape, S.K.; Geological Association of Canada Special Paper 33, 1987. p. 1–3.

Abstract

Quantifying the reservoirs and fluxes of the most volatile and soluble phases of the earth is one of the major tasks of modern geochemistry. As well as playing a leading role in influencing the stability and changes of the environment, such fluids are significant in the mechanics of the lithosphere and influence deep electrical properties and the fluids and their pathways influence the propagation of seismic waves. Deep recycling of fluids is more important than previously recognized as such recycling is demanded by studies in deep trenches of the Japan type where even sediments are missing. The inventory of volatile-soluble species is thus important to our understanding of many processes from lithosphere dynamics to toxic waste disposal.

Résumé

L'évaluation quantitative des aquifères et de l'écoulement des phases les plus volatiles et les plus solubles de la terre est un des principaux objets d'étude de la géochimie moderne. En plus de jouer un rôle déterminant sur la stabilité ou les changements de l'environnement, ces fluides exercent une influence appréciable sur les propriétés mécaniques de la lithosphère et, dans une moindre mesure, sur les propriétés électriques en profondeur. De plus, la trajectoire des fluides influe sur la propagation des ondes sismiques. Le recyclage des fluides en profondeur est un phénomène beaucoup plus important qu'on ne l'avait d'abord cru puisqu'il est d'un intérêt capital dans le cadre d'études effectuées dans les fosses sous-marines profondes et dépourvues de sédiments, du genre de celles du Japon. Par conséquent, l'inventaire des phases volatiles ou solubles est une étape importante de la compréhension de nombreux processus allant de la dynamique de la lithosphère à l'enfouissement des déchets toxiques.

It was a great honour to be invited to write an overview of some aspects of this symposium. Over the past two decades there have been very important advances in our knowledge of the interactions between the hydrosphere (and in part atmosphere and biosphere) and the earth's crust and even with the deeper mantle. In a sense we live on a water cooled planet. We know very well that in regions where the thermal gradient much exceeds 50°C km^{-1}, hot spring activity appears and as Straus and Schubert (1977) showed, the adiabatic gradient for convective flow in a porous-cracked medium is exceeded even at average crustal thermal gradients. In particular the study of cooling processes at ocean ridges has been shown to involve fluid transport of a very significant fraction of the thermal energy of the earth. Overall, perhaps something like a third of the total global energy production is carried out of the crust by fluid motions. This process loads the outer 10 km or so of the crust with fluids and volatile-bearing mineral phases and the total volatile inventory in the solid crust is similar to that of the great ocean reservoir.

As studies of the structure of continental crust have accelerated (e.g. Oliver, 1982; Mathews, 1982) the influence of fluids even in deep crust increasingly attracts attention. At a recent meeting on the lower crust held in Durham

England (1984) seismologists reported on deep rocks with "rather low" seismic velocities. Prof. William Brace of M.I.T. discussed evidence on deep permeability and concluded that it was possibly much higher than we imagined. The deep Kola bore in the U.S.S.R. has revealed the same surprise (Kozlovsky, 1984). And those who work with deep electrical conductivity have shown that frequently deep crust is highly conductive and saline fluids or small amounts of partial melt are suspected (Gough, 1981). Further the seismologists are worred about the nature of the reflectors they see in vibroseis techniques and in particular the so-called "bright spots" that may indicate concentrations of trapped fluids.

One of the great tasks of modern earth science is to quantify the geochemical processes which control the more mobile or soluble components and the classic volatile species of the planet (CO_2, N_2, H_2, C-H gases, S gases, NH_3, rare gases, halogen salts, sulphates, etc.). The unsatisfactory state of present knowledge is clearly shown in Holland's (1984) recent synthesis. The quantities and fluxes of such species between major geospheres are often poorly quantified and before there are known we will not understand many of the significant perturbations in the surface environment.

The chemistry of deep fluids is often a little surprising. Often there seems to be too much CO_2-CH_4 or salt and strange concentration of isotopes such as ^3He. It is partly because of such observations that there is increasing interest in the necessity of deep drilling for scientific purposes and in part the growing necessity of finding sites for toxic waste disposal in the crust has accelerated the search for *exact* knowledge on the nature and flow of fluids in the crust. During the recent Alfred Wegener conference on continental drilling (Seeheim, Germany, 1985) we were shown results from initial drilling in two possible German sites for deep holes in crystalline rocks. What was to me impressive, was the frequency of discontinuities in these rocks; shear zones, vein swarms, gas breccias, etc. The volumes of isotropic domains were small. Deep drilling in the Cyprus ophiolite has revealed the same features with alteration and wide aperture veins being almost as common in the deepest parts of the complex as near the surface. There is thus great need, for both practical and scientific reasons, to improve our knowledge of solid-fluid behaviour in the crust, the major theme of this symposium, and we must start near the more accessible surface.

FLUID-SOLID EARTH INTERACTIONS AND TECTONICS

As the processes of plate tectonics are better understood through studies of structure and heat flow in the crust, the role of volatile phases in the processes has steadily increased. We have learned that in ridge cooling processes the entire ocean mass is processed through the basaltic crust in a few million years modifying the chemistry of both oceans and crust (Wolery and Sheep, 1976; Fyfe and Lonsdale, 1981). The process loads the mantle-derived crust with volatiles such as H_2O, CO_2, S on a large scale and possibly even with salt. Recently it has even been suggested that calcium cycling in this process may significantly perturb atmospheric CO_2 and contribute to "greenhouse" effects (Owen and Rea, 1985).

One of the most important sea floor processes in terms of the water inventory may be serpentinization. Decades ago, Hess (1962) pointed out that serpentinites or partially serpentinized peridotites are one of the ubiquitous materials of sea floor rocks. New seismic observations in the sea floor (Sharman, 1985) and studies on the rate process (MacDonald and Fyfe, 1985) indicate that serpentinites may be common down to the oceanic Moho. Observations from the Cyprus drilling project will certainly confirm such an idea. Thus the water content of the ocean crust may be much greater than once thought. But there is another most important aspect of this process. The serpentine minerals which are clay like in structure also act as good semipermeable membranes for salt so a process may occur:

$$\text{Peridotite} + \text{sea water} \rightarrow \text{serpentinite} + \text{salt}\ ,$$

brines may be produced. J. Vibetti et al. (1985) has found very saline (50% salt) brines in some of the fluid inclusions from deeper alteration zones in Cyprus. This process suggests a much more general phenomenon. Whenever mildly saline waters interact with rocks which are in the process of hydration (e.g. dilute solution plus gabbro → greenschist + salt) saline brines with very low water fugacity must result unless the salt can be accomodated in appropriate minerals (e.g. halogen fixation in amphiboles, etc.). One would expect that deep waters, particularly those where an original marine or mixed source is possible, would tend to become more saline as slow hydration reactions occur. Does this process contribute to some of the high deep conductivity anomalies?

Peridotites are very common rocks of oceanic and crustal lithosphere and their hydration products are equally common. Along with clay rich sediments they are some of the most simple water carriers of the outer crustal layers. Once hydrated, various P-T changes may lead to their dehydration which occurs over very narrow intervals because of chemical and mineralogical simplicity. When such processes occur, because the fluid release is so large, highly porous or conductive zones may be created over a very wide range of depths. Processes in such materials may produce many of the discontinuities observed at depth (Fyfe, 1985).

One of the laws of plate tectonics is that the new lithosphere created at ridges is almost quantitatively subducted (see Fyfe, 1983). This means that the volatile loaded spilitic-serpentinite crust moves back into the mantle. There is also increasing evidence that a considerable volume of pelagic sediments may be subducted (see Hilde and Uyeda, 1983) and recent observations in the Japan trench (Kaiko Project Staff, 1985) show that not only are sediments subducted but the slumped trench wall debris is subducted. Thus volatiles are being recycled towards the mantle. How much, how far, is not well understood but few would doubt that the genesis of andesite magmas is related to such deep volatiles (see Thorpe, 1982). Canadian lithoprobe studies on the Vancouver Islands, will eventually provide more precise information on this process (see Clowes et al., 1984). It is interesting to note that in his recent synthesis, Holland (1984) does not consider subductive recycling in any serious manner (see p. 180) even though the supporting facts are growing. Even the classic primoidial component ^3He is perhaps suspect as Ozima et al. (1984), Sano et al. (1985) have shown that this species from extraterrestrial sources is concentrated in pelagic clays and Fe-Mn particles. Present evidence indicates that the ocean mass, and oceanic sulphate and bicarbonate, is probably recycled in less than a billion years.

Another process of profound importance involves continental collision of the Himalayan type, a process which modern crustal seismic studies steadily shows may have been one of the most common in building continents by sweeping together pieces of light crust. In the Himalayan collision a continental area comparable in size to Australia has been underthrust to form a great band of thick, 60 km + crust. The fluids released and mobilized in such a process must be enormous. The process in a sense is like taking a large continental mass, heating it by about 300°C and compressing it by loading with 30 km of overburden. As we have indicated (Fyfe and Kerrich, 1985; Fyfe, 1986), the H_2O mass mobilized will be similar to that of the ice caps. But if large masses of carbonates or salt are involved in the underthrust stratigraphy, CO_2-NaCl release may be massive and the same is true for H-C-N organics. Such fluids will rise into the overthrust by hydraulic fracture (shear zone) mechanisms and while some may escape, some may load the entire overthrust with volatiles. We do not understand such processes but when thrust processes are considered, the initial fluid state of normal high level crust is a fundamental piece of knowledge required. It is such fluids that will be moved easily in any tectonic process.

There are other common processes that may mobilize crustal fluids on large scales (e.g. magma underplating, Herzberg et al., 1983; Fyfe, 1986; strike slip faulting with override, Allis, 1981).

What is important is that when a crust with volatiles, fixed in mineralogy or in pores and fractures, is perturbed by a pressure or temperature change or change in regional stress patterns, some mobilization of fluids is certain. In terms of the outer fluid-saturated layers this is of great importance when one considers any long term waste disposal scenario and, at extreme latitudes, even ice loading must be considered as a key variable in the dynamics. At this time our knowledge of the inventory, and our knowledge of volatile dynamics is not adequate.

REFERENCES

Allis, R.G., 1981. Continental underthrusting beneath the Southern Alps of New Zealand: Geology, v. 9, p. 303-307.

Clowes, R.M., Green, A.G., Yorath, C.J., Kanasewich, E.R., West, G.F. and Garland, G.D., 1984. Lithoprobe – a natural program for studying the third dimension of geology; Journal of the Canadian Society of Exploration Geophysicists, v. 20, p. 23-39.

Fyfe, W.S. and Kerrich, R., 1985. Fluids and thrusting: Chemical Geology, v. 49, p. 353-362.

Fyfe, W.S. and Lonsdale, P., 1981. Ocean floor hydrothermal activity: in The Oceanic Lithosphere, The Sea vol. 7 (C. Emiliani, editor), John Wiley, p. 589-638.

Fyfe, W.S., 1983. Subduction and the geochemical cycle: Tectonophysics, v. 99, p. 271-277.

Fyfe, W.S., 1985. Fluids, tectonics and crustal deformation: Tectonophysics, v. 119, p. 29-36.

Fyfe, W.S., 1986. Fluids in deep continental crust: Journal of Geophysical Research (in press).

Gough, D.I., 1981. Magnetomer arrays and geodynamics: In Evolution of the Earth's crust (R.J. O'Connell and W.S. Fyfe, editors), Geodynamics Series, v. 5, American Geophysical Union, p. 87-95.

Herzberg, C.T., Fyfe, W.S. and Carr, M.J., 1983. Density constraints on the formation of the continental Moho and crust: Contributions to Mineralogy and Petrology, v. 84, p. 1-5.

Hess, H.H., 1962. Serpentinites, orogeny and epeirogeny: In Crust of the Earth (A. Poldervaart, editor), Geological Society of America, Special Paper v. 62, 762 p.

Hilde, T.W.C. and Uyeda, S., 1983. Convergence and subduction: Tectonophysics special volume, v. 99, p. 85-400.

Holland, H.D., 1984. The chemical evolution of the atmosphere of oceans: Princeton University Press, 582 p.

Kaiko Project Staff, 1985. Japanese deep-sea trench survey: Nature, v. 313, p. 432-433.

Kozlovsky, Y.A., 1984. The world's deepest well: Scientific American, v. 251, no. 6, p. 98-104.

MacDonald, A.H. and Fyfe, W.S., 1985. Rate of serpentinization in seafloor environments: Tectonophysics, v. 116, p. 123-135.

Matthews, D.H., 1982. BIRPS: deep seismic reflection profiling around the British Isles: Nature, v. 298, p. 709-710.

Oliver, J., 1982. Changes at the crust-mantle boundary: Nature, v. 299, p. 398-399.

Owen, R.M. and Rea, D.K., 1985. Sea-floor hydrothermal activity links climate to tectonics: the Eocene carbon dioxide greenhouse: Science, v. 227, p. 166-169.

Ozima, M., Takayanagi, M., Zashu, S. and Amari, S., 1984. High $^3He/^4He$ ratio in ocean sediments: Nature, v. 311, p. 448-450.

Sano, Y., Yoyoda, K. and Wakita, H, 1985. $^3He/^4He$ ratios of marine ferromanganese nodules: Nature, v. 317, p. 518-520.

Sharman, F.G., 1985. Oceanic lithosphere: origin structure and dynamics: Special Volume Tectonophysics, 116.

Straus, J.M. and Schubert, G., 1977. Thermal convection of water in a porous medium: effect of temperature and pressure dependent thermodynamic and transport properties: Journal of Geophysical Research, v. 82, p. 325-333.

Thorpe, R.S., 1982. Andesites: orogenic and related rocks: Wiley, New York, 724 p.

Vibetti, N.J., Kerrich, R. and Fyfe, W.S., 1985. Presence of Ca-Na dominated hypersaline brines in quartz-calc silicate veins of Troodos ophiolite (Cyprus): December meeting, American Geophysical Union San Francisco, abstract.

Wolery, T.J. and Steep, N.H., 1976. Hydrothermal circulation and geochemical flux at mid-ocean ridges: Journal of Geology, v. 84, p. 249-275.

Normative Salt Characterization of Natural Waters

Blair F. Jones
U.S. Geological Survey, 432 National Center, Reston, Virginia 22092, U.S.A.

Marc W. Bodine, Jr.
U.S. Geological Survey, MS 939, Denver Federal Center, Denver, Colorado 80225, U.S.A.

Jones, Blair F. and Bodine, Marc W. Jr.; Normative salt characterization of natural waters; in Saline water and gases in crystalline rocks, Editors: Fritz, P., and Frape, S.K.; Geological Association of Canada Special Paper 33, 1987. p. 5–18.

Abstract

The salt norm is the quantitative idealized equilibrium salt assemblage that would crystallize if a natural water evaporated to dryness at 25°C and 1 bar pressure under atmospheric partial pressure of CO_2*. The salt norms for a wide variety of natural waters were calculated using the new computer program SNORM. SNORM distributes solutes into stable normative salt associations based on the Gibbs Phase Rule, available free energy values, and observed low-temperature mineral associations.*

Although most natural water compositions represent multiple solute origins, three major categories can readily be identified: meteoric or weathering norms that are characterized by alkali-bearing sulfate and carbonate salts; marine-like norms with the familiar seawater-type halite-bischofite-carnallite-kieserite-anhydrite association; and diagenetic norms that yield salts, such as Ca-bearing chlorides (antarcticite and tachyhydrite) and sylvite, which suggest solute alteration by secondary mineral reactions. The solute source or derivation process within each of the above categories commonly is indicated by the presence or absence of diagnostic normative salts and their relative abundance in the normative salt assemblage. For example, salt norms may (1) identify lithologic source; (2) identify the relative roles of carbonic- and sulfuric-acid hydrolysis in the evolution of weathering waters; (3) identify the origin of connate waters from normal marine, hypersaline, or evaporite salt resolution processes; and (4) distinguish between dolomitization and silicate hydrolysis or exchange for the origin of diagenetic waters.

Résumé

La norme saline est la représentation quantitative théorique de l'assemblage salin qui cristalliserait à l'équilibre durant l'évaporation complète d'une eau naturelle à 25 °C, sous une pression de 1 bar et une pression atmosphérique partielle de CO_2*. La norme saline d'une grande variété d'eaux naturelles a été calculée à l'aide du nouveau logiciel SNORM. Celui-ci distribue les solutés en associations stables de sels normatifs en se basant sur la Loi des phases de Gibbs, les valeurs d'énergie libre disponible et les associations minéralogiques observées à faible température.*

Bien que la plupart des eaux naturelles renferment des solutés d'origine diverse, il est possible de distinguer trois catégories principales: (1) les normes météoriques ou d'altération, caractérisées par la présence de sulfates alcalins et de sels de carbonates; (2) les normes de type marin, caractérisées par l'association halite-bischofite-carnallite-kieserite-anhydrite propre à l'eau de mer et (3) les normes diagénétiques libérant des sels, tels que les chlorures de calcium (antarcticite et tachyhydrite) et la sylvite qui associent l'altération des solutés à des réactions minéralogiques secondaires. La présence ou l'absence de sels normatifs caractéristiques et leur abondance relative dans l'assemblage salin normatif indiquent générale-

ment la source des solutés ou le processus de dérivation de chacune de ces catégories. On peut, par exemple, utiliser les normes salines pour: (1) identifier la lithologie de la source; (2) identifier les rôles respectifs de l'hydrolyse acide de nature carbonique ou sulfurique dans l'évolution des eaux d'altération; (3) distinguer l'origine des eaux connées normales de celle des eaux marines, sursalées ou des processus de redissolution des sels évaporitiques et (4) choisir entre la dolomitisation et l'hydrolyse, ou l'échange, des silicates pour expliquer l'origine des eaux diagénétiques.

INTRODUCTION

Chemical composition plays a central role in the classification of natural waters and interpretation of their source and evolution. Water chemistry has become increasingly important to our understanding of a broad spectrum of hydrologic and geologic processes from flow regimes at waste disposal sites and aquifer dynamics in shallow environments, to diagenesis, ore transport and deposition, and migration and maturation of fluid hydrocarbons in deep-basin environments.

In this paper we offer a quantitative chemical-mineralogical method of characterizing natural waters based on a recasting of the solutes of a conventional water analysis into an idealized 25°C equilibrium assemblage of mineral salts. We suggest that characterization of a water's composition as an assemblage of salts leads to a more detailed chemical classification of natural waters and hopefully provides indications of solute source and subsequent interactions.

Dissolved constituents in natural waters expressed as salt abundances are not new. In the nineteenth century, it was a common practice to report water compositions in terms of abundances of simple salts. As Hem (1970) points out, this practice predated the acceptance of the Arrhenius concept of dissociated ions, and, although it also attempted to express water composition in terms of the salts produced upon evaporation, it was actually more closely related to the classical gravimetric analytical procedures than to the natural occurrence of mineral salts. Such schemes relied chiefly on forming simple salts with little regard to salt speciation and association. It is not surprising, then, that this form of expressing water compositions diminished in the twentieth century with only occasional use for specific purposes in the more recent literature (see for example, Rankama and Sahama, 1950, p. 318; Lambert, 1978.)

Comparison and study of both marine and nonmarine evaporite deposits, on the other hand, have demonstrated some striking differences in the associated salt mineral assemblage (Stewart, 1963; Braitsch, 1971; Jones, 1970; Eugster, 1980; Sonnenfeld, 1984), that commonly are indicative of both lithologic origin and subsequent geochemical evolution of the dissolved constituents (Hardie and Eugster, 1970; Eugster and Hardie, 1978; Eugster and Jones, 1979; Eugster and others, 1980). Thus, it would seem potentially informative to attempt reconstruction of the solute content of a natural water into the equilibrium salt assemblage expected were the solution evaporated to dryness under earth-surface conditions.

This salt assemblage is analagous to the CIPW norm (Cross and others, 1902), which is an idealized equilibrium assemblage of igneous minerals calculated from the rock's chemical composition. The CIPW norm has proven useful in igneous petrology for characterizing and classifying igneous rocks and providing invaluable quantitative data for rigorous interpretation of the origin and evolution of an igneous complex. We suggest a parallel application of the idealized equilibrium salt assemblage – the salt norm, which is calculated from a water's chemical composition. Indeed, its similar construction and association permits the general classification of any natural solution based on saline mineral assemblages. Such a classification is more detailed and suggestive of solute origin and subsequent interaction than major cation-anion predominance currently, the most commonly used system of hydrochemical classification.

Transformation of a standard water analysis into the normative salt assemblage is performed by the new computer program SNORM. Preliminary results from SNORM for a wide variety of water compositions suggest that a few key normative salts are diagnostic of the solute source, such as meteoric as opposed to connate marine waters, and the relative abundance of selected normative salts are indicative of the character of water-rock interaction in subsurface environments, such as silicate hydrolysis in a crystalline host as opposed to dolomitization in a carbonate matrix. Examples of salt norms and their interpretation will be presented and discussed following a brief outline of the SNORM program. A complete description and rationale for SNORM is given elsewhere (Bodine and Jones, 1986).

SNORM

SNORM is a computer program written specifically to compute the normative salt assemblage from the solute concentration data in a conventional water analysis. The program transforms 18 solutes (table 1) into a normative salt assemblage after appropriate adjustments. The assemblage is constituted from a listing of 63 salts (table 2) that are permissable normative salts. Although SNORM computes the normative assemblage directly from the solutes without proceeding along a succession of reaction paths and salt precipitation, the salt norm can be visualized as the solid residuum that coexists with an infinitely small, last vestige of solution upon evaporation of the sample to dryness at 25°C.

Table 1

Solutes in water analyses that SNORM recasts into the normative salt assemblage

	CATIONS		ANIONS	
Major				
	Mg^{2+}	Na^{+}	Cl^{-}	HCO_3^{-}
	Ca^{2+}	K^{+}	SO_4^{2-}	CO_3^{2-}
Minor				
	Li^{+}	Sr^{2+}	Fl^{-}	NO_3^{-}
	NH_4^{+}	Ba^{2+}	Br^{-}	B*
			I^{-}	PO_4^{3-}

* Boron assigned as borate with a charge of $-7/12$ (-0.583) per boron or the charge per boron that is appropriate for the borate stoichiometry of the normative salt in the assemblage.

Table 2

Abbreviations of normative salts in SNORM and their chemical formulae

		(Bi)carbonates			
cal	Calcite	$CaCO_3$	kal	Kalicinite	$KHCO_3$
mag	Magnesite	$MgCO_3$	lca	–	Li_2CO_3
dol	Dolomite	$CaMg(CO_3)_2$	tes	Teschemacherite	NH_4HCO_3
tro	Trona	$Na_3H(CO_3)_2{\cdot}2H_2O$	str	Strontionite	$SrCO_3$
pir	Pirssonite	$Na_2Ca(CO_3)_2{\cdot}2H_2O$	wit	Witherite	$BaCO_3$
		Sulfates			
anh	Anhydrite	$CaSO_4$	blo	Bloedite	$Na_2Mg(SO_4)_2{\cdot}4H^2O$
gyp	Gypsum	$CaSO_4{\cdot}2H_2O$	leo	Leonite	$K_2Mg(SO_4)_2{\cdot}4H_2O$
kie	Kierserite	$MgSO_4{\cdot}H_2O$	pic	Picromerite	$K_2Mg(SO_2)_2{\cdot}6H_2O$
eps	Epsomite	$MgSO_4{\cdot}7H_2O$	aph	Aphthitalite	$K_3Na(SO_4)_2$
arc	Arcanite	K_2SO_4	lso	–	Li_2SO_4
the	Thenardite	Na_2SO_4	mas	Mascagnite	$(NH_3)_2SO_4$
mir	Mirabilite	$Na_2SO_4{\cdot}10H_2O$	cel	Celestite	$SrSO_4$
gla	Glauberite	$Na_2Ca(SO_4)^2$	bar	Barite	$BaSO_4$
syn	Syngenite	$K_2Ca(SO_4)_2{\cdot}H_2O$	bur	Burkeite	$Na_6CO_3(SO_4)_2$
pol	Polyhalite	$K_2Ca_2Mg(SO_4)_4{\cdot}2H_2O$			
		Chlorides			
ant	Antarcticite	$CaCl_2{\cdot}6H_2O$	sal	Salammoniac	NH_4Cl
bis	Bischofite	$MgCl_2{\cdot}6H_2O$	sca	–	$SrCl_2{\cdot}2H_2O$
tac	Tachyhydrite	$CaMg_2Cl_6{\cdot}12H_2O$	sch	–	$SrCl_2{\cdot}6H_2O$
syl	Sylvite	KCl	bca	–	$BaCl_2{\cdot}H_2O$
car	Carnallite	$KMGCl_3{\cdot}6H_2O$	bch	–	$BaCl_2{\cdot}2H_2O$
hal	Halite	$NaCl$	kai	Kainite	$KMgClSO_4{\cdot}3H_2O$
lic	–	$LiCl_2{\cdot}H_2O$			
		Nitrates			
nic	Nitrocalcite	$Ca(NO_3)_2{\cdot}4H_2O$	nil	–	$LiNO_3{\cdot}3H_2O$
nim	Nitromagnesite	$Mg(NO_3)_2{\cdot}6H_2O$	nia	Ammonia niter	NH_4NO_3
nit	Niter	KNO_3	nis	–	$Sr(NO_3)_2$
sod	Soda niter	$NaNO_3$	nib	Nitrobarite	$Ba(NO_3)_2$
		Borates			
iny	Inyoite	$Ca_2B_6O_{11}{\cdot}13H_2O^+$	bor	Borax	$Na_2B_4O_7{\cdot}10H_2O^+$
ind	Indirite	$Mg_2B_6O_{11}{\cdot}15H_2O^+$	ule	Ulexite	$NaCaB_5O_9{\cdot}8H_2O^+$
		Fluorides			
flu	Fluorite	CaF_2	vil	Villiaumite	NaF
sel	Sellaite	MgF_2	lif	–	LiF
		Phosphates			
hap	Hydroxyapatite	$Ca_5(PO_5)_3OH$§	fap	Fluorapatite	$Ca_5(PO_5)^3F$
mgp	–	$Mg_3(PO_4)_2$	wag	Wagnerite	Mg_2PO_4F
nap	–	Na_3PO_4			

§ Normative hydroxyapatite calculated as $Ca_{4.75}(PO_4)_{3.17}$.

$^+$ If calculated with average borate charge of $-7/12$ per boron; mineral compositions calculated as: inyoite, $Ca_{1.79}B_{6.14}O_{11}{\cdot}13H_2O$; ulexite, $Ca_{0.98}Na_{0.98}B_{5.02}O_9{\cdot}8H_2O$; indirite, $Mg_{1.79}B_{6.14}O_{11}{\cdot}15H_2O$, and borax, $Na_{2.28}B_{3.91}O_7{\cdot}10H_2O$.

SNORM performs two major tasks: (1) It reads solute concentration values from a conventional water analysis into the program, recasting these values into other appropriate concentration units, and adjusts the solute concentrations to yield cation-anion charge balance; and (2) determines the equilibrium normative salt assemblage based on the principles of phase equilibria, and assigns quantitatively all solutes into the salts of the determined assemblage.

Solutes

The solutes (Table 1) that SNORM distributes into normative salts (Table 2) were selected to comply with two criteria: (1) The solute, at least occasionally, occurs in natural waters in more than negligible (trace) concentrations; and (2) the solute forms its normative salt(s) through direct combination with other solutes in the analysis without interacting with the aqueous environment or requiring any chemical or charge modification. This latter criterion was adopted to exclude solutes that require interaction with water (hydrolysis, a pH effect; or H_2O dissociation, a redox effect), or a chemical and charge modification, because the character of the solute in the analysis differs from its character in the resultant solid phase. Thus, neither pH nor redox conditions enter directly into SNORM computations and cation-anion charge balance of the adjusted solute concentrations is preserved in the normative salt assemblage.

Normative salts

SNORM assembles the salt norm for each water analysis from a list of 63 normative salts (table 2). The list contains the necessary array of salts to permit representation of the composition of any known natural water by an idealized equilibrium assemblage.

Whenever possible, the normative salts have been assigned compositions that occur naturally as minerals. However, several of the salts in table 2 do not occur naturally, but only as normative salts, because SNORM allows no interaction with host-rock salts, allows no consumption of solutes through neoformation or transformation of clays and other silicates, allows no solid solution in the normative salts except for bromide and iodide in chlorides, and, unlike most natural occurrences, "evaporates" the water to total dryness.

A wide variety of compound (double) salts are present in evaporite and other low-temperature environments; these salts range in abundance from the common dolomite, fluorapatite, and carnallite, to the rare tachyhydrite, kalistrontite, and pirssonite. The occurrence, genesis, and stability of many compound salts are well documented (see, for example, Braitsch, 1971; Eugster, 1980; Eugster and others, 1980; Harvie and others, 1984) and these must be included when forming the normative assemblage. Therefore, we have included all identifiably stable-compound salt minerals that are composed solely of major solutes – that is, compound salts in the Ca-Mg-K-Na-Cl-SO_4-HCO_3-CO_3-H_2O system, as well as several of the common compound-salt minerals, fluorapatite, wagnerite, and ulexite, that contain minor solutes. Of this latter group, apatite and wagnerite are SNORM's only normative calcium and magnesium phosphate minerals.

We have, however, excluded all compound salts that are not found as minerals, and most compound salt minerals that contain a minor solute, such as barytocalcite, $CaBa(CO_3)_2$, neighborite, $NaMgF_3$, and kalistrontite, $K_2Sr(SO_4)_2$, among others. This is done partially to reduce the number of phases in the listing of normative salts, but more importantly, to eliminate uncertainties about these rare minerals. inconclusively indicative of their the stabilities and permissable mineral associations. Furthermore, standard chemical potentials of most compound salts are unknown. A principal reason for referring to the salt norm as a *idealized* equilibrium assemblage of salts is because these compound salts were omitted from the listing; some of these salts are undoubtedly stable at surface conditions.

The salt norm for each sample then consists of the equilibrium assemblage of normative salts from table 2, the abundances of which are quantitatively equivalent to the adjusted solute abundances (concentrations). Assignment of the correct salts to the norm occurs in three steps: (1) The group of salts from table 2 that are eligible candidates based on solute species in the analysis and gross solute concentration relationships are defined; (2) the Gibbs phase rule is applied to determine the number of salts that comprise the assemblage; and (3) the single assemblage of eligible salts that contains only permissable salt associations and is quantitatively equivalent to solute concentrations is identified.

INTERPRETATION OF SALT NORMS

Attempts to interpret the origin of the principal constituents dissolved in natural waters frequently have used ion ratios and graphical techniques. These considerations commonly rely on diagnostic relations between individual solutes, which remain relatively unaffected or follow known trends through subsequent geochemical evolution of the water mass. Such trends and relationships accompanying the extensive concentration of natural surface and subsurface waters have been generalized by Hardie and Eugster (1970), Eugster and Hardie (1978), Carpenter (1978), Eugster and Jones (1979), and Al Droubi and others (1980). These papers emphasize the control of major-element distribution in waters by precipitate mineral mass balances (particularly calcite, gypsum, and halite), and the potential of using constituents conserved in solution for tracing the geochemical history of any water.

The law of mass action requires that cations and anions be added or subtracted from solution in stoichiometric proportion by the congruent dissolution or precipitation of a mineral. Thus, molar proportions of ions in a given water reflect the composition of original simple salt minerals or the products of nonsalt mineral dissolution by natural acid (carbonic from the hydrolysis of CO_2 or sulfuric from oxidation and hydrolysis of sulfides). The computational association of cations and anions into a salt norm, such as is done in SNORM, quantitatively projects an eventual disposition of the dissolved constituents according to the most stable salt assemblage at 25°C and one bar pressure.

Outside a system closed to all components except H_2O, the computation does not directly address the geochemical evolution of a natural water, but much may be inferred from some knowledge of relative mineral solubility sequence, precipitation kinetics, and geologic setting. Such interpretation must allow for the differences in environmental influences likely to accompany the range in solubility between alkaline-earth carbonate, calcium sulfate, and alkali sulfate or carbonate, a spread in solute concentration of three orders of magnitude. Perhaps the most profound consequence of this exercise will be the prediction of further evolution of solute composition out of contact with initial precipitates, particularly those containing more sparingly soluble constituents. However, interpretation becomes questionable for very dilute waters in which solute sources are dominated by aerosol dissolution, biologic degradation, or anthropogenic contaminants. Some effects of such initial contributions may remain despite extensive subsequent reaction in the normal weathering environment.

Hardie and Eugster (1970) have underscored the importance of the calcite and gypsum solubility "divides" in the subsequent solute evolution of concentrating waters. The required removal of Ca and carbonate or sulfate in equal proportions at the time of precipitation assures the continued dominance of whichever of the two constituents is most abundant at that point. As Ca must exceed total CO_3 and SO_4 to prevail at both junctures, it is easy to understand why calcium-dominated chloride waters are relatively uncommon. In fact, the clear excess of equivalent sulfate over calcium in seawater leads to its clear alkali-magnesium, sulfo-chloride character with evaporative concentration. The importance of the two common, relatively insoluble

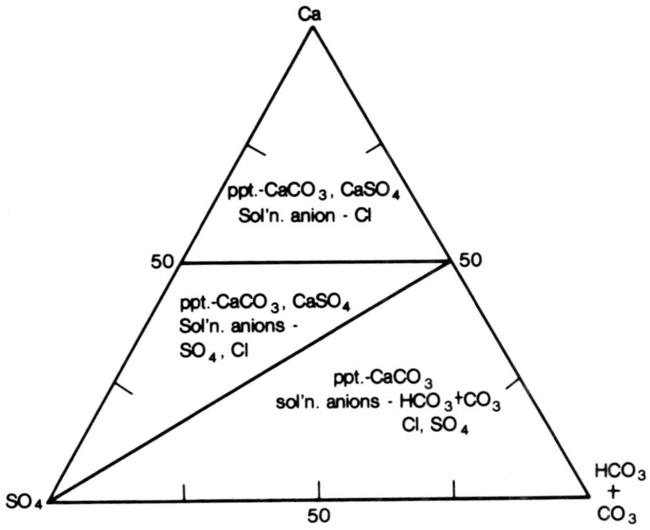

Figure 1. *Relative proportions of calcium, carbonate species, and sulfate in natural waters and character of precipitated salts (from R.J. Spencer, personal communication, 1982)*

calcium salts to the compositional evolution of all natural waters is illustrated by a trilinear diagram (fig. 1) taken from R.J. Spencer (Univ. of Calgary, personal commun. 1982). The plot of water compositions on this diagram will qualitatively predict the major-solute distribution in all saline waters. The calculation of salt norms permits these predictions to be made in a more quantitative way.

Readily discernible differences of genetic importance can be illustrated in the comparison of salt norms (table 3) for seawater and for the average world river compositions given by Livingstone (1963). The abundance of alkaline-earth carbonates, as represented by calcite and dolomite, the presence of alkali-bearing double sulfates glauberite and syngenite, and the relatively minor amount of halite as the sole chloride in the river-water norm differ markedly from the chloride-rich, carbonate-poor seawater assemblage. The contrast illustrates the fundamental difference in ultimate anionic source for reacting solutions (H_2CO_3 vs. HCl), and the more geochemically evolved state of the oceans as compared to most continental waters (MacKenzie and Garrels, 1966). It generally is agreed that the oceans achieved their present chloride concentration very early in earth history by degassing at HCl/H_2O ratios very close to present crustal values (Holland, 1984). Subsequent additions of primordial chloride to the hydrosphere

Table 3

Major solute concentrations and major-ion normative salts for mean compositions of seawater and world river waters
(Seawater from Riley and Chester, 1971; river waters from Livingstone, 1963)

			River waters*				
		Seawater	North America	South America	Europe	Australia	WORLD
Solutes (mg/kg)							
	Mg	1,293.	5.	1.5	5.6	2.7	4.1
	Ca	411.	21.	7.2	31.1	3.9	15.
	Na	10,760.	9.	4.	5.4	2.9	6.3
	K	399.	1.4	2.	1.7	1.4	2.3
	Cl	19,350.	8.	4.9	6.9	10.	7.8
	HCO_3	143.	68.	31.	95.	31.6	58.4
	SO_4	2,709.	20.	4.8	24.	2.6	11.2
Total Dissolved Solids	(TDS)	35,150.	133.4	56.1	173.4	55.2	106.1
Normative salts, anhydrous (wt. %)							
Calcite	$CaCO_3$	–	17.7	29.5	25.0	–	18.2
Dolomite	$CaMg(CO_3)_2$	–	36.9	29.4	34.2	48.6	40.8
Magnesite	$MgCO_3$	0.3	–	–	–	4.9	–
Anhydrite	$CaSO_4$	4.0	6.8	–	24.2	8.3	3.2
Kierserite	$MgSO_4$	6.1	–	–	–	–	–
Syngenite	$K_2Ca(SO_4)_2$	–	2.8	–	–	–	8.7
Glauberite	$Na_2Ca(SO_4)_2$	–	20.5	2.9	2.9	–	10.1
Aphthitalite	$K_3Na(SO_4)_2$	–	–	11.6	–	–	–
Thenardite	Na_2SO_4	–	–	4.3	–	–	–
Bischofite	$MgCl_2$	6.5	–	–	–	–	–
Carnallite	$KMgCl_3$	4.9	–	–	–	8.4	–
Sylvite	KCl	–	–	–	–	5.1	–
Halite	NaCl	78.1	13.6	19.5	9.1	24.6	16.8
	Sum†	99.9	98.3	97.2	95.4	99.9	97.8

* Africa and Asia not listed because potassium not given (Africa) or combined with sodium (Asia).

† TDS includes minor solutes; salt norms total <100% because minor-solute normative salts not listed (chiefly indirite and celestite in seawater, and niter and soda niter in river waters).

Table 4

Salt norms (anhydrous weight percent) of representative water samples
(salts of minor solutes, except total of nitrate salts, omitted)

Sample number		1	2	3	4	5	6	7	8	9	10
TDS Total Diss. Solids (mg/l)		332*	243	229*	634*	3,641*	567*	1,323*	1,052*	148*	113*
Calcite	$CaCO_3$	88.4	53.7	–	–	–	–	0.3	2.3	17.2	7.0
Dolomite	$CaMg(CO_3)_2$	4.7	20.3	71.8	57.0	13.5	10.4	1.7	0.4	30.7	42.8
Magnesite	$MgCO_3$	–	–	1.0	0.2	72.4	65.2	–	–	–	–
Trona	$Na_3H(CO_3)_2{\cdot}2H_2O$	–	–	7.4	–	–	–	89.1	65.9	–	–
Pirssonite	$Na_2Ca(CO_3)_2$	–	–	–	–	–	–	0.6	4.7	43.3	25.1
Burkeite	$Na_6CO_3(SO_4)_2$	–	–	–	–	–	–	–	–	–	–
Anhydrite	$CaSO_4$	–	5.0	–	24.7	1.6	10.8	–	–	–	–
Kieserite	$MgSO_4{\cdot}H_2O$	–	–	–	–	–	–	–	–	–	–
Thenardite	Na_2SO_4	–	–	–	–	–	–	–	–	–	–
Aphthitalite	$K_3Na(SO_4)_2$	0.2	–	1.8	–	–	–	0.5	0.4	0.2	12.1
Glauberite	$Na_2Ca(SO_4)_2$	–	14.0	–	12.6	–	–	–	–	–	–
Syngenite	$K_2Ca(SO_4)_2{\cdot}H_2O$	–	2.6	–	3.7	–	–	–	–	–	–
Polyhalite	$K_2MgCa_2(SO_4)_4{\cdot}2H_2O$	–	–	–	–	–	–	–	–	–	–
Bloedite	$Na_2Mg(SO_4)_2{\cdot}4H_2O$	–	–	–	–	–	–	–	–	–	–
Leonite	$K_2Mg(SO_4)_2{\cdot}4H_2O$	–	–	–	–	–	–	–	–	–	–
Kainite	$KMg(SO_4)Cl{\cdot}3H_2O$	–	–	–	–	–	–	–	–	–	–
Halite	$NaCl$	5.9	4.4	–	1.1	4.6	2.2	5.0	27.5	3.7	8.3
Sylvite	KCl	0.4	–	5.3	–	–	–	1.4	0.5	1.8	1.5
Carnallite	$KMgCl_3{\cdot}6H_2O$	–	–	–	–	3.9	–	–	–	–	–
Bischofite	$MgCl_2{\cdot}6H_2O$	–	–	–	–	–	–	–	–	–	–
Tachyhydrite	$CaMg_2Cl_6{\cdot}12H_2O$	–	–	–	–	–	–	–	–	–	–
Antarcticite	$CaCl_2{\cdot}6H_2O$	–	–	–	–	–	–	–	–	–	–
Nitrate salts		–	–	–	–	3.8	9.8	–	–	–	–

* mg/kg   + Epsomite ($MgSO_4{\cdot}7H_2O$) rather than kieserite.   × Picromerite ($K_2Mg[SO_4]_2{\cdot}6H_2O$) rather than leonite.

1. Groundwater from Castle Hayne Limestone, New Bern, North Carolina (White and others, 1963, table 6).
2. Groundwater from Lillehammer sandstone and conglomerate, SE Norway (Englund, 1983, table 1).
3. Groundwater from Columbia River basalt, Farmington, Oregon (White and others, 1963, table 2).
4. Groundwater from Sioux Quartzite, Sioux Falls, South Dakota (White and others, 1963, table 9).
5. Groundwater from serpentine, Baltimore, Maryland (White and others, 1963, table 2).
6. Groundwater from Peebles Dolomite, Bainbridge, Ohio (White and others, 1963, table 7).
7. Groundwater from Homestake Mine, Lawrence Co., South Dakota (White and others, 1963, table 24).
8. Groundwater from Benton Shale, La Prele, Wyoming (White and others, 1963, table 5).
9. Groundwater from rhyolite, Mebane, North Carolina (White and others, 1963, table 1).
10. Groundwater from granite, McCormack, South Dakota (White and others, 1963, table 1).

compared to the mass of Cl in the oceans or sedimentary rocks can be considered negligible.

The extent of variance in salt norms for continental waters in table 3 can be seen through the differences in assemblage, as well as amounts, for the average river waters of some continents. Note the more sodic North American waters (led by the Mississippi River system) are characterized by abundant glauberite, the more potassic waters from South America (dominated by the Amazon) produce aphthitalite, and the more chloridic Australian waters (principally the River Murray) make normative sylvite. Despite a variance of 3 times in carbonate-species concentration, the four average river norms all contain between 50 and 60 weight percent carbonate minerals. All the river norms contain halite (<20 percent) and only the South America average lacks anhydrite (all sulfates of this assemblage are alkali-bearing).

The contrast in salt norms not only between the average river waters and seawater but also between the river waters themselves, suggests the usefulness of the salt norm in identifying solute sources despite the lack of direct information on the geochemical evolution of the waters. Except for dilute waters deriving their principal solutes from surficial sources, major-solute categories can most readily be related to principal lithologies (Jones, 1966; Garrels, 1967; Hem, 1970; and Drever, 1982). Thus, for the major anions, carbonate waters are associated with the dissolution of limestones or silicate hydrolysis. Sulfate waters reflect the oxidation of reduced sulfur and/or the dissolution of gypsum. Chloride waters are most likely associated with an ultimate solute source in marine salts or hydrothermal systems, though commonly they are recycled and modified in continental basins, such as the Great Salt Lake (Spencer and others, 1985). At the same time, for the major cations, the alkalis are prominent in waters associated with siliceous crystalline rocks, pyroclastics, and the sediments derived therefrom; calcium is dominant in waters from carbonate or plagioclase-rich rocks; magnesium is characteristic of waters interacting with mafic rocks and marine muds; and all the major cations can be important in waters associated with mixed clastic lithologies. Such considerations, particularly for saline waters, can point to normative salt assemblages characteristic of genetic end members.

To test the interpretive value of the normative analysis and to establish the most diagnostic assemblages, we have computed salt norms for more than 500 analyses of fresh and saline waters from a wide variety of environments worldwide. From the final tabulations, we have selected 50 examples (table 4) to illustrate a broad range of normative attributes. Because solute derivation is most readily related to mineral reaction and tends to be of much interest in salinity problems, particular attention was given to various

Table 4 continued.

Sample number		11	12	13	14	15	16	17	18	19	20
TDS (mg/l)		477*	4,111*	1.023*	563*	8,710	371,700	428	905*	3,960	115
Calcite	$CaCO_3$	–	–	0.1	–	–	–	–	7.4	–	1.8
Dolomite	$CaMg(CO_3)_2$	0.9	8.7	24.6	24.3	–	–	–	4.5	5.0	22.9
Magnesite	$MgCO_3$	0.4	–	–	7.8	0.6	0.4	3.4	–	0.1	–
Trona	$Na_3H(CO_3)_2{\cdot}2H_2O$	60.2	15.1	–	–	–	–	–	–	–	–
Pirssonite	$Na_2Ca(CO_3)_2$	–	6.3	–	–	–	–	–	–	–	–
Burkeite	$Na_6CO_3(SO_4)_2$	31.4	30.2	–	–	–	–	–	–	1.5	–
Anhydrite	$CaSO_4$	–	–	70.6	56.8	3.0	1.7	40.8	–	–	8.9
Kieserite	$MgSO_4{\cdot}H_2O$	–	–	–	–	35.8	76.6	28.6⁺	–	–	–
Thenardite	Na_2SO_4	–	–	–	–	–	–	–	52.9	45.6	–
Aphthitalite	$K_3Na(SO_4)_2$	0.3	5.1	–	–	–	–	–	–	0.5	–
Glauberite	$Na_2Ca(SO_4)_2$	–	–	–	6.2	–	–	–	24.9	–	46.0
Syngenite	$K_2Ca(SO_4)_2{\cdot}H_2O$	–	–	–	4.0	–	–	–	4.7	–	8.8
Polyhalite	$K_2MgCa_2(SO_4)_4{\cdot}2H_2O$	–	–	–	–	2.4	–	4.5	–	–	–
Bloedite	$Na_2Mg(SO_4)_2{\cdot}4H_2O$	–	–	–	–	–	–	18.8	–	–	–
Leonite	$K_2Mg(SO_4)_2{\cdot}4H_2O$	–	–	–	–	–	–	–	–	–	–
Kainite	$KMg(SO_4)Cl{\cdot}3H_2O$	–	–	–	–	–	–	–	–	–	–
Halite	$NaCl$	6.5	33.2	3.5	0.6	55.2	13.2	3.7	4.7	46.6	11.3
Sylvite	KCl	–	–	0.3	–	–	–	–	–	–	–
Carnallite	$KMgCl_3{\cdot}6H_2O$	–	–	0.7	–	2.9	7.7	–	–	–	–
Bischofite	$MgCl_2{\cdot}6H_2O$	–	–	–	–	–	0.4	–	–	–	–
Tachyhydrite	$CaMg_2Cl_6{\cdot}12H_2O$	–	–	–	–	–	–	–	–	–	–
Antarcticite	$CaCl_2{\cdot}6H_2O$	–	–	–	–	–	–	–	–	–	–
Nitrate salts		–	–	–	–	–	–	–	–	–	–

Sample number		21	22	23	24	25	26	27	28	29	30
TDS (mg/l)		1,107*	1,886*	475*	3,264*	324,7*	339,300	239,200	643*	82	1,713
Calcite	$CaCO_3$	–	–	34.8	–	–	–	–	–	1.7	–
Dolomite	$CaMg(CO_3)_2$	15.6	32.4	11.5	–	–	–	–	39.1	23.2	–
Magnesite	$MgCO_3$	10.9	0.2	–	2.9	2.3	0.07	0.1	–	–	8.0
Trona	$Na_3H(CO_3)_2{\cdot}2H_2O$	–	–	–	–	–	–	–	4.4	–	–
Pirssonite	$Na_2Ca(CO_3)_2$	–	–	–	–	–	–	–	15.1	–	–
Burkeite	$Na_6CO_3(SO_4)_2$	–	–	–	–	–	–	–	17.6	–	–
Anhydrite	$CaSO_4$	–	23.9	24.4	1.8	31.8	–	–	–	–	31.4
Kieserite	$MgSO_4{\cdot}H_2O$	–	–	–	–	–	–	4.4⁺	–	–	7.5
Thenardite	Na_2SO_4	3.7	–	–	–	–	–	–	–	4.2	–
Aphthitalite	$K_3Na(SO_4)_2$	–	–	–	–	–	0.2	–	19.2	18.7	–
Glauberite	$Na_2Ca(SO_4)_2$	64.7	5.3	15.8	–	21.5	–	–	–	21.3	–
Syngenite	$K_2Ca(SO_4)_2{\cdot}H_2O$	–	17.4	10.8	2.7	–	–	–	–	–	–
Polyhalite	$K_2MgCa_2(SO_4)_4{\cdot}2H_2O$	–	–	–	61.5	3.1	0.2	1.6	–	–	7.5
Bloedite	$Na_2Mg(SO_4)_2{\cdot}4H_2O$	–	–	–	–	39.3	37.6	0.8	–	–	–
Leonite	$K_2Mg(SO_4)_2{\cdot}4H_2O$	–	–	–	–	–	9.2ˣ	7.1	–	–	–
Kainite	$KMg(SO_4)Cl{\cdot}3H_2O$	–	–	–	–	–	–	–	–	–	–
Halite	$NaCl$	1.1	17.8	2.3	31.2	2.0	52.7	86.1	3.5	30.8	41.8
Sylvite	KCl	–	–	–	0.03	–	–	–	–	–	–
Carnallite	$KMgCl_3{\cdot}6H_2O$	–	–	–	–	–	–	–	–	–	3.5
Bischofite	$MgCl_2{\cdot}6H_2O$	–	–	–	–	–	–	–	–	–	–
Tachyhydrite	$CaMg_2Cl_6{\cdot}12H_2O$	–	–	–	–	–	–	–	–	–	–
Antarcticite	$CaCl_2{\cdot}6H_2O$	–	–	–	–	–	–	–	–	–	–
Nitrate salts		–	–	–	–	–	–	–	1.0	–	–

* mg/kg ⁺ Epsomite ($MgSO_4{\cdot}7H_2O$) rather than kieserite. ˣ Picromerite ($K_2Mg[SO_4]_2{\cdot}6H_2O$) rather than leonite.

11. Groundwater from Navajo Sandstone, Mexican Water, Arizona (White and others, 1963, table 4).
12. Doughty Springs, Delta Co., Colorado (White and others, 1963, table 23).
13. Groundwater from Camillus-Vernon Shale, Syracuse, New York (White and others, 1963, table 5).
14. Groundwater from Pahasapa Limestone, Rapid City, South Dakota (White and others, 1963, table 6).
15. Interstitial brine #47, Chott el Jerid, Tunisia (Gueddari and others, 1983, table 1).
16. Surficial brine LS10 from Salicor playa, LaMancha region, Spain (Soriano and others, 1977, table 1).
17. Groundwater from Sudbury area, Ontario, Canada (Frape and others, 1984, table 1).
18. Groundwater from the Chester Granite, Virginia (White and others, 1963, table 11).
19. Groundwater from Enoch's Pond well, Balley Co., Texas (Chemerys, J.C. and Wood, W.W., pers. commun.).
20. Groundwater from Sudbury area, Ontario, Canada (Frape and others, 1984, table 1).
21. Groundwater from glacial deposits, Malcolm, Iowa (White and others, 1963, table 11).
22. Mammoth Springs, Yellowstone Park, Wyoming (White and others, 1963, table 25).
23. Groundwater from Siamo Slate, Morris Mine, Michigan (White and others, 1963, table 10).
24. Interstitial fluid #20, Chott el Jerid, Tunisia (Gueddari and others, 1983, table 1).
25. Groundwater from Pierre Shale, Langdon, North Dakota (White and others, 1963, table 5).
26. Salt Lake, Roosevelt Co., New Mexico (Chemerys, J.C. and Wood, W.W., pers. commun.).
27. Interstitial brine sample 47-27-022, Salt Lake playa, Hudspeth Co., Texas (Boyd, 1982, table 7).
28. Groundwater from gabbro, Waterloo, Maryland (White and others, 1963, table 2).
29. Groundwater from morraine, Moss area, SE Norway (Englund and Myhrstad, 1980, table 1).
30. Deep well, Ouargla Oasis #3, Tunisia (Clarke and Jones, 1972, table 2).

Table 4 continued.

Sample number		31	32	33	34	35	36	37	38	39	40
TDS (mg/l)		103,800	135,700	331,200	310,900	389,200	253,300*	266,400	907	4,308	271,300
Calcite	$CaCO_3$	–	–	–	–	–	0.2	–	–	–	–
Dolomite	$CaMg(CO_3)_2$	–	–	–	–	–	–	–	3.4	–	0.08
Magnesite	$MgCO_3$	0.3	–	–	0.02	–	–	0.02	–	2.1	–
Trona	$Na_3H(CO_3)_2{\cdot}2H_2O$	–	–	–	–	–	–	–	–	–	–
Pirssonite	$Na_2Ca(CO_3)_2$	–	–	–	–	–	–	–	–	–	–
Burkeite	$Na_6CO_3(SO_4)_2$	–	–	–	–	–	–	–	–	–	–
Anhydrite	$CaSO_4$	–	0.04	1.1	1.6	–	<0.01	0.5	31.1	28.5	0.2
Kieserite	$MgSO_4{\cdot}H_2O$	6.9	–	–	0.07	–	–	5.2	–	3.7	–
Thenardite	Na_2SO_4	–	–	–	–	–	–	–	–	–	–
Aphthitalite	$K_3Na(SO_4)_2$	–	–	–	–	–	–	–	–	–	–
Glauberite	$Na_2Ca(SO_4)_2$	–	–	–	–	–	–	–	–	–	–
Syngenite	$K_2Ca(SO_4)_2{\cdot}H_2O$	–	–	–	–	–	–	–	–	–	–
Polyhalite	$K_2MgCa_2(SO_4)_4{\cdot}2H_2O$	7.4	–	–	0.2	–	–	–	–	–	–
Bloedite	$Na_2Mg(SO_4)_2{\cdot}4H_2O$	–	–	–	–	–	–	–	–	–	–
Leonite	$K_2Mg(SO_4)_2{\cdot}4H_2O$	–	–	–	–	–	–	–	–	–	–
Kainite	$KMg(SO_4)Cl{\cdot}3H_2O$	1.9	–	–	–	–	–	–	–	–	–
Halite	NaCl	80.2	84.3	97.9	97.4	48.1	52.7	80.2	30.8	45.5	48.7
Sylvite	KCl	–	–	0.02	0.8	15.9	13.6	–	–	–	–
Carnallite	$KMgCl_3{\cdot}6H_2O$	3.3	1.0	0.1	0.8	25.4	0.2	12.3	7.7	3.5	7.2
Bischofite	$MgCl_2{\cdot}6H_2O$	–	–	–	–	–	–	1.8	24.8	16.8	–
Tachyhydrite	$CaMg_2Cl_6{\cdot}12H_2O$	–	2.1	–	–	–	–	–	2.2	–	43.7
Antarcticite	$CaCl_2{\cdot}6H_2O$	–	12.2	0.7	–	9.2	31.1	–	–	–	0.05
Nitrate salts		–	–	–	–	–	–	–	–	–	–

Sample number		41	42	43	44	45	46	47	48	49	50
TDS (mg/l)		363,400	241,000	400,300	510	72*	898	534	252	709	1,472*
Calcite	$CaCO_3$	–	–	–	10.5	–	–	36.2	–	–	–
Dolomite	$CaMg(CO_3)_2$	0.2	–	–	26.0	5.2	0.3	21.8	6.6	39.7	–
Magnesite	$MgCO_3$	–	–	–	–	21.8	–	–	–	–	6.8
Trona	$Na_3H(CO_3)_2{\cdot}2H_2O$	–	–	–	–	–	–	–	–	–	–
Pirssonite	$Na_2Ca(CO_3)_2$	–	–	–	–	–	–	–	–	–	–
Burkeite	$Na_6CO_3(SO_4)_2$	–	–	–	–	–	–	–	–	–	–
Anhydrite	$CaSO_4$	<0.01	0.08	0.01	23.4	26.0	10.4	29.9	6.2	23.2	36.3
Kieserite	$MgSO_4{\cdot}H_2O$	–	–	–	–	–	–	–	–	–	5.2
Thenardite	Na_2SO_4	–	–	–	–	–	–	–	–	–	–
Aphthitalite	$K_3Na(SO_4)_2$	–	–	–	–	–	–	–	–	–	–
Glauberite	$Na_2Ca(SO_4)_2$	–	–	–	–	–	–	–	–	–	–
Syngenite	$K_2Ca(SO_4)_2{\cdot}H_2O$	–	–	–	–	–	–	–	–	–	–
Polyhalite	$K_2MgCa_2(SO_4)_4{\cdot}2H_2O$	–	–	–	–	–	–	–	–	–	–
Bloedite	$Na_2Mg(SO_4)_2{\cdot}4H_2O$	–	–	–	–	–	–	–	–	–	–
Leonite	$K_2Mg(SO_4)_2{\cdot}4H_2O$	–	–	–	–	–	–	–	–	–	–
Kainite	$KMg(SO_4)Cl{\cdot}3H_2O$	–	–	–	–	–	–	–	–	–	–
Halite	NaCl	4.3	19.0	19.0	33.0	20.0	66.3	9.4	8304	27.0	40.9
Sylvite	KCl	–	0.09	–	–	–	–	0.4	–	–	–
Carnallite	$KMgCl_3{\cdot}6H_2O$	23.0	0.03	10.0	1.7	9.9	5.3	2.3	1.5	0.9	4.5
Bischofite	$MgCl_2{\cdot}6H_2O$	–	–	–	–	–	–	–	–	4.0	3.8
Tachyhydrite	$CaMg_2Cl_6{\cdot}12H_2O$	48.0	–	17.0	–	–	3.4	–	2.5	0.4	–
Antarcticite	$CaCl_2{\cdot}6H_2O$	22.8	79.7	53.0	5.2	–	14.2	–	2.5	–	–
Nitrate salts		–	–	–	–	14.1	–	–	–	4.8	2.3

* mg/kg [+] Epsomite ($MgSO_4{\cdot}7H_2O$) rather than kieserite. [×] Picromerite ($K_2Mg[SO_4]_2{\cdot}6H_2O$) rather than leonite.

31. Interstitial brine sample 48-24-021, Salt Lake basin playa, Hudspeth Co., Texas (Boyd, 1982, table 7).
32. Average of 40 formation waters, Louisiana Gulf Coast (Dickey and others, 1972).
33. Sample AV12, seepage in Avery Island salt dome, Louisiana (Kumar, 1983, table 2).
34. Caspian Gates Spring, Shiraz basin, Iran (Zak and Gat, 1975, table 1).
35. Sample AV36, seepage in Avery Island salt dome, Louisiana (Kumar, 1983, table 2).
36. Salton Sea geothermal well, I.I.D. no. 1, California (White, 1968, table 1).
37. Interstitial fluid #111, Chott el Jerid, Tunisia (Gueddari and others, 1983, table 1).
38. Corral Rubio playa LP31, LaMancha region, Spain (Marfil and others, 1975, table 4).
39. Feiran well #9, southern Sinai desert (Starinsky and others, 1983, table 2).
40. Formation water from well Zohar #6, Sherif Fm., Jurassic, southern Israel (Fleischer and others, 1977, table 1).
41. Oil field brine, Delhi-Taylor #2, Seven Mile Fm., Grand Co., Utah (Mayhew and Heylmun, 1965, no. 4).
42. Waters from North Mine, Sudbury district, Ontario, Canada (Frape and Fritz, 1982, table 1).
43. Oil field brine, Niagaran Fm., Ingham Co., Minnesota (Carpenter, 1979, table 5).
44. Groundwater from granite, well D647, SW Sinai area (Gat and Issar, 1974, table 2).
45. Groundwater from Baltimore Gneiss, Bucks Co., Pennsylvania (White and others, 1963, table 10).
46. Groundwater from gneiss and amphibolite, As, Moss areas, SE Norway (Englund, 1983, table 1).
47. Groundwater from Grimsby area, NE England (Howard and Lloyd, 1983, table 3).
48. Groundwater from Quaternary marine deposits, SE Norway (Englund, 1983, table 1).
49. Groundwater from Conasauga Limestone, Birmingham, Alabama (White and others, 1963, table 6).
50. Deep well, Rahmat 2, Tunisia (Clarke and Jones, 1972, table 2).

ground waters, seawater mixtures, and saline basin waters. In table 4, normative salt minerals are arranged according to major anion groups in order of carbonates, then sulfates, then chlorides, with mixed anion phases intermediate. Within each anion group, the sequence generally is from simple, single cation-anion phases to more complex double salts. With minor exceptions, this system corresponds to increasing single-phase aqueous solubility. Our attempt to associate increasing salt complexity with increasing solubility produces a different cation sequence for the chlorides than for the carbonates and sulfates.

We tried to choose sample salt norms for table 4 so as to have two examples with abundant amounts of each major solute phase. Some additional examples were taken from special lithologic environments or from unusual normative salt assemblages. From the results of all the salt-norm computations, a diagnostic chart was derived (fig. 2). Although many natural waters reflect multiple solute origins, salt norms basically fall into three major categories: meteoric, marine, and diagenetic. These general categories are based on qualitative similarity of complete normative assemblage rather than on detailed characteristics of constituent phases.

Meteoric norms

Salt norms for meteoric waters deriving their solutes principally from rock weathering reflect the principal agent promoting dissolution in the pedogenic environment. In addition to direct aqueous dissolution of mineral salts, such as the carbonate minerals, carbonic acid and sulfuric acid in meteoric waters each generate their characteristic solute compositions. A distinctive feature of the normative assemblages from meteoric waters is the absence of the alkaline-earth-bearing chlorides.

Carbonic acid hydrolysis of rock-forming minerals produces a salt norm dominated by carbonate minerals (table 4). The very high abundance in the norm from the Tertiary Castle Hayne Limestone (#1) of North Carolina or the calcic Permian clastics of SE. Norway (#2), and the abundant normative dolomite in waters from the Mississippian Pahasapa Limestone of South Dakota (#14) and the Cambrian Conasuaga Limestone of Alabama (#49) are examples of straightforward carbonate-rock dissolution. The occurrence of normative alkali carbonate phases (trona or kalicinite), on the other hand, is diagnostic of silicate hydrolysis. Examples of silicate hydrolysis are given in table 4 in the trona-bearing salt norms for ground waters from siliceous crystalline or clastic rocks, for example, the granite of the Black Hills, South Dakota (#7), the Cretaceous Benton Shale of Wyoming (#8), or the lower Mesozoic Navajo sandstone, Arizona (#11).

In the absence of carbonate rock dissolution, contrasts in amounts of trona, alkali alkaline-earth double carbonate (pirssonite), and alkaline-earth carbonate in the norm reflect variations in the alkali/alkaline-earth ratios in silicate minerals (feldspar and pyroxene or amphibole) in the associated lithologies. Increased importance of Ca-plagioclase and/or mafic minerals is reflected in the norms for ground water from the granite of McCormack, South Carolina (table 4, #10), or metamorphosed volcanic rocks at Mebane, North Carolina (table 4, #9). The norms for waters associated with normal carbonic acid weathering of mixed alkali-bearing silicate rocks commonly contain sylvite (table 4, #3, 7, 8, 9, 10). The lower solubility of the trona-sylvite pair compared to that of the kalicinite-halite pair under atmospheric conditions leaves potassium associated with chloride in the normative calculation. Normative kalicinite appears only in salt norms for ground waters where potash feldspar is the dominant solute source. Large amounts of normative dolomite can result from hydrolysis of mixed Ca-Mg silicates, such as the plagioclase-pyroxene mineralogy of basalt or gabbro (table 4, #3 and 28). Abundant normative magnesite, on the other hand, requires hydrolysis of dominantly magnesium silicate, such as serpentine (table 4, #5), biotite (table 4,#45), or even Mg clays in carbonate rock (table 4, #6).

Meteoric waters aided in weathering by sulfuric acid derived from the oxidation of sulfides (perhaps cycling through acid rain; table 4, #2 and #29) produce salt norms characterized by sulfate minerals. Predominant anhydrite in the norm can, of course, simply reflect dissolution of original evaporitic $CaSO_4$, such as in waters from Silurian gypsiferous red beds of central New York (table 4, #13), and the gypsiferous Pahasapa Limestone of South Dakota (#14). In addition, abundant normative potassium-magnesium-mixed sulfates can be characteristic of residual meteric fluids following fractional precipitation of gypsum and/or carbonates (table 4, #27 and #31).

The presence of either of the pure alkali sulfates, thenardite (Na_2SO_4) and aphthitalite ($K_3Na(SO_4)_2$), or of the alkali sulfo-carbonate, burkeite ($Na_6CO_3(SO_4)_2{\cdot}2H_2O$), or of the alkali-calcium double salts, glauberite ($Na_2Ca(SO_4)_2$) and syngenite ($K_2Ca(SO_4)_2{\cdot}H_2O$) can be considered diagnostic of sulfatic-weathering solutions. The alkali sulfates or sulfo-carbonate can be expected from waters associated with the weathering of siliceous crystalline rocks and their clastic derivatives, whereas the sulfate double salts reflect more calcic lithologies. Solute magnesium derived from sulfuric-acid hydrolysis of mafic lithologies can appear in salt norms as the monohydrate sulfate, keiserite, or its more hydrous equivalent, epsomite. Usually, however, the magnesian component of dissolution or hydrolysis appears in the salt norm as dolomite or as the alkali-bearing complex salts bloedite ($Na_2Mg(SO_4)_2$), polyhalite ($K_2MgCa_2(SO_4)_4{\cdot}2H_2O$), leonite ($K_2Mg(SO_4)_2{\cdot}4H_2O$), and kainite ($KM_g(SO_4)Cl{\cdot}3H_2O$). The occurrence of such alkali magnesium sulfates in the salt norm indicates sulfuric-acid hydrolysis of mixed-mafic silicate, either from crystalline rocks (table 4, #17) or fine clastic sediments (table 4, #25). Examples of sulfate-rich norms spanning the whole range of crystalline and clastic lithologies are given in table 4 for waters associated with granite in Virginia (#18), granitic basement derived glacial deposits in Iowa (#21) or gabbro at Waterloo, Maryland (#28), basic plutonic rocks at Sudbury, Ontario (#17), and calcic micaceous slate from northern Michigan (#23).

Marine norms

Seawater-type norms dominated by sulfate and chloride salts are particularly characterized by magnesium-associated chloride. This association cannot be obtained from hydrothermal sources because of the aqueous insolubility of Mg-silicates at elevated temperature. Such an association in the salt norm from other than ultimately marine sources would require dissolution of chloride-bearing mafic minerals.

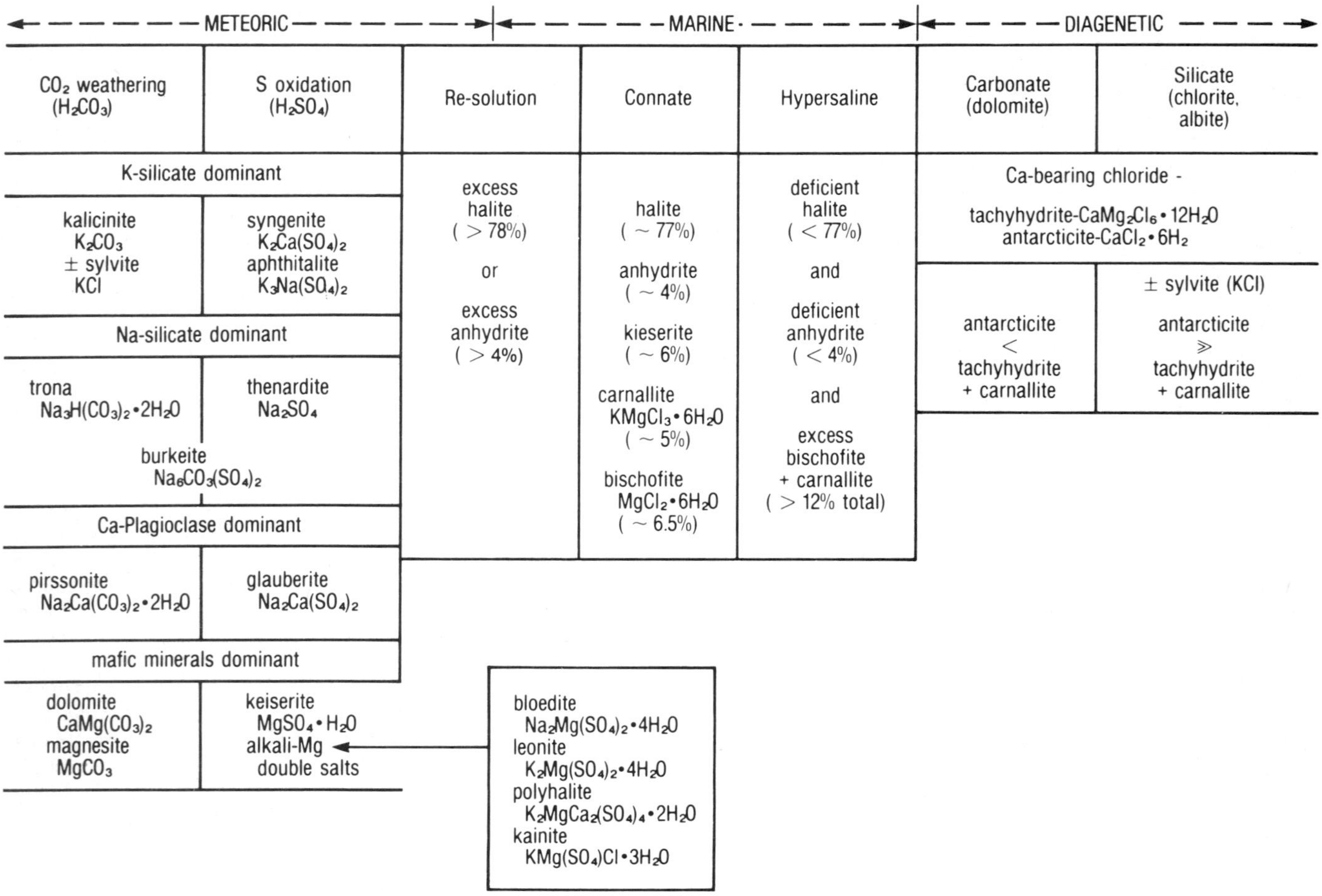

Figure 2. *Diagnostic chart for normative salts*

Normative diagnosis of marine solute sources is complicated not only by mixing or continental re-cycling, but also by solute redistribution within and marginal to the marine environment. Three subdivisions can readily be recognized: connate, resolution, and hypersaline. These categories are based on abundance, as well as the presence or absence of key phases in the salt norm. Thus, any marine connate fluid or simple seawater dilution should give the seawater salt norm (table 3) with the characteristic assemblage halite-anhydrite-kieserite-carnallite-bischofite in the proportions given (as also shown in fig. 2). A solute matrix derived from resolution of the most common evaporite minerals, halite and gypsum, is derived in the same way as other meteoric waters; the matrix will be enriched in NaCl and $CaSO_4$, relative to the seawater norm, and depleted in the other constituents. In contrast, the solutes of a hypersaline marine bittern will be depleted in NaCl and $CaSO_4$ because of fractional precipitation of those salts, but the salt norm will include excess magnesium and, most likely, potassium-bearing chloride salts. This is because the molar ratio of chloride to sodium in seawater is greater than unity (the molar value is 1.17), so that the next most abundant cations, Mg and K, both appear as chlorides in the salt norm. Changes in proportions of the major normative salts corresponding to the evaporative concentration of seawater are given in table 5, as calculated from the analyses of hypersaline liquors in coastal marine-salt evaporation ponds.

Of all the samples for which salt norms were calculated, only a few computations produced results similar to those expected for simple connate seawater dilution. This is not unexpected, as mixing, continental cycling, or mineral-water interaction of any sort could be expected to produce differences from seawater proportions. Because of the relative stabilities of the minerals considered by SNORM, additions of calcium, carbonate, or sulfate to a seawater solute matrix will appear in the salt norm as excess magnesite and anhydrite. This also can eliminate bischofite and introduce sylvite into the norm as Mg becomes associated with carbonate or sulfate rather than with chloride. Additions of sodium will be seen as NaCl, with an associated redistribution of other mineral proportions, depending on the accompanying anion. Thus, an excess of magnesite, anhydrite, or halite over that in the seawater norm appears indicative of mixing or continental cycling. The effects of marine aerosols may be seen in the dilute waters of some areas subject to coastal precipitation patterns (table 4, #45 and #47).

The waters that most closely approach the seawater norm are from the continental saline basins of northern Algeria (table 4, Chott el Jerid, #37) and western Australia (table 6). The former has an overall excess of chlorides and a deficiency of sulfate, indicating hypersaline conditions. However, the excess of halite compared to the seawater norm suggests some resolution as well. The Yilgarn examples include (1) groundwater norms with definite, though small, deficiencies of halite and anhydrite along with sufficient bischofite to indicate clear hypersaline affinities (table 6, #500 and #518), and (2) surface water with normative proportions about as close to seawater as might be imagined in continental environs. Most of the Yilgarn salt norms, however, show a telltale excess of halite resulting from resolution. In fact, the majority of the Yilgarn salt norms (table 6) produce enough excess halite to suggest recycling of marine solutes with little contribution from direct weathering of the granitic crystalline basement of the region (unless dominated by chloride-bearing mafic minerals). However, the normative salt abundances for some Yilgarn samples (table 6, #65 and #243) and most other closed basin waters can be interpreted as the result of mixing of meteoric weathering solutions with a recycled marine-solute matrix. Well water from the north Sahara (table 4, #50) produced a complete seawater assemblage in the salt norm but with halite and bischofite quantities much reduced and anhydrite and magnesite increased over a seawater norm, primarily because of addition of sulfate and carbonate solutes from continental sources. Similarly, well water of the southern Sinai (table 4, #39), is deficient in normative halite and has an excess of magnesium chloride, suggesting a hypersaline solute origin as supported by strontium-isotope evidence (Starinsky and others, 1983). Excess normative anhydrite and magnesite, however, suggest other (meteoric) solute contributions.

A key normative mineral indicator of meteoric additions to a marine-type, sulfo-chloride anion matrix is the unusual mineral, kainite $KMg(SO_4)Cl{\cdot}3H_2O$. This phase appears in the norms of samples from the edge of the Yilgarn block and from the Salt Lake playa of West Texas (table 4, #31), which has derived the bulk of its solutes from weathering of the evaporitic Permian strata of the area. These salt norms represent the addition of excess sulfate to a marine solute matrix, while the proportions of alkalies and Mg were maintained.

A less diagnostic mineral commonly appearing in salt norms for modified marine waters is sylvite. If a water containing alkaline-earth or sodium chloride is mixed with a water containing potassium sulfo-carbonate from weathering, the salt norm calculation will associate the carbonate and sulfate with the other cations. After consuming available Mg in polyhalite, kainite, or carnallite, KCl is left as a distinct phase. Mixing of any other type of water with a residual marine bittern merely increases the likelihood of sylvite occurring as a single phase in the salt norm. Thus, additional information besides the presence of sylvite in the norm is needed to suggest solute affinities. The appearance of sylvite in normative alkali sulfo-carbonate assemblages attests to its compatibility with weathering solutions; however, its presence within other chloride-dominant normative assemblages bespeaks its involvement in different phenomena, which will be discussed below.

Another normative mineral that can serve as an indicator of mixing of weathering solutions with a seawater-type solute matrix is kieserite. The abundance of normative kieserite can be reduced by addition of redissolved halite and/or calcium sulfate. Normative kieserite will be increased in a solute matrix with additions of sulfate from weathering, regardless of whether subsequent sequential evaporative precipitation of $CaSO_4$ and/or NaCl occurs (table 4, #15 and #16). The addition of dilute calcium sulfo-carbonate solutes typical of meteoric waters to oxic, gypsum-saturated, hypersaline fluids also will deplete normative kieserite through the loss of sulfate with calcium and of magnesium with carbonate (the latter will also result from sulfate reduction). Without the low values for activity of water generated by coexistant hydroscopic chlorides, kieserite is unstable relative to its more hydrous equivalent, epsomite ($MgSO_4{\cdot}7H_2O$; table 4, #27). Kieserite occurs in the norm of all but 4 of 48 analyses from the Yilgarn block, western Australia (table 6).

Waters illustrating the extreme proportions of solute NaCl that can result from halite resolution in the subsurface include a salt dome mine seepage (table 4, #33) and a surface spring associated with a shallow subsurface dome (#34). The clear indication that excess halite from salt

Table 5

Salinity, major anion concentration, and salt norms for hypersaline marine liquors from surface evaporation ponds

	–Concentration (mg/kg)–				Salt norm* (anhydrous weight percent) [Abbreviations of normative phase names from table 2]				
Sample	Salinity	Chloride	Sulfate	E_{Br}†	hal	anh	kie	car	bis
Black Sea ponds, Ukraine (Zherebtsova and Volkova, 1966)									
gypsum ppt.	170,000	95,310	11,760	5.9	79.0	1.8	7.1	5.0	7.1
halite ppt.	270,000	149,400	22,580	15.0	71.5	0.3	10.3	7.25	10.7
Mg-sulfate ppt.	309,200	148,100	63,920	44.2	30.8	–	26.1	19.3	23.7
K-salt ppt.	314,600	173,600	43,490	71.0	13.8	–	17.5	27.7	40.9
>K-salt ppt.	329,800	192,700	41,120	90.2	5.2	–	15.8	18.7	60.2
Portoroz salt ponds (Herrmann and others, 1973)									
pan 12	137,300	76,800	10,130	3.8	77.1	3.2	6.4	4.8	8.2
pan 13c	252,200	143,900	15,600	7.1	78.7	0.6	7.2	5.0	8.2
pan 13e	274,100	151,100	29,900	12.5	61.0	0.2	13.4	7.7	17.3
pan 13h	273,100	149,000	33,000	15.8	51.2	0.2	15.0	10.2	23.0

* Normative carbonate salts not reported; aqueous carbonate species negligible (Portoroz) or not reported (Black Sea).

† Degree of evaporation of brine compared to normal seawater (after Levy, 1977) based on relative Br content; value a minimum beyond halite saturation because of neglect of Br uptake by chloride salts.

Table 6

Salt norms for selected analyses of groundwaters from the salt lake areas of the Yilgarn Block, Western Australia; samples chosen from data of Mann (1983) to illustrate complete compositional range in normative marine salts and are presented in order of increasing halite

[Abbreviations of normative phase names from table 2]

Sample	Location (part of block)	Salinity (mg/l)	hal	anh	mag	–Salt norm (anhydrous weight percent)– kie	pol	leo	blo	kai	car	bis
Marine assemblages												
500	SW corner	323,300	71.9	0.3	<0.1	8.4	–	–	–	–	2.2	17.1
518	WSW edge	147,100	74.6	3.0	<0.1	5.0	–	–	–	–	2.0	15.4
62	N. edge	48,200	77.0	5.9	0.1	12.0	–	–	–	–	3.1	1.8
70	NNW edge	23,600	77.8	15.5	0.4	3.7	1.0	–	–	–	1.6	–
506	S. part	142,300	81.2	0.9	<0.1	5.4	–	–	–	–	4.5	5.9
509	S. edge	217,200	83.2	1.8	–	2.9	–	–	–	–	3.7	8.4
393	center	151,900	84.6	2.9	<0.1	4.2	–	–	–	–	0.6	7.5
403	W. part	153,200	85.5	3.5	<0.1	3.3	–	–	–	–	4.8	2.8
399	N. central	230,900	91.0	1.0	<0.1	4.0	–	–	–	–	2.9	1.0
Mixed marine-meteoric assemblages												
65	N. part	21,840	81.3	–	0.9	9.1*	6.5	1.5	0.7	–	–	–
243	NE edge	22,700	88.7	–	<0.1	4.7*	2.2	2.5	–	1.8	–	–

* Normative epsomite rather than kieserite.

resolution predominates over solute diagenesis, as suggested by Hanor (1984), also is indicated in the salt norm for the average of 40 formation waters from the Louisiana Gulf Coast (table 4, #32).

The chloride minerals characteristic of hypersaline marine fluids or similar continental and modified sedimentary basin brines are bischofite ($MgCl_2{\cdot}6H_2O$) and the K-Mg double halide, carnallite ($KMgCl_3{\cdot}6H_2O$). Small quantities of these minerals also appear in the norm of relatively dilute ground waters associated with diverse rock types, suggesting the presence of seawater-type solutes and possibly aerosols; in ground waters from coastal sediments (table 4, #47), this is most likely the case. Ground waters in crystalline rock terranes of the United States, Canada, and Scandinavia (table 9, #5, #45, and #46), however, may contain solutes of diagenetic origin. Substantial normative amounts of bischofite, carnallite, and/or sylvite are computed for brines of clearly saline strata or dilutions thereof (table 4, #35, #37, #38, #39, #41). Normative abundance of carnallite and bischofite more than the amounts found in the seawater norm appear to be clearly associated with hypersaline waters.

Diagenetic norms

The calcium-bearing chloride minerals, antarcticite ($CaCl_2{\cdot}6H_2O$) and tachyhydrite ($CaMg_2Cl_6{\cdot}12H_2O$), indicate solute diagenesis and the most highly altered fluid compositions with respect to earth-surface conditions. Yet, calcium chloride-bearing normative assemblages were computed for a number of analyses representing diverse conditions. Nearly all such salt norms were obtained from subsurface fluids associated with deep sedimentary basins, highly saline strata, or crystalline shield areas (table 4, #32-33, #35-36, #40-44, #46). Notable exceptions occur in a few salt norms from coastal marine sediments (table 4, #48). Assemblages containing both calcium-bearing chlorides are most readily interpreted in terms of the diagenetic alteration of a residual marine-solute matrix dominated by magnesium salts through exchange of Mg and Na in solution for Ca in solids – that is, carbonate, clay minerals or feldspar. Such exchange in silicates is promoted by elevated temperature through decrease in the solubility of magnesium silicate phases. The resulting fluid can be expected to produce an antarcticite-sylvite norm (table 4, #36). The blending of marine bittern with a deep sedimentary basin brine or hydrothermal fluid is the process postulated by Knauth and others (1980) for the Louisiana salt dome brines. Addition of exchanged solute calcium to a marine bittern would all but eliminate sulfate from the fluid by precipitation of anhydrite. In variable proportions, exchange of other ions for calcium is characteristic of many highly concentrated solutions encountered in the subsurface.

Diagenetic salt norms can apparently be subdivided on the basis of the phase ratios of calcium chloride to magnesium chloride. The development of calcium chloride brines in the Bardawil Lagoon area of north Sinai by evaporative concentration, by mixing with shallow meteoric ground waters, and by dolomitization of carbonate sediments has been documented by Levy (1977). Salt norms for representative analyses from the Bardawil Lagoon area are given in table 7. In no instance does the Ca/Mg concentration (or salt norm antarcticite/tachyhydrite) ratio exceed unity. However, the decreasing ratio of activity coefficients for Ca/Mg with increasing fluid concentration permits an increasing molar Ca/Mg ratio in equilibrium with dolomite. The ratio to be found in many natural solutions, however, will differ with respect to the solute matrix and the structural state of the Ca and Mg carbonates. Although it is impossible to determine an exact limit for Ca/Mg, above which dolomitization alone cannot account for the relative normative abundances, the ratio probably is from 4 to 5 (Stoessel and Moore, 1984).

Table 7

Salt norms for representative brines from sabkhas south of the Bardawil Lagoon, north Sinai (data from Levy, 1977); samples listed according to increasingly greater distance from normal Meditteranean seawaters of the principal lagoon

[Abbreviations of normative phase names from table 2]

Sample	Sabkha	Salinity (mg/kg)	E_{Mg}*	–Salt norm (anhydrous weight percent)– hal	anh	mag	dol	kie	car	bis	tac	ant
1083	Hawash N.	135,800	3.7	79.8	5.7	0.07	–	5.7	5.4	6.2	–	–
1082	Hawash N.	280,900	7.9	81.0	1.1	0.06	–	5.7	5.4	6.7	–	–
1083A	Hawash S.	278,700	13.2	70.2	0.8	0.09	–	8.0	6.5	14.3	–	–
1107	Hafiordim	271,400	4.5	88.3	1.0	0.02	–	3.0	3.3	4.3	–	–
1120	Hafiordim	270,000	5.8	86.7	0.8	0.01	–	0.7	4.3	7.5	–	–
1144	Hagvaot N.	269,300	8.1	80.6	0.5	0.03	–	5.2	4.9	8.8	–	–
1114	Hagvaot S.	271,700	6.1	85.6	1.8	0.02	–	0.3	2.6	9.7	–	–
1166	Hagvaot S.	262,700	16.1	63.9	1.3	–	0.04	–	5.9	25.7	3.2	–
1111	Hayareah	260,700	5.0	82.0	0.4	–	0.01	–	0.5	–	15.2	1.8
1113	Hayareah	100,900	3.6	59.5	2.1	–	0.04	–	0.5	–	28.5	9.4
1160	Hayareah	275,500	20.2	35.0	0.2	–	0.02	–	0.1	–	58.9	5.8

* Degree of evaporation of brine compared to normal seawater (after Levy, 1977) based on relative Mg content; computation neglects loss of Mg to diagenetic reactions reflected in normative occurrences of $CaCl_2$ phases.

Very large ratios of normative antarcticite to magnesium chloride salt ratios clearly are associated with silicate authigenesis, – most specifically, formation of chlorite and albite. The best examples in table 4 are in the crystalline shield brine from Canada (#42) and the sedimentary basin brine of Michigan (#43). Additions of minor amounts of these solute matrices to alkali carbonate or sulfate-bearing waters can readily obliterate the $CaCl_2$ signature (as perhaps in table 4, #39 and #45), but salt norm examples are found where small amounts of antarcticite suggest such a contribution (table 4, #44 and #46). Although calcium is usually strongly preferred on clay-mineral cation-exchange sites, the possibility of displacing some calcium to the solution by high solute concentrations of other cations must, however, be considered.

SUMMARY

Examination of more than 500 salt norms calculated from water analyses representing a wide range of concentrations and hydrochemical settings has suggested that certain key phases or assemblages are characteristic of solute origin. In addition, it appears that the interpretation of the salt norm can depend on the relative abundance of diagnostic phases as well as their presence or absence. Normative assemblages calculated by SNORM seem to fall into three major categories: meteoric, marine, or diagenetic. Although the presence of only alkali-bearing carbonates and sulfates in the salt norm seems to distinguish meteoric waters and their principal supplemental weathering agent, differentiation of types of marine or diagenetic fluids requires comparison of relative abundances of normative salts. The salt norms of resolution, connate, or hypersaline fluids of marine or continental recycled seawater type are distinguished by the relative amounts of halite, anhydrite, and Mg-bearing chlorides, whereas different principal diagenetic processes operating on subsurface fluids are reflected in alkaline-earth chloride ratios. Although the detailed criteria have been based on the most indicative data sets known to the authors, the tests so far have hardly been exhaustive.

REFERENCES

Al Droubi, Abdallah, Fritz, Bertrand, Gac, Jean-Ives and Tardy, Yves, 1980. Generalized residual alkalinity concept; application to prediction of the chemical evolution of natural waters by evaporation: American Journal of Science, v. 280, p. 560-572.

Bodine, M.W. and Jones, B.F., 1986. The Salt Norm: A quantitative chemical-mineralogical characterization of natural waters: U.S. Geological Survey Water Resources Investigations 86-4086, 116 p.

Boyd, F.M., 1982. Hydrogeology of the Northern Salt Basin of West Texas and New Mexico: Ph.D. thesis, University of Texas at Austin, p. 61-71.

Braitsch, Otto, 1971. Salt Deposits, their Origin and Composition: Springer-Verlag, New York, 297 p.

Carpenter, A.B., 1978. Origin and chemical evolution of brines in sedimentary basins: Oklahoma Geological Survey Circular 79, p. 60-78.

______, 1979. Interim report on lead and zinc in oil-field brines in the central Gulf Coast and in southern Michigan: AIME Annual Meeting, New Orleans, La., Report 79-95, 15 p.

Clarke, F.E. and Jones, B.F., 1972. Significance of ground-water chemistry in performance of North Sahara tube wells in Algeria and Tunisia: U.S. Geological Survey Water Supply Paper 1757-M, p. 1-39.

Cross, C.W., Iddings, J.P., Pirsson, L.V. and Washington, H.S., 1902. A quantitative chemico-mineralogical classification and nomenclature of igneous rocks: Journal of Geology, v. 10, p. 555-690.

Dickey, P.A., Collins, A.G. and Fajardo, I., 1972. Chemical composition of deep formation waters in southwestern Louisiana: American Association of Petroleum Geologists Bulletin, v. 56, p. 1530-1533.

Drever, J.I., 1982. The Geochemistry of Natural Waters, Prentice Hall, Englewood Cliffs, N.J., 388 p.

Englund, Jens-Olaf, 1983. Chemistry and flow patterns in some groundwaters of southeastern Norway: Norges Geologiske Undersokelse, offprint 380, p 221-234.

Englund, Jens-Olaf and Myhrstad, J.A., 1980. Groundwater chemistry of some selected areas in southeastern Norway: Nordic Hydrology, v. 11, p. 33-54.

Eugster, H.P., 1980. Geochemistry of evaporitic lacustrine deposits: Annual Review of Earth and Planetary Science, v. 8, p. 35-63.

Eugster, H.P. and Hardie, L.A., 1978. Saline lakes *in* Lerman, A., ed., Lakes: Chemistry, Geology, Physics: Springer-Verlag, New York, p. 237-293.

Eugster, H.P., Harvie, C.E. and Weare, J.H., 1980. Mineral equilibria in the six-component sea water system, Na-K-Mg-Ca-SO_4-Cl-H_2O, at 25°C: Geochimica et Cosmochimica Acta, v. 44, p. 1355-1348.

Eugster, H.P. and Jones, B.F., 1979. Behavior of major solutes during closed-basin brine evolution: American Journal of Science, v. 279, p. 609-631.

Fleischer, E., Goldberg, M., Gat, J.R. and Magaritz, M., 1977. Isotopic composition of formation waters from deep drillings in southern Israel: Geochimica et Cosmochimica Acta, v. 41, p. 511-525.

Frape, S.K. and Fritz, Peter, 1982. The chemistry and isotopic composition of saline groundwaters from the Sudbury Basin, Ontario: Canadian Journal of Earth Sciences, v. 19, p. 645-661.

Frape, S.K., Fritz, Peter and McNutt, R.H., 1984. Water-rock interaction and chemistry of groundwaters from the Canadian Shield: Geochimica et Cosmochimica Acta, v. 48, p. 1617-1627.

Garrels. R.M., 1967. Genesis of some ground waters from igneous rocks *in* Abelson, P.H., ed., Researches in Geochemistry: John Wiley, New York, v. 2, p. 405-421.

Gat, J.R. and Issar, A., 1974. Desert isotope hydrology: water sources of the Sinai Desert: Geochimica et Cosmochimica Acta, v. 38, p. 1117-1131.

Gueddari, Moncef, Monnin, Christophe, Perret, Dominique, Fritz, Bertrand and Tardy, Yves, 1983. Geochemistry of brines of the Chott el Jerid in southern Tunisia – application of Pitzer's equations: Chemical Geology, v. 39, p. 165-178.

Hanor, J.S., 1984. Variation in the chemical composition of oil-field brines with depth in northern Louisiana and southern Arkansas: implications for mechanism and rates of mass transport and diagenetic reaction: Transactions – Gulf Coast Association of Geological Societies, v. 34, p. 55-61.

Hardie, L.A. and Eugster, H.P., 1970. The evolution of closed-basin brines: Mineralogical Society of America, Special Publication 3, p. 273-290.

Harvie, C.E., Eugster, H.P. and Weare, J.H., 1982. Mineral equilibria in the six-component seawater system, Na-K-Mg-Ca-SO_4-Cl-H_2O at 25°C. II: Compositions of the saturated solutions: Geochimica et Cosmochimica Acta, v. 46, p. 1603-1618.

Hem, J.D., 1970. Study and interpretation of the chemical characteristics of natural waters: U.S. Geological Survey, Water-Supply Paper 1473, 363 p.

Herrmann, A.G., Knake, Doris, Schneider, Jurgen and Peters, Heide, 1973. Geochemistry of modern seawater and brines from salt pans: main components and bromine distribution: Contributions to Mineralogy and Petrology, v. 40, p. 1-24.

Holland, H.D., 1984. The chemical evolution of the atmosphere and oceans: Princeton University Press, Princeton, N.J., 528 p.

Howard, K.W.F. and Lloyd, J.W., 1983. Major ion characterization of coastal saline ground waters: Ground Water, v. 21, p. 429-437.

Jones, B.F., 1966. Geochemical evolution of closed basin water in the western Great Basin, *in* Rau, J.L., ed. 2nd Symposium on Salt, Northern Ohio Geological Society, Cleveland, Ohio, v. 1, p. 181-200.

———, ed., 1970. Symposium on the mineralogy and geochemistry of non-marine evaporites: Mineralogical Society of America, Special Paper 3, p. 213-319.

Knauth, L.P., Kumar, M.B. and Martinez, J.D., 1980. Isotope geochemistry of water in Gulf Coast salt domes: Journal of Geophysical Research, v. 85, p. 4863-4871.

Kumar, M.B., 1983. Character of meteoric leaks in the salt mines of southern Louisiana, USA: Journal of Hydrology, v. 66, p. 351-363.

Lambert, S.J., 1978. Geochemistry of Delaware Basin ground waters: *in* Austin, G.S., ed., Geology and Mineral Deposits of Ochoan Rocks in Delaware Basin and Adjacent Areas: New Mexico Bureau of Mines and Mineral Resources, Circular 159, p. 33-38.

Levy, Yitzhak, 1977. The origin and evolution of brine in coastal sabkhas, northern Sinai: Journal of Sedimentary Petrology, v. 47, p. 451-462.

Livingstone, D.A., 1963. Chemical composition of rivers and lakes, *in* Fleischer, Michael, ed., Data of Geochemistry [6th ed.]: U.S. Geol. Survey Professional Paper 440-G, 64 p.

Mackenzie, F.T. and Garrels, R.M., 1966. Chemical mass balance between rivers and oceans: American Journal of Science, v. 264, p. 507-525.

Mann, A.W., 1983. Hydrogeochemistry and weathering on the Yilgarn Block, Western Australia – ferrolysis and heavy metals in continental brines: Geochimica et Cosmochimica Acta, v. 47, p. 181-190.

Marfil, Rafaela, Bermejo, E.P. and de la Pena, J.A., 1975. Sedimentacion salina actual en las lagunas de la zona Corral Rubio-La Higuera (provincia de Albacete): Estudios Geologicos, v. 31, p. 543-552.

Mayhew, E.J. and Heylmun, E.B., 1965. Concentrated subsurface brines in the Moab Region, Utah: Utah Geological and Mineralogical Survey, Special Studies 13, 28 p.

Rankama, Kalervo and Sahama, Th. G., 1950. Geochemistry: University of Chicago Press, Chicago, 912 p.

Sonnenfeld, Peter, 1984. Brines and Evaporites: Academic Press, New York, 613 p.

Soriano, Jose, Marfil, Rafaela and de la Pena, J.A., 1977. Sedimentacion salina actual en las lagunas del norte de Alcazar de San Juan (Cuidad Real): Estudios Geologicos, v. 33, p. 123-130.

Spencer, R.J., Eugster, H.P., Jones, B.F. and Rettig, S.L., 1985. Geochemistry of Great Salt Lake, Utah I: hydrochemistry since 1850, Geochimica et Cosmochimica Acta, v. 49, pp. 727-737.

Starinsky, A., Bielski, M. Ecker, A. and Steinitz, G., 1983. Tracing the origin of salts in groundwater by Sr isotopic composition (the crystalline complex of the Southern Sinai, Egypt): Isotope Geoscience, v. 1, p. 257-267.

Stewart, F.J., 1963. Marine evaporites, *in* Fleischer, Michael, ed., Data of Geochemistry (6th ed.): U.S. Geological Survey Professional Paper 440-Y, 53 p.

Stoessel, R.K. and Moore, C.H., 1983. Chemical constraints and origins of four suggested groups of Gulf Coast Reservoir fluids: American Association of Petroleum Geologists Bulletin, v. 67, p. 896-906.

White, D.E., 1968. Environments of generation of some base-metal ore deposits: Economic Geology, v. 63, p. 301-335.

Zak, Israel and Gat, J.R., 1975. Saline waters and residual brines in the Shiraz-Sarvistan Basin, Iran: Chemical Geology, v. 16, p. 179-188.

Zherebtsova, I.K. and Volkova, N.N., 1966. Experimental study of behavior of trace elements in the process of natural solar evaporation of Black Sea water and Sasyk-Sivash brine: Geochemistry International, v. 3, p. 656-670.

Geochemical Trends for Groundwaters from the Canadian Shield

S.K. Frape and P. Fritz

Department of Earth Sciences, University of Waterloo, Waterloo, Ontario, Canada, N2L 3G1

Frape, S.K. and Fritz, P., Geochemical trends for groundwaters from the Canadian Shield; in Saline water and gases in crystalline rocks, Editors: Fritz, P., and Frape, S.K.; Geological Association of Canada Special Paper 33, 1987. p. 19–38.

Abstract

Deep groundwaters from the Canadian Shield are highly saline Ca-Na-Cl brines. The stable isotopic signature of the most concentrated brines, found at sites from across the Shield, is very similar. The hydrogen and oxygen stable isotope signature of these brines is very different from saline waters or brines collected in sedimentary or hydrothermal environments. The water component of the Shield brines may have a very different origin than the majority of the solute component. Most of the salinity associated with these brines appears to be derived from intense water-rock interaction with local host rocks, rather than being derived from external sources. Although major ion ratios such as Ca/Cl are uniform from site to site across the Canadian Shield; many minor and trace element ratios reflect differences in host chemistry between sites. It is proposed that the similar major element chemistries for brines could only be achieved if the same primary and secondary mineral equilibrium controls were to occur in each area. Such water-rock interaction, coupled with solubility controls of most ionic species in Ca-Na-Cl brines, exert the major influence on the evolution of these waters.

The stable isotopic signature of shallow fresh and brackish groundwaters is meteoric in origin. These waters derive the majority of their solutes from intense rock-water interaction. As most sites sampled are disturbed environments, the influence of these shallower groundwaters is seen in the form of mixing or dilution trends with the more concentrated brines.

Résumé

Les eaux souterraines profondes du bouclier canadien sont des solutions sursalées riches en Ca-Na-Cl et, dans l'ensemble du bouclier, la concentration des isotopes stables des eaux les plus concentrées est très semblable. Celle de l'hydrogène et de l'oxygène diffère fortement de ce qui est mesuré dans les eaux salées et sursalées en milieu sédimentaire ou hydrothermal. La composante aqueuse des saumures du bouclier pourrait avoir une origine très différente de celle de la plupart des composants dissous. La majeure partie du contenu salin de ces saumures semble provenir d'une interaction intense entre l'eau et les encaissants locaux, plutôt que de sources externes. Bien que les rapports d'ions majeurs, comme Ca/Cl, soient semblables d'un site à l'autre dans l'ensemble du bouclier, ceux d'un grand nombre d'éléments mineurs et d'éléments traces reflètent une variation de la composition chimique des encaissants. On a suggéré que la similitude des caractéristiques chimiques des éléments majeurs de ces saumures n'était possible que si chaque secteur était régi par des contraintes d'équilibre minéralogique primaires et secondaires identiques. Une interaction de ce genre entre l'eau et l'encaissant, à laquelle s'ajoutent les conditions de solubilité de la plupart des ions des saumures riches en Ca, Na et Cl, sont les principaux facteurs influant sur l'évolution de ces eaux.

La concentration des isotopes stables dans les eaux souterraines douces et saumâtres de faible profondeur a une origine météorique. La plupart des composants dissous de ces solutions dérivent d'une interaction intense entre l'eau et la roche. Comme la plupart des échantillons proviennent d'environnements perturbés, l'influence des eaux souterraines moins profondes prend la forme d'un mélange avec les eaux plus concentrées ou de leur dilution dans des porportions variables.

INTRODUCTION

The earliest recorded occurrences of saline groundwaters on the Canadian Shield are found in the Annual Report of the Geological Survey of Canada for 1877/78 (Anonymous, 1889). This report contains a description of combustible gases and Ca-Cl brines issuing from a borehole in the Silver Islet mine near Thunder Bay. In fact, the mining industry was to provide the main source of information about saline water in Shield rocks for the next one hundred years. Reports by Lane in 1908, 1914, and 1927 describe the occurrence and chemistry of brines found in association with copper mines on the Keweenaw Peninsula, Michigan. Still later, in 1931, Tanton gave a detailed description of the brines, gases and the associated mining problems from the Silver Islet mine. Descriptions of evaporite minerals found in crystalline rocks and believed to be associated with saline water (approximately 8000 mg l^{-1} of Cl) were provided by Langford and Hancox (1936) describing occurrences in the McIntyre Mine, Ontario.

A common theme throughout the early literature, which describes saline waters from the Canadian Shield, was the association with gold and silver mines. Examples are the work of Bruce (1941) on saline waters in the Sturgeon River Gold Mines and the extensive work of Boyle (1961, 1979) dealing with gold mines from across the Canadian Shield. It should be noted that some of the initial thoughts on the origin of the brines and saline waters were first proposed by Boyle (1961) in describing the Yellowknife gold mines. His postulate that the waters represented residual ore forming fluids is shared by others.

In the late 1970's the Canadian committment to research into disposal of nuclear fuel waste in stable plutonic rock lead to numerous reconnaissance and detailed groundwater surveys on the Canadian Shield. In 1979, Fritz and Reardon recorded the occurrence of Ca-Na-Cl brines within mines of the Sudbury Basin. Later detailed sampling in the Sudbury Basin and a number of other sites across the Canadian Shield (Figure 1) has shown such brines to be common at depth in almost all areas (Frape and Fritz, 1981, 1982; Fritz and Frape, 1982; Frape, et al., 1984; and McNutt, et al., 1984). During the same period Atomic Energy of Canada Limited (A.E.C.L.) in drilling several non-mineralized plutons (granite and gabbroic rocks) encountered saline waters in deep drill holes (Bottomley and Graham, 1980; Davison, 1984; Gascoyne et al., this volume). Therefore, as a result of the programs funded by A.E.C.L., a relatively detailed picture of the occurrence, chemistry, hydrogeology and evolution of Canadian Shield groundwaters has been developed in the last decade. This research has lead to a variety of related studies and a new interest and awareness of groundwaters in crystalline rock environments.

HYDROGEOLOGY

Groundwaters across the Canadian Shield occur in a variety of crystalline rock types. In almost all cases active hydrogeological flow systems containing generally fresh waters (dissolved loads <1000 mg l^{-1}) exists at depths of less than about 200 meters. These waters occur in numerous faults and joint systems, which are found to be highly interconnected when detailed hydrogeological testing is performed (Dickin, et al., 1985). Evidence from mining environments supports the above statements as many mines have a considerable flow of fresh meteoric groundwaters into the shallow mine workings. Of course, mines represent disturbed environments which most likely had very different hydrogeologic regimes before the start of blasting and mine dewatering operations.

With increasing depth at most mine sites the volumes of groundwater entering the workings and the apparent hydrologic interconnectivity of structures appears to decrease. Evidence from non-mining sites and boreholes would suggest that large water filled faults or shear zones do exist at greater depths (below 200 meters) in barren plutons; but these structures do not appear to be as common as in the shallow environments (Davison, 1985).

In the deep Canadian Shield (below 1000 meters) some large vertical or nearly vertical fault systems exist and often in active mining zones these contain fresh to brackish groundwaters (TDS <10000 mg l^{-1}). Examples are the Bob's Lake Fault, Sudbury or the Negus and Pud faults, Yellowknife. Due to mine drainage of the rock mass, it has been possible to show that these faults are hydraulically connected to surface as evidenced by the appearance of young tritiated meteoric waters at depth and the progressive dilution of deep saline waters over a period of years at some sites (Frape et al., 1984). With time these structures allow fresh meteoric groundwaters to enter deeper zones in the Shield which are adjacent to the fault. In further discussion of geochemical trends, within the groundwaters of the Canadian Shield, it is very important to remember that many of the brackish and saline groundwaters are mixtures of brine and dilute meteoric water. However, at depths as great as 2000 meters large horizontal or subhorizontal faults have been encountered during drilling. These fault systems contain considerable volumes (100's litres per minute of flow) of very concentrated Ca-Na-Cl brines (usually >250 g l^{-1}). Some boreholes encountering these structures often cease flowing after variable periods of time (days to years). In many instances the boreholes cease flowing as a result of plugging due to the precipitation of secondary minerals such as gypsum. Little or no dilution of the brine issuing from these boreholes has been observed. However, the majority of deep boreholes, initially containing very saline waters, do show a decrease in geochemical concentration with time. There is no evidence at present to suggest that these deep groundwater systems associated with subhorizontal faults are hydrogeologically active, unless disturbed by mining or drilling operations.

GEOCHEMISTRY

The geochemical aspects pertaining to studies on the Canadian Shield were initiated in order to describe the chemical composition of groundwaters found in crystalline rocks. Continuing surveys very rapidly became focussed on the occurrence, and extent of saline waters and brines within the crystalline rocks of the Shield. This was in great part due to the unexpected occurrence of waters of such chemical composition at depth in the Canadian Shield and it was felt that a much greater understanding of these fluids was necessary in the context of nuclear fuel waste disposal in plutonic rocks. Finally, the origin and/or evolution of these waters has recently become a major concern of many investigators, both those interested in waste disposal and those concerned with the evolution of ore deposits.

In all cases important geochemical questions which require answers are (1) the source of the salinity in deep groundwaters systems, (2) the extent of water-rock interaction and its influence on modifying groundwater chemistry,

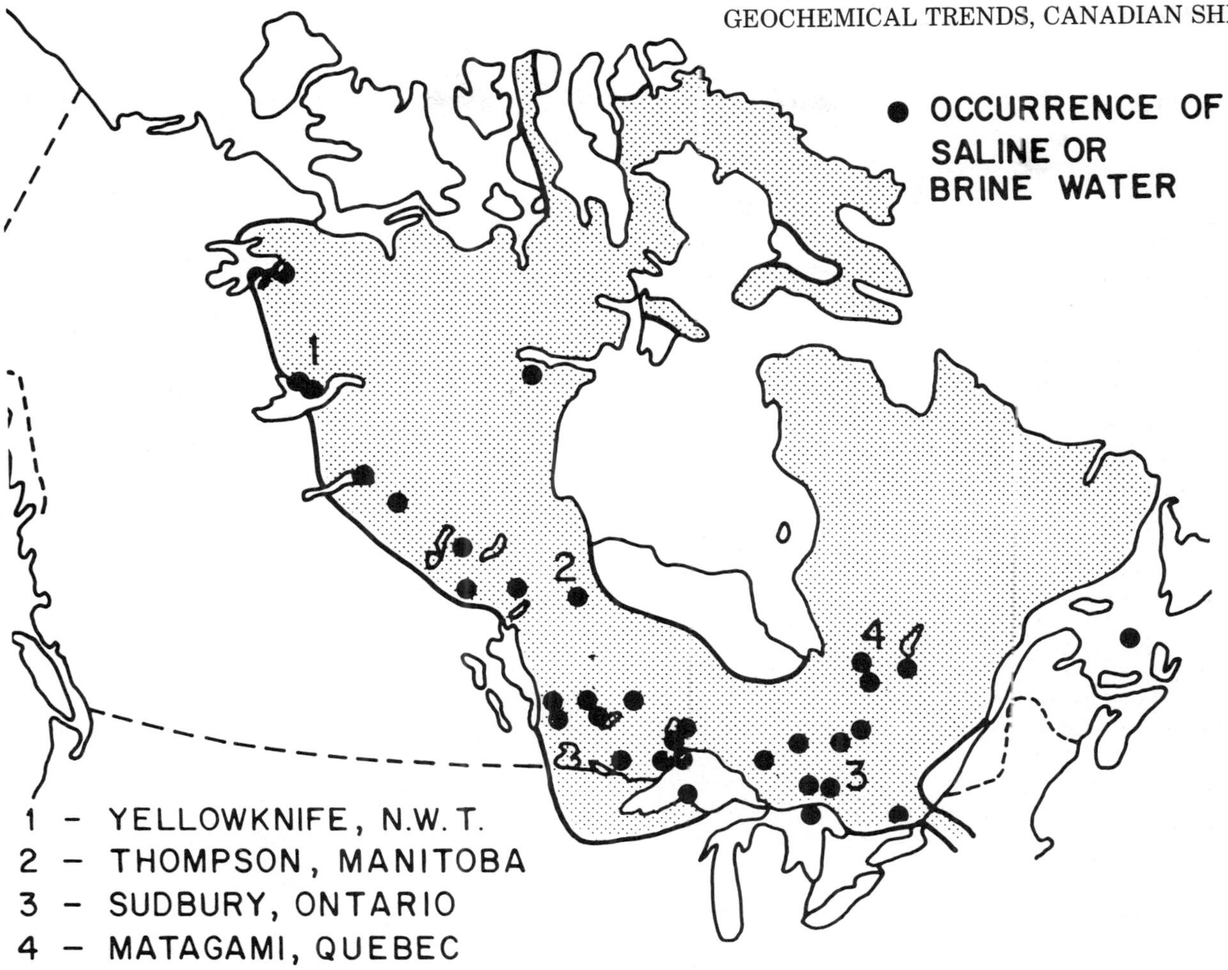

Figure 1. *Location of sampling sites on the Canadian Shield (numbers refer to the main sites discussed in the text).*

and (3) the age, direct or indirect, of groundwaters found within the Canadian Shield.

The object of this paper is to provide a summary of the geochemistry of Canadian Shield groundwaters. A portion of this geochemical information has been presented in previous papers by the authors and additional information is presented in several other papers contained within this volume. Although key points of earlier works will be presented and summarized, new information on many minor and trace elements is presented and some additional aspects of rock-water interaction will be discussed.

Table 1 modified from Frape et al., (1984) contains representative major, minor and trace element data for groundwater samples from the Canadian Shield. Mine sites from which samples were obtained are shown in Figure 1. As mentioned in previous publications the deep saline groundwaters from the Shield have a relatively uniform composition. Thus, the relationship of density to total dissolved solids (TDS) is very consistent for waters from all sites (Figure 2a). Typical geochemical trends with increasing depth are shown in Figures 2b to 2d. The total dissolved load increases relatively rapidly with increasing depth, particularly below 1000 meters (Figure 2b). As the brines are primarily composed of the ions Ca-Na-Cl the same depth trend as TDS is reflected in the calcium and chloride concentration (Figures 2c and 2d). It should be noted that the depth refers to the sampling depth. In many cases, particularly in deeper areas of most mines, the actual water producing zone is often much deeper than the borehole collar. Unfortunately, drill records are usually not detailed enough for accurate estimates of true sample depth and therefore collar depth was chosen to provide some uniformity to the data base.

DISCUSSION

Origin of the water component

Figure 3 shows the relationship between oxygen and hydrogen isotope contents in waters in the crystalline rocks of the Canadian Shield. Shallow groundwaters reflect local climatic regimes and plot on local Meteoric Waterlines which usually are close to the global line, occasionally very shallow groundwater samples show a component of evaporated surface waters. The second major group of groundwater samples are brines; the most concentrated of which have an almost identical isotopic signature for all sites. Many samples reflect mixing between the brines and shallow meteoric groundwaters.

Computing best fit regression lines for the most saline data points at each site leads to the shaded area as shown in Figure 3. A similar observation is made by Pearson (this volume) who shows that if chemical compositions are extrapolated to halite saturation, all brines have very similar

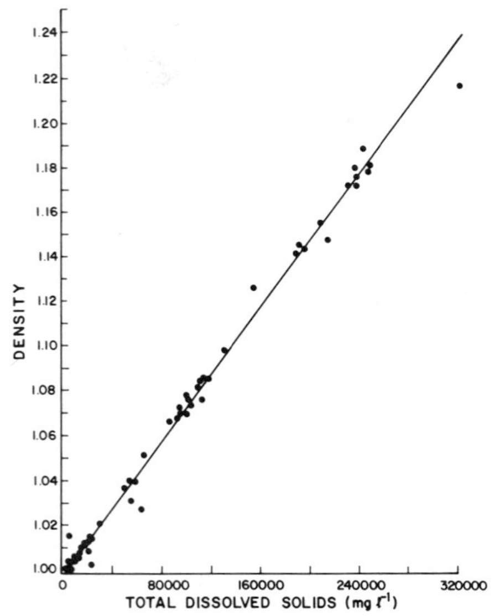

Figure 2a. *The relationship of total dissolved solids and density for groundwaters from the Canadian Shield.*

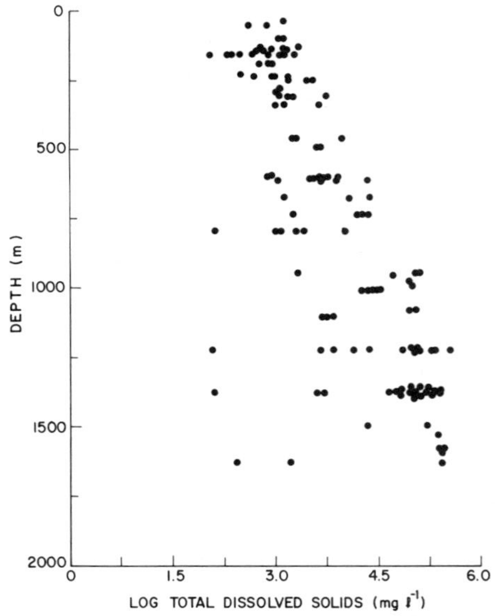

Figure 2b. *The relationship of depth and total dissolved solids for groundwater from the Canadian Shield.*

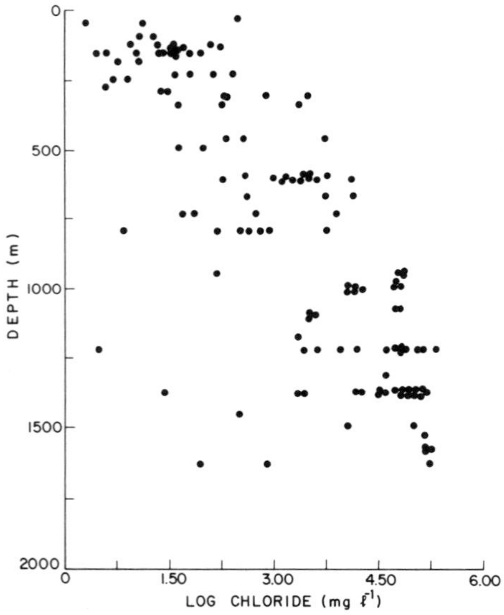

Figure 2c. *The relationship of depth and chloride concentration for groundwaters from the Canadian Shield.*

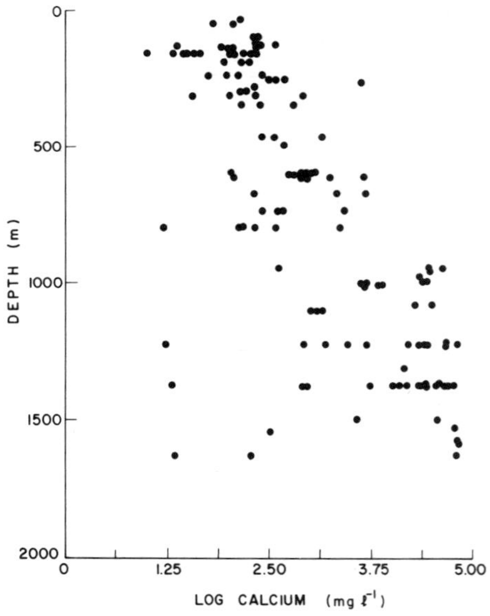

Figure 2d. *The relationship of depth and calcium concentration for groundwaters from the Canadian Shield.*

trapolated to halite saturation, all brines have very similar extrapolated isotopic contents. This area was chosen because the data from three of four major sites intercept here. The very low intercept of the Thompson site with Matagami and Sudbury and lack of intercept with Yellowknife is puzzling and is heavily influenced by a few recent samples from this area. The deuterium value of the brine at the intercept point would be approximately −20 to 0°/oo SMOW and the oxygen-18 value approximately −10 to −7.0°/oo SMOW. As postulated in Frape et al.(1984) this may indicate a common source brine at great depth across the Canadian Shield.

However, the origin of the water component of these most concentrated brines is still in great debate. The isotopic signature is unusual compared to fluids from deep sedimentary or hydrothermal environments; but may be common for deep brines from crystalline environments.

As indicated by the nature of the hydrogeology and the disturbance by mining activity, the majority of saline waters and brines which are reported in this paper are mixtures of a fresh or brackish end member with the very

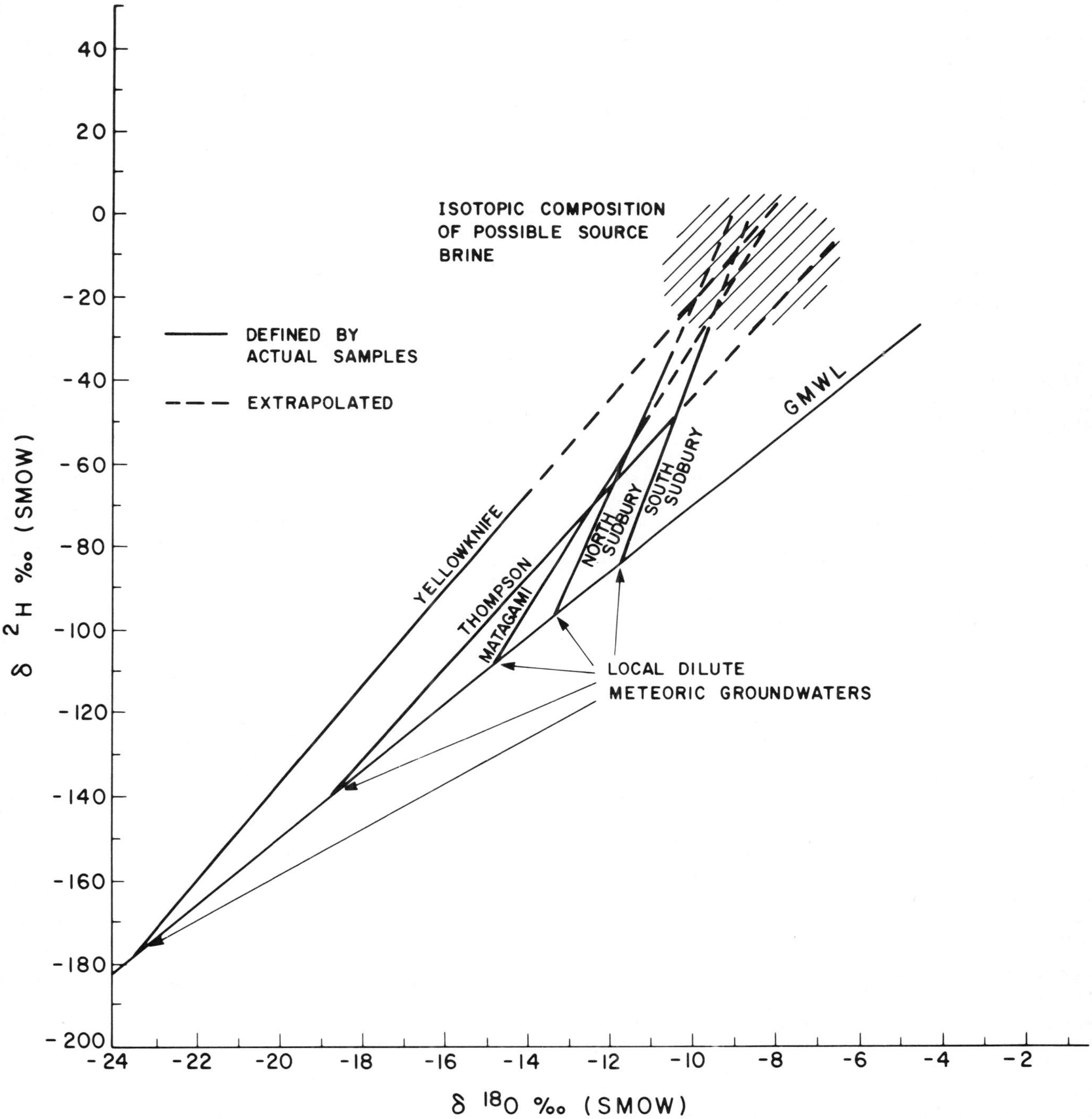

Figure 3. *Oxygen-18 and deuterium plot for various groundwaters found on the Canadian Shield. Regression lines are drawn in order to extrapolate the possible isotopic composition of a common source brine or brines for major sites on the Shield.*

concentrated brines. This mixing has been discussed in detail in several previous papers (Frape et al., 1984; Fritz and Frape, 1982). In describing the geochemical nature of saline groundwaters on the Canadian Shield it is very important to be aware of this mixing and the effect this may have on any interpretation of the naturally occurring geochemical trends.

It is important to note that due to the disturbed nature of most mining environments some fresh groundwaters with low TDS values occur at great depths. If the tritium concentration (^{3}H as Tritium Units) for these samples is considered (Figure 4) it is apparent that a substantial portion of many samples is composed of relatively young (<45 years) groundwaters. This "leakage" of young water along structural flowpaths to great depths as a result of drilling and mine dewatering is seen in the data compiled in Table 2.

Origin of the saline component

Almost all recent geochemical studies in crystalline rock environments have realized a very important characteristic of the solute component of groundwater samples. That is the primary origin of the salt is very difficult or impossible to identify because it is readily masked by the intensity of rock-water interaction in crystalline rock systems. Therefore, in any scheme that addresses the origin or evolution of the saline or brine component of a deep groundwater system we must attempt to assess the degree of rock-water interaction.

Table 1

Geochemistry of representative groundwater samples from sites on the Canadian Shield.

Site	Sample	depth (m)	pH	temp Celsius	Ca mg/L	Na mg/L	Mg mg/L	K mg/L	Sr mg/L	Li mg/L	Cl mg/L	Br mg/L	I mg/L	F mg/L	HCO_3 mg/L	SO_4 mg/L
BRINE																
Thompson	4000-4	1500	5.00	18	64000	45000	5100	199	1080	0.07	207000	1760	—	0.05	19	284
Sudbury	N3644b	1650	6.90	21.5	62000	22000	48.0	320	1700	0.7	168000	1530	—	0.14	—	213
Sudbury	N3646A	1600	6.55	22	63800	18900	78.0	430	1580	0.81	162700	1250	7.0	7.0	58	223
Sudbury	N3651	1600	6.80	—	65000	16880	12.0	122	1390	0.6	156000	1090	—	—	—	138
Yellowknife	4500-6C	1372	5.55	22.5	57300	32600	920	495	1640	1.67	142000	1520	0.2	26.9	2	1
Matagami	R36	1800	6.70	9.5	61300	14800	3400	338	1460	0.25	143581	1785	—	—	101	405
Yellowknife	YK2041	1372	—	20.5	44700	30000	890	430	950	0.74	115000	1010	—	—	—	57
Thomson	4000-5	1220	6.25	20	46300	17000	1960	126	910	0.09	115000	1110	—	0.2	9	107
Yellowknife	YK2042	1372	—	20.5	39300	21700	820	164	910	0.72	109000	935	—	—	—	54
Norita	4E-85 #1	—	—	—	29100	13300	4200	335	960	0.28	90000	1134	—	—	75.9	5
Yellowknife	YK4500-6A	1372	—	—	28600	15000	550	118	530	0.58	73600	667	—	—		219
Thompson	4000-6	1220	6.30	21	26800	8670	637	59.9	558	0.12	63800	793	1.33	5	15	2
SALINE																
Sudbury	CCS3250B	991	6.60	16	27100	8350	105	36.6	351	0.2	58300	579	4.06	11.0	48	123
Sudbury	N3645	1100	—	—	19300	7130	80.5	122	490	0.56	55600	451	0.4	7	25	116
Matagami	R46	1300	—	—	22600	4850	1500	62.5	547	0.13	50828	628	—	—	72	5
Thompson	4000-3	1220	6.00	22	15900	6800	549	45.7	277	0.1	41400	405	1.01	3.68	10	2
Yellowknife	4500-3	1372	6.60	23	12900	10900	268	111	671	0.78	37100	358	1.01	5.74	28	746
Yellowknife	4500-2	1372	6.35	23	15700	9420	250	110	314	0.4	31900	324	0.92	5	69	198
Norita	UN 242	—	7.30	8	5700	2200	1000	35.0	201	0.13	17905	234	—	—	29	5
Thompson	2200-1	671	6.40	13.5	4840	2930	233	92.6	98.3	0.05	13700	115	0.29	1.4	28	444
Yellowknife	4900-2	1494	—	—	3740	3200	67.5	230	65.1	0.3	11520	767	0.2	15.7	141	391
Thompson	T3-2000	610	—	6.5	4540	2740	160	32.1	79.7	0.05	12600	111	0.29	1.34	55	1
Sudbury	FR33192B	1006	8.47	—	4470	2300	2.0	23.1	72	0.14	11300	78.7	1.06	1.9	15	533
Norita	UN 249	—	7.20	7.5	2640	1050	525	16.5	107	0.1	7600	118	—	—	38	5
BRACKISH																
Yellowknife	5	457	6.65	—	1400	1600	174	29.8	47.8	0.06	5250	96.8	—	—	280	—
Thompson	1002-2	305	6.30	9.5	820	908	145	7.8	14.6	0.07	3120	34.4	0.35	0.67	263	28
Sudbury	G-1600	488	—	—	480	180	330	23.5	4	0.19	97	—	—	—	213	2950
Yellowknife	4	335	6.95	—	648	757	115	18.8	—	—	2360	36.3	—	—	243	45
Matagami	R42	1100	—	—	370	110	130	3.1	6.35	0.05	287	3.6	—	0.42	161	1160
Yellowknife	YK2388	550	—	—	264	290	55.0	3.2	9.1	0.04	542	4.9	—	0.24	—	602
FRESH																
Yellowknife	14	130	7.40	—	80.0	107	42.3	4.3	—	—	51	0.41	—	—	341	227
Yellowknife	6	175	7.25	—	103	26	45.9	5.7	—	—	4	0.1	8	—	350	197
Thompson	400-1	100	7.50	4	23.6	119	21.6	6.2	0.6	0.08	123	1.4	0.23	0.71	264	9
Sudbury	DOW0002	152	—	—	21.0	30	6.0	1.8	<0.1	—	63	—	—	—	23	25
Sudbury	L26000	793	—	—	16.0	12	2.9	2.2	<0.1	—	7	—	—	—	32	43
Thompson	4000-7	1220	6.20	10	16.5	3.5	4.1	1.1	—	<0.01	3	0.18	0.06	0.07	27	51

In assessing the origin of the salt component, graphical comparisons have been made between major or minor ions in solution and dissolved load (Frape et al., 1984) or with an ion such as chloride which is assumed to act in a conservative manner (Nordstrom et al. 1985; Edmunds et al. 1985). Such plots are highly useful in discussing the interrelationships of many cations and anions with the major anionic component (Cl) of most brines. However, as pointed out by Edmunds et al. (1985) in referencing other elements to Cl we must remember that Cl is usually being added to solution by rock-water reactions and only once it is in solution does it pursue a conservative geochemical role. Therefore Cl is an excellent indicator of mixing and dilution processes in most crystalline environments because of its conservative nature, but not for the description of geochemical processes which generated the observed brines.

Major and minor cations

Typical major element plots of calcium and magnesium versus chloride in Canadian Shield brines are shown in Figure 5a and 5b. The nature of these plots is quite different, probably representing very different geochemical controls on the ions involved. At low concentration of chloride such as would be found in shallow, fresh groundwater systems, the figures are very similar with considerably

Table 1 continued.

Sample	Si mg/L	B mg/L	Al mg/L	Fe mg/L	Mn mg/L	Cr mg/L	Co mg/L	Ni mg/L	Cu mg/L	Zn mg/L	Pb mg/L	Cd mg/L	TDS mg/L
BRINE													
4000-4	4.8	—	—	—	—	—	—	—	—	—	—	—	324500
N3644b	—	—	—	—	—	—	—	—	—	—	—	—	255800
N3646A	9.6	6.2	1.6	2.07	4.57	<0.4	19.9	3.31	0.33	0.52	3.03	0.97	249100
N3651	—	—	—	—	—	—	—	—	—	—	—	—	240700
4500-6C	8.0	9.19	17.5	18.6	21.8	0.5	5.8	12.7	0.81	0.59	2.42	1.02	237100
R36	—	—	—	—	—	—	—	—	—	—	—	—	227200
YK2041	—	—	—	—	—	—	—	—	—	—	—	—	193100
4000-5	7.5	—	—	—	—	—	—	—	—	—	—	—	182600
YK2042	—	2.0	—	—	—	—	—	—	—	—	—	—	172900
4E-85 #1	—	—	—	—	—	—	—	—	—	—	—	—	139100
YK4500-6A	—	—	—	—	—	—	—	—	—	—	—	—	119300
4000-6	8.5	0.74	1.44	1.81	8.0	0.79	1.15	1.05	0.38	0.22	1.98	0.25	101400
SALINE													
CCS3250B	6.0	5.88	1.63	1.87	1.11	0.78	1.26	1.05	0.39	0.15	1.90	0.24	95040
N3645	9.4	2.3	0.71	1.63	1.74	0.23	9.4	1.50	1.18	0.25	1.39	0.45	83370
R46	—	—	—	—	—	—	—	—	—	—	—	—	81090
4000-3	6.1	1.37	0.91	1.18	2.39	0.54	0.83	0.70	0.27	0.15	0.18	0.16	65440
4500-3	8.0	1.58	4.74	9.16	3.04	0.5	1.0	3.0	0.50	0.30	2.42	0.30	63140
4500-2	10.0	2.1	5.1	4.91	2.40	0.5	1.0	3.0	0.50	0.30	0.50	0.30	58340
UN 242	—	—	—	—	—	—	—	—	—	—	—	—	27300
2200-1	5.3	0.72	0.36	0.53	0.75	0.21	0.26	0.34	0.11	0.070	0.44	0.05	22490
4900-2	3.0	0.49	0.5	6.18	0.04	0.5	1.0	3.0	0.50	0.30	1.6	0.30	20480
T3-2000	8.5	1.25	0.36	0.26	0.61	0.19	0.26	0.22	0.09	0.04	0.44	0.05	20340
FR33192B	5.8	2.0	0.3	0.27	0.11	0.22	0.20	0.20	0.11	0.03	0.31	0.04	18810
UN 249									—	—	—	—	12090
BRACKISH													
5	—	—	—	—	—	—	—	—	—	—	—	—	8883
1002-2	6.4	1.29	0.36	0.37	0.29	<0.04	<0.07	<0.09	<0.01	<0.01	<0.12	<0.01	5371
G1600	—	—	—	—	—	—	—	—	—	—	—	—	4277
4	—	—	—	—	—	—	—	—	—	—	—	—	4223
R42	—	—	—	—	—	—	—	—	—	—	—	—	2231
YK2388	—	—	—	—	—	—	—	—	—	—	—	—	1770
FRESH													
14	—	—	—	—	—	—	—	—	—	—	—	—	853
6	—	—	—	—	—	—	—	—	—	—	—	—	739
400-1	4.4	0.62	0.36	0.07	0.02	<0.04	<0.07	<0.09	<0.01	<0.01	<0.12	<0.01	579
DOW0002	15.7	—	—	0.05	0.05	—	—	—	—	—	—	—	185
L26000	6.1	—	—	5.00	0.11	—	—	—	—	—	—	—	126
4000-7	10.0	0.07	0.36	0.11	0.02	<0.04	<0.07	<0.09	<0.01	<0.01	<0.12	<0.01	117

Table 2

Results of analyzes for sequential water samples from several boreholes at Con Mine, Yellowknife and Lockerby Mine, Sudbury.

Sample	Ca mg/L	Na mg/L	Mg mg/L	K mg/L	Cl mg/L	Br mg/L	SO_4 mg/L	HCO_3 mg/L	2H (SMOW)	18O (SMOW)	3H (TU's)
Sudbury Lockerby L40140											
1979	50000	12000	10.2	83	65000	540	46	68	–46	–11.3	8
1980	4980	2310	5.2	25	15500	117	46	113	–79	–12.1	70
Yellowknife Con Mine DDH 3457											
1980	12900	10900	268	111	37100	358	746	28	–163	–22.8	3
1981	10500	8050	209	53	29400	304	715	34	–162	–22.3	34
1985	4500	3200	67	22	18100	176	723	35	–159	–21.4	37

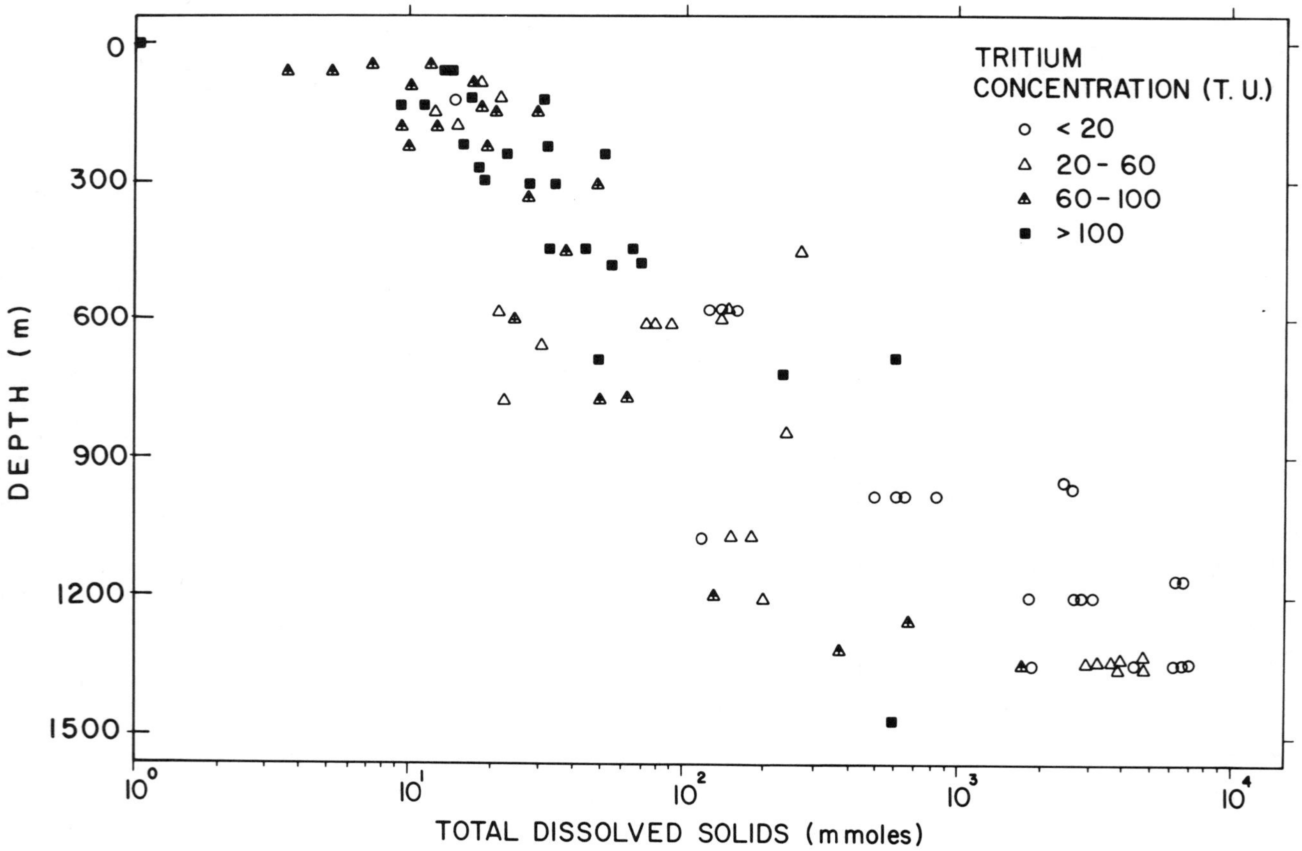

Figure 4. *The tritium content of groundwater samples for the Canadian Shield expressed in a plot of total dissolved solids versus depth.*

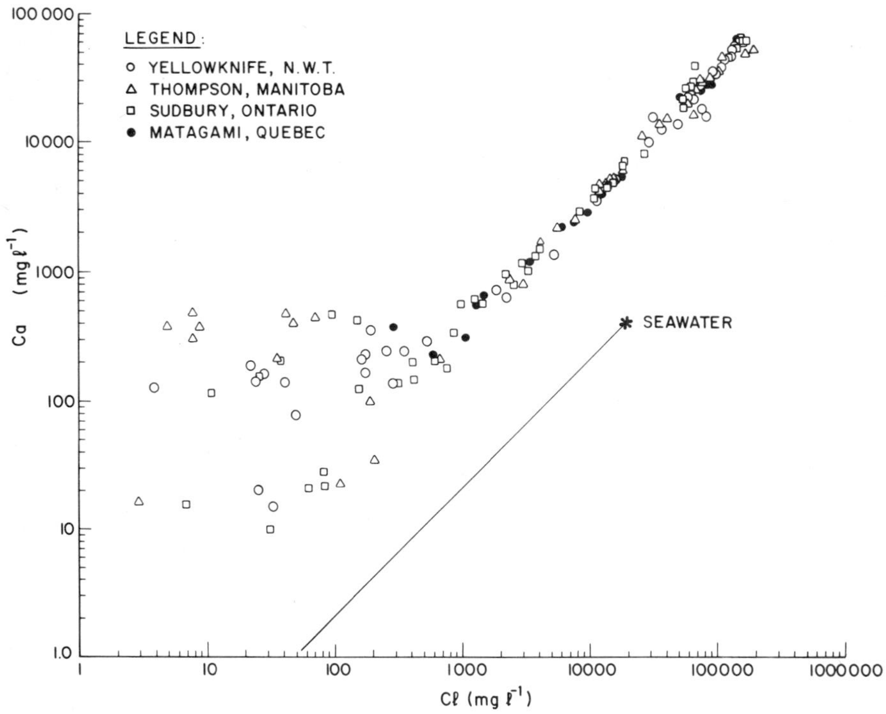

Figure 5a. *The relationship of calcium concentration and chloride concentration for groundwaters from the Canadian Shield relative to the dilution line for seawater.*

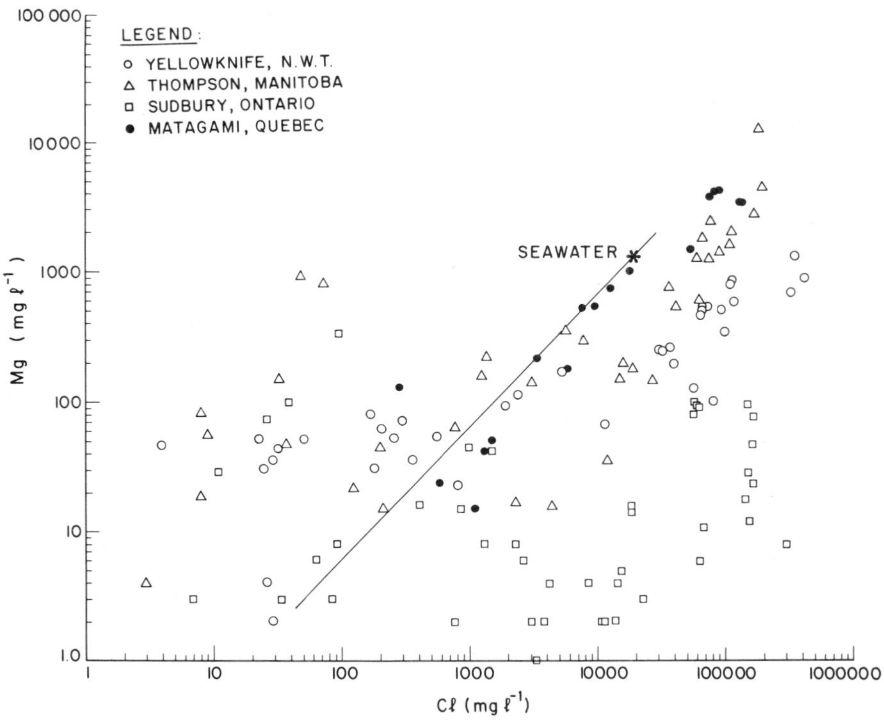

Figure 5b. *The relationship of magnesium concentration and chloride concentration for groundwaters from the Canadian Shield.*

scatter among the plotted elements. At higher concentrations of chloride two very different trends are apparent. Calcium shows a very linear relationship with chloride at elevated concentrations. This trend is due to mixing of shallow waters with brines. The uniform Ca content of the most concentrated brines may be primarily due to equilibration of Ca with secondary minerals such as calcite and gypsum which have been identified as major fracture fillings at most locations on the Canadian Shield (Frape et al. 1984) or the formation of a Ca-Cl mineral such as antarcticite.

Magnesium exhibits a very different trend which appears

to be independent of chloride even at higher concentrations (Figure 5b). In earlier papers the control of Mg concentration was attributed to two mechanisms. The first was source or composition of the rock. It was found that magnesium rich rocks contained waters with elevated Mg levels, an observation also made by Garrels (1967, 1976), Tamari et al. (1983) and others. The other major control on Mg has been postulated to be the incorporation of Mg into the secondary mineral chlorite or other silicates (Edmunds et al. 1985; Fritz and Frape, 1982). It thus, displays a much more 'labile' geochemistry which does not lead to uniform concentrations, across the Shield but rather only at individual sites. Of interest in Figure 5b, therefore is the division by various sites: Thompson and Matagami waters, both dilute and concentrated, have much higher Mg levels per concentration of chloride, reflecting the very Mg rich rocks at these sites versus the relatively Mg poor rocks at Sudbury. As well mixing lines can be seen in the data plotted on Figure 5b for the more concentrated samples at Thompson, Matagami and Yellowknife and may reflect a strong brine control or Ca-Cl solubility control on Mg at very high concentrations of Cl.

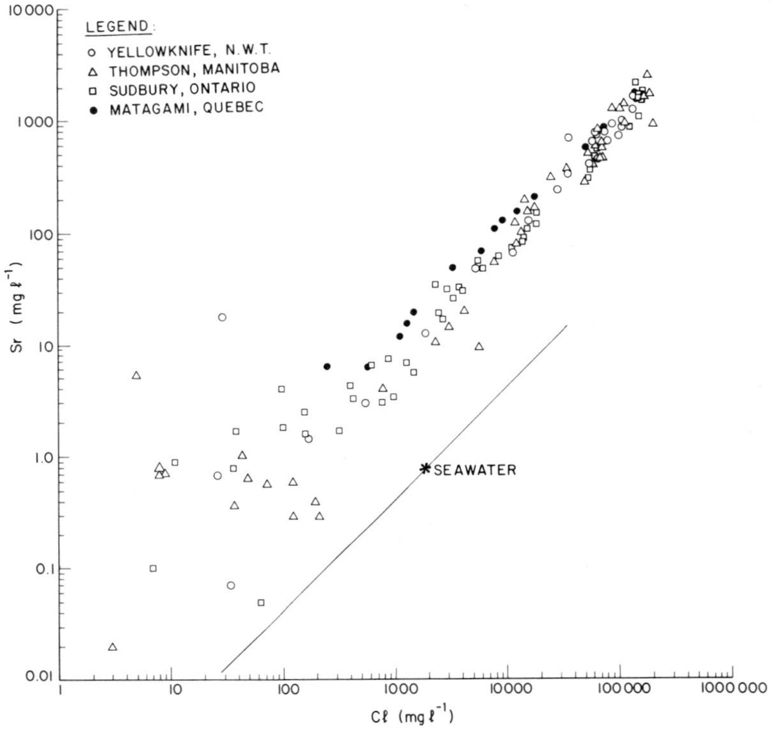

Figure 6a. *The relationship of strontium concentration and chloride concentration for groundwaters from the Canadian Shield relative to the dilution line for seawater.*

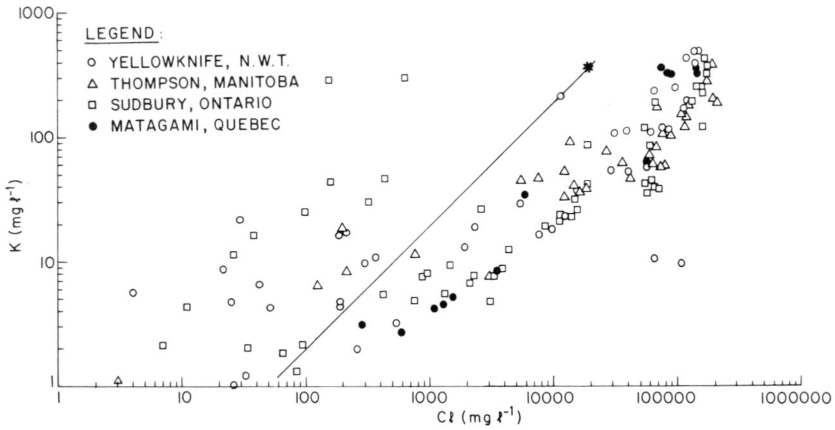

Figure 6b. *The relationship of potassium concentration and chloride concentration for groundwaters from the Canadian Shield.*

Sodium follows trends similar to that of calcium; although some influence of the rock chemistry was noted, especially at sites with a dominance of Na rich felsic rock types (Frape et al., 1984). This rock dependence thus shows some resemblance to Mg, although in a much less pronounced manner.

Other important cations which may be useful for assessing the source of salts and the degree of rock-water interaction are strontium, potassium and lithium. These elements are plotted against chloride in Figures 6a, b and c. Generally in felsic rocks strontium would have a very strong relationship with Ca, while this relationship is often poor in mafic rocks (Turekian, 1956). In the saline groundwaters and brines of the Canadian shield strontium levels certainly correlate well with calcium concentration and Figure 6a is very similar in appearance to the Ca versus Cl plot (figure 5a). This would suggest that much the same controls exist for strontium concentration as for calcium concentration. It is important to note that the Sr/Cl ratio is very similar for all sites across the Shield.

The value of strontium as a rock water indicator is primarily in the use of the $^{87}Sr/^{86}Sr$ isotopic ratio to measure the degree of equilibrium of the water and rock, as well as assessing the degree of mixing of waters having different isotopic ratios. The importance of strontium isotopes in assessing the degree of water-rock interaction for Canadian Shield groundwater has been discussed in detail by McNutt (this volume) and McNutt et al. (1984).

Potassium forms a major cationic component of rock forming minerals such as K-feldspar, micas and many secondary fracture filling clays. Potassium is easily dissolved from rocks during weathering, but it also tends to be easily removed from solution into secondary minerals such as the interlayer positions of phyllosilicates (Goldschmidt, 1958). As with Mg there is considerable scatter in the Cl versus K plot (Figure 6b). However, there are no apparent trends with rock type either in the dilute or concentrated groundwaters. In many respects this scatter of data may reflect the

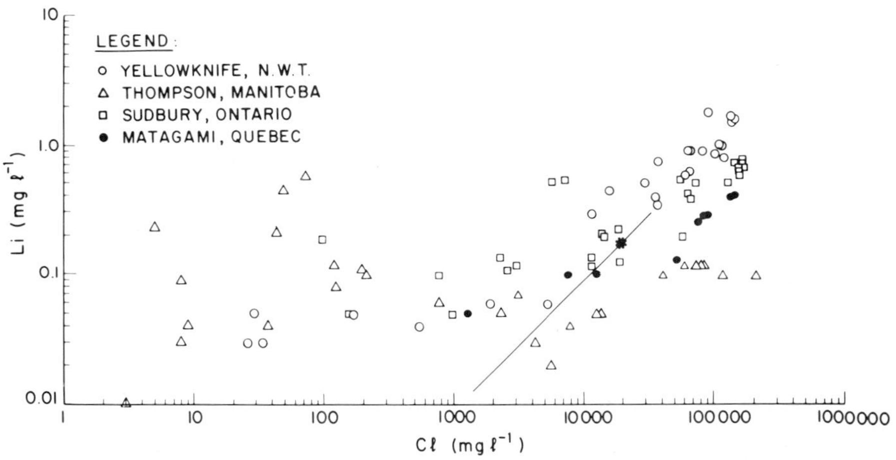

Figure 6c. *The relationship of lithium concentration and chloride concentration for groundwaters from the Canadian Shield.*

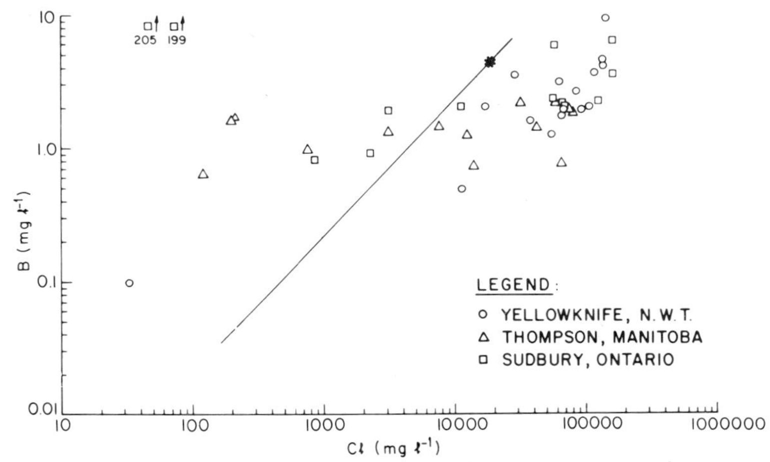

Figure 6d. *The relationship of boron concentration and chloride concentration for groundwaters from the Canadian Shield.*

variability in weathering susceptibility of major K host minerals such as feldspars and phyllosilicates; as well as the complex nature of K uptake in secondary clay minerals and the geochemical controls on such mineral forming reactions.

Of the minor element cations studied, lithium shows some of the strongest trends related to source rock and to intensity of rock-water interaction. In general, granitic rocks have higher lithium levels than mafic rocks (Horstman, 1957), often associated with minerals such as muscovite; although as pointed out by Edmunds et al. (this volume) lithium is usually found in the structure of minerals such as biotite in the granites they have studied. The geochemical behaviour of lithium and sources of this element are reflected in the distributions shown on Figure 6c. At all sites, with the exception of Thompson, the lithium concentrations are greater in the brines than the more dilute waters similar to the observations of Edmunds et al (1985). As can be seen on Figure 6c individual mixing or dilution lines exist for each site. In the most concentrated brines mafic rock dominated sites have the lowest Li values as would be predicted by Horstman (1957). At the Thompson site the Li values are relatively constant and may even show a slight decrease with increasing dissolved load. At the present we have no readily available explanation for this trend although one possibility is that many of the more concentrated samples issue from boreholes dominantly in intermediate to mafic rocks which would have lower Li levels. At sites such as Yellowknife where intermediate to felsic rocks occur the Li/Cl ratios are much higher than at the more mafic dominated sites such as Thompson and Matagami. The work of Edmunds et al (this volume) and Fuge and Power (1969) suggests that once a groundwater has acquired lithium it does not readily lose the ion to secondary minerals or exchange reactions. This would imply that lithium may continue to accumulate in solution with time and depending on the availability may be an excellent indicator of the degree of water-rock interaction.

Boron concentrations in Canadian Shield groundwaters increase with increasing salinity or Cl concentration (figure 6d). Edmunds et al. (1984) assume that the lithium and boron concentration of the groundwaters are derived from weathering of biotites, and other micas. The observed trends and relationship of both elements with chloride are very similar. However, in the Shield there are no obvious trends with rock type or site for boron concentration. This may indicate several possible sources and controls of chemistry, similar to potassium. Most groundwaters sampled on the Canadian Shield have boron concentrations below 4 mg l^{-1}, although some very concentrated brines have values approaching 10 mg l^{-1}. The highest values recorded are from two very acidic samples (pH <4) taken in the ore zone of the Fox Mine, Lynn Lake, Manitoba. These are brackish samples (TDS <10000 mg l^{-1}) with boron levels of approximately 200 mg l^{-1}. Although boron concentrations of this magnitude have been recorded in marine clays and some evaporites (Goldschmidt, 1954); it is probable that they only occur in Canadian Shield Waters which have a low pH.

Halogens

The plots of bromide, fluoride and iodide against chloride concentration provide insight into rock-water interaction within the Canadian Shield (Figures 7a, b and c).

Fluoride concentrations are usually below 10 mg l^{-1} in fresh and brackish groundwaters found at shallow depths, but levels can reach 25 mg l^{-1} in deep brines. Figure 7a shows that the concentration of F to Cl is highly variable at sites across the Canadian Shield. This is not surprising as Cl and F have very different chemical behaviour in rock forming processes (Kamineni, this volume). The F concentrations in some Yellowknife brines are slightly higher than at other sites and a slight positive correlation between F and Cl levels does exist at this site. Fluoride does not appear to be strongly related to the geochemistry of chloride, although increasing F concentration accompanies in-

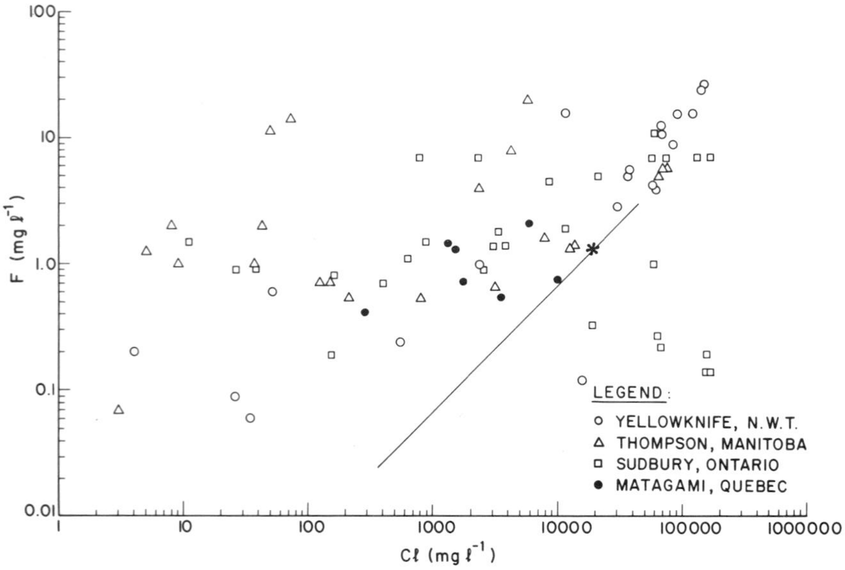

Figure 7a. *The relationship of fluoride concentration and chloride concentration for groundwaters from the Canadian Shield.*

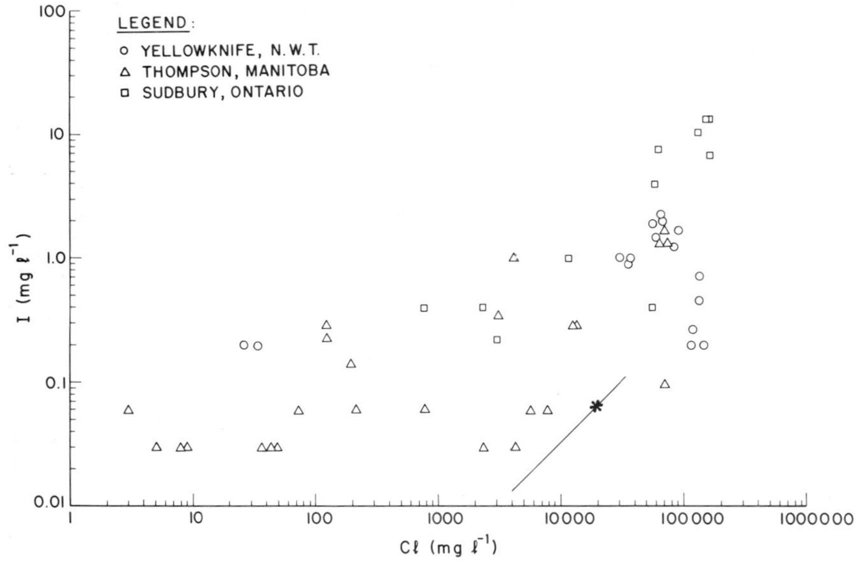

Figure 7b. *The relationship of iodide concentration and chloride concentration for groundwaters from the Canadian Shield.*

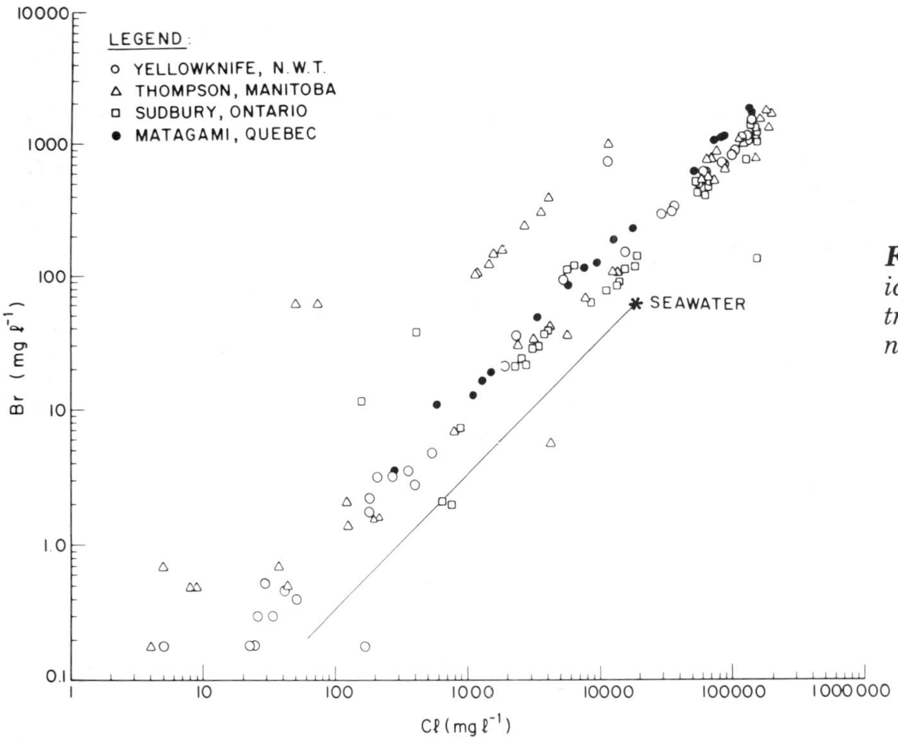

Figure 7c. *The relationship of bromide concentration and chloride concentration for groundwaters from the Canadian Shield.*

creasing salinity. Similar trends have been noted by Nordstrom et al. (1985) and have been attributed to a mineral reaction-controlling process provided by the formation or dissolution of the mineral fluorite. Fluorite is a common fracture filling at some sites on the Fennoscandian Shield, and future work may show that similar mineralogies and controls exist for the Canadian Shield as well. In general fluoride is the most reactive of the halogen ions and has a much greater geochemical ability to participate in rock forming and weathering reactions in crystalline rocks. It also appears that F has a strong geochemical relationship with Li. A plot of these two elements shows a relatively strong correlation for waters from Yellowknife where felsic rocks are more common (Figure 8). It is uncertain whether this trend is due to a common source and mixing with a strong solubility control of both elements in the more concentrated brines, or post mixing interaction with fracture minerals.

A stronger correlation exists between iodide and chloride (Figure 7b), although the variable iodide levels in many of the most concentrated brines would suggest several possible geochemical controls may be active. Slightly higher concentrations of iodide occur in some Sudbury Basin brines (maximum 13.8 mg l^{-1}) than at other Canadian Shield sites. In all samples with chloride concentrations less than seawater, iodide is considerably enriched relative

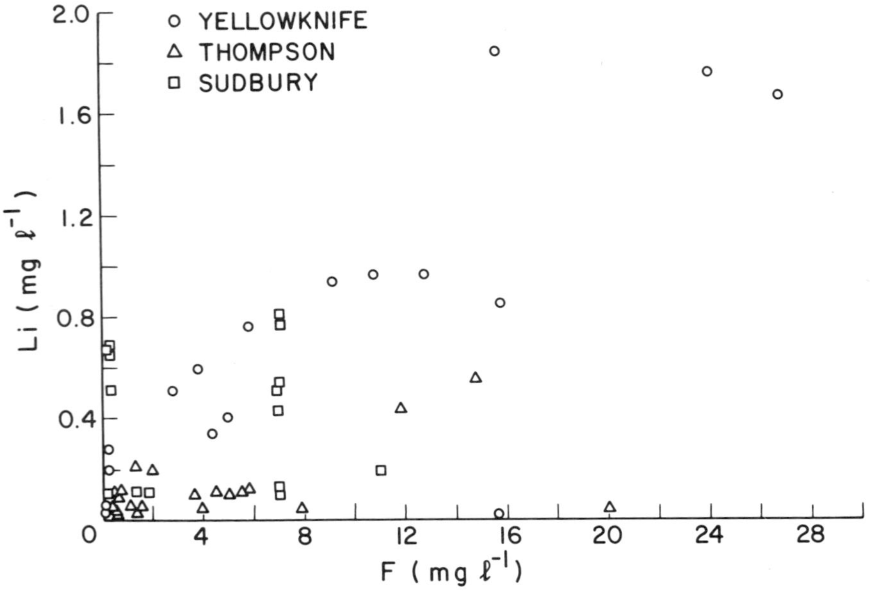

Figure 8. *The relationship of fluoride concentration and lithium concentration for groundwaters from the Canadian Shield.*

to this potential source. As well, iodide concentrations in Shield brines are much higher than the values reported for crystalline rocks (approximately 0.2 ppm) by Fuge et al., (1978) or (0.5 ppm) by Becher et al. (1968). As iodide is very water soluble (Fuge et al., 1978) and like bromine does not readily substitute into mineral phases because of its large ionic size, it will tend to concentrate preferentially in the aqueous phase during rock-water interaction. Lloyd et al. (1982) postulate that iodide occurs at crystal boundaries or in fluid inclusions. In the former it would probably be readily leachable.

A very strong correlation of Cl to Br exists in groundwaters from across the Canadian Shield. In leaching studies of volcanic rocks from Japan, Yoshida et al. (1971) concluded that Br and Cl chemistries were controlled by the same factors and therefore should show strong correlations. However, only poor correlation exists for I or F with either Br or Cl, and therefore these halogens are controlled by different geochemical mechanisms.

Numerous authors have pointed out the very conservative nature of Br and Cl in aqueous solutions and brines (Rittenhouse, 1967; Collins, 1975; Carpenter, 1978; Edmunds et al., 1985). In a previous paper the authors have used the Br/Cl ratio to demonstrate that Shield brines are highly enriched in bromide, or depleted in chloride, relative to seawater (Frape et al., 1984).

At many Canadian Shield sites the fresh groundwaters appear to be gaining Br preferentially over Cl. Data plotted on Figure 7c shows a very interesting trend between Br and Cl. Some Thompson samples have much higher Br/Cl ratios (~0.1) than some brines collected less than 10 km away or than most Canadian Shield groundwaters. This suggests that the concentrations of one or both elements may be significantly controlled by rock sources. Furthermore, the second parallel mixing or dilution line suggests that uniform Br/Cl ratios do not exist for the most concentrated brines found across the Canadian Shield. The significance of the differences is important in evaluating evolutionary schemes for the origin of Shield brines based on the conservative nature of the Br/Cl ratio (Fritz and Frape, 1982), and possible sources such as sedimentary basin brines or fossil seawater. Here conservative is used to imply that little addition or subtraction from solution occurs and therefore the Br/Cl ratio can be used to indicate a possible source of the brine.

The available data suggest that no single source exists and that the variability over short distances and the slightly higher Br/Cl ratios in more mafic rocks are probably related to source controls in the crystalline host rock. For instance Duce (1972) found that Br levels were slightly higher in mafic rocks (~2.5 ppm) over granitic rocks (~1.6 ppm). Nordstrom et al., (1985) in leaching experiments of granites has found Br/Cl ratios of around 0.010. It is interesting to note that the authors have found saline groundwater springs issuing from a very Na rich felsic granite in the Thunder Bay, Ontario area which have Br/Cl ratios similar to seawater (.0033) (Frape et al., 1984a). Such ratios in a groundwater discharging from crystalline rocks on the Canadian Shield further complicates any scheme using these elements as conservative ions to predict brine origin.

Thus, even if the sole source of the Br and/or Cl is not the crystalline rocks of the Canadian Shield, it becomes apparent that the concentrations of even these "conservative" ions are influenced by the rock mass and the intensity of rock-water interaction.

Heavy metals

It has often been assumed that the heavy metal concentrations in solution are heavily dependent upon the availability of metals in the host rock. At many mine sites because of the proximity of sulphide ore bodies and the oxidation of sulphides by fresh groundwaters, this availability is potentially very high. A plot of total metals versus sulphate (Figure 9a) does not reveal any obvious relationships between sulphate and total metal concentration. Therefore, as shown by sulphate concentration, the oxidation of sulphides, although a possible source, may not be a key control on the levels of metal ions in these groundwaters. The only exceptions may be a group of very acidic

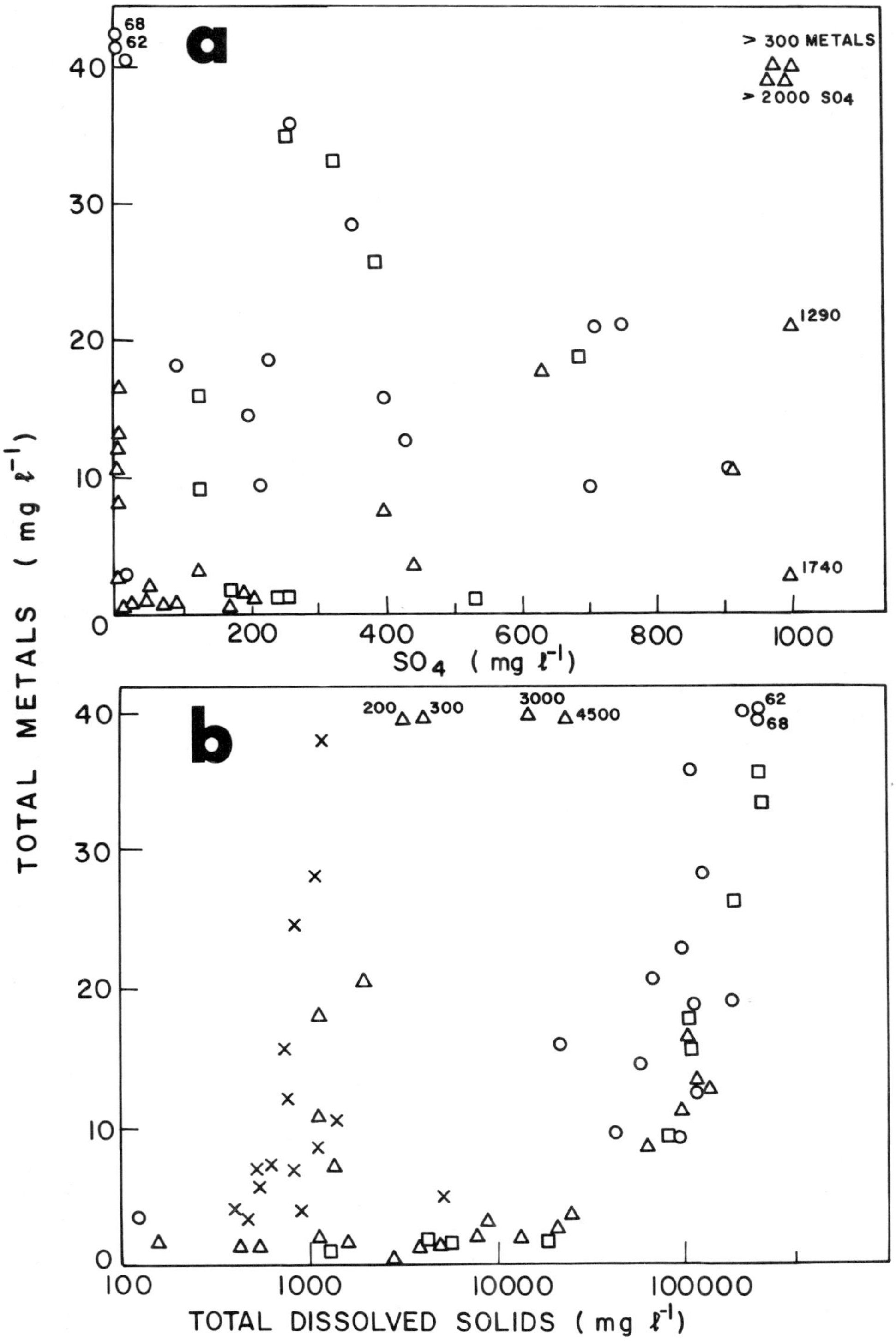

Figure 9. The sum of total heavy metal concentration plotted against (a) sulphate concentration and (b) against total dissolved solids for Canadian Shield groundwaters (total heavy metal axis was cut off at 40 mg l^{-1} in both figures although samples as shown exist above this value).

samples (pH <4) from the Thompson site which have total metal values in excess of 300 mg l^{-1} and SO_4 values in excess of 2000 mg l^{-1}. However, for the majority of groundwaters sampled other controls such as solubility of metals in high ionic strength solutions or alternate sources such as silicates-carbonates must dominate.

It appears that the concentration and thus solubility of some heavy metals such as Fe, Mn and Zn are much higher in the low TDS waters (< 10000 mg l^{-1}) where redox processes, pH controls and oxy-hydoxide complexes would likely dominate. Therefore, the majority of the total heavy metals in low TDS samples are usually combinations of these three elements (Figure 9b). At very high TDS values (> 50000 mg l^{-1}) Fe and Mn concentrations are often elevated, but many of the other heavy metals, with the exception of Zn, show concentration increases of several orders of magnitude. Some elements such as Co, Ni and Pb are often the dominant heavy metals found in brine samples. It is important to note that Cu concentrations are usually very low in the brines. This coupled with high Ni and Co levels poses a problem in interpreting the controls on metal species in such solutions. Dissociation constant data would predict that Cu concentrations should be very high as Cu complexes readily with Cl; whereas Ni and Co complex weakly with Cl. It would be convenient to assume that metal chloride complexes account for the increased

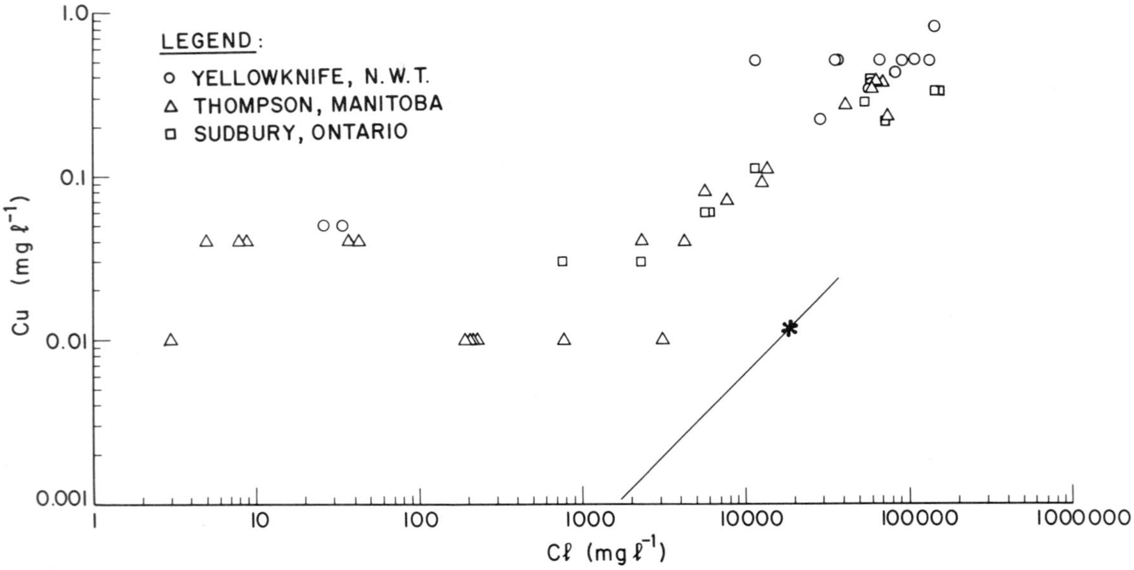

Figure 10a. *The relationship of copper concentration and chloride concentration for groundwaters from the Canadian Shield.*

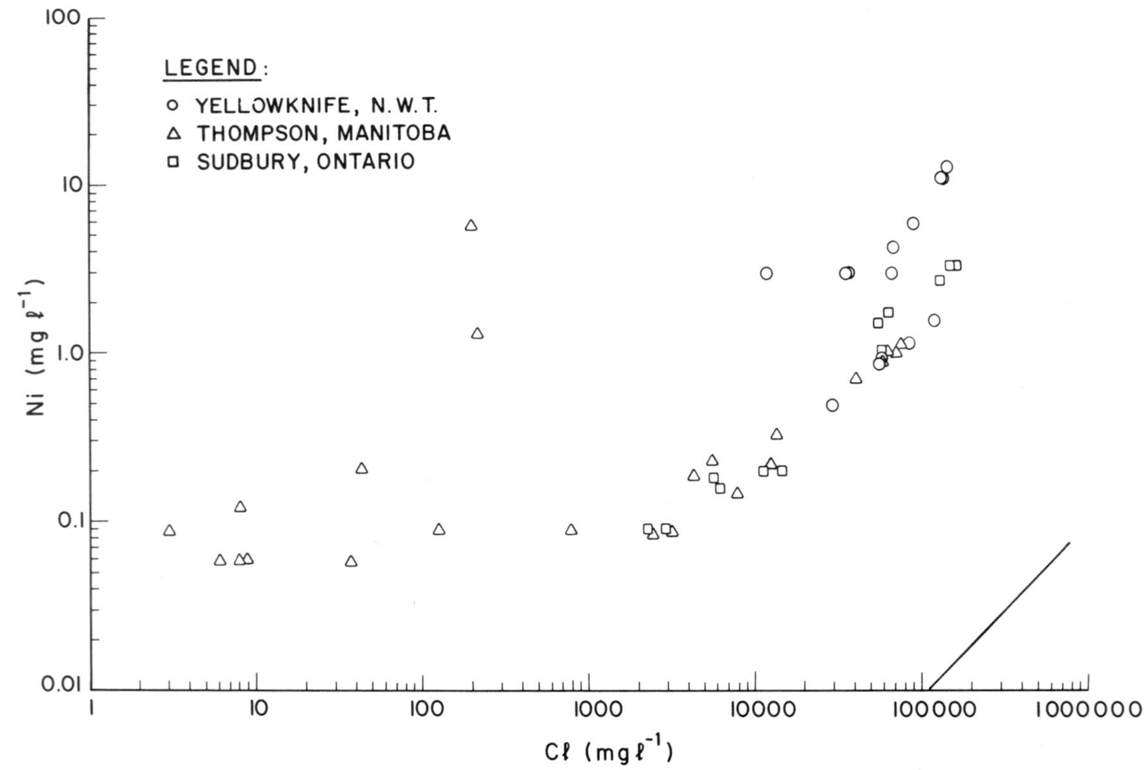

Figure 10b. *The relationship of nickel concentration and chloride concentration for groundwaters from the Canadian Shield.*

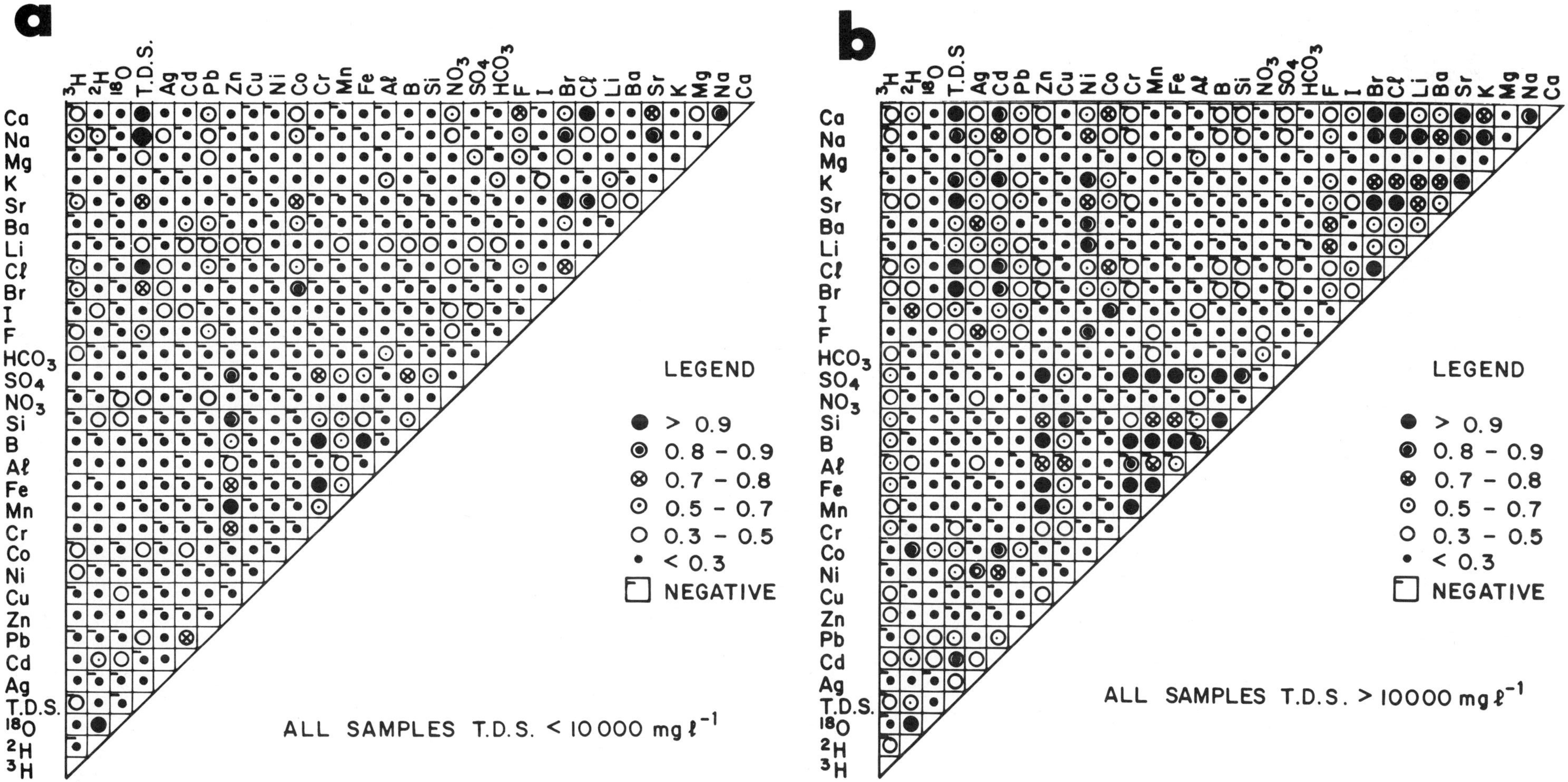

Figure 11. *Correlation matrix for parameters from Canadian Shield groundwaters with (a) TDS < 10000 mg l*$^{-1}$ *and (b) TDS > 10000 mg l*$^{-1}$.

concentrations of some heavy metal species in the brine; but this may not be geochemically the most correct explanation.

It should be noted that almost all plots of metal versus TDS show a similar low concentration of total metals in intermediate concentration waters (5000-50000 mg l^{-1} TDS) (Figure 9b). Exceptions are very acidic waters found at some mine sites. Obvious reasons for these low concentrations of metals may be related to the instability of metal compounds in waters which are mixtures of very concentrated metal rich chloride brines and dilute brackish or fresh waters. Many intermediate waters have approximately equal concentrations of Cl, SO_4 and HCO_3 and the low solubility of metal compounds in such solutions or the formation of meta-stable phases may result in the precipitation of metal compounds as part of the dominant fracture fillings found in these zones. Much more detailed experimental and field sampling would be required to test any of these hypothesis.

Plots of Cu and Ni versus chloride concentration are typical of trends for groundwaters from the Shield (Figures 10a and 10b). Copper concentration in the most saline brines are very uniform at a maximum value of 1.0 mg l^{-1}, which suggests a strong solubility control in Ca-Na-Cl solutions (Figure 10a). Clear trends with rock type were not noted. This is also the case for Ni which shows maximum concentration of ~ 13 mg l^{-1} (Figure 10b). Elevated Ni levels (> 1 mg l^{-1}) occur in several more dilute samples (Cl ~ 250 mg l^{-1}) that have very low pH values (<4). It is interesting to note that brines from massive sulphide nickel mines at Thompson and Sudbury have lower Ni/Cl ratios than brines from the Yellowknife gold mines. Major aqueous geochemical controls on nickel are pH (soluble in acidic solutions) and the coprecipitation with Fe-Mn oxy-hydroxides (Avis, 1972). It is very apparent that a strong geochemical control, is influencing the concentrations of Cu, Ni and most metals in the brines. As metal-chloride complexing is probably not the major control we must look for other possibilities. Perhaps the coprecipitation with Fe-Mn complexes or equilibrium with dominant secondary mineral phases would help to control metal levels in the brines. At this time it is unclear what the major controls on metal concentrations may be for these Ca-Na-Cl brines.

Correlation matrices

In an attempt to further explore possible geochemical relationships within the Canadian Shield data set, correlation matrices were generated for the fresh and brackish groundwaters (TDS < 10000 mg l^{-1}) and the saline waters and brines (TDS > 10000 mg l^{-1}) (Figures 11a and b). These plots were generated using a Spearman correlation program from the SAS statistics package. Many very strong correlations exist in both of the data sets and in consequence the division of the r values was chosen to best distinguish the better correlations. The fresh and brackish groundwater data set contained approximately 75 samples for which most parameters were analyzed; whereas the saline water and brine data set is based on approximately 150 samples which have almost complete chemistries.

Many of the strong correlations with Chloride, which have been shown already in previous figures, appear using this approach. However, a number of high correlations exist which were not readily apparent. As an example both data sets show very strong relationships between boron, aluminum and several of the heavy metals such as chromium or iron. It is not certain what these particular correlations represent, although the influence of the weathering of mafic or ultramafic rocks or dykes may play a role. Likewise both data sets show strong correlations between many of the heavy metals as well as individual metal species and sulphate. The figures represent a readily understandable summary of the geomechanical interrelationships in these data sets. Inorder to fully understand all of the strong correlation presented considerably more experimental water-rock interaction studies on crystalline rock will be required.

SUMMARY

It is very apparent that the source of the salinity in deep Canadian Shield groundwaters is not readily discernable. The dominant theories for the origin of salinity in crystalline rocks have been summarized and argued by several of the authors in this volume and other publications. There is a school strongly supported by Kelly et al (1986) and Guha (this volume) which would argue for the emplacement of basin brines or modified Paleozoic seawater into the Canadian Shield. Thus the salinity would have an allocthonous origin and we need only account for the uniform isotopic signature and geochemical similarity of brines from all locations (Frape and Fritz, 1982). Indeed some evidence would suggest that sedimentary basin brines may be entering marginal areas of the Shield (Gascoyne et al (this volume)).

The second major school of thought would suggest that the high salinities of typical Shield brines occur as a product of the leaching of fluid inclusions or intense reactions with certain mineral phases such as micas or amphiboles. (Nordstrom et al., 1985; Gascoyne et al., this volume; Kamineni, this volume; Edmunds et al, this volume, 1984, 1985). In all cases whether fluid inclusions, grain boundary salts or weathered mineral phases are the source, the origin of the salts is autochthonous and therefore the degree of rock-water interaction becomes critical.

One indication of the degree of rock-water interaction in the Canadian Shield may be the maximum levels or uniformity of concentration of certain elements in the most saline solutions. Many cations and anions have very uniform values. Any disturbance or mixing-dilution of these brines would probably result in a non-equilibrium situation. For most of the heavy metals and elements such as boron, potassium and fluorine this does result in strong post mixing reaction. Other elements such as, strontium, sodium, and calcium show a very conservation behaviour which leads to well defined mixing lines. Elements such as potassium and magnesium may have several competing mineral phases active over the complete salinity range. Thus, the concentration of these elements in solution can be variable over a very wide range of salinities. As well, the concentrations of many elements such as Mg, Li, F and to a lesser extent Br appear to be directly related to the composition of the rocks at different sites.

It appears that the extent of rock water interaction is extreme because of the similarities of Shield brines from a wide range of locations. This similarity could only be achieved if the same primary and secondary mineral controls were to occur in each area, as well as the same solubility controls dominating for individual elements or complexes in Ca-Na-Cl fluids.

The age of the solute is probably that of the rock, as much of the solid component of the groundwaters appears to come from or exchange with the host rock. A large portion of the water may have a separate origin from the solute phase which, because of its very different isotopic composition in the most concentrated brines, presents a very difficult fluid to define either for age or origin.

ACKNOWLEDGEMENTS

The authors wish to express their appreciation to Atomic Energy of Canada Limited for financial and analytical support from 1979-1986. Without this assistance the advances and interest in groundwaters within the Canadian Shield would not have occurred. In particular we wish to acknowledge the special help of M. Gascoyne and R. Pearson during this period. As well, financial assistance was provided by NSERC to the authors. Finally, the help of most mining Companies on the Canadian Shield has been invaluable, but we particularly wish to thank the personnel of Cominco Limited, Giant Yellowknife Mines, INCO metals, Falconbridge Nickel Mines Limited and Noranda Mines Limited. The paper benefited from readings by E. Reardon, M. Gascoyne and several anonymous readers.

REFERENCES

Anonymous, 1889. Annual Report 1887-88: Geological Natural History Survey of Canada, v. 3, p. 28T-29T.

Avias, J., 1972. Nickel: Element and geochemistry: In the Encyclopedia of Geochemistry and Environmental Sciences IVA, Ed. R.W. Fairbridge, Van Nostrand Reinhold Pub., New York.

Becher, V., Bennet, J.H. and Manvel, O.K., 1968. On iodine and uranium in ultrabasic rocks and carbonatites: Earth Planterary Science Letters, v. 4, p. 357-362.

Bottomley, D.J. and Graham, B.W., 1980. Progress Report on hydrogeochemical studies in the hydrogeologic research programme 1978/79. Atomic Energy of Canada Limited, Technical Record TR-28, 29 p.

Boyle, R.W., 1961. Geology, geochemistry and origin of the gold deposits of the Yellowknife district, Northwest Territories: Geological Survey of Canada Memoir 310, 193 p.

______, 1979. The geochemistry of gold and its deposits: Geological Survey of Canada, Bulletin 280, 584 p.

Bruce, E.L., 1941. Concentrated saline water from the Sturgeon River Gold Mines: Transactions of the Royal Society of Canada, v. 35, p. 25-29.

Carpenter, A.B., 1978. Origin and chemical evolution of brines in sedimentary basins: Oklahoma Geological Survey Circular 79, p. 60-77.

Collins, A.G., 1975. Geochemistry of oil field waters: Elsevier, Amsterdam, 485 p.

Davison, C.C., 1984. Hydrogeological characterization at the site of Canada's Underground Research Laboratory: Groundwater Monitoring Review 4, p. 95-102.

Davison, C.C., 1985. Hydrogeological characterization of the URL site: Proceedings of the Seventeenth Information Meeting of the Nuclear Fuel Waste Management Program, Atomic Energy of Canada Limited, Technical Record, 299, p. 231-253.

Dickin, R.C., Frape, S.K., Fritz, P., Leech, R.E.J. and Pearson, R., 1985. Groundwater chemistry to depths of 1,000 m in low permeability granitic rocks of the Canadian Shield: International Symposium on Groundwater Resources Utilization and Contaminant Hydrogeology, I.A.H., Montreal, May, 1984.

Duce, R.A., 1972. Bromine: Element and geochemistry: In the Encyclopedia of Geochemistry and Environmental Sciences IVA, Ed. R.W. Fairbridge, Van Nostrand Reinhold Co., New York.

Edmunds, W.M., Andrews, J.N., Burgess, W.G., Kay, R.L.F. and Lee, D.J., 1984. The evolution of saline and thermal groundwaters in the Carnmenellis granite: Mineralogical Magazine, v. 48, p. 407-424.

Edmunds, W.M., Kay, R.L.F. and McCartney, R.A., 1985. Origin of saline groundwaters in the Carnmenellis granite: Natural processes and reaction during Hot Dry Rock reservoir circulation: Chemical Geology, v. 49, p. 287-302.

Frape, S.K. and Fritz, P., 1981. A preliminary report on the occurrence and geochemistry of saline groundwaters on the Canadian Shield: Atomic Energy of Canada Limited, Technical Record 136, 71 p.

______, 1982. The Chemistry and isotopic composition of saline groundwaters from the Sudbury basin, Ontario: Canadian Journal of Earth Sciences, v. 19, p. 645-661.

Frape, S.K., Fritz, P. and Blackmer, A.J., 1984a. Saline groundwater discharges from crystalline rocks near Thunder Bay, Ontario, Canada: International Symposium on Hydrochemical Balances of Freshwater Systems, Uppsala, Sweden, September, 1984.

Frape, S.K., Fritz, P. and McNutt, R.H., 1984. The role of water-rock interaction in the chemical evolution of groundwaters from the Canadian Shield: Geochimica et Cosmochimica Acta, v. 48, p. 1617-1627.

Fritz, P. and Reardon, E.J., 1979. Isotopic and Chemical Characteristics of mine water in the Sudbury area: Atomic Energy of Canada Limited, Technical Report, 35, 37 p.

Fritz, P. and Frape, S.K., 1982. Saline groundwaters in the Canadian Shield – a first overview: Chemical Geology, v. 36, p. 179-190.

Fuge, R., 1974. Chlorine. In: K.H. Wedepohl (editor), Handbook of Geochemistry, Springer, Berlin, Ch. 17.

Fuge, R., 1977. On the behaviour of fluorine and chlorine during magnatic differentiation: Contributions to Mineralogy and Petrology, v. 61, p. 245-249.

Fuge, R., 1979. Water-soluble chlorine in granitic rocks: Chemical Geology, v. 25, p. 169-174.

Fuge, R. and Power, G.M., 1969. Chlorine and fluorine in granitic rocks from SW England: Geochimica et Cosmochimica Acta, v. 33, p. 888-893.

Fuge, R., Johnson, C.C. and Phillips, W.J., 1978. Iodine in granitic and associated rocks: Chemical Geology, v. 22, p. 347-352.

Garrels, R.M., 1967. Genesis of some ground waters from igneous rocks: In Researches in Geochemistry, v. 2, (P.H. Abelson ed.), p. 405-420. John Wiley, New York.

______, 1976. A survey of low temperature water-mineral relations: In Interpretation of Environmental Isotope and Hydrochemical Data in Groundwater Hydrology. I.A.E.A. Special Volume, Vienna, Austria, 1976, p. 65-84.

Gascoyne, M., Davison, C.C., Ross, J.D. and Pearson, R., (this volume).Saline groundwaters and brines in plutons in the Canadian Shield.

Goldschmidt, V.M., 1954. Geochemistry: Oxford, Clarendon Press, 730 p.

Guha, J., Leroy, J. and Guha, D., 1979. Significance of fluid phases associated with shear zone Cu-Au mineralization in the Dore Lake Complex, Chibougamou, Quebec: Bulletin Minéralogie, v. 102, p. 569-576.

Horstman, E.L., 1957. The distribution of lithium, rubidium and calcium in igneous and sedimentary rocks: Geochimica et Cosmochimica Acta, v. 12, p. 1-28.

Kamineni, D.C., (this volume). Halogen-bearing minerals in plutonic rocks: a possible source of chlorine in saline groundwater in the Canadian Shield.

Kelly, W.C., Rye, R.O. and Livnat, A., 1986. Saline minewaters of the Keweenaw Peninsula, Northern Michigan: Their nature, origin and relation to similar deep waters in precambrian

crystalline rocks of the Canadian Shield: American Journal of Science, v. 286, p. 281-308.
Lane, A.C., 1908. Mine waters and their field assay: Geological Society of America, Bulletin, v. 19, p. 501-512.
______, 1914. Mine water composition, an index to the course of ore-bearing currents: Economic Geology, v. 9, p. 239-263.
______, 1927. Calcium chloride waters, connate and diagenetic: American Association of Petroleum Geologists Bulletin, v. II, no. 12, p. 1283-1305.
Langford, G.B. and Hancox, E.G., 1936. Hypogene anhydrite from the McIntyre mine, Porcupine district, Ontario: Economic Geology, v. 31, p. 600-609.
Lloyd, J.W., Howard, K.W.F., Pacey, N.R. and Tellam, J.H., 1982. The value of iodide as a parameter in the chemical characterization of groundwaters: Journal of Hydrology, v. 57, p. 247-265.
McNutt, R.H., Frape, S.K. and Fritz, P., 1984. Strontium isotopic composition of some brines from the Precambrian Shield of Canada: Isotope Geoscience, v. 46, p. 205-216.
Nordstrom, D.K., Andrews, J.N., Carlsson, L., Fontes, J.C., Fritz, P., Moser, H. and Olssen, T., 1985. Hydrogeological and hydrogeochemical investigations in boreholes – final report of the phase I geochemical investigations of the Stripa groundwaters: SKB Stripa Project, Swedish Nuclear Fuel and Waste Management Company, Report 85-06, 250 p.
Rittenhouse, G., 1967. Bromine in oil-field waters and its use in determining possibilities of origin of these waters: American Association of Petroleum Geologist Bulletin, v. 51, p. 2430-2440.
Rittenhouse, G., Fulton, R.B., Grabowski, R.J. and Bernard, J.L., 1969. Minor elements in oil-field waters: Chemical Geology, v. 4, p. 189-209.
Tamari, Y., Tsuji, H. and Kusaka, Y., 1983. Influence of geology on the chemical compositions of groundwaters – Analysis of Na, K, Ca, Mg, and Sr utilizing a chemical leaching method: Proc. International Symposium Water-Rock Interaction. Misasa, Japan, September, 1983, Institute of Thermal Spring Research, Okayama University, p. 472-474.
Tanton, T.L., 1931. Fort William and Port Arthur, and Thunder Cape Map areas, Thunder Bay District, Ontario: Geological Survey of Canada, Memoir 167, p. 222.
Turekian, K., 1956. The geochemistry of strontium: Geochimica et Cosmochimica Acta, v. 10, p. 145-196.
Yoshida, M., Takahashi, K. and Yonehara, N., 1971. Fluorine, chlorine, bromine and iodine in volcanic rocks: Volcanological Society of Japan Bulletin, v. 16, p. 43-44.

Models of Mineral Controls on the Composition of Saline Groundwaters of the Canadian Shield

F.J. Pearson, Jr.
Intera Technologies, Inc., 6850 Austin Center Blvd., Suite 300, Austin, Texas 78731, U.S.A.

Pearson, F.J., Models of mineral controls on the composition of saline groundwaters of the Canadian Shield; in Saline water and gases in crystalline rocks, Editors: Fritz, P., and Frape, S.K.; Geological Association of Canada Special Paper 33, 1987. p. 39–51.

Abstract

Geochemical modelling of brines from the crystalline rocks of the Canadian Shield strongly suggests that low temperature reactions between silicate minerals and water control the geochemical evolution of these brines. This argument is further substantiated by stable isotope analyses on the brines. The samples collected are typically mixtures with dilute meteoric waters, but if halite saturation is assumed for endmember brines then extrapolation of both ^{18}O and ^{2}H data is possible. The $\delta^{18}O$ and $\delta^{2}H$ values estimated for these source brines range from −8.3 to −9.1 and +10 to −10 permille respectively. These values could result from an approach to water-rock equilibrium at temperatures close to about 25°C.

Résumé

La modélisation géochimique des eaux sursalées des roches cristallines du bouclier canadien semble démontrer que des réactions à faible température entre l'eau et les minéraux silicatés régissent l'évolution géochimique des saumures. Des analyses d'isotopes stables effectuées sur les saumures appuient cette hypothèse. Habituellement, les échantillons recueillis sont des mélanges comprenant une part d'eau météoritique mais, en admettant que les saumures initiales soient saturées en halite, l'extrapolation de la teneur en ^{18}O et ^{2}H est possible. L'estimation de la teneur en $\delta^{18}O$ et $\delta^{2}H$ de ces saumures originelles varie respectivement de −8,3 à −9,1 et +10 à −10 pour mille. Ces valeurs pourraient résulter de conditions voisines du point d'équilibre eau-roche à des températures d'environ 25°C.

INTRODUCTION

Groundwaters with salinities of up to 325 grams per liter (g/ℓ) have been found at depth within the Canadian Shield. Such groundwaters have been sampled in mines in the Sudbury, Thompson and Yellowknife mining districts by staff of the Department of Earth Sciences at the University of Waterloo (Fritz and Reardon, 1979; Fritz and Frape, 1980 and 1982; Frape and Fritz, 1981a; 1981b and 1982; Frape and others, 1984; McNutt and others, 1984, and in these proceedings). Samples of saline waters from boreholes have been collected at two AECL research areas by AECL staff and contractors as described by Gascoyne and others (these proceedings). This paper proposes likely compositions for source brines in the mining districts and examines the state of geochemical and isotopic equilibrium that exists between these brines and their host rocks.

In papers by Frape and Fritz and their co-authors, it was first pointed out that a group of samples collected from a given mine or district have the following characteristics:

- While the waters have salinities ranging from less than one to several hundred g/ℓ, the proportion of the various dissolved species stays relatively constant;
- The major anion is chloride and the main cation is calcium. Sodium (in one mining district, magnesium) is the next most prominent cation;

- The concentration of the major ions sodium, calcium and chloride tends to increase with depth.

As Frape, Fritz and their co-authors have also pointed out, the stable isotopic composition of the groundwaters also varies systematically both within groups of samples from a single mine or mining district, and from one district to another. The $\delta^{18}O$ and δD values of the least saline samples fall on the normal meteoric water line, which describes the world wide variation of these values in precipitation and other non-evaporated surface waters. With increasing salinity, both the $\delta^{18}O$ and δD values become more positive and describe straight lines, but with slopes different from that of the normal meteoric water line. The systematic variation of the isotopic composition of these waters suggests that the samples represent mixtures of local meteoric water with a highly saline fluid. Further, this variation provides a tool for estimating the properties of the highly saline source brine.

In a number of groundwater systems it has been possible to evaluate quantitatively the controls exerted by minerals of the aquifer on the chemistry and isotopic composition of the water (Plummer and others, 1983). Most such studies have been carried out on sedimentary aquifers or, if in crystalline rocks, only in groundwaters at temperatures approaching or exceeding 100°C. Mineral controls are not as easily demonstrated in lower temperature groundwaters in crystalline rock aquifers because of slow reaction rates between water and rock forming silicate minerals.

Frape and Fritz (1982, for example) present a number of hypothesis on the origin of the saline waters of the shield. Regardless of their origin, the most saline waters and the source brines from which they are derived by mixing with local meteoric waters, are likely to have had a long residence time in the crystalline rock of the shield. This paper tests the hypothesis that, because of this great residence time, even very slow silicate reactions have influenced the chemical and isotopic composition of the source brines. These reactions could have been extensive enough to obscure if not obliterate evidence on the origin of the water of the brines.

Several difficulties attended the testing of this hypothesis. First, to model the solution chemistry required the application of an equation for activity coefficients for dissolved species which is suitable to waters of high ionic strength. Second, thermodynamic data for rock forming silicate minerals are usually determined in high temperature experiments and extrapolated to lower temperatures with considerable uncertainty. Finally, water-rock isotopic fractionation factors at low temperatures are also extrapolated from high temperature measurements. The uncertainties in low-temperature fractionation factors are even higher than those for mineral thermodynamic data.

ESTIMATION OF CHEMISTRY OF SOURCE BRINES

All the samples of saline ground waters so far reported seem to represent mixtures of meteoric water with a highly concentrated brine. Two lines of evidence were used to estimate the composition of the as yet unsampled source brine. The first takes advantage of the systematic relationship between the isotopic composition of the water samples and their chemical composition. The second considers the range of possible source brine chemistries as limited by reactions with minerals of the shield.

Chemical and isotopic composition of mixed samples

The data on water chemistry and isotopic composition display two types of systematic behavior which are useful for estimating the composition of the source brines. The first regularity is in the variation of the isotopic composition of the samples with their location. The second is the regular variation of the salinity of the samples with their isotopic composition.

Variation of isotopic composition with location

The pattern of variation of the stable isotopic composition of samples from each mining district is similar. Figure 1 is a graph of $\delta^{18}O$ for one of the mines sampled near Yellowknife, NWT. The solid line on this figure is the normal meteoric water line:

$$\delta D = 8\ \delta^{18}O + 10.$$

The data from each mine describe two straight lines which intersect the meteoric water line at about the same point. One has a slope of less than 8 and the other a slope of greater than 8. The slopes of the less steep lines of each of the sample groups ranges from 3.4 to 5.2 with a mean of 4.5 (Pearson, 1986, Table 2-1). This slope is typical of meteoric waters which have undergone some evaporative concentration (Gat and Gonfiantini, 1981, Chap. 3). The groups of samples describing lines of low-slope include some samples of moderate salinity, but all the samples of highest salinity are in the other group. The salinities observed in the low-slope group could well result from simple evaporative processes.

The lack of samples of compositions intermediate between the two groups on Figure 1 is remarkable, and suggests that two separate groundwater systems may be present. Most mine samples were collected where the original groundwater flow system has been disturbed by the mines themselves. Thus, it is not generally possible to determine which part of the original flow system was the source for a given sample, and explore the implications of the lack of intermediate samples in any detail.

The samples on which the remainder of this paper will concentrate are those whose isotopic compositions describe lines of slopes greater than 8. These groups include the most saline samples taken at each location. The slopes of these lines differ from location to location but, when plotted together, they tend to converge. This convergence was pointed out by Frape and Fritz and their colleagues, who also noted the existence of a group of samples with slopes of less than 8, like those of evaporating surface water, discussed above.

Figure 2 shows the trends of the isotopic compositions of samples from the several mining districts, it is based on Figure 2-11 of Pearson (1985) and is similar to Figure 1 of Frape and others, 1984. The slopes of the lines are not greatly different and they are uncertain because of scatter in the data on which they are based. Thus, the region of convergence can be considered only very broadly defined by this figure. However, it is clear that the common meeting point has a δD value of about zero and a $\delta^{18}O$ value of perhaps minus 8 or 9.

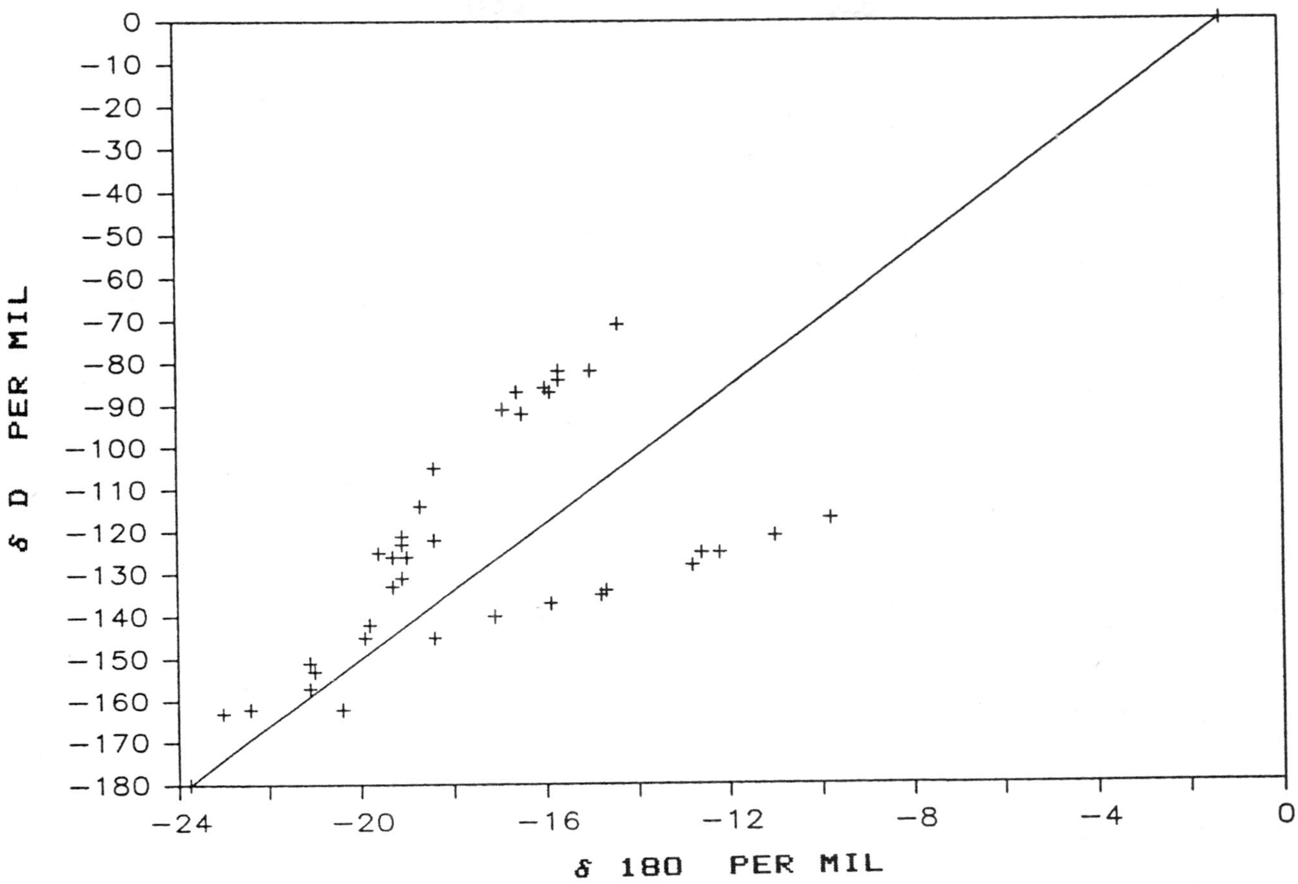

Figure 1. *Plot of $\delta^{18}O$ vs. δD for samples from the Con Mine Yellowknife, NWT.*

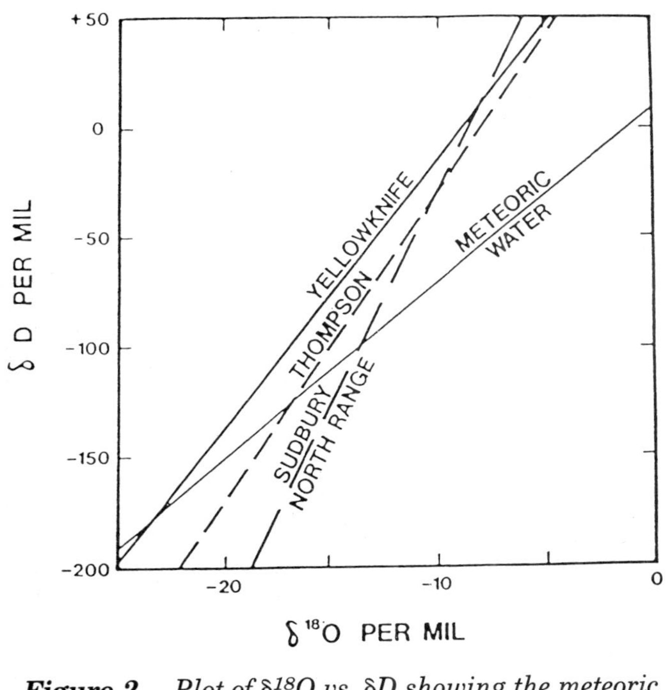

Figure 2. *Plot of $\delta^{18}O$ vs. δD showing the meteoric water line and regression lines for sample groups from Yellowknife, Thompson and Sudbury.*

This is an unusual isotopic composition, because most natural waters fall either on or to the lower right of the meteoric water line, while this composition is to the upper left of that line. As discussed below, this composition could be developed by isotopic exchange at relatively low temperatures between water of almost any initial isotopic composition and silicate rock.

Variation of salinity with isotopic composition

Samples with isotopic compositions to the upper left of the meteoric water line show regularly increasing salinity with the departure of their isotopic composition from the meteoric water line. Figure 3 shows the salinity (total dissolved solids) and chloride concentrations of the same group of samples shown in Figure 1 plotted against their $\delta^{18}O$ values. The relationship between increasing concentrations of chemical species and increasing oxygen 18 concentrations (indicated by less negative $\delta^{18}O$ values) is clear. Figure 4 shows the same behavior for the sodium and calcium concentrations of the same group of samples.

Graphs such as these for groups of analyses from each of the districts sampled have been used to estimate the composition of the source brines or end member brines which, when mixed with local meteoric water, give rise to the samples of varying salinity. Initial estimates of source brine composition were made by extrapolating concentrations to $\delta^{18}O$ values of minus 8 per mil, a value chosen as

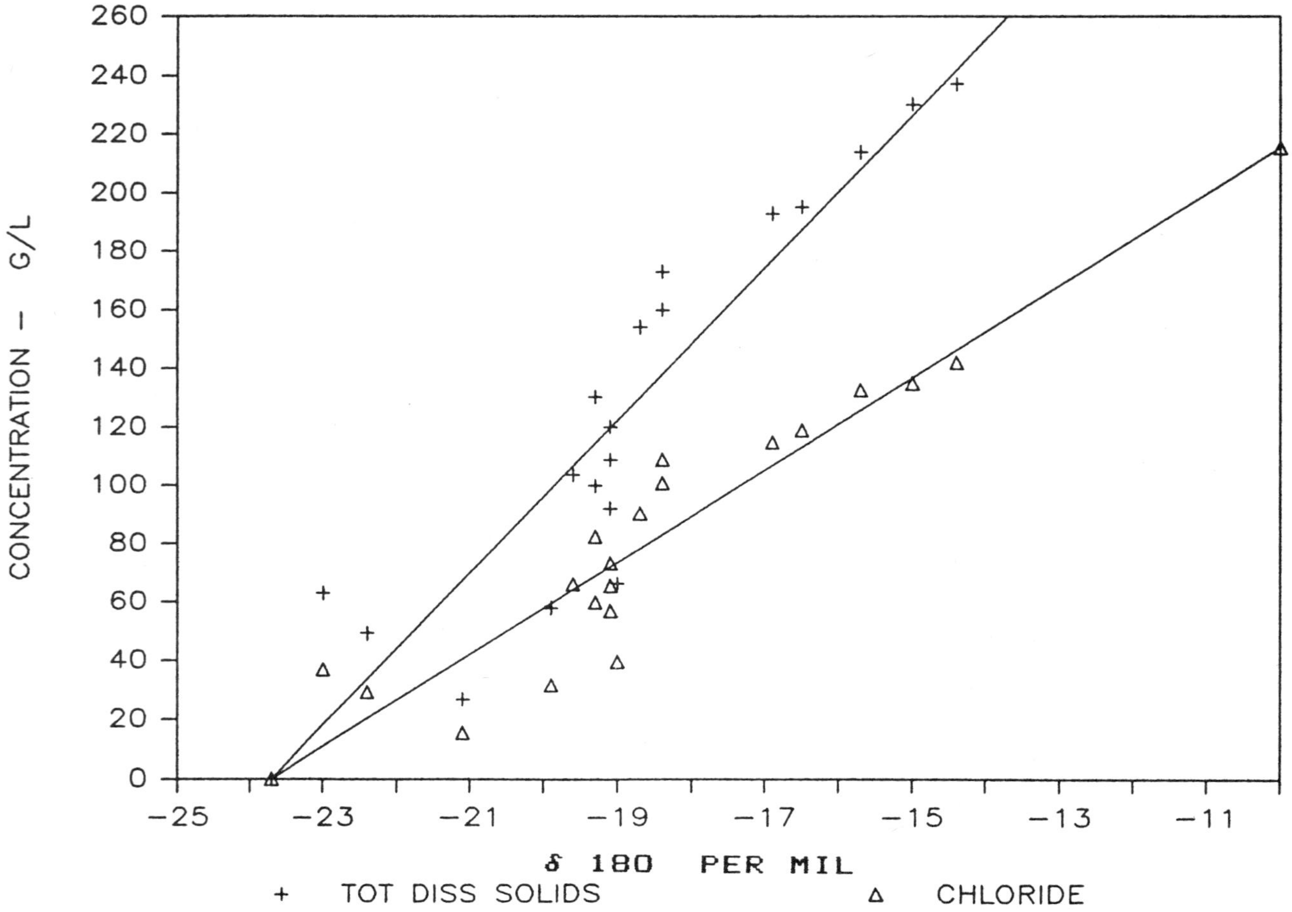

Figure 3. *Plot of total dissolved solids and chloride concentrations vs.* $\delta^{18}O$ *for samples from the Con Mine Yellowknife, NWT.*

approximately that of the intersection of the three lines shown on Figure 2.

Mineralogic constraints on salinity of source brine

An initial estimate of the major ion concentrations of the source brines was made by extrapolating the concentration vs. isotopic composition relations, like those shown in Figures 2 and 3. Mineral-water equilibrium calculations as discussed in Section 3. below, showed that the compositions estimated were close to or exceeded saturation with respect to halite.

Regardless of their source, the concentrations of sodium and chloride cannot exceed the solubility product of the mineral halite. Thus, in making the final estimates, source brine compositions, were constrained to be just at halite saturation. This was done by adjusting the concentrations of all ions proportionately until the ion activity product of sodium and chloride (calculated as described below) in the resulting solution just equalled the equilibrium constant for halite.

Figure 5 is a bar diagram which compares the major ion concentrations of the source brines calculated for each of the three mining districts. The source brines for each district seems to have essentially the same composition. It is difficult to be certain whether the differences between the calculated compositions represent real differences in the brine chemistry from district to district, or are simply artifacts of the sampling and extrapolation procedure or local effects from the mining environment in which they are found. A single brine composition may be valid for all districts.

Some further discussion on the use of halite as a constraint on the composition of the source brines is warranted. There can be little question that the brines could not exceed the solubility of halite. However, if the brines are approaching halite saturation there should be some evidence for the presence of halite in the shield. This point was addressed by J. Guha during discussions at the brine symposium. He observed that when it was possible to observe freshly opened vugs in mines, actual halite crystals were sometimes found. Most study of the mineralogy of shield rocks is done by examination of thin sections. In preparing a thin section, any halite likely to be present would certainly be dissolved. Its presence however, could be inferred from square holes in the thin section. At the same symposium several petrographers observed that they have in fact seen such square holes. Thus, the actual presence of mineral halite in the crystalline rock of the shield is not out of the question.

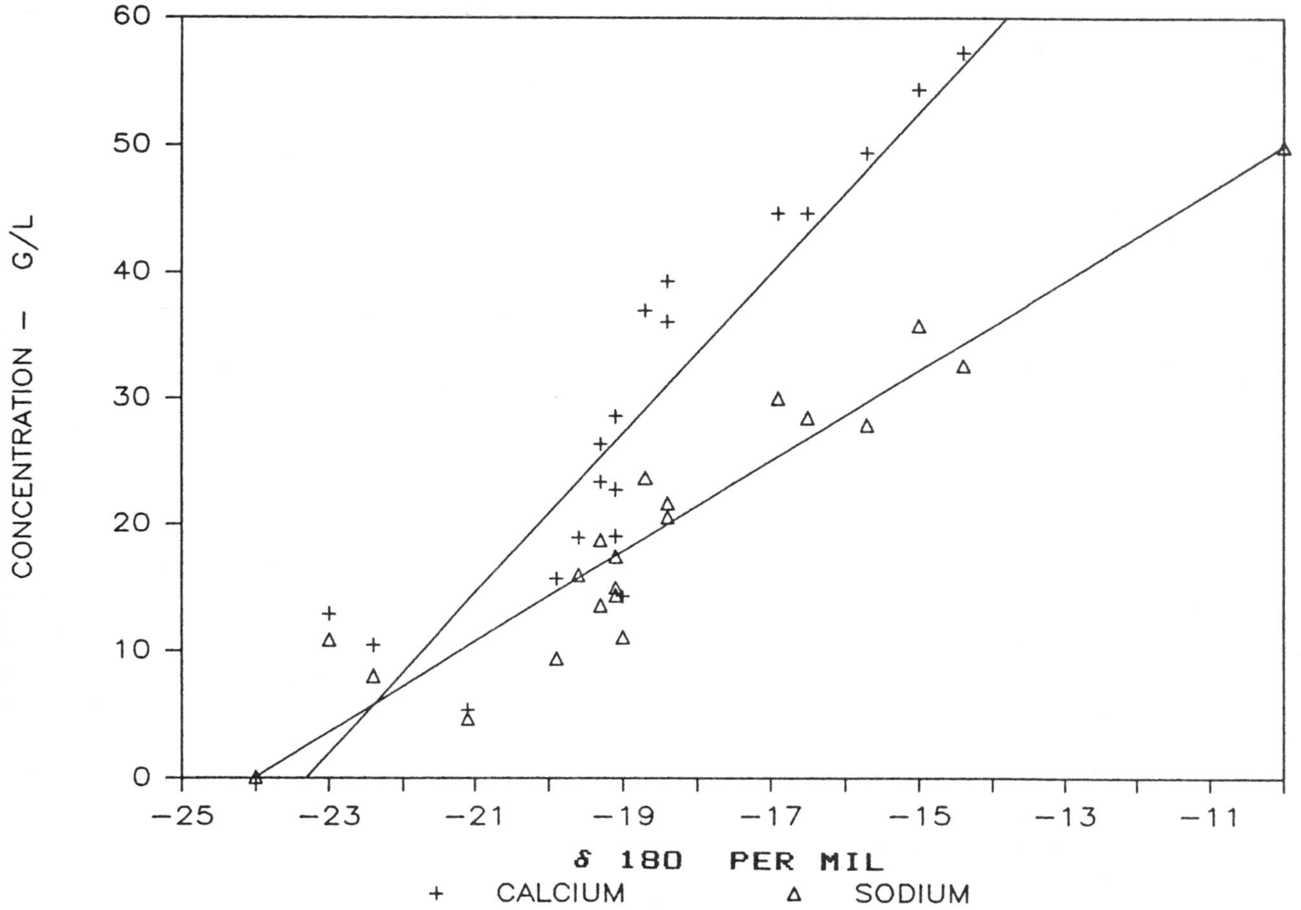

Figure 4. *Plot of calcium and sodium Concentrations vs.* $\delta^{18}O$ *for Samples from the Con Mine Yellowknife, NWT.*

It is likely that a number of other rapidly reacting minerals influence the chemistry not only of the mixtures which have been sampled, but probably also of the source brines as well. Effects of such minerals, including the sulfates and carbonates, can be best examined in concert with studies of the stable isotopic composition of dissolved material in the brines. Work of this type is in progress by Frape and Fritz and does not form a part of this paper.

Rock forming silicates are the minerals present in the largest amounts in the shield and therefore are those with the potential to have the greatest influence on the chemistry of brines. The influence of these minerals is the subject of the modeling described in the next section.

MODELS OF MINERAL CONTROLS ON BRINE CHEMISTRY

Two aspects of the chemistry of the brines in the shield have been examined quantitatively and are discussed in this section. The first is the stable isotopic composition of the water of the brines. The second includes the ratios of the concentrations of the major ions dissolved in the brines. A third aspect – the assumption that halite solubility provides an upper limit to the salinity of the brine – was discussed in the previous section. The suggestion that low-temperature rock-water equilibrium, or reactions leading to such an equilibrium, influence or control the chemistry of the shield brines has been made by Frape, Fritz and their colleagues. Recently, McNutt and others (1984) have shown that the strontium isotope ratios of saline waters from several mining districts are the same as those of the rock from those districts. They take this as evidence of rock-water equilibrium and suggest further that the isotopic composition of the water molecules of the brines could also result from an approach to such equilibrium.

Stable isotope composition of water

At temperatures of 600 to 800°C and above there is little or no isotopic fractionation between silicate minerals and water. Thus, it is to be expected that waters associated with silicate melts will have the same hydrogen and oxygen isotopic composition as the melts themselves and so will have a range of $\delta^{18}O$ values from +5 to +10 per mil and of δD values from −40 to −80 per mil (see for example Taylor, 1979). Also, at high temperatures, reactions occur rapidly so water in high temperature systems can rapidly reach isotopic equilibrium with its containing rock. At lower temperatures, isotopic fractionation between the solid and liquid phases becomes important, and generally, the amount of fractionation increases with decreasing temperature.

The temperature of the source brines during most of their

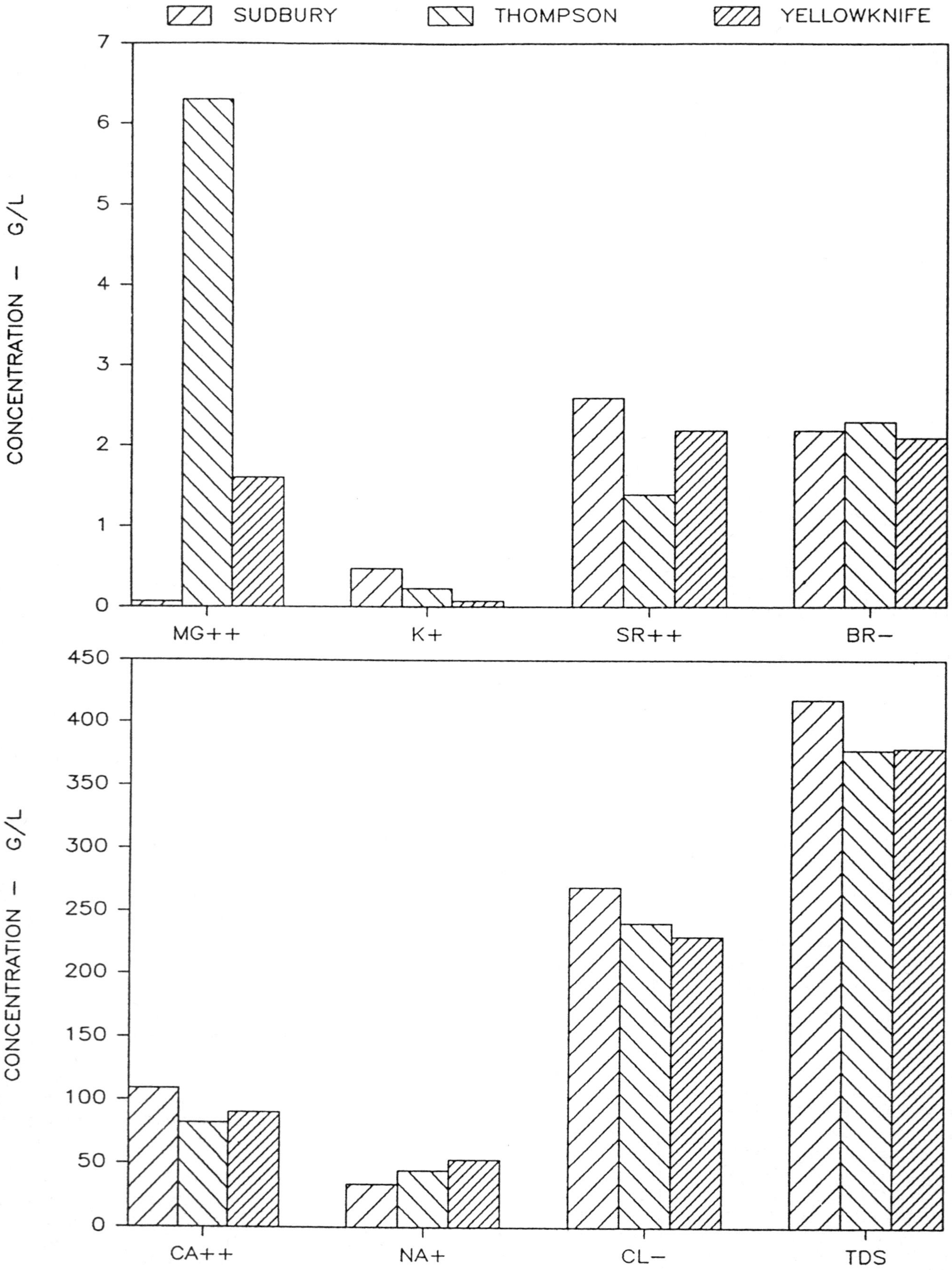

Figure 5. *Bar graphs comparing estimated compositions of source brines.*

residence time is likely to have been relatively low. From measurements reported by Jessop and Judge (1971) and Jessop and Lewis (1978), the thermal gradient of the shield can be taken as about 13°C per kilometer. If the average annual temperature over the surface of the shield is taken as 8°C the brines would have a temperature of 25°C at a depth of 1.3 kilometers or of about 60°C at 4 kilometers. It is unlikely that higher gradients were active for significant parts of the recent geologic history of the shield, and thus the brines are not likely to have been exposed to temperatures very much higher than at present for any considerable length of time.

Oxygen isotopes

The $\delta^{18}O$ values for whole rock samples and silicate minerals in the shield are reported to range between +6 and +13 per mil (Ding and Schwarcz, 1984; Kerrich, 1984). At high temperatures these rocks would be in equilibrium with the waters of the isotopic composition suggested by Taylor (1979) for magmatic water.

Oxygen isotope fractionation factors for the important shield minerals quartz and feldspar have been measured only at high temperatures (O'Neil and Taylor, 1967; Clayton and others, 1972). As used here, the expression, "fractionation factor" means δ(rock) – δ(water), in per mil, which is virtually the same as the value of 1000 $\ell n\,\alpha$ in the equations below. (Fritz and Fontes, 1980, Introduction). Quartz data extend down to only 200°C while feldspar data have not been collected below 350°C. Estimation of fractionation factors at lower temperatures requires extrapolation of the high temperature data. The correct form of the equation which should be used to make these extrapolations is uncertain. The commonly used expression is (see for example Friedman and O'Neil, 1977):

$$1000\ \ell n\ \alpha = A + B/T^2$$

Extrapolations based upon this equation lead to fractionation factors at 25°C of about 36 per mil for quartz, 30 per mil for alkali feldspar and about 20 per mil for plagioclase (Friedman and O'Neil, 1977 Figure 18).

Bottinga and Javoy (1973) point out that there is some theoretical justification for preferring an equation of the form

$$1000\ \ell n\ \alpha = A/T + B/T^2$$

to express the temperature dependence of isotope fractionation factors. They demonstrate that this equation fits the feldspar experimental data of O'Neil and Taylor (1967) as well as does the more commonly used expression. Using Bottinga and Javoy's equation to extrapolate the feldspar data leads to a fractionation factor at 25°C of +47 per mil for alkali feldspar (Pearson, 1985, Figure 2-12).

The calculated $\delta^{18}O$ values of halite-saturated source brines for the mixed samples are from −8.3 to −9.1 per mil. Taking the shield rock $\delta^{18}O$ values as between +6 and +13 per mil leads to rock-water isotopic fractionation factors of 14 to 22 per mil. Using quartz fractionation factors extrapolated with the usual equation, fractionation of 14 to 22 per mil corresponds to equilibrium temperatures between 100 and 200°C. Using alkali feldspar factors conventionally extrapolated it corresponds to temperatures between 70 and 160°C, while based on plagioclase fractionation similarly extrapolated, it corresponds to a range of 15 to 70°C. Based on the alkali feldspar data extrapolated using the Bottinga and Javoy equation this fractionation corresponds to a temperature between 120 and 210°C.

Because of the uncertainties in the isotopic fractionation data between water and silicate minerals at low temperatures it is not possible to draw definite conclusions about whether the brines and the shield rock are in isotopic equilibrium. However, it is clear that low temperature reaction between the brines and the rock could have brought about the unusually negative $\delta^{18}O$ values of the brines (relative to their δH values). An approach to a further examination of the isotopic relations between these brines and the shield rock would be to measure the oxygen isotopic composition of albite crystals which appear from petrographic examination to have been formed by a process of recent albitization.

Hydrogen isotopes

The hydrogen isotopic fractionation factors between water and rock forming silicate minerals are even less well known than are those for the oxygen isotopes. However, a number of studies have been made of silicate minerals formed at low temperatures as a result of weathering processes and which therefore, are assumed to be in equilibrium with local meteoric water. These studies suggest that hydrogen isotope fractionation is in the opposite sense to oxygen isotope fractionation in that the heavier isotope is concentrated in the fluid rather than in the solid phase. As Savin (1980, Figure 8-7) shows, the hydrogen isotopic composition of clay minerals formed during weathering can be as much as 40 per mil more negative than that of the water from which they are presumed to have been formed.

The calculated δH values of the halite saturated source brines for the mixtures is between −10 and +10 per mil. Nothing is known about the hydrogen isotopic composition of the shield rocks themselves but, based on Taylor's (1979) estimate, they may be in the range of −40 to −80 per mil. If Savin's low temperature fractionation factors for clay minerals are also appropriate to the shield brine system, low temperature equilibrium could produce fluids with δD values as much as 40 per mil more positive than those of the rock. Although this is a very tenuous line of argument because of the lack of information about fractionation factors and of the actual hydrogen isotopic composition of the shield rock, it nonetheless suggests that the estimated hydrogen isotopic composition of the brines could be caused by low temperature equilibrium with the hydrogen in the shield rock itself.

Major Species Concentrations

The possible influence of water-rock interactions on the chemistry of both the source brines and the mixed samples has been examined by geochemical modeling. Before discussing the results a brief review of the geochemical modeling process itself is warrented.

Geochemical Modeling

There are three requirements before one can perform meaningful water-rock equilibrium calculations. First, one must have reliable data on the chemistry of the water being examined. Second, one must have available a conceptual model of solution behavior which will permit the calculation of thermodynamic activities of dissolved species. The conceptual model must be appropriate to the types of waters being studied. Third, one must have thermodynamic data on the water-rock reactions of interest at the temperatures of concern.

Water Chemical Data: The concentrations of the major ions in the various saline water samples which have been reported appear to be quite reliable. The compositions of the source brines, however, depend on extrapolation from the composition of the sampled mixtures. Most of the major ions correlate well with isotopic composition of the mixtures, and thus, extrapolation to their composition in the brines seems reasonably secure. Some constituents however, such as the alkalinity, sulfate and the magnesium concentration in the Sudbury samples correlate so poorly with the isotopic composition of the mixtures that their extrapolated composition in the source brines is quite unreliable. These are the same constituents which Frape and Fritz (1982, Figure 5) show correlate poorly with sample depth, and thus, are clearly influenced by processes such as the formation of acids from sulfide oxidation (Frape and others, 1984) associated with mining activities or other non-surface effects.

The analytical data on minor species in the mixture may also not be as reliable as the major ion data. Aluminum, for example, may be influenced by the inclusion of fine particulate or colloidal aluminum-bearing substances in the samples, no matter how carefully they were filtered. In addition, aluminum analyses themselves are difficult under the best of circumstances and would be particularly so in such brines as these. Thus, the geochemical calculations described here are limited to those invloving the major ions.

Activity coefficient model: To perform useful water-rock equilibrium calculations requires a technique for translating the analytical data on concentrations (m) of species in solution to their thermodynamic activities (α). The activity coefficient ($\gamma = \alpha/m$) is used to make this translation.

A conceptual model for finding activity coefficients which is in wide use in geochemical computer codes combines the Debye-Hueckel model for ion activity coefficients with the ion-pair model of solution behavior (Garrels and Thompson, 1962; Nordstrom and others, 1979). This conceptual model does not reproduce the behavior of complex solutions which are much more concentrated than sea water which has an ionic strength of 0.7. Because the most concentrated mixed samples and the source brines themselves have ionic strength in excess of 8, the conventional model of solution behavior is not applicable to them.

Instead the ion-interaction conceptual model has been used. This model describes the activity coefficients of dissolved species as explicit functions of the concentrations of all other ions in solution. The model has been brought to its present state of development by Pitzer and his students. A review by Pitzer (1979) provides an overview of the model itself. That the model is also applicable to systems as complicated as natural brines has been shown by the work of, for example, Harvie and others (1984).

Thermodynamic Data for Reactions: The dominant minerals in the shield system are silicates. Because such minerals react only slowly with water at low temperatures, most experimental data on mineral water equilibria have been taken at high temperatures. Low temperature values required for the calculations of this study are generally obtained by extrapolation from the experimental, high-temperature data. This will certainly give rise to some uncertainties in the thermodynamic data used at low temperatures. The size of the errors are not known, but because they are based on a larger body of experimental data, they will certainly be relatively smaller than those for the low temperature isotope fractionation data described above. Nonetheless the possibility that the results of the calculations to be described may be influenced by errors in the thermodynamic data used cannot be overlooked.

The second difficulty is that many of the silicate minerals of interest in this sytem – e.g., plagioclase – are solid solutions. Solid solution properties are not well enough known to treat their effects on solution behavior satisfactorily. As will be discussed below, it is quite possible that the sodium to calcium ratio of the brines, which is one of their most striking features, could be brought about by reactions with plagioclase minerals. In the absence of sufficient solid solution data this hypothesis cannot be tested.

Silica

The least soluble silica mineral is quartz. Because the rate of quartz precipitation is so low, the silica content of low-temperature groundwaters is commonly controlled by another metastable polymorph such as chalcedony.

The silica content in the analyzed samples of mixed waters is between 4 and 10 mg/ℓ. The sample of lowest silica concentration (4.1 mg/ℓ) would be in equilibrium with quartz at about 20°C. This temperature is close to that of collection of most of the samples and is a temperature appropriate to the depth at which the undisturbed brines may be found in the shield.

The higher silica concentrations correspond to quartz saturation at temperatures as high as 50°C. These higher concentrations may reflect mixing with shallower waters of higher silica content or reaction of the waters sampled with rock flour generated during mining or the drilling of boreholes. The hypothesis that the silica content of the brines is controlled by low-temperature equilibrium with quartz is in keeping with conclusions which can be drawn from other mineral equilibria considerations as well, as discussed below.

Cations

Cations of compositional significance in the rock-forming minerals of the shield include: sodium, potassium and calcium in the feldspar and micas, and magnesium in the dark minerals and micas. The magnesium concentrations vary among the samples from the several mining districts and in the source brines calculated from them. This may be a result of differences in local rock types or it may result from differences in the environment of the mines themselves as suggested by Frape and others (1984). Magnesium is not examined further here.

Strontium has higher concentrations in most samples and the source brines calculated from them than does potassium or magnesium (see Figure 5). Strontium however, does not have a separate controlling mineral phase as do the other cations mentioned. McNutt and others (1984) have shown that the strontium concentration of the waters is strongly correlated with their calcium concentrations and suggest that both have the same origin because of their

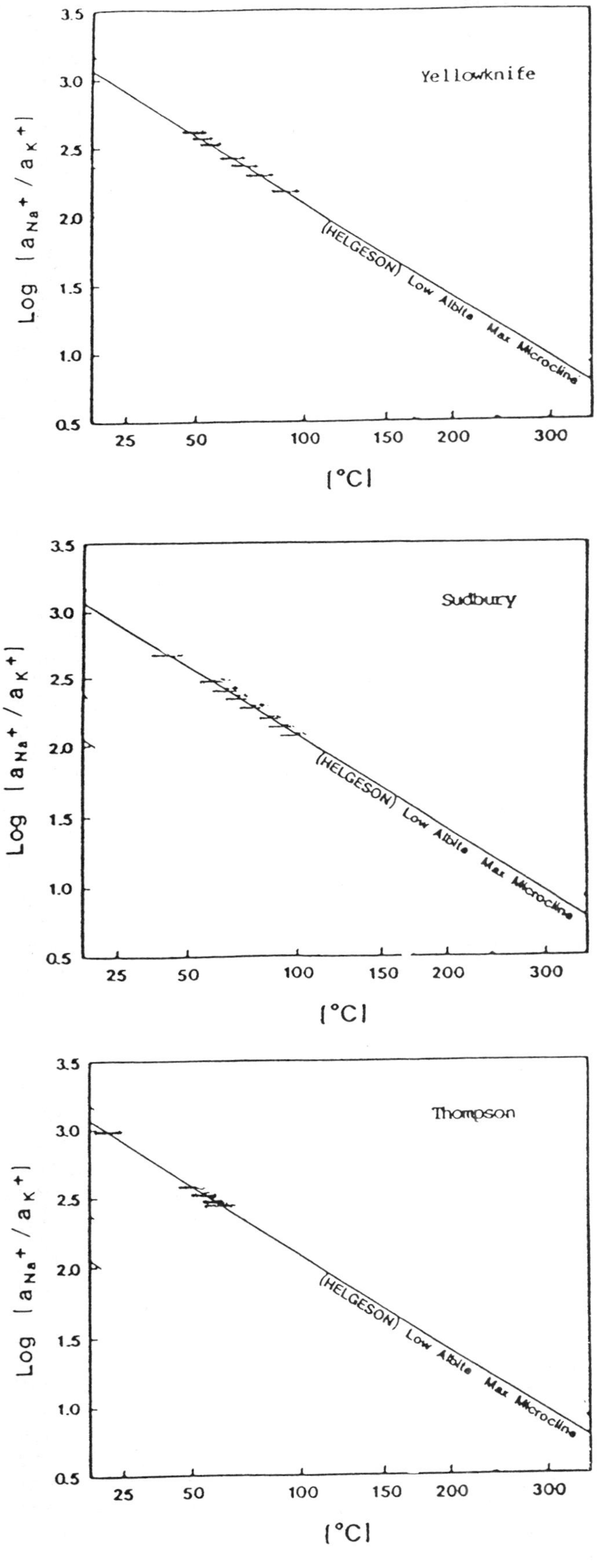

Figure 6. *Graphs of the logarithm of the ratio of sodium activity to potassium activity vs. temperature for samples from Yellowknife, Sudbury and Thompson districts.*

similar geochemical behavior. This origin is presumably in the plagioclase feldspars. McNutt and others also show strontium equilibrium between rock and water based on strontium isotope ratios. Because of the strong calcium-strontium correlation it may mean that water-rock equilibrium exists with respect to calcium as well.

Sodium/Potassium Ratios: The ratio of sodium to potassium is used in studies of geothermal systems to indicate water-rock equilibrium temperatures (Fournier, 1981). The controlling reaction is thought to be:

$$NaAIS_3O_8 + K^+ = KAISi_3O_8 + Na^+$$

Where the solids are the alkali feldspars, albite and microcline. The equilibrium constant for this reaction is:

$$K = \alpha_{Na^+}/\alpha_{K^+}$$

which is a function of temperature. α, the thermodynamic activity of the species, is related to its concentration (m) as measured in the samples or calculated for the source brines by the activity coefficient (γ). Activity coefficients for the major ions dissolved in the source brines and most concentrated samples were calculated using an ion-interaction model for solution behavior, the conceptual model appropriate for such highly saline waters as these. It is necessary to make such calculations before attempting to understand possible water-rock interactions controlling water chemistry. As the following shows, conclusions drawn from considerations only of analyzed molalities are considerably at variance from those drawn from the calculated ion activities.

Figure 6 shows graphs of the logarithm of the ratios of the activity of sodium to potassium against temperature. The lines on these graphs were calculated for equilibrium between alkali feldspars with the thermodynamic properties of low albite and maximum microcline as given by Helgeson and others (1978). The calculated activity ratios are plotted on these lines. Within each group of samples the activity ratio increases regularly with salinity. An immediate, though improbable, interpretation of this observation might be that the more dilute samples equilibriated at higher temperatures than did the more concentrated samples.

When geothermal waters are plotted on diagrams of the type shown in Figure 6, the vertical axis is often the log ratio of molalities rather than the log ratio of activities (see, for example, Fournier 1981, Figure 4-7). Because the ratios of molalities are not affected by dilution, this procedure can be used successfully even on mixed samples. Furthermore, the use of mole ratios in place of activity ratios is not seriously in error in waters of relatively low salinity where the ratio of the activity coefficient of sodium to that of potassium is not greatly different from one.

In waters of such high salinity as those modeled here, this assumption of nearly equal activity coefficients may not be valid, so that ratios of activity and of molality may not be equivalent. If so, it would also be true that activity ratios

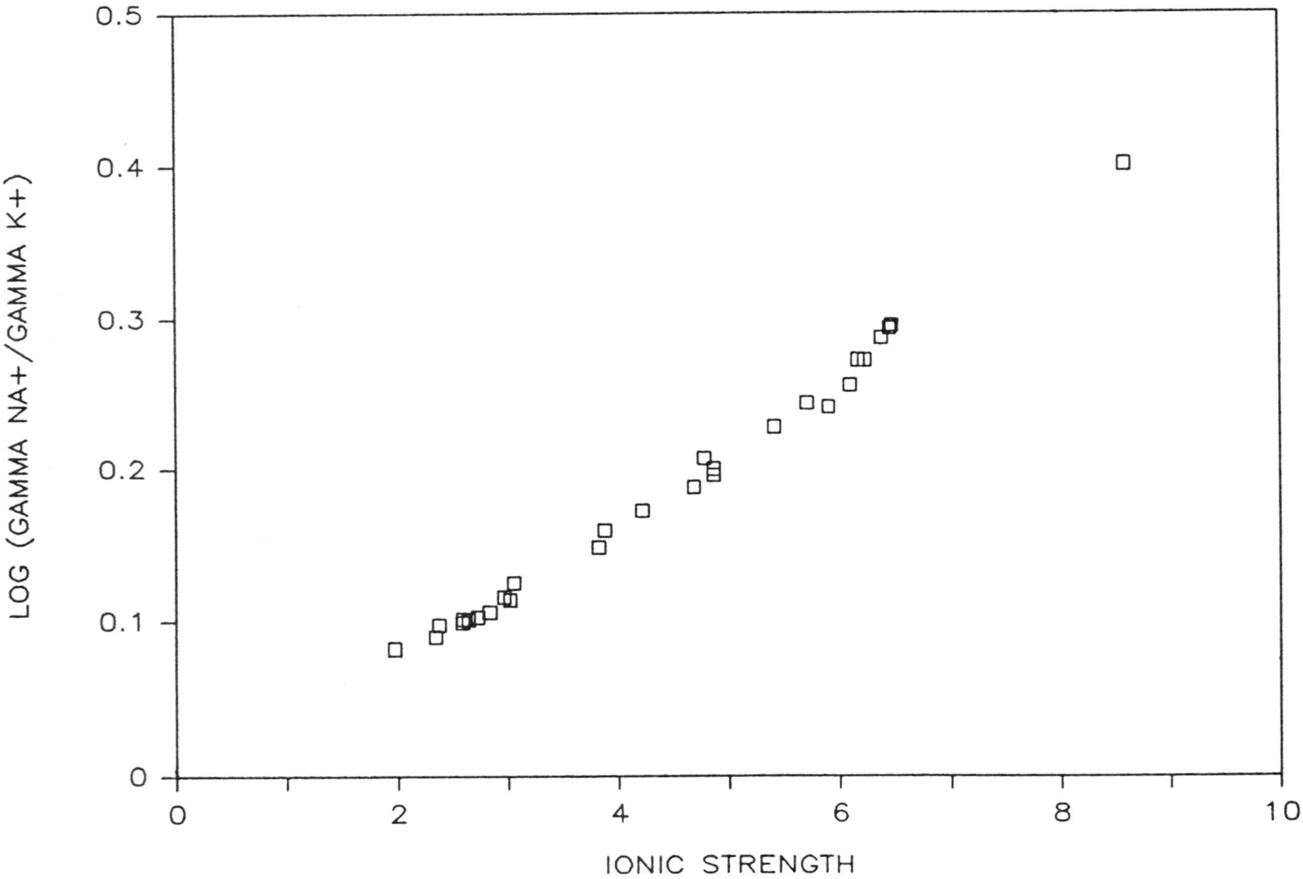

Figure 7. *Graph of the log of the ratio of the activity coefficient of sodium to the activity coefficient of potassium versus ionic strength for samples of saline water from the Canadian shield.*

would change with dilution and could not be used to estimate an equilibrium temperature of an undiluted brine unless corrected to the salinity of that brine.

The activity coefficients calculated as part of the aqueous modeling of the mixed samples have been used to examine the dependence of the activity ratio of sodium to potassium on salinity. Figure 7 is a graph of the log activity coefficient ratio of sodium to potassium versus ionic strength for the mixed samples. This graph shows a steady increase in the activity coefficient ratio from about 0.1 in the most dilute samples modeled (80-100 g/ℓ total dissolved solids) to greater than 0.4 in the most concentrated sample (325 g/ℓ). The ionic strength of the source brines is about 10 which corresponds to a log activity coefficient ratio of about 0.5.

The ratios plotted in Figure 6 must be adjusted upward by factors of 0.1 to 0.4 to correct for the effects of dilution on the sodium and potassium activity coefficients. When this is done the corrected log activity ratios cluster at values between about 2.7 to 3.0. These values correspond to equilibrium between fluid and sodium and potassium feldspars at about 25°C. This is virtually the same temperature as suggested by quartz saturation in the water of lowest measured silica content and is concordant with the depths at which the brines may occur. Thus for purposes of further calculations in this report, the temperature of the undiluted brines is assumed to be 25°C.

Hydrogen Ion (pH): Reactions with silicate minerals can also exert a control on the pH of ground waters. The system K_2O-Al_2O_3-SiO_2-H_2O is commonly used to demonstrate this control. Muscovite ($KAl_3Si_3O_{10}(OH)_2$), kaolinite ($Al_2Ai_2O_5(OH)_4$), and potassium feldspar are the minerals involved. Stability relationships in such systems are usually displayed in activity ratio diagrams. Drever (1982) illustrates such diagrams and describes their construction.

Stability diagrams for the systems including alkali feldspars, muscovite and kaolinite are shown in Figure 8. These diagrams were also constructed for 25°C and used the thermodynamic data of Helgeson and others (1978).

Samples from each district are plotted on two diagrams. One was constructed for a log SiO_2 activity of -3.6 and the other for a log activity of -4.0. A log activity value of -3.6 is about 15 mg SiO_2/ℓ, while a log activity of -4 corresponds to quartz saturation at 25°C and equals about 5 mg SiO_2/ℓ.

Points for samples from the Sudbury and Yellowknife districts fall parallel to the K-feldspar-albite line but are displaced from it by perhaps ½ log unit into the K-feldspar or muscovite fields. This is the same magnitude as the displacement of the sodium to potassium ratio toward erroneously low values as a result of dilution. If adjusted for the effects of dilution, these points would fall essentially on the line separating the K-feldspar and albite fields.

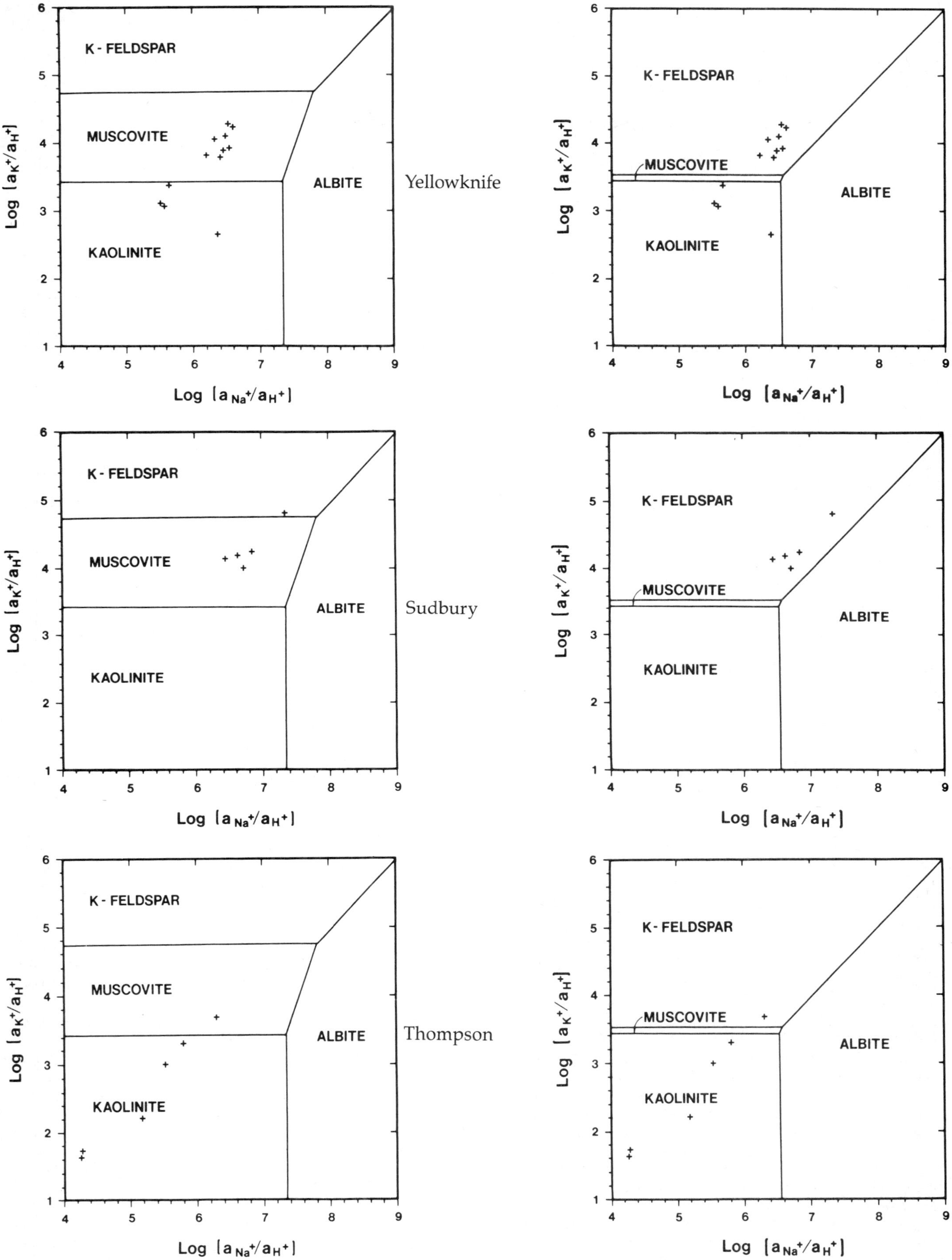

Figure 8. *Plot of activity ratios of saline water samples on a stability diagram for alkali feldspars, muscovite and kaolinite. Figures to the left are for log $\alpha SiO_2 = -4.0$, and to the right for log $\alpha SiO_2 = -3.6$.*

Samples from the Sudbury and Yellowknife districts fall either in the feldspar or the muscovite field depending on which silica activity is assumed. Because the silica activity of the source brines is not likely to be lower than that set by quartz saturation (log $^{\alpha}SiO_2 = -4$) and could well be slightly higher, it would appear that the pH of the source brines may well be controlled by a feldspar-muscovite equilibrium.

Samples from the Thompson district are displaced toward lower sodium and potassium concentrations. As Frape, and others, 1984, point out this behavior may be, like the high magnesium content of the Thompson samples, related to the presence of magnesium-rich ultramafic rock.

Calcium-sodium ratios: A striking aspect of the chemistry of the saline samples and source brines is their relatively high calcium to sodium ratio. Because plagioclase feldspar is the dominant source of calcium and sodium in most shield rock, if water-rock equilibrium does control the chemistry of these waters, reactions with plagioclase are likely to control the dissolved concentrations of these species. As pointed out above, the lack of thermochemical data on plagioclase solid solution precludes equilibrium calculations like those made for potassium and pH. However, there are qualitative indications that an approach to equilibrium with plagioclase would develop a solution like the source brines.

When water reacts with fresh crystalline rock, during weathering for example, it develops a calcium to sodium ratio with about the same value as that of the plagioclase feldspars in the rock (Garrels, 1967). From feldspars typical of granitic rocks this would lead to calcium to sodium ratios of less than one, considerably lower than those observed in the mixed samples, or calculated in the source brines.

There is evidence in many igneous rocks including those in the shield that albitization takes place as a result of low temperature water-rock interaction. The formation of secondary albite would lead to an enrichment of calcium in the solution from which it was precipitating.

Garrels (1967) notes that Headden (1903) found high calcium to sodium ratios in laboratory solutions of high CO_2 contents which he allowed to react with alkali feldspars. Headden suggested that this might be due to differential release of calcium from sodic plagioclase. Whether the process is called secondary albitization or differential calcium solution the result appears to be that with time waters tend to become enriched with calcium relative to the feldspars with which they are in contact.

SUMMARY AND CONCLUSIONS

Samples of saline water from mines and boreholes in the Canadian shield are mixtures of local meteoric water, and a more highly saline brine.

There is strong correlation between the stable isotopic composition of samples from various mines and mining districts and their chemical composition. These correlations make it possible to estimate the composition of source brines. These estimates have been refined by assuming that the salinity of the source brines are saturated with halite.

The $\delta^{18}O$ and δD values estimated for these source brines range from -8.3 to -9.1 per mil and from $+10$ to -10 per mil respectively. These values could result from an approach to water-rock isotopic equilibrium at temperatures of about 25°C.

Geochemical modeling of the activities of the major dissolved species in the saline water samples have been carried out. The calculations employed the ion-interaction (Pitzer) conceptual model of solution behavior, which is appropriate for waters of such high ionic strengths.

The calculations support the hypothesis that the sodium to potassium ratio of the source brines is controlled by low temperature (25°C) equilibrium with alkali feldspars, and that pH may be set by additional equilibrium with respect to muscovite. Calculation of controls on the calcium concentrations is not possible, but there is qualitative evidence that supports the hypothesis that the calcium to sodium ratio of the brines results from an approach to equilibrium with plagioclase feldspars, such as a albitization.

ACKNOWLEDGMENTS

Shaun Frape and Peter Fritz of the Department of Earth Sciences at the University of Waterloo were most gracious in providing ready access to their data and ideas on the brines in the Canadian shield.

This paper is based on a report prepared by INTERA Technologies, Inc. for Atomic Energy of Canada, Ltd., Whiteshell Nuclear Research Establishment, Pinawa, Manitoba.

REFERENCES

Bottinga, Y. and Javoy, M., 1973. Comments on oxygen isotope geothermometry: Earth and Planetary Science Letters, v. 20, p. 250-265.

Clayton, R.N., Friedman, I., Graf, D.L., Mayeda, T.K., Meents, W.F. and Shimp, N.F., 1966. The origin of saline formation waters, 1 Isotopic composition: Journal of Geophysical Research, v. 71, p. 3869-3882.

Ding, T.P. and Schwarcz, H.P., 1984. Oxygen and isotopic chemical composition of rocks in the Sudbury Basin, Ontario: Canadian Journal of Earth Sciences, v. 21, p. 305-318.

Drever, J.I., 1982. The geochemistry of natural waters: Prentice-Hall Inc., Englewood Cliffs, N.J., 388 p.

Fournier, R.O., 1981. Application of water geochemistry to geothermal exploration and reservoir engineering: *in* Rybach, L. and Muffler, L.J.P., *eds.*, Geothermal Systems: Principles and Case Histories: John Wiley and Sons, Ltd., Chichester, p. 109-143.

Frape, S.K. and Fritz, P., 1981a. A preliminary report on the occurrence and geochemistry of saline groundwaters on the Canadian shield: Atomic Energy of Canada Ltd., Pinawa, Manitoba, Technical Record TR-136, 68 p.

Frape, S.K. and Fritz, P., 1981b. The occurrence and the geochemistry of groundwaters on the Canadian shield: Final report to Atomic Energy of Canada Ltd., Pinawa, Manitoba, Project No. 002-08-18, 74 p.

Frape, S.K. and Fritz, P., 1982. The chemistry and isotopic composition of saline groundwaters from the Sudbury Basin, Ontario: Canadian Journal of Earth Sciences, v. 19, p. 645-661.

Frape, S.K., Fritz, P. and McNutt, R.H., 1984. Water-rock interaction and chemistry of groundwaters from the Canadian Shield: Geochimica et Cosmochimica Acta, v. 48, p. 1617-1627.

Friedman, Irving and O'Neil, J.R., 1977. Compilation of stable isotope fractionation factors of geochemical interest: Chapter KK of; Data of Geochemistry, Sixth Edition, U.S. Geological Survey Professional Paper 440-KK, p. KK1-KK12.

Fritz, P. and Fontes, J. Ch., 1980. Handbook of Environmental Isotope Geochemistry, vol. 1, The Terrestrial Environment, a: Elsevier Scientific Publishing Co., Amsterdam, 545 p.

Fritz, P. and Reardon, E.J., 1979. Isotopic and chemical characteristics of mine water in the Sudbury area: Atomic Energy of Canada Ltd., Pinawa, Manitoba, Technical Record TR-35, 37 p.

Fritz, P. and Frape, S.K., 1980. Comments of the ^{18}O , ^{2}H and chemical composition of saline groundwaters on the Canadian shield: *in* Isotope Studies of Hydrologic Processes, Proceedings of an American Geophysical Union Symposium, University of Northern Illinois, DeKalb, Illinois. p. 57-63.

Fritz, P. and Frape, S.K., 1982. Saline groundwaters in the Canadian shield – a first overview: Chemical Geology, v. 36, p. 179-190.

Garrels, R.M., 1967. 0Genesis of some groundwaters from igneous rocks. *in* Abelson, P.H., *ed.* Researches in Geochemistry, v. 2, John Wiley and Sons, Inc., New York, p. 405-420.

Garrels, R.M. and Thompson, M.E., 1962. A chemical model for sea water at 25°C and one atmosphere total pressure: American Journal of Science, v. 260, p. 57-66.

Gat, J.R. and Gonfiantini, R. *eds.*, 1981. Stable Isotope Hydrology, Deuterium and Oxygen-18 in the Water Cycle: International Atmomic Energy Agency, Vienna, Technical Reports Series 210, 339 p.

Harvie, C.E., Moller, Nancy, and Weare, J.H., 1984. The prediction of mineral solubilities in natural waters: The Na-K-Mg-Ca-H-Cl-SO_4-OH-HCO_3-CO_3-CO_2-H_2O system to high ionic strengths at 25°C: Geochimica et Cosmochimica Acta, v. 48, p. 723-751.

Headden, W.P., 1903. Significance of silica acids in waters of mountain streams. American Journal of Science, v. 166, p. 169-186.

Helgeson, H.C., Delany, J.M., Nesbitt, H.W. and Bird, D.K., 1978. Summary and critique of the thermodynamic properties of rock-forming minerals: American Journal of Science, v. 278A, p. 1-229.

Jessop, A.M. and Judge, A.S., 1971. Five measurements of heat flow in Southern Canada: Canadian Journal of Earth Science, v. 8, p. 711-716.

Jessop, A.M. and Lewis, Trevor, 1978. Heat flow and heat generation in the Superior Province of the Canadian shield: Tectonophysics, v. 50, p. 55-77.

Kerrich, R., 1984. Chronology and ambient temperature/pressure conditions of fluid flow through the Eye-Dashwa Lakes Pluton based on the $^{18}O/^{16}O$ ratio and fluid inclusions: Atomic Energy of Canada Limited, Pinawa, Manitoba, Technical Record TR267, 32 p.

McNutt, R.H., Frape, S.K. and Fritz, Peter, 1984. Strontium isotopic composition of some brines from the Pre-Cambrian shield of Canada: Isotope Geoscience, v. 2, p. 205-215.

Nordstrom, D.K., Plummer, L.N., Wigley, Tm.L., Wolery, T.J., Ball, J.W., Jenne, E.A., Bassett, R.L., Crerar, D.A., Florence, T.M., Fritz, B., Hoffman, M., Holdren, G.R., Jr., Lafon, G.M., Mattigod, S.V., McDuff, R.E., Morel, F., Reedy, M.M., Sposito, G., Thrailkill, J., 1979. A comparison of computerized chemical models for equilibrium calculations in aqueous systems: *in* Jenne, E.A. *ed.*, Chemical modeling in aqueous systems: American Chemical Society, Washington, ACS Symposium Series 93, p. 857-892.

O'Neill, J.R. and Taylor, H.P., Jr., 1967. The oxygen isotope and cation exchange chemistry of feldspars: The American Mineralogist, v. 52, p. 141-1437.

Pearson, F.J., Jr., 1985. Geochemical and isotopic evidence on the origin of brines in the Canadian shield: Report by INTERA Technologies, Inc., Texas for Atomic Energy of Canada Limited, Pinawa, Manitoba. (in press)

Pitzer, K.S., 1979. Theory: Ion interaction approach: *in* Pytkowicz, R.M., *ed.*, Activity Coefficients in Electrolyte Solutions, v. 1, CRC Press Inc., Boca Raton, Florida, p. 157-208.

Plummer, L.N., Parkhurst, D.L. and Thorstenson, D.C., 1983. Development of reaction models for groundwater systems: Geochimica et Cosmochimica Acta. v. 47, p. 666-686.

Savin, S.M., 1980. Oxygen and hydrogen isotope effects in low-temperature mineral-water interactions: *in* Fritz, P. and Fontes, J. Ch. *eds.*, Handbook of Environmental Isotope Geochemistry, v. 1 The Terrestrial Environment A. Elsevier Scientific Publishing Co., Amsterdam, p. 283-327.

Taylor, H.P., Jr., 1979. Oxygen and hydrogen isotope relationships in hydrothermal mineral deposits: *in* Barnes, H.L., *ed.*, Geochemistry of Hydrothermal Ore Deposits: John Wiley and Sons, N.Y., p. 236-277.

Saline Groundwaters and Brines in Plutons in the Canadian Shield

M. Gascoyne, C.C. Davison, J.D. Ross, R. Pearson
Atomic Energy of Canada Limited, Applied Geoscience Branch, Whiteshell Nuclear Research Establishment, Pinawa, Manitoba, Canada R0E 1L0

Gascoyne, M., Davison, C.C., Ross, J.D. and Pearson, R., Saline groundwaters and brines in plutons in the Canadian Shield; in Saline water and gases in crystalline rocks, Editors: Fritz, P., and Frape, S.K.; Geological Association of Canada Special Paper 33, 1987. p. 53–68.

Abstract

The geochemistry of groundwaters in plutons of the Canadian Shield is being investigated as part of the hydrogeological activities of the Canadian Nuclear Fuel Waste Management Program. Two granitic plutons (Lac du Bonnet batholith, Manitoba and Eye-Dashwa Lakes pluton, Atikokan, Ontario), a gabbroic pluton (East Bull Lake, Massey, Ontario) and a gneissic/metagabbroic pluton (Chalk River, Ontario) are being characterized in terms of groundwater chemistry and its relation to geology and hydrogeology.

Brackish and saline groundwaters have been found below 300 m at three of the four research areas, ranging from 3 g/L TDS in gabbro, to 40 g/L in granite at Atikokan. The presence of tritium in some groundwater samples indicates dilution by recent water, probably introduced by drilling or before borehole isolation. Projection to zero tritium for the Atikokan samples suggests the presence of brines at 1 km depth, with TDS in excess of 140 g/L. Dominant ions are Na, Ca, Cl, with some SO_4.

Groundwater salinity and chlorinity appear to be broadly related to hydraulic permeability and, hence, depth, for samples from plutons and mines (Frape et al. 1984a). A model is proposed of advective transport and mixing above 300 m depth with molecular diffusion below 300 m to account for the observed concentration gradients.

Low Br/Cl and high Na/Ca ratios for the Lac du Bonnet and East Bull Lake groundwaters may be due to mixing with adjacent sedimentary basin brines. Low Na/Ca and high Br/Cl ratios for the Atikokan samples compare closely with values for mine brines and Stripa (Sweden) groundwaters and are thought to indicate a fluid inclusion – mineral alteration origin for the observed salinity. Stable isotope analyses show all groundwaters to have a meteoric origin and significant deuterium enrichment is not observed, although Atikokan data suggest ^2H-rich waters may be present if effects of dilution by surface water are removed. Chemical equilibrium modelling of Lac du Bonnet groundwaters indicates calcite saturation at all depths, gypsum undersaturation above ~800 m and stability with respect to kaolinite. Corresponding abundances of calcite and gypsum are observed as fracture infillings from core log data.

Résumé

L'étude géochimique des eaux souterraines des plutons du bouclier canadien est un des aspects hydrogéologiques reliés au programme de gestion des déchets nucléaires. L'étude comprend la détermination de la composition chimique de l'eau souterraine et de son influence sur la géologie et l'hydrogéologie. Elle porte sur deux plutons granitiques (batholite du lac du Bonnet, Manitoba et pluton des lacs Eye-Dashwa, Atikokan, Ontario), un pluton gabbroïque (lac East Bull, Massey, Ontario) et un pluton métagabbroïque/gneissique (Chalk River, Ontario).

Dans trois des quatres secteurs étudiés, on a trouvé à plus de 300 m de profondeur des eaux souterraines saumâtres et salées, dont la teneur de l'ensemble des solides dissous (TDS) variait de 3 g/l dans le gabbro à 40 g/l dans le granite d'Atikokan. La présence de tritium dans certains échantillons d'eau souterraine indique une dilution par de l'eau récente, probablement introduite par les forages ou antérieure au colmatage des trous. Si la teneur en tritium des échantillons d'Atikokan est ramenée à zéro, on peut soupçonner, à 1 km de profondeur, la présence d'eaux sursalées dont la teneur en solides dissous excède 140 g/l. Les principaux ions sont Na, Ca, Cl et une certaine quantité de SO_4.

La salinité et la teneur en chlore de l'eau souterraine d'échantillons provenant de plutons et de mines semble dépendre fortement de la perméabilité hydraulique, et par conséquent, de la profondeur (Frape et al. 1984a). Le modèle proposé, transport et mélange advectifs à moins de 300 m de profondeur et diffusion moléculaire à plus de 300 m, tient compte des gradients de concentration mesurés.

Le faible rapport Br/Cl et le fort rapport Na/Ca des eaux souterraines des lacs du Bonnet et East Bull pourraient résulter d'un mélange avec les eaux sursalées d'un bassin sédimentaire adjacent. Le faible rapport Na/Ca et le fort rapport Br/Cl des échantillons d'Atikokan sont très semblables à ceux des saumures de mines et de l'eau souterraine de Stripa (Suède) et pourraient relier l'origine de la salinité aux inclusions fluides et aux altérations minéralogiques. Des analyses d'isotopes stables révèlent l'origine météorique de toutes les eaux souterraines ainsi que l'absence d'enrichissement en deutérium, quoique d'après les données d'Atikokan, la présence d'eaux riches en ^{2}H *serait possible en éliminant les effets de la dilution par les eaux superficielles. La modélisation de l'équilibre chimique des eaux souterraines du lac du Bonnet met les aspects suivants en relief: saturation de la calcite à toutes les profondeurs, sous-saturation du gypse à moins de 800 m environ et stabilité de la kaolinite. L'abondance de calcite et de gypse se retrouve dans le matériel remplissant les fractures des carottes de sondage.*

INTRODUCTION

The occurrence of saline waters, at depth, in the Canadian Shield has long been known from mining activities (Lane, 1914). This knowledge has become widespread in the last five years due to reconnaissance studies for the Canadian Nuclear Fuel Waste Management Program (CNFWMP, summarized by Dixon and Rosinger, 1984). This program, managed by Atomic Energy of Canada Limited (AECL), includes detailed studies of the hydrogeology and hydrogeochemistry of igneous and high-grade metamorphic rock types in the Canadian Shield, to characterize groundwater flow paths and chemistries to depths of over 1 km. The information will be used in assessing the feasibility of isolation of nuclear fuel waste by a multi-barrier system which prevents or retards subsequent leakage of radionuclides to the biosphere. The most likely mechanism for transporting radionuclides to the surface is dissolution or suspension in flowing groundwater. It is, therefore, necessary to characterize and understand the flow regimes of groundwater in plutonic rocks, both by observing existing occurrences (in mines, for instance), and by accessing sources in undisturbed rock formations, using boreholes and down-hole sampling systems.

In the first part of this program, researchers at the Department of Earth Sciences, University of Waterloo, obtained support through the CNFWMP to determine the occurrence and chemistry of saline waters in mines in the Canadian Shield. Their results form an important part of the current state of knowledge of the major-element chemistry of brines in crystalline rocks (Frape et al., 1984a) and the isotopic characteristics and possible origins of these brines (Fritz and Frape, 1982, McNutt et al., 1984). Recently, this work has been extended to (1) an examination of surface discharges of saline waters ("moose licks") in northwest Ontario (Frape et al., 1984b) and (2) measurements of redox potential and solubility limits of minerals in brine solutions to determine more precisely the properties of the observed brine compositions (Frape and Fritz, 1985).

The second part of the program has been performed by researchers at, or contracted by, AECL's Whiteshell Nuclear Research Establishment (WNRE) in Manitoba. Boreholes have been drilled to depths of up to 1 km in four plutons across the Shield (Figure 1). Various methods of borehole drilling have been used and most have been completed using multi-level packer systems (Table 1) for long-term monitoring purposes. These allow access to discrete zones for hydraulic head measurements and sampling for chemical analysis using downhole pumps or wire-line "bomb" samplers.

This paper describes results obtained in characterizing the water chemistry of the four research areas listed in Table 1 and compares the data with results obtained from the Shield mines and plutonic rocks of other countries.

HYDROGEOLOGY OF THE RESEARCH AREAS

The subsurface hydrogeology of all four plutons investigated in the research areas listed in Table 1 is controlled largely by major sub-horizontal and/or sub-vertical fracture zones. In addition, some areas are bounded on at least one side by major structural or stratigraphic features (e.g. faults, dikes, Paleozoic cover, etc.). The main characteristics of the individual areas are summarized below.

Chalk River, Ontario

The Chalk River research area (Figure 1) is located about 200 km northwest of Ottawa in the Grenville Structural Province of the Canadian Shield. The area is lithologically and structurally complex. Bedrock is comprised mainly of monzonitic orthogneiss and metagabbro containing numerous diabase and pegmatite veins. It is highly fractured, with a major regional fault zone (the Ottawa-Bonnechere graben system) striking northwesterly through the region. Surficial deposits of Pleistocene glaciofluvial or lacustrine materials are generally less than 3 m thick with frequent bedrock outcroppings. Calcite is the most common fracture filling mineral at the Chalk River research area (Bottomley et al., 1984a).

The hydrogeology of the Chalk River research area is controlled mainly by a network of vertical and sub-horizontal fractures. Many of these have been demonstrated to be permeable and interconnected during borehole pumping and tracer tests (Raven et al. 1984). Most Chalk River boreholes have been drilled for geological or hydraulic characterization. One borehole, CR-13, was drilled by air percussion methods, specifically for hydrogeochemical characterization (Bottomley et al., 1984a). The results for

Table 1

Research area and borehole characteristics forming part of the hydrogeological work in the Canadian Nuclear Fuel Waste Management Program.

Research Area	Rock Type	Designation	Number of Holes	Diameter and Type	Max Depth (m)	Completion System
Chalk River, Ont.	Quartz monzonite gneiss	FS	15	165 → 152 mm	100	multi-level
		CR	21	apd	610	pvc casing + Westbay
East Bull Lake, Massey, Ont.	Layered gabbro/ anorthosite	P	14	150 mm apd	100	none
		EBL	4	76 mm dd	875	Westbay
Eye-Dashwa Lakes, Atikokan, Ont.	Granite	FL	21	76 mm dd	40	water-table piezometer
		ML	16	76 mm dd	70	multi-level pvc casing
		ATK	8	76 mm dd	1050	PIP/Westbay
Whiteshell (WN)/ Underground Research Laboratory (URL) Lac du Bonnet, Man.	Granite	B	33	76 mm dd	15	slotted well/ piezometer
		B	10	155 mm apd	50	multi-level pvc casing
		M	14	155 mm apd	400	multi-level pvc casing
		URL	11	76 mm dd	1090	Westbay
		WN	10	76 mm dd + 152 mm apd	1000	Westbay + multi-level pvc casing

dd = diamond drilled borehole
apd = air percussion drilled borehole
PIP = pressure inflated packer

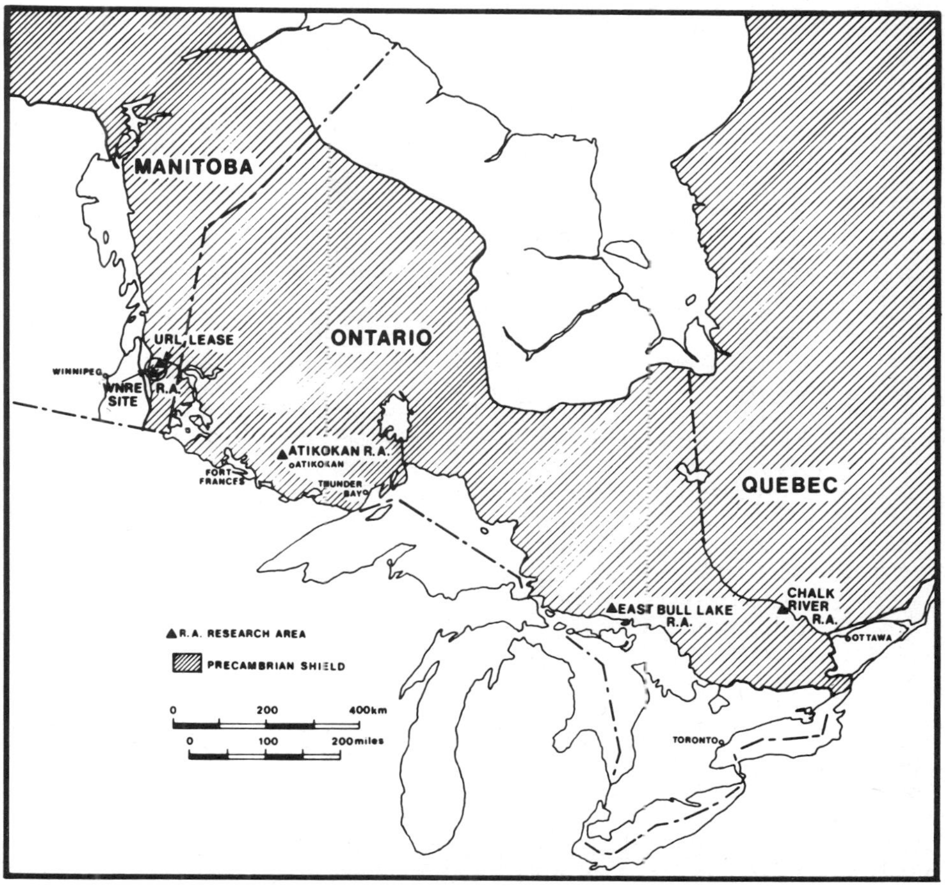

***Figure 1.** Location of research areas in plutonic rocks on the Canadian Shield.*

the Chalk River gneiss, described here, are from measurements made in this borehole. Sampling of groundwater in the metagabbro (below 400 m depth) has been hampered by the presence of tritium-rich, casing-installation water and the low hydraulic conductivity of the zone. The presence of the high total dissolved solids (TDS) waters found at all other research areas has not been confirmed at the Chalk River research area.

The East Bull Lake Research Area

The East Bull Lake Research Area is located near Massey, Ontario (Figure 1) and lies close to the southern margin of the Superior Structural Province of the Canadian Shield. The Grenville Structural Province is 80 km to the southeast. The East Bull Lake pluton occupies the central part of the study area, and is coeval with the Huronian Supergroup that overlies Archean rocks, including granites emplaced at least 2500 Ma ago. The elongate pluton comprises various gabbroic rock units and two gabbro dykes. Based on recent drilling, the pluton is about 760 m deep and is underlain by granodiorite.

The major geochemical field studies at East Bull Lake have concentrated on the identification and chronological interpretation of fracture filling materials within the body and on the characteristics of groundwater chemistry to a depth of 850 m.

Fracture filling minerals are abundant at all depths in the East Bull Lake gabbro. These include serpentinite in rubble and fracture zones and various low temperature minerals such as iron oxyhydroxides, clays, calcite and laumontite in smaller scale fractures and lineaments (McCrank et al., 1984, Kamineni, pers. comm.). Hydrogeological work at East Bull Lake has attempted to define the

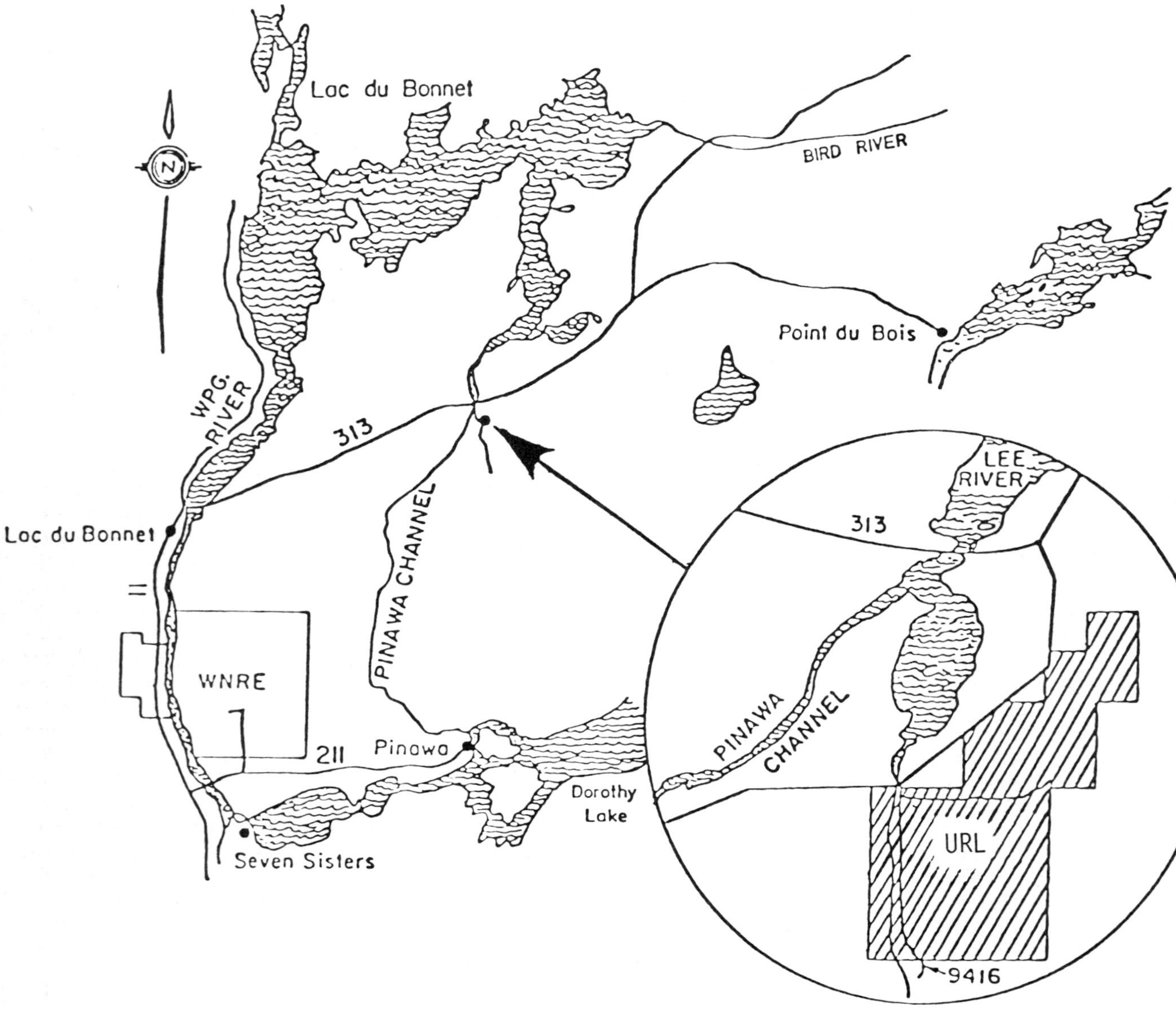

Figure 2. Regional setting for the Whiteshell Nuclear Research Establishment and Underground Research Laboratory lease area, together forming part of the Whiteshell Research Area.

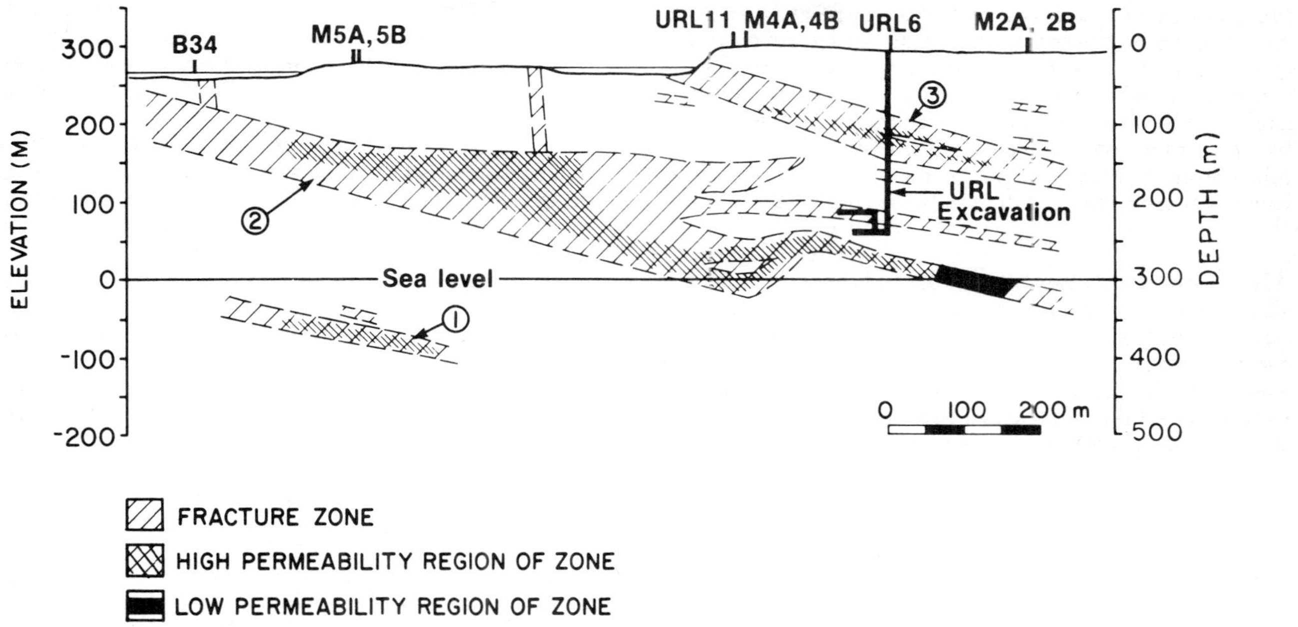

Figure 3. *Schematic representation of fracture zones (numbered) in granite at the URL location showing relative hydraulic permeabilities and boreholes which access the zones. The section runs from north (at B34) to south (at M2B).*

groundwater flow system in the central 2 × 2 km plateau of the gabbroic body, to depths of 800 m.

Two of the deeper boreholes were found to contain pronounced, high hydraulic conductivity fracture zones at about 200 m and 400 m depth. Open fracture zones were encountered to depths of 775 m suggesting that high hydraulic conductivities may also occur below 400 m. Available data suggest that recharge conditions exist in these boreholes.

Research on the hydrogeologic and hydrochemical properties of the groundwater at East Bull Lake is continuing and only preliminary data are included in this paper.

Eye-Dashwa Lakes, Atikokan

The Atikokan research area is centered on the Eye-Dashwa Lakes granite pluton in north-west Ontario (Figure 1). The pluton is 2500 Ma old and lies within the Superior Structural Province of the Canadian Shield. The granite is more fractured than the Lac du Bonnet batholith (see following section) and contains an extensive suite of secondary minerals, that fill fractures to depths of over 1 km. The correlation between fractures and mineralogies, and their relative chronology, has been described in detail by Stone and Kamineni (1982). Inclined fracture zones, observed in cored boreholes up to 1200 m in length, appear to control the groundwater hydrogeology of the area. The presence of the Bartlay Fault to the south-east and the 350 m deep open pit of the Steep Rock iron mine, also to the south-east, undoubtedly influence the regional hydrogeology, but their effect on the research area flow systems is unclear at present. Core and acoustic televiewer (ATV) logs of two deep 76 mm holes drilled in 1979 and three similar holes drilled in 1984/85 indicate the presence of inclined fracture zones or faults that have highly permeable sections. In one instance, a fracture zone 150 m below the surface gives artesian conditions at the borehole surface. Shallow bedrock holes, drilled in nearby topographic highs and lows indicate localized recharge to groundwater from hills, with discharge in valleys (Ridgway and Pearson, 1985). More extensive regional flow patterns are thought to control deep groundwater flow, as indicated by the variations in hydraulic heads below 300 m.

Whiteshell Research Area

The Whiteshell (WN) and the Underground Research Laboratory (URL) areas near Lac du Bonnet, Manitoba, are part of the Whiteshell Research Area, located on the western portion of the Lac du Bonnet granite batholith (Figures 1 and 2). The batholith is 2500 Ma old and also lies within the Superior Structural Province of the Canadian Shield. Both pink and grey granite are found at the WN/URL locations with grey predominating below 200 m. Secondary minerals are found at all depths in the rock, ranging from high temperature, deuteric fillings of epidote, quartz and chlorite, to low temperature, geologically-recent deposits of calcite, goethite and clays. The geology and fracture-filling mineralogy is described in detail by Kamineni et al. (1984).

In the WN/URL area, the batholith is relatively undisturbed by major tectonic features. In conjunction with core and ATV logs, pumping tests and hydraulic head measurements show that the hydrogeology is controlled by subhorizontal fracture zones, known to extend to at least 350 m depth (Davison, 1984a). These zones are partially interconnected by near-vertical fractures, particularly close to the surface. The location and relative permeabilities of the zones are shown in Figure 3.

Permeabilities in the fracture zones are as high as 10^{-4} m/s, but in the unaltered, unfractured rock, values range from 10^{-9} to 10^{-13} m/s. Permeability of the unfractured rock generally decreases with increase in depth, such that in the few boreholes in the Whiteshell Area that penetrate below 700 m, fractures are uncommon and the permeability of the unfractured rock does not exceed 10^{-12} m/s.

Examination of cores and observations in the URL access shaft have shown that the fracture zones are characterized by a pink to red coloration of the granite, tending to a beige colour in the more permeable sections. Larger abundances of low temperature alteration minerals are also found here and give the rock a loose and crumbly texture.

To the west of the Whiteshell Research Area, the batholith is covered by fluvio-glacial gravels, sands and clay of Pleistocene age but records of private wells in the area show that sub-horizontal fracture zones in the granite are still present and act as major sources of groundwater. Groundwater flow directions in the area are thought to be from the more elevated land in the southeast portion of the Whiteshell Research Area (305 m a.s.l.), to the Winnipeg and Lee Rivers in the west (248 m a.s.l.). The hydrogeology further west is unclear at present, and is likely to be complicated by the presence of overlying Paleozoic rocks (of the Western Canada Sedimentary Basin) which contain brines derived from evaporitic sediments.

GROUNDWATER SAMPLING AND ANALYSIS

Many techniques have been evaluated for sampling groundwaters from boreholes in plutonic rocks, as part of the Canadian Nuclear Fuel Waste Management Program. A major objective has been to obtain samples that are representative of the groundwaters present in the rock prior to intervention. This has been generally done as follows.

After drilling, the boreholes are cleared of drill waters, iron particles, rock chips and rock flour from the drilling operation using air-lifting or swabbing techniques. These involve the rapid and vigorous lifting of water through a steel drill rod to the surface by down-hole air injection, or by a mechanical piston on a wire line. Drilling water is sometimes tagged by rhodamine WT prior to use and is detected by fluoresence analysis during subsequent sampling operations. After initial clean-up, sections of the hole, chosen after examination of fracture logs, are isolated by inflatable packers, such as in the Westbay system, and specific zones are sampled. An air-pressure operated ('squeeze') pump or electric pump is used to sample the zone via a stand-pipe connected to the upper packer in the zone. In the Westbay multi-interval casing system, samples are taken by a pressure-actuated valve and cylinder system, which is lowered to the packed-off interval on an armoured cable. Access to the interval can also be made through a sliding-sleeve port using a special open/close tool. Full details of the types and operation of borehole casing systems are given by Davison (1984a). General sampling methodologies are reviewed in detail by Bottomley et al. (1984a).

Most sampling in this study has been done by swabbing and pumping from borehole zones isolated by packers. In some cases, long-term pumping and monitoring of discrete, permeable zones has been accomplished using a mobile straddle-packer/downhole sampling and sensor system (Bottomley et al., 1984a). In most other cases, pumped water was passed into a flow cell at the surface which contained Eh, pH and specific conductance sensors. A bypass to the cell, containing an in-line 0.45 μm filter, allowed samples to be taken for analysis once flow cell measurements stabilized. If the flow cell was not used, samples were taken and filtered as soon as possible in the laboratory. When

Table 2

Summary of samples taken, pretreatment methods and analyses made for groundwater from a given zone in a research area borehole.

Sample	Species/Element	Container	Volume	Filtered?	Preservative	Analytical Methods
Anions	HCO_3,SO_4,Cl,Br, F,B,PO_4,NO_3	HDPE[1]	0.12	yes	none	filtration, ion liquid chromatography, colorimetry
Cations	Na,Ca,Mg,K,Sr,Si	HDPE	0.06	yes	4 mL/L HCl	Atomic Absorption
Trace	Li,Fe,Mn,V and others	HDPE	0.06	yes	8 mL/L HNO_3	inductively coupled plasma spectrometry
Isotopes	^{2}H,^{3}H,^{18}O	HDPE	0.06	no	none	mass spectrometry, liquid scintillation counting
C,S	^{13}C,^{14}C,$^{34}SO_4$,$S^{18}O_4$	HDPE	~50	no[2]	none[3]	mass spectrometry, liquid scintillation counting
Sr	$^{87}Sr/^{86}Sr$	HDPE	0.25	yes	8 mL/L HNO_3	mass spectrometry
U	U,$^{234}U/^{238}U$,^{226}Ra	HDPE	4	yes	6 mL/L HNO_3	alpha spectrometry
Rn	^{222}Rn	glass vessel	0.05	no	none	gas scintillation counting
DG	H_2,He,O_2,N_2,CO_2 CH_4,Ar,H_2S	steel cylinder	0.05	no	none	mass spectrometry
DNG	He,$^{3}He/^{4}He$, Ne isotopes	Cu tube	0.02	no	none	mass spectrometry

1. high density polyethylene bottle
2. unless containing significant sediment
3. precipitating reagents may be added

possible, alkalinity was determined in the field. For a given zone, a suite of samples were taken for laboratory analysis and a summary of the samples taken and parameters determined is given in Table 2. Most laboratory analyses were performed by the Analytical Science Branch, WNRE, Pinawa. Stable isotopic and tritium analyses of water were determined at the University of Waterloo or the Chalk River Nuclear Laboratories using standard mass spectrometry and direct scintillation counting methods. Other isotopic and noble gas analyses were determined by University laboratories, involved in the Nuclear Fuel Waste Management Program.

RESULTS

Compositions of the groundwaters encountered at the four research areas are shown in Table 3. For all data shown, the sum of cations equals the sum of anions to within ± 10% (each expressed as milliequivalents per litre). Considerably more data than shown in Table 3 have been obtained for these research areas but most results apply to near-surface, fresher waters. Only the most representative and the most saline samples are reported here. Data for the Chalk River research area are taken from Bottomley et al. (1984a).

Many of the analyses are for groundwaters that have a component of drilling water as shown by the presence of rhodamine WT (East Bull Lake). In the Chalk River case, this is indicated by high tritium levels, derived from surface water containing about 2000 T.U. The presence of low, but finite, tritium concentrations (10 to 120 T.U.) in groundwater (see Table 3) indicates that dilution has occurred probably, by either un-tagged drill water or surface waters that entered the sample zone during periods when the borehole zones were not isolated by packers. Such intra-borehole flow may occur when hydraulic heads are not the same throughout the hole, and is most frequently observed when a fracture zone at depth has a much lower hydraulic head than zones above or below it.

DISCUSSION

Variations in Salinity and Chlorinity

In all research areas groundwater salinity (i.e. total dissolved solids, TDS) and chlorinity were found generally to increase with depth. The most representative values are summarised in Table III. Highest concentrations have been

Table 3

Chemistries of saline groundwater samples from boreholes in plutonic rocks in four research areas on the Canadian Shield. All ionic concentrations are in mg/L. Dashes indicate no analyses were made.

Research Area	Bore-hole	Depth (m)	Na	K	Ca	Mg	Sr	Cl	SO_4	HCO_3	Si	F	Br	pH	Eh (mv)	Rhoda-mine Wt. (μg/L)	^3H (±8TU)
Chalk	CR-13	341	209	1.1	35.1	2.4	1.7	325	54	50	3.5	2.5	3.0	7.9	+330	–	0
River,	CR-13	423	70	2.6	15.3	2.8	0.5	37	22	155	3.5	0.7	0.3	7.3	+325	–	316
Ontario	CR-13	486	92	5.8	8.1	1.5	0.3	48	8.2	171	3.5	0.7	0.5	8.3	−100	–	615
	CR-13	576	53	2.8	18.6	2.8	0.6	29	8.8	159	6.0	0.7	<0.1	8.1	0	–	512
East Bull	EBL-1	405	242	0.8	6.0	0.1	0.07	328	16	65	16	0.3	1.5	10.2	+130	23	13
Lake,	EBL-4	443	846	2.8	279	2.0	3.50	1743	77	71	6.1	3.5	5.7	8.9	–	2	24
Ontario	EBL-2	538	584	2.1	293	1.7	2.90	1350	13	30	1.5	2.0	2.1	7.4	–	13	62
Eye-Dashwa	ATK-1	790	613	2.7	978	2.4	24	2380	80	13	2.1	0.8	17.9	6.0	+220	–	106
Lakes,	ATK-1	880	1830	9.8	11900	3.8	308	25400	430	45	3.9	4.5	141	7.3	+10	–	86
Atikokan,																	
Ontario	ATK-5	680	86	1.5	940	1.1	24	1860	91	<2	2.7	0.5	16.6	6.6	+100	–	112
	ATK-5	700	116	2.8	1240	1.4	31	2560	114	12	<1.2	0.6	19.3	5.7	+150	–	86
	ATK-5	980	816	7.4	8980	0.9	220	18100	263	<2	2.4	3.1	137	5.5	+160	–	62
	ATK-5	1050	779	8.5	7460	1.4	205	12200	186	21	1.7	2.3	119	5.5	−210	–	72
Lac du	WN-1	416	1490	8.4	1600	64	–	4600	914	62	–	–	–	–	–	–	56
Bonnet	WN-4	395	316	12.1	472	40.3	–	1380	310	237	–	–	–	8.0	–	–	–
Batholith	WN-4	465	818	24.4	921	65.5	–	3050	496	164	–	–	–	7.8	–	–	–
(WN/URL),	WN-4	487	2510	26.7	1480	24.1	–	6350	1090	72	–	–	–	7.6	–	–	–
Manitoba	WN-4	513	1530	20.7	1060	25.9	–	4000	631	208	–	–	–	8.1	–	–	–
	WN-4	568	457	12.6	424	23.2	–	1440	243	358	–	–	–	8.1	–	–	–
	WN-4	577	3080	28.9	1890	24.9	–	7740	1240	14	–	–	–	6.9	–	–	–
	WN-4	720	3130	12.9	1860	24.0	28	7410	1180	115	5.1	1.4	25.1	–	–	–	12
	WN-4	800	4170	13.5	1680	23.1	21	9570	1210	42	5.8	<0.2	23.0	–	–	–	12
	URL-2	874	50	3.2	1000	8.7	2	1755	22	50	7	0.5	13	7.5	–	–	64
	M3A	300	1550	7.0	510	1.5	4	2980	604	49	4	2.2	6.2	7.4	–	–	16
	M3A	375	1700	5.3	510	6.0	5	3020	693	<2	4	2.8	<0.1	6.1	–	–	18
	M5A	322	810	3.9	620	5.6	4	1880	473	72	6	2.9	5.7	7.2	–	–	50
	M7	390	1900	10.0	3600	16	40	8000	720	25	6	3.4	51	7.1	−140	–	0
	M10	400	1270	6.7	1440	10.9	12	3830	753	30	2	1.9	29	7.5	–	–	17
	M13	335	799	4.3	126	6.2	1	856	504	36	2	1.9	6.8	7.7	–	–	9
	M14	340	1700	5.7	2450	17.5	25	5970	956	30	6	3.1	39	7.1	–	–	0

found at 1 km depth at Atikokan (40 g/L) but these values are considered minima because of the presence of drilling water and/or shallow groundwater components in the samples. Values of tritium content for local surface waters in the area are typically 30-120 T.U. The minimum salinity of the undiluted deep groundwater at Atikokan can be calculated assuming this water has negligible tritium. If the conservative case is considered whereby the freshwater or drill water component contains 120 T.U., then for sample ATK 1 (86 T.U.), a potential salinity of at least

$$\frac{120}{120-86} \cdot 40 = 140 \text{ g/L}$$

is calculated for the undiluted groundwater.

Projections for the tritium-containing waters of the WN/URL area (Table 2) suggest that no greater salinity than 15-20 g/L will be found at the depths examined. These salinities are, in fact, observed in two tritium-free samples shown in Table 3.

The results from the research areas may be compared with data obtained for the Canadian Shield mines (Frape et al., 1984a) if local differences in source and rock type are ignored. Variation in TDS and chlorinity with depth, are shown in Figures 4a, b, using the data from Frape et al. (1984a), Table 3 and shallow groundwater chemistries for the WN/URL area (Gascoyne and Elliott, 1985). High TDS concentrations are found in shallow groundwaters (Figure 4a) due to bicarbonate and sulphate derived from overburden materials. The TDS is seen to decrease rapidly with depth and then increase equally rapidly at about 300 m depth. The continuing increase in TDS then proceeds exponentially with linear increase in depth. In contrast, groundwater chlorinity (Figure 4b) shows a continuous increase throughout the depth range. In both diagrams a change of slope occurs at about 300-400 m depth. The rate of increase in TDS and Cl below this depth is reasonably constant (both rates are about 37% per 100 m) if the waters contaminated with recent surface water (shown as large circles in Figures 4a, b) are rejected or their analyses corrected to compensate for the presence of this surface water.

Projected concentrations of the undiluted Shield brines have been determined by Pearson (1985) using $\delta^{18}O$, δ^2H and major ion concentrations of the diluted brines sampled from the Yellowknife, Thompson and Sudbury mines. By regressing each isotopic or ionic component, against $\delta^{18}O$ content, the concentration of that component, at halite saturation, was obtained for the undiluted brine. Maximum concentrations for each mine region are listed in Table 4. Compositions of these 'parent' brines are shown as arrows in Figures 4a, b. It can be seen that the compositions of many of the Shield mine samples lie close to those calculated for the parent brine. These samples, when not showing dilution by recent tritium-rich water presumably are diluted only by zero-tritium groundwaters derived from shallower, less saline flow paths. Extrapolation of the data in Figures 4a, b to estimated parent brine compositions suggests that these parent brines may be encountered below about 1700 m in most areas.

The relatively constant rate of increase of TDS and chlorinity below 300 m shown in Figures 4a and b implies that there is an overriding control on salinity that is applicable to the Shield, irrespective of variations in local topography, distribution of recharge/discharge zones and rock type. These latter factors probably account for the scatter in data between 0 and 300-400 m depth, resulting from dominantly advective processes in the hydrogeological regimes. Below this depth, it appears that an extensive and more uniform situation exists, controlling the groundwater salinity. This is ascribed tentatively to molecular diffusion rather than advection. In other words, the mixing of essentially stagnant, saturated brines, with the water from shallower, lower salinity, active or paleo-flow systems appears to take place by diffusion below 300 m and advection above this level. Diffusion, driven solely by chemical differences between zones of parent brine and the lower limit of penetration of advective processes, is one process that might account for such a distinctive relationship. An alternative explanation of the change in slope could be that a mixing line for relatively local groundwater flow systems intersects a mixing line for deep regional flows, at about 300 m depth. This may be a valid process, and can permit advective mixing at all depths. However, the coincidence of data points from widespread areas on the Shield around a single line, for depths greater than 300 m (Fig. 4b), suggests that advective mixing of deep flows is not occurring in these localities.

The main control over the depth to which advective processes penetrate (assuming relatively low hydraulic gradients) may be rock permeability. The change in hydraulic permeability for the Atikokan granite, with depth, is

Table 4

Estimated compositions (in g/L) of parent brines from three mines in the Canadian Shield (from Pearson, 1985).

Location	Ca	Na	Mg	K	Sr	Cl	Br	TDS
Sudbury, Ontario	109	34	0.068	0.47	2.6	270	2.2	418
Thompson, Manitoba	82.2	44.7	6.3	0.23	1.4	241	2.3	378
Yellowknife, N.W.T.	90.6	53	1.6	0.07	2.2	230	2.1	380

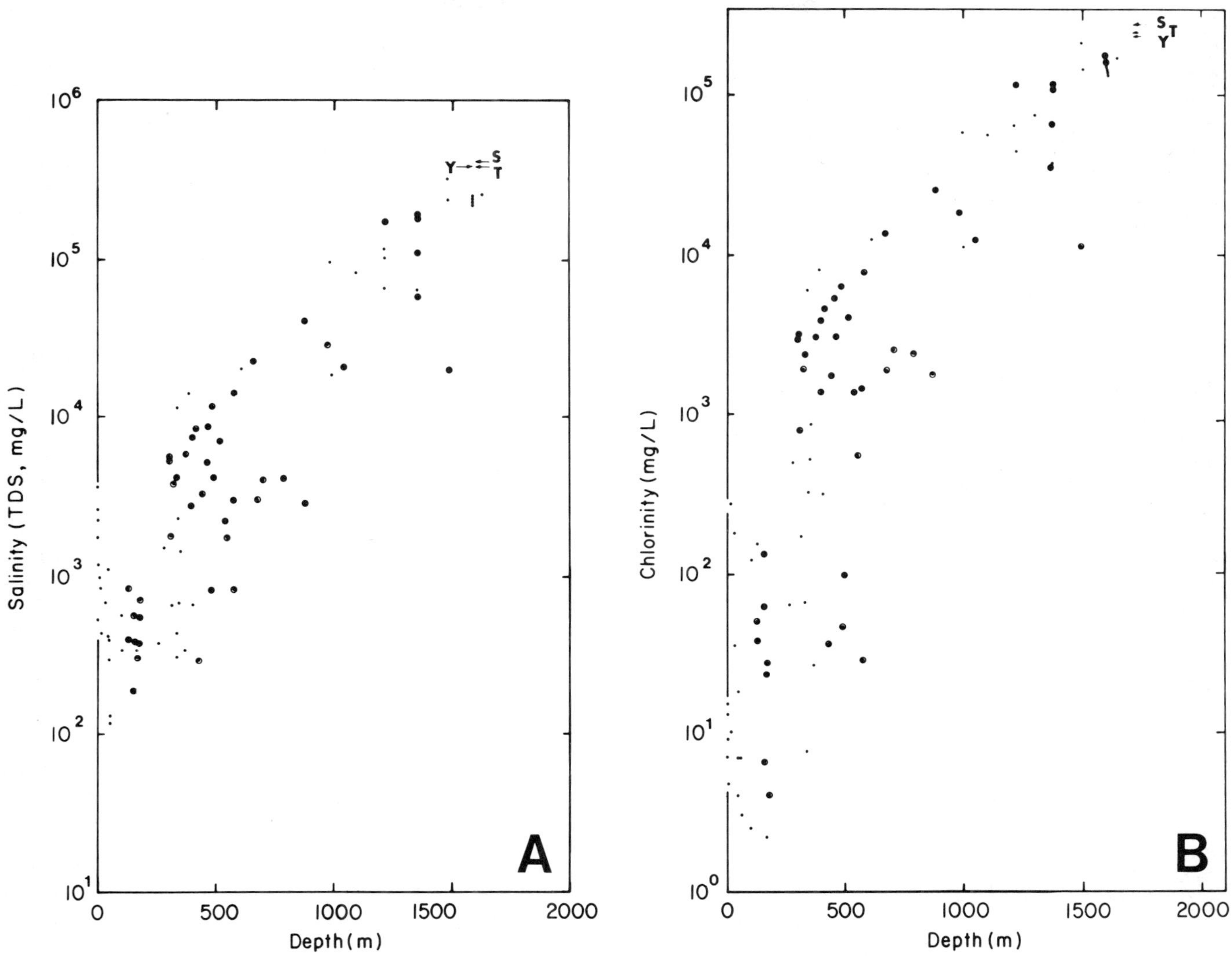

Figure 4. *Variation of a) total dissolved solids (TDS) and b) chlorinity with depth for research area and Shield mine waters. Large circles indicate dilution by fresh waters (from presence of ≥15 tritium units) for samples deeper than 100 m. Arrows show projected parent brine compositions for Sudbury (S), Thompson (T) and Yellowknife (Y) samples (see text).*

shown in Figure 5. Again, a break in slope is seen in the depth range 300-500 m with relatively constant low permeabilities (10^{-13} to 10^{-11} m/s) below this depth. It should be noted, however, that discrete zones of higher permeability (up to at least 10^{-10} m/s) do exist at depths of at least 1000 m and these are usually the fracture zones from which many of the waters shown in Figures 4a, b are sampled. Nevertheless, it is proposed here that diffusion of dissolved salts is the main process occurring in these higher permeability zones. Advection will tend to occur in these zones only if high hydraulic gradients exist – a situation unlikely to be found in low relief Shield terrain.

Situations exist where advective processes may be less important in diluting saline waters as they migrate towards the surface. The 'moose licks' of northern Ontario (Frape et al., 1984b) may be such an example, where saline waters (up to 10 g/L) discharge at the surface, controlled, presumably, by diffusion and minor advection through an isolated fracture network extending to considerable depth. Conversely, the recent observation of relatively dilute waters at ~800 m depth in the Atikokan research area may indicate an example of advection to considerable depth (the presence of a major sub-vertical fault zone at this depth undoubtedly assists the advective process, however).

Some advection of dissolved solids at depth may also be occurring due to long term-geological movements, such as isostatic rebound following deglaciation. Crustal depression and relaxation by ice loading is recognised as an important method of inducing flow at depth in fractured rocks but it is unclear how significant this is in aiding transport of saline waters to the surface.

Variation in Groundwater Chemistry

Major Ions. Groundwater compositions in all research areas are found to change in character with increasing depth, either as a result of rock-water interaction or be-

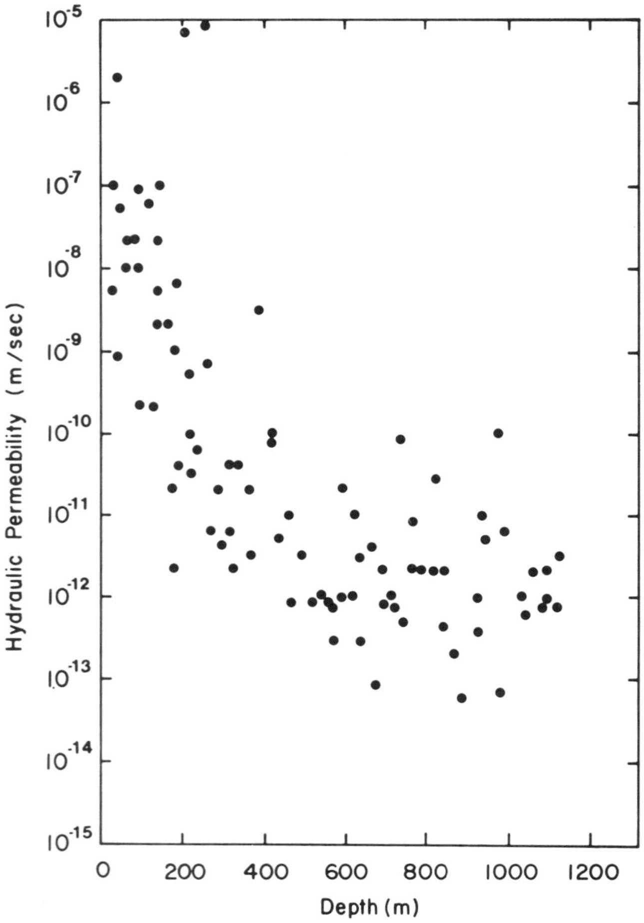

Figure 5. *Variation of permeability with depth at the Atikokan research area.*

cause of mixing with deeper waters of different chemical signatures. The composition of shallow groundwaters is described elsewhere (Bottomley et al., 1984a, Gascoyne and Elliott 1985, Bottomley et al., 1985). All surface and shallow groundwaters are dominated by bicarbonate and, in some cases, contain considerable sulphate. This is a result of dissolution of carbonates, sulphates and sulphides in the overburden, by slightly acidic, CO_2- and O_2-rich recharge waters. The pH values attained in shallow groundwaters in the gabbro at East Bull Lake, are somewhat higher than those of groundwater from granitic rocks in the other areas (pH varies between 8.5 and 10 above 50 m depth, compared with 7.0 to 8.5 in granites) indicating that hydrolysis of the mafic minerals, with limited CO_2- buffering, is the dominant reaction.

Shallow groundwaters are generally Na-Ca-(Mg) dominated and molar Na/Ca ratios range between 0.2 and 165 for gabbro and between 0.2 and 4 for granite, at the East Bull Lake and WN/URL locations, respectively. With increasing depth, Mg and SO_4 concentrations decrease. Sulphate concentration rises sharply at higher salinities in both rock types but Mg only increases in WN/URL groundwaters. The decreases are probably due to Mg-exchange with clay minerals and less pyrite oxidation to form sulphate. Increases in Mg and SO_4 concentrations may be caused by increasing stability as associated complexes form in the saline waters. Calcium, Na and Cl also increase markedly at greater depths and eventually become responsible for over 90% of the salinity. The transition in chemistry from shallow to deep groundwater is also noted by the loss of HCO_3 (Table 3). This may be due to precipitation of calcite induced by the increasing Ca concentrations.

The variation of chemistry from shallow to deep regions within the same pluton is shown schematically in Figure 6a for the sub-horizontal fracture zones that dominate groundwater flow at the URL location (from Davison, 1984b). Hydraulic head and density data obtained from the numerous instrumented boreholes at the location (Davison, 1984a) permit flow directions and areas of recharge and discharge to be determined and these are indicated on Figure 6a. The chemistry of these groundwaters is illustrated in the modified Piper diagram (Figure 6b). To show their relation to the hydrogeological environment, waters of similar chemistry are identified as one of six types. Shallow waters, of low to moderate salinity (small circles, types III-3, IV, V and VI), lie in the Ca-Na/HCO_3 corners. As depth increases, the chemistry evolves towards the upper left, Na/Cl-SO_4, corner (larger circles, types III-2 and III-1). Below 400 m, Ca concentrations increase and waters move towards the upper centre of Figure 6b (large circles, and hexagons, type II).

One groundwater, from a low permeability zone at 1000 m depth, indicates how far the evolution towards Ca-rich chemistries may go. The sample from borehole URL-2 (Table 3), although still diluted by considerable amounts of drilling water, is almost a pure solution of calcium chloride (Na/Ca molar ratio is <0.1). In contrast, however, the Na/Ca ratio at the WN site shows the opposite trend, i.e. one of increasing value with depth (see data in Table 3). Variations in molar Na/Ca ratio for deep zones in borehole WN-4 range from 1.2 to 4.3. Unfortunately no other groundwaters from similar depths at either location have been obtained so far, to determine which is the more common characteristic.

The data from Atikokan (Table 3) indicate that the calcium concentration relative to that of sodium, increases at greater depths (Na/Ca changes from 1.2 to 0.15). In contrast with the URL, however, Ca is usually the dominant cation at Atikokan, throughout the depth range.

Evolution of a groundwater chemistry towards calcium enrichment relative to sodium is an expected characteristic for a granitic environment, based on thermodynamic information. Albitization of plagioclase and alteration of hornblende in granite that is in contact with sodium-rich groundwaters will occur over geologic time to remove sodium from solution and release calcium. This has been observed in altered zones at the URL (Kamineni et al., 1984). Mixing with Na-Ca brines from a sedimentary origin may obscure the effects of this process, however, and could explain the high Na/Cl ratios in some WN/URL groundwaters. This possibility is considered in greater detail in the following section.

Br/Cl Ratios

Differences in composition between groundwater samples from the research areas and those from the mines are further illustrated by the variation of the molar ratios of Ca/Mg relative to Br/Cl, shown in Figure 7. Distinct regions on the diagram are seen for each research area and the mine samples, although there is considerable overlap be-

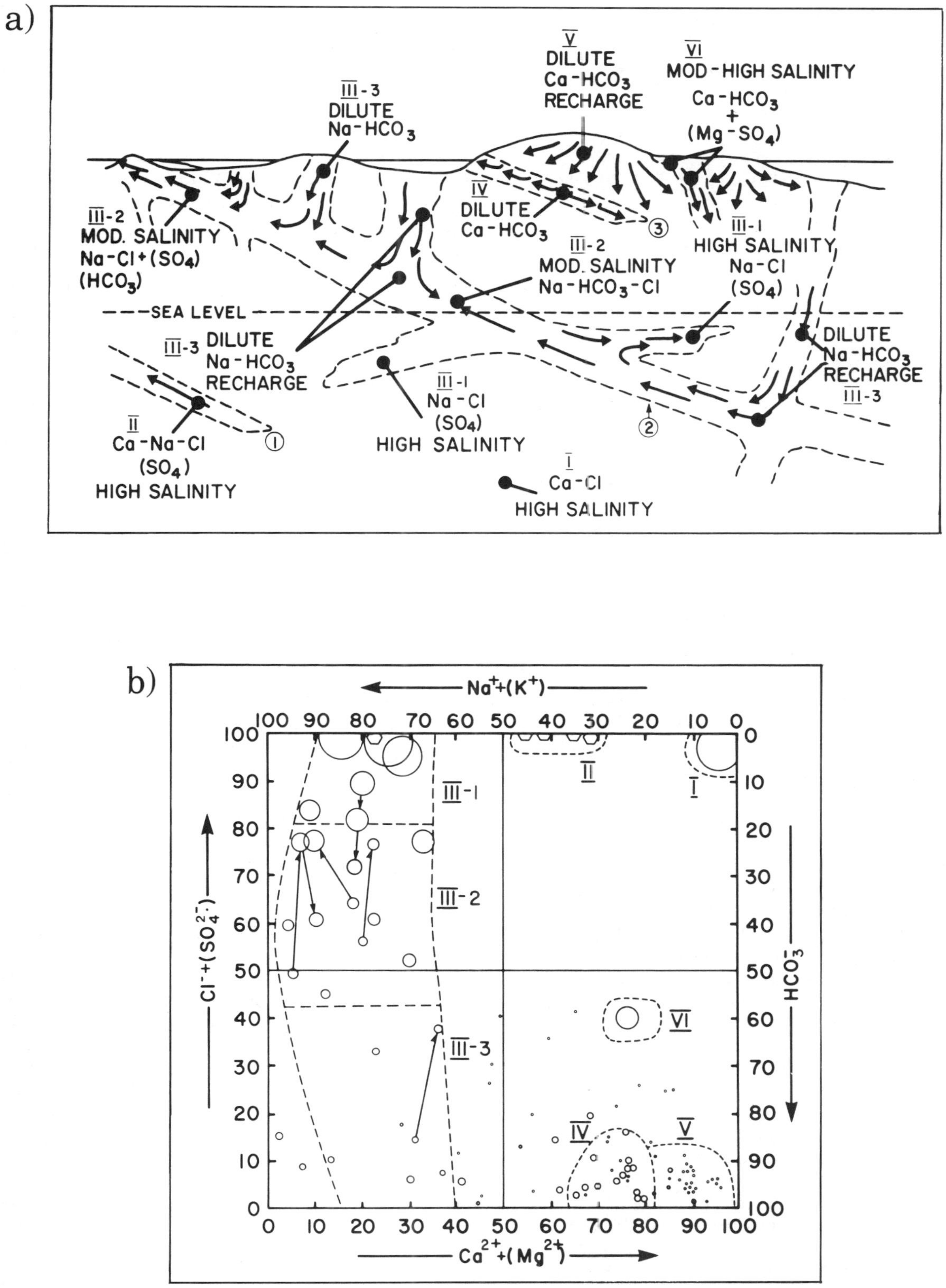

Figure 6. *Schematic representation of groundwater compositions at the URL lease area: a) showing groundwater flow paths and geochemical patterns in fracture zones (see Figure 3) and b) modified Piper diagram showing major ion chemistry of groundwaters. The size of circle indicates salinity of the water (hexagons refer to highest salinities), Roman numerals indicate water chemistry type shown in Figure 6a, and arrows show variation in chemistry during a pump test.*

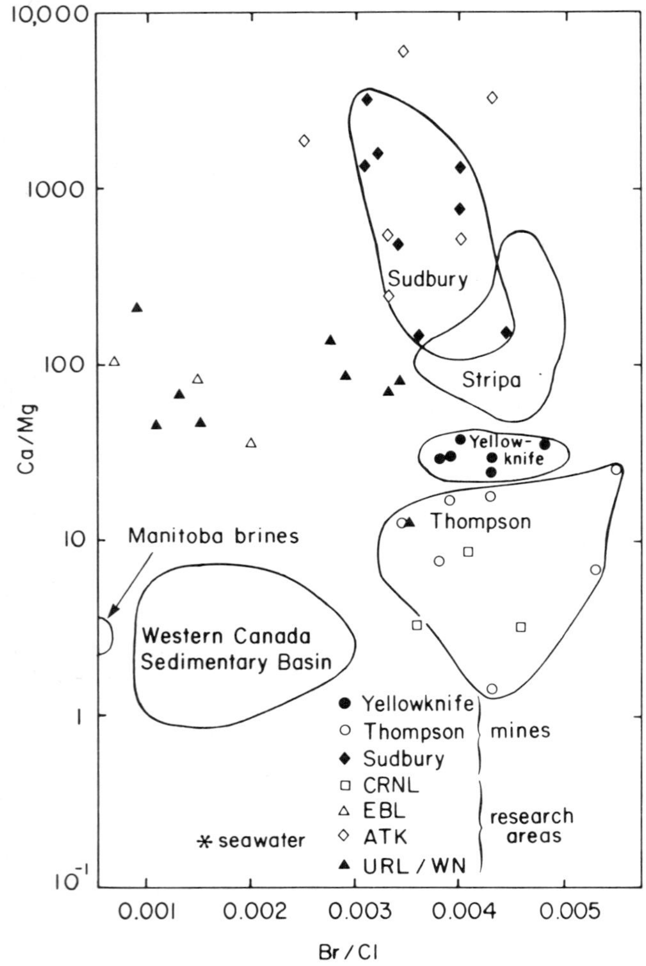

Figure 7. *Variation of Ca/Mg with Br/Cl (expressed as molar ratios) for research area saline groundwaters showing relationship to Shield mine waters and groundwaters from Stripa (from Nordstrom 1983), the Western Canada Sedimentary Basin (from Hitchon et al., 1971) and Manitoba Paleozoic rock brines (Betcher, personal communication, 1985).*

tween some of these. Data for groundwaters from the Stripa granite, Sweden (from Nordstrom 1983) are also shown in Figure 7 and fall in the same region as those from the mine/Atikokan samples.

Of significance is the relationship between the Br/Cl ratios for the saline waters and that of seawater. Several workers have argued that the Br/Cl ratio of saline groundwaters is a useful indicator of marine or non-marine origins of the water, because dilution or concentration (by evaporation) of seawater does not markedly affect the ratio of these conservative ions (Carpenter 1978, Nordstrom 1983, Edmunds et al., 1984, Frape et al., 1984a). High Br/Cl ratios are thought to indicate leaching of fluid inclusions, 'intergranular salts', or structurally-bound halogens that have a magmatic or hydrothermal origin and are, therefore, distinct from seawater. Nordstrom (1983) has observed that the Br/Cl ratios for leached, crushed granite at Stripa are identical to those in the deeper groundwaters and proposes that the salinity of Stripa groundwaters can be accounted for largely by leaching of fluid inclusions. The Br/Cl ratios at Stripa are very similar to those observed in Shield mine and Atikokan groundwaters (Figure 7), thereby suggesting a similar origin for the salinity.

The ratios in samples from the Chalk River research area also fall in the mine/Atikokan grouping and thus conflict with the proposal by Bottomley et al. (1984a) that the origin of the Na-Cl in groundwater at about 350 m depth at Chalk River, is post-glacial intrusion of Champlain seawater. Based on Cl concentrations, only a maximum of 6% of the groundwater can be intruded seawater and this would have a pronounced lowering effect on the Br/Cl ratio if mixed with 'fresh' groundwaters that have a Br/Cl molar ratio of about 0.004 (Figure 7, other data in Bottomley et al., 1984a).

For the same rock type (granite) there is a considerable difference between Atikokan and WN/URL grouping of Br/Cl ratios. Many WN/URL samples in the depth range 300-800 m, have Br/Cl values similar to those for Western Canada Sedimentary Basin brines (Figure 7, based on data from Hitchon et al., 1971 and Betcher, pers. comm.). These data lie close to the Br/Cl ratio for seawater indicating a marine sedimentary origin for the fluids. Some data points for the WN/URL area, however, lie closer to the mine-Atikokan data fields giving a much broader span of Br/Cl ratios than seen in any other area grouping. One possible explanation of this span is that some intrusion of adjacent

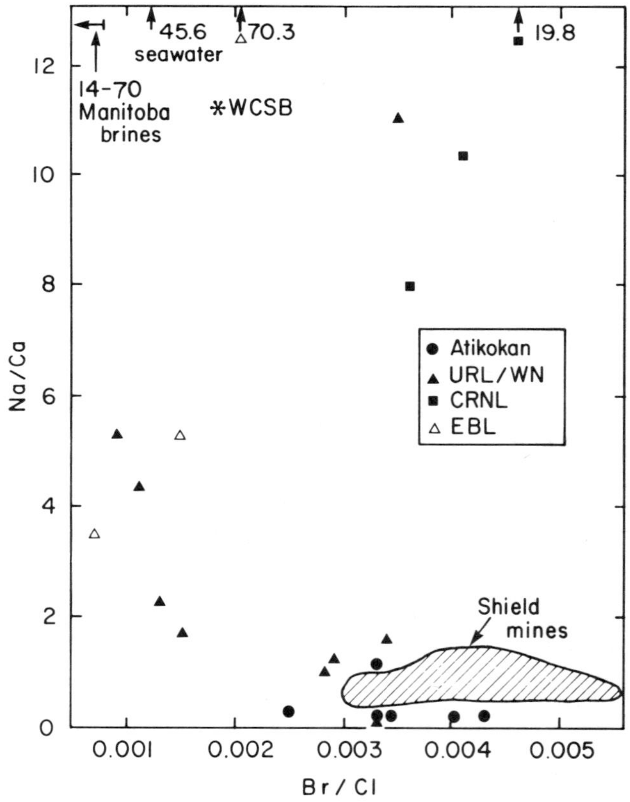

Figure 8. *Relationship between Na/Ca and Br/Cl (expressed as molar ratios) for research area groundwaters, as in Figure 7.*

sedimentary brines in the Winnipeg Formation (40 km to the west of the WN/URL area) may be occurring, or more likely, may have occurred, during Pleistocene times, or earlier, when the Paleozoic rocks may have covered the present edge of the Shield. The WN/URL samples with high Br/Cl may then be relatively unmixed and more characteristic of evolution in a granitic rock environment alone.

Further indication of mixing with basin brines comes from the high Na/Ca ratios for most of the WN/URL samples that also have low Br/Cl ratios (Figure 8). Atikokan and mine groundwaters, in contrast, have low Na/Ca and high Br/Cl ratios.

Saline groundwaters from the East Bull Lake research area also have Br/Cl ratios that are low compared with other Shield data and fall in the data field for the WN/URL location (Figure 7). These results may be explained by groundwaters in the gabbro pluton mixing with saline groundwaters from the Proterozoic sedimentary formations 40 km south of East Bull Lake, a situation similar to that of WN/URL. Sodium/Ca ratios are also high for the East Bull Lake samples (Figure 8) and this is further supportive of mixing with a NaCl-rich groundwater.

Isotopic Variations

The determination of ^{2}H and ^{18}O in groundwaters at the four research areas has been performed along with chemical analyses although results for East Bull Lake and the URL samples are not yet available. In earlier work, Bottomley et al. (1984a, b) described the variation in ^{2}H and ^{18}O at the Chalk River and Whiteshell (WN) areas and have shown how groundwater at different depths has discrete isotopic characteristics. All waters plot on, or close to, the global meteoric water line of Craig (1961). In both areas, however, shallow groundwaters (<50 m) tend to be enriched in ^{2}H and ^{18}O relative to samples of older (tritium-free) water from greater depths (Figure 9). The more saline samples from the Whiteshell site (up to 17 g/L TDS, Table III) are different again and plot slightly above the global meteoric water line. This deuterium enrichment is observed most clearly in Shield brines (Frape et al. 1984a) and the WN results may indicate some similarity to brines in this respect.

Although not a clear ^{2}H-enrichment, a further suggestion of this trend is observed in isotopic data for deep groundwaters at Atikokan. The variation of $\delta^{2}H$ and $\delta^{18}O$ isotopic data is shown in Figure 10, for surface waters, and shallow and deep groundwaters at Atikokan. The meteoric water line for Atikokan ($\delta D = 7.8\delta^{18}O + 7.9$) is shown for reference. Also included in Figure 10, for comparison purposes, are trends observed for deep mine groundwaters (Frape et al., 1984a). The Figure shows that Atikokan surface waters are affected by evaporation and with increasing depth the groundwaters follow this trend to about 500 m. This indicates either dilution of formation water with surface drill water, or an active, recent flow system, or a mixture of both. Between 500 m and 700 m there is a shift in this trend with increasing depth. This shift is further substantiated by the fact that, where two consecutive samples are taken from the same depth, the second sample, assumed to be more representative of formation water, also follows the trend of groundwaters indicated in Figure 10. The shift in trend below 500 to 700 m could be extrapolated above the meteoric water line, assuming that the deeper groundwaters are still considerably diluted by drilling water. This extrapolation would indicate that these deeper groundwaters are consistent with the deep mine groundwaters in terms of isotopic content. Current sampling work at Atikokan may resolve this question.

Solubility Controls and Rock-water Interaction

Any attempt at interpreting the geochemistry of deep groundwaters in the research area plutons must account for the presence or absence of dissolved ions, preferably with data from the rock mass to support the proposed rock-water interactions. Data for the Chalk River and Atikokan research areas are insufficiently abundant to attempt a detailed correlation and results from East Bull Lake are still in preparation. Considerably analytical data for groundwaters at the URL area have been obtained, however, and

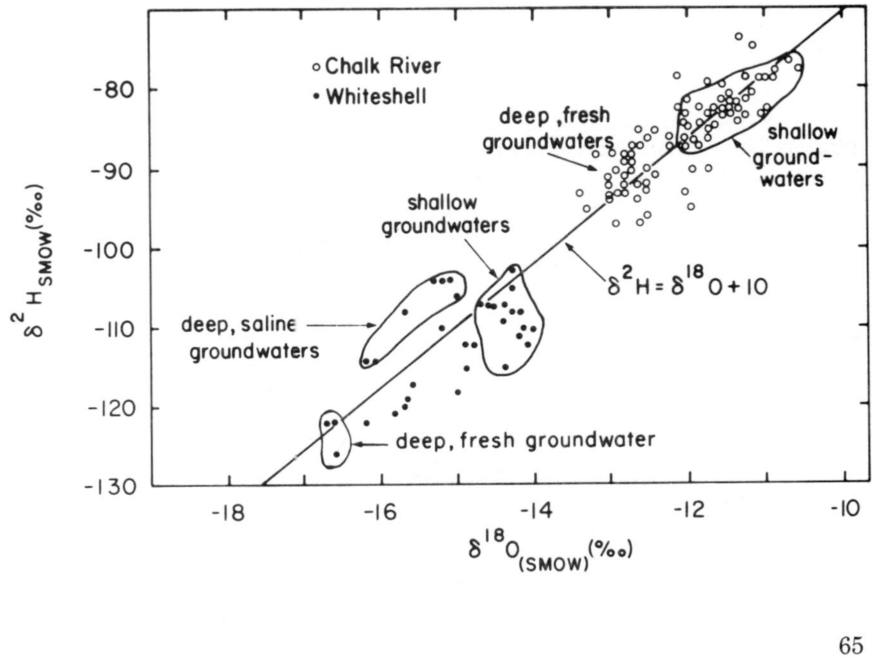

Figure 9. The stable isotopic composition of fresh and saline groundwaters at the Chalk River and WN research areas (after Bottomley et al., 1984a,b), showing the meteoric water relationship of Craig (1961), relative to Standard Mean Ocean Water (SMOW).

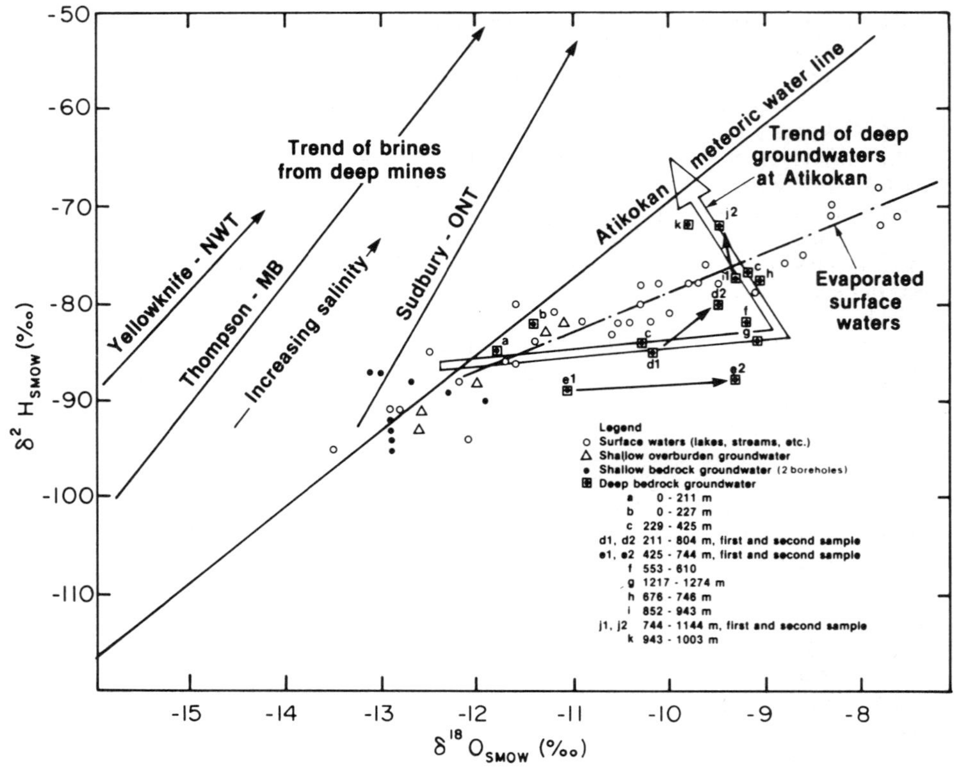

Figure 10. *Variation of stable isotopic composition of Atikokan groundwater showing relationships to depth of sample and trends observed in mine groundwaters (from Frape et al., 1984a).*

these are used, below, to test for relationships between modern water chemistry and existing fracture filling mineralogy.

Dissolved carbonate – calcite equilibria. A prominent characteristic of saline groundwaters in plutonic rocks is the depletion of bicarbonate with increasing salinity and depth. The fact that most of these waters are undersaturated with respect to atmospheric CO_2 indicates that HCO_3 has been lost via a process involving either reduction (to CH_4, for instance) or precipitation of carbonate minerals. Calcite is ubiquitous as a secondary fracture-infilling mineral in all four research areas and presumably occurs by precipitation from groundwater under conditions that were either hydrothermal (i.e., relatively ancient) or low-temperature (geologically recent).

Saturation indices with respect to calcite have been determined for surface, shallow and deep groundwaters at the

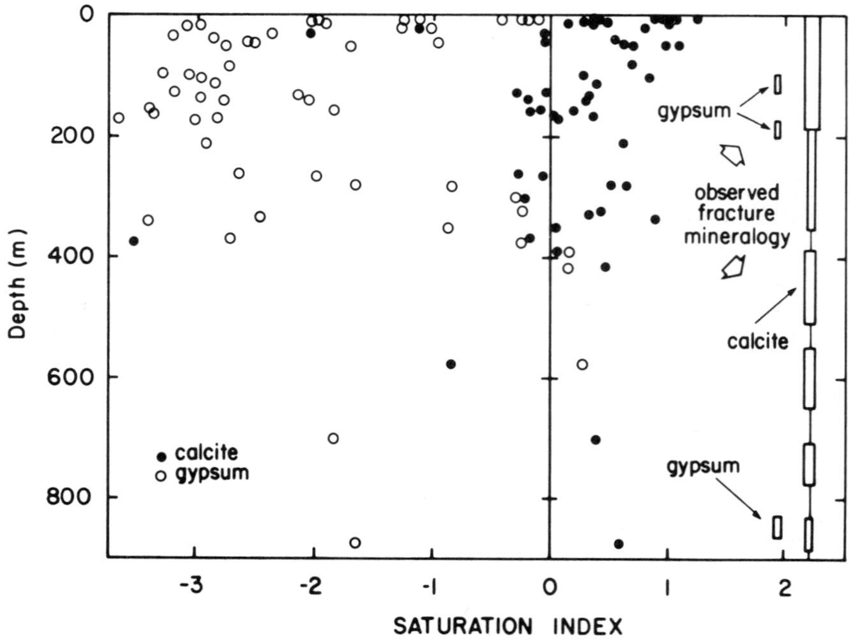

Figure 11. *Calculated saturation indices with respect to calcite and gypsum for groundwaters from the URL area, showing schematic representation of mineral occurrences in rock cores.*

URL, using an interactive version of the geochemical code PHREEQE (Gascoyne and Elliot 1985). These are plotted against depth and correlated with the relative abundance of calcite in fracture filling minerals in Figure 11. It can be seen that most waters shallower than 100 m are strongly supersaturated with respect to calcite (saturation index >0.3), and even at greater depths many waters are still supersaturated. These results suggest that calcite will be a stable phase at most depths in the granite and this is, in fact, observed throughout the sections of cores from these boreholes.

Sulphate-gypsum equilibria. Data for the saturation indices of waters with respect to gypsum are also plotted in Figure 11, together with relative abundances of gypsum in fractures for the URL. In contrast to calcite, gypsum is considerably undersaturated in most waters to depths of at least 800 m. This suggests that gypsum is unlikely to be observed as a fracture-filling mineral in contact with modern groundwaters at the URL. In fact, the only occurrences of gypsum in the granite are traces in one core between 100 and 200 m depth and a more prominent occurrence at about 830 m depth in borehole URL-2.

In contrast, at Atikokan, gypsum is a more common fracture-filling mineral, especially below 500 m depth. Calculations suggest that deep Atikokan groundwaters are saturated or supersaturated with respect to gypsum but little reliance should be placed on these data because of the high dilution of these samples by surface or drilling water. The higher calcium concentrations that exist in Atikokan relative to WN/URL groundwaters, however, argue for greater stability of gypsum fracture-filling minerals in the granite at Atikokan.

Silicate-clay mineral equilibria. Clay minerals and iron oxhydroxides are also common fracture-filling minerals in plutonic rocks in the four research areas. It is difficult to determine the state of equilibrium between the groundwater and iron oxyhydroxide minerals because (1) steel rods and packers are or have been present in the boreholes, (2) accurate analysis of the Fe^{2+}/Fe^{3+} ratio and Eh in the groundwaters is problematic and (3) it is difficult to identify and fully characterize the iron-containing mineral controlling the groundwater chemistry.

The relationship between clay minerals and water chemistry can be examined using major cation and dissolved silica concentrations and the results expressed on stability field diagrams. Figure 12 shows the relationship between cation/hydrogen ion ratios and silica, calculated for WN/URL groundwaters (Gascoyne and Elliott, 1985) and the stability field diagrams determined by Nesbitt (1985). Most groundwaters lie well within the kaolinite field for Na^+, Ca^{2+} and Mg^{2+} indicating that clay mineralogy does not significantly control the concentration of these ions in groundwaters. The positions of some waters in the muscovite/illite fields indicates that alteration of muscovite to illite and illite to kaolinite may control the K^+ concentrations of these groundwaters.

Although kaolinite is predicted to be the stable mineral phase in contact with groundwaters at the WN/URL area, Kamineni et al. (1984) report that none has so far been observed in fracture-filling minerals. Illite is observed, however, and this agrees with the location of data on the potassium diagram (see Figure 12).

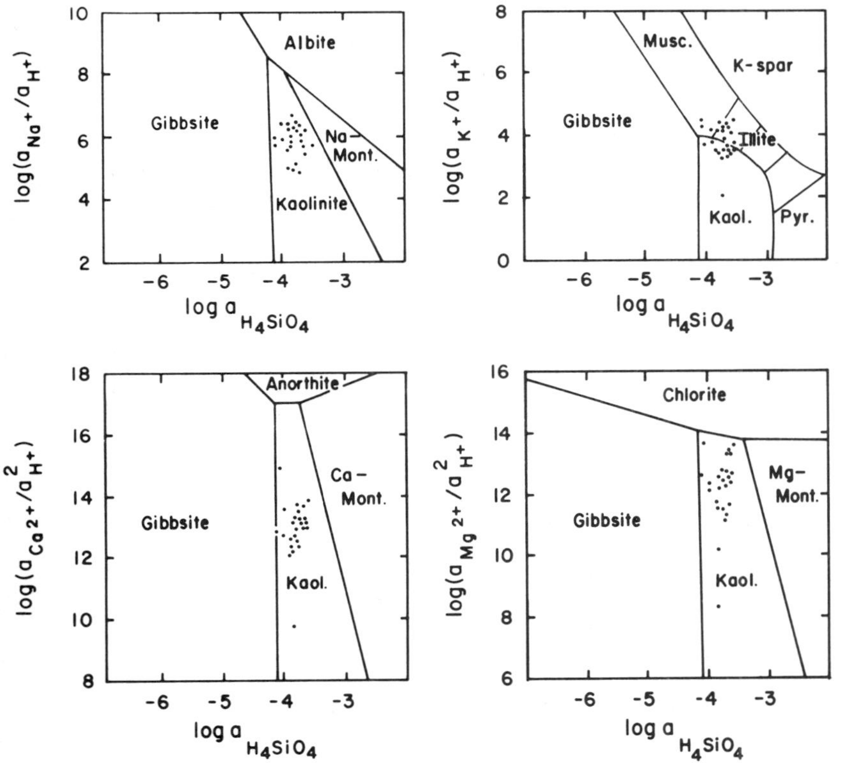

Figure 12. *Relationship of groundwater chemistries to low temperature silicate minerals using data from groundwaters at the WN/URL research area and stability field diagrams of Nesbitt (1985).*

SUMMARY AND CONCLUSIONS

The composition of groundwaters in three rock types (granite, gabbro and gneiss) for four widely spaced research areas across the Canadian Shield has been determined as part of the Canadian Nuclear Fuel Waste Management Program. Although results for the gabbro and granite areas are preliminary and await further sampling, the deeper, saline groundwaters are uniformly found to be sodium-, calcium- and chloride-rich. Salinities no greater than 40 g/L have been measured at present, but the existence of salinities of over 140 g/L is possible if the diluting component of surface water is removed.

The distinct correlation between salinity and chlorinity of Shield groundwater with depth in bedrock is interpreted as advective mixing of groundwaters to depths of 300 to 400 m driven by surface recharge-discharge controls. Below this depth a more stagnant zone of higher salinity water is postulated, in which advection is minimal. Transport of dissolved salts upwards from this zone is thought to occur mainly by diffusion rather than advection, and is driven only by the chemical gradient existing between fresh waters at about 300 m depth and saturated brines below 1700 m depth. The existence of surface 'moose licks' and low salinity waters in a deep fault zone at Atikokan are notable exceptions to this situation.

Distinctly different Br/Cl and Na/Ca ratios in groundwater are observed between research areas; those with low Br/Cl and high Na/Ca may be influenced by mixing with sedimentary basin brines within 50 km of the research area, and those with high Br/Cl and low Na/Ca are more typical of results obtained for saline groundwaters in mines on the Canadian Shield and in other granitic bodies. The latter group are thought to obtain their salinity largely from dissolution of fluid inclusions in the rock and alteration of halogen-bearing minerals.

Reasonable agreement is found between the secondary minerals present in fractures in the bedrock and those predicted from geochemical modelling of the water chemistry data, although the absence of kaolinite in one granitic research area, the URL, is anomalous.

Ongoing work, involving continued sampling of deep groundwaters and further geochemical modelling, should allow a better comparison between rock and water chemistries. The possibility of sedimentary brine intrusion into plutonic rocks in two research areas will also be examined in more detail by future sampling work and by analysis of other indicative parameters, e.g., ^{34}S in dissolved sulphate.

Characterization of the composition of deep groundwaters in relatively undisturbed plutons, coupled with an understanding of the flow paths and transit times of these waters, will assist in the design of a vault for disposal of nuclear fuel waste.

REFERENCES

Bottomley, D.J., Ross, J.D., and Graham, B.W., 1984a. A borehole methodology for hydrogeochemical investigations in fractured rock: Water Resources Research, v. 20, p. 1277-1300.

Bottomley, D.J., Ross, J.D., and Clarke, W.B., 1984b. Helium and neon isotope geochemistry of some groundwaters from the Canadian Precambrian Shield: Geochimica et Cosmochimica Acta, v. 48, p. 1973-1985.

Bottomley, D.J., Gascoyne, M., Ross, J.D., and Ruttan, T., 1985. Hydrogeochemical investigations in the East Bull Lake gabbro: Atomic Energy of Canada Limited Technical Record, TR-382.

Carpenter, A.B., 1978. Origin and chemical evolution of brines in sedimentary basins: Oklahoma Geological Survey Circular v. 79, p. 60-77.

Craig, H., 1961. Isotopic variations in meteoric waters: Science v. 133, p. 1702-1703.

Davison, C.C., 1984a. Monitoring hydrogeological conditions in fractured rock at the site of Canada's Underground Research Laboratory: Monitoring Groundwater, v. 4, p. 95-102.

Davison, C.C., 1984b. Hydrogeological characterization at the site of Canada's Underground Research Laboratory: *in* Proceedings of the International Groundwater Symposium On Groundwater Resources Utilization and Contaminant Hydrogeology, Vol. II, Montreal, Que., May 1984, p. 310-335.

Dixon, R.S. and Rosinger, E.L.J., 1984. The Canadian Nuclear Fuel Waste Management Program, 1983, Annual Report, Atomic Energy of Canada Limited, AECL-7811.

Edmunds, W.M., Andrews, J.N., Burgess, W.G., Kay, R.L.F., and Lee, D.J., 1984. The evolution of saline and thermal groundwaters in the Carnmenellis granite: Mineralogical Magazine, v. 48, p. 407-424.

Frape, S.K. and Fritz, P., 1985. Unpublished progress report, University of Waterloo, Waterloo, Ontario.

Frape, S.K., Fritz, P., and McNutt, R.H., 1984a. Water-rock interaction and chemistry of groundwaters from the Canadian Shield: Geochimica et Cosmochimica Acta, v. 48, p. 1617-1627.

Frape, S.K., Fritz, P., and Blackmer, A.J., 1984b. Saline groundwater discharges from crystalline rocks near Thunder Bay, Ontario, Canada. *In* Hydrochemical Balances of Freshwater Systems, (ed.) E. Eriksson, International Assocociation for Hydrological Sciences, Publication No. 150, p. 369-379.

Fritz, P. and Frape, S.K., 1982. Saline groundwaters in the Canadian Shield – a first overview: Chemical Geology, v. 36, p. 179-190.

Gascoyne, M. and Elliott, L.C.M., 1985. A simple well-water dilution model. Atomic Energy of Canada Limited, Technical Record, TR-391.

Hitchon, B., Billings, G.K., and Klovan, J.E., 1971. Geochemistry and origin of formation waters in the western Canada sedimentary basin. II. Factors controlling chemical composition: Geochimica et Cosmochimica Acta, v. 35, p. 567-598.

Lane, A.C., 1914. Mine water composition, an index to the course of ore bearing currents: Economic Geology, v. 9, p. 239-262.

Kamineni, D.C., Dugal, J.J.B., and Ejeckam, R.B., 1984. Geochemical investigations of granitic core samples from boreholes at the Underground Research Laboratory site near Lac du Bonnet, Manitoba: Atomic Energy of Canada Limited Technical Record TR-221.

McCrank, G.F.D., Stone, D., Ejeckam, R.B., Kamineni, D.C., Sikorsky, R.I., and McEwen, J., 1984. Surface and subsurface geological investigations of the East Bull Lake pluton, Atomic Energy of Canada Limited Technical Record, TR-299, in press.

McNutt, R.H., Frape, S.K., and Fritz, P., 1984. Strontium isotopic composition of some brines from the Precambrian Shield of Canada: Isotope Geoscience, v. 2, p. 205-215.

Nesbitt, H.W., 1985. A chemical equilibrium model for the Illinois Basin formation waters: American Journal of Science, v. 285, p. 436-458.

Nordstrom, K., 1983. Conceptual framework for the chemical proccesses in the Stripa groundwaters: SKB Stripa Project Report 83-01, p. 105-115.

Pearson, F.J. Jr., 1985. Geochemical and isotopic evidence on the origin of brines in the Canadian Shield: Report by Intera Techn. Inc., Texas, for Atomic Energy of Canada Limited, WNRE, Pinawa, Manitoba.

Raven, K.G., Novakowski, K.S., and Bottomley, D.J., 1984. Hydrogeologic research at the Chalk River research area: Atomic Energy of Canada Limited Technical Report, TR-218, p. 11-29.

Ridgway, W.R. and Pearson, R., 1985. Regional characterization studies in support of a major hydrogeological investigation at Atikokan, Northwest Ontario: International Symposium on Management of Hazardous Chemical Waste Sites, October 9-10, 1985.

Stone, D. and Kamineni, D.C., 1982. Fractures and fracture infillings of the Eye-Dashwa Lakes pluton, Atikokan, Ontario: Canadian Journal of Earth Sciences, v. 19, p. 789-803.

Halogen-Bearing Minerals in Plutonic Rocks: A Possible Source of Chlorine in Saline Groundwater in the Canadian Shield

D. Choudari Kamineni

Atomic Energy of Canada Ltd. on attachment to Geological Survey of Canada,
601 Booth Street, Ottawa, Ontario, Canada K1A 0E8

Kamineni, D.C., Halogen-bearing minerals in plutonic rocks: a possible source of chlorine in saline groundwater in the Canadian Shield; in Saline water and gases in crystalline rocks, Editors: Fritz, P., and Frape, S.K.; Geological Association of Canada Special Paper 33, 1987. p. 69–79.

Abstract

A literature search indicates that the halogens, chlorine and fluorine, can concentrate in significant amounts in certain minerals occurring in plutonic rocks. The chlorine-bearing minerals in these rocks can be categorized as: (1) hydroxysilicate minerals, where the hydroxyl is substituted by chlorine, (2) silicates that contain chlorine as an essential constituent, and (3) secondary minerals such as compounds of metal hydroxides. Degradation and dissolution of chlorine-bearing minerals during water-rock interaction can release significant amounts of chlorine, particularly near fractures in plutonic rocks.

Chemical analysis of rock samples and microprobe study of alteration minerals from the East Bull Lake anorthosite-gabbro complex, in Northeastern Ontario, showed that chlorine is concentrated in some bulk rock samples, and in the minerals hornblende and biotite. Near open fractures, chlorine-bearing hornblende and biotite are replaced by low-grade metamorphic, chlorine-free minerals such as prehnite, laumontite, clays and gypsum. The chlorine content of bulk rock samples shows a corresponding decrease in these zones. The breakdown of chlorine-bearing minerals may account for the high concentrations of chlorine in groundwater encountered at depth in plutonic rocks.

Résumé

Une étude documentaire indique que le chlore et le fluor peuvent se concentrer en quantité appréciable dans certains minéraux des roches plutoniques. Dans ce type de roche, les phases chlorurées peuvent être ainsi classées: (1) minéraux hydroxysilicatés dans lesquels l'hydroxyle est remplacé par le chlore, (2) silicates dans lesquels le chlore constitue un composant essentiel et (3) minéraux secondaires, comme les composés d'hydroxydes de métaux. Durant l'interaction entre l'eau et la roche, la dégradation et la dissolution des minéraux chlorurés peuvent dégager des quantités appréciables de chlore, particulièrement à proximité des fractures des roches plutoniques.

L'analyse chimique d'échantillons du complexe anorthositique-gabbroïque du lac East Bull, au nord-est de l'Ontario, et l'étude à la microsonde électronique de minéraux d'altération provenant du même secteur ont révélé une concentration importante de chlore dans certains échantillons mégascopiques ainsi que dans la hornblende et la biotite. À proximité des fractures ouvertes, la hornblende et la biotite qui renfermaient du chlore sont remplacées par des minéraux de faible degré métamorphique et dépourvus de chlore, tels que la préhnite, la laumontite, les argiles et le gypse. La teneur en chlore des échantillons mégascopiques décroît également dans ces zones. La dégradation des minéraux chlorurés pourrait expliquer la forte concentration du chlore dans les eaux souterraines s'écoulant en profondeur dans les roches plutoniques

INTRODUCTION

The origin and evolution of groundwater in plutonic environments have been major research topics in recent years. This is mainly because of the fact that plutonic rocks in the Canadian Shield are being investigated as potential hosts for the disposal of high-level nuclear fuel waste. Consequently, the chemical characteristics of groundwater in plutonic rocks are important as they may play a significant role in mobilization and dispersion-sorption of radionuclides from a waste disposal vault.

Frape and Fritz (1982) reported that the compositions of groundwater in the Canadian Shield changes with depth. The waters above 650 m are fresh or brackish, whereas those below 650 m are saline or briny (Frape et al., 1984). Since the latter type of waters appears to be concentrated at the proposed vault depths (i.e.$\approx$1 km), an understanding of the origin and chemical characteristics of this type of water is essential.

Three theories that are feasible have been proposed for the origin and/or source of chlorine in saline groundwaters. They are: (1) leaching of chlorine-bearing biotites (Edmunds et al., 1984, 1985), (2) release of fluid inclusions due to fracturing (Nordstrom, 1983 and Lindblom, 1984) and (3) equilibration of trapped and concentrated sea water with clay minerals (Nesbitt, 1985). But investigations from the Canadian Shield suggested rock-water interactions can have a significant role in the origin and evolution of saline groundwaters and brines.

A study of the relationships among rock alteration, bulk rock chemistry and groundwater composition in the Eye-Dashwa lakes pluton, in Northwestern Ontario, led Kamineni and Dugal (1982) to conclude that groundwater chemistry below 500 m can, in general, be accounted for by rock-water interaction. Detailed studies by Frape et al. (1984) and McNutt et al. (1984) also suggested this mechanism for the origin of saline waters and brines in the Canadian Shield. In this report, the source of chlorine, which is an important component in these waters, is discussed. In this context, the nature and occurrence of halogen-bearing minerals in plutonic rocks are explored and a case history of an anorthosite-gabbro layered complex from the Canadian Shield where chlorine-bearing amphiboles and biotites can be linked to source of salinity in groundwaters are presented.

HALOGENS IN PLUTONIC ENVIRONMENTS

Halogens, particularly fluorine and chlorine, are important constituents in magmatic and metamorphic rocks of plutonic environments. Correns (1956) reviewed the halogen contents of various plutonic rocks and concluded that none of the halogens correlated with SiO_2 content. Johns and Huang (1967) studied the distribution of chlorine in terrestrial rocks and estimated the abundance of chlorine in the Earth's crust and in the upper mantle to be 180μg/g and 100μg/g, respectively. In the crustal materials, halogens have three modes of occurrence (Correns, 1956): (1) incorporation in mineral lattices, (2) in aqueous solutions or as gases in mineral inclusions, and (3) dissolved in solid glasses. In this report, the first mode is examined in detail.

The first two modes of occurrence listed above are generally confined to plutonic environments, and they appear to be complementary. Halogens participating in magmatic and metamorphic processes would be either incorporated into suitable mineral lattices or occur as fluid-gas inclusions in most minerals. Fluorine and chlorine compounds have high and low solubilities, respectively, in magmas (Koster van Groos and Wyllie, 1968 and 1969). Hence, these elements behave differently in magmatic systems; fluorine is retained by the magma, occurring in late differentiates, whereas chlorine is expelled (Fuge, 1977). The expelled chlorine may form immiscible liquids in association with alkali metals and produce alkaline rocks (Roedder and Coombs, 1967; Fuge, 1977). According to Carmichael et al. (1974), if hydroxyl or fluorine are present, chlorine is rejected by hydroxyl minerals and expelled from the magma. Only in magmas of lower silica activity where sodalite is generally crystallized, chlorine is retained.

Some aspects of the theoretical concepts outlined above have been confirmed in studies carried out on various rock suites. Greenland and Lovering (1966) determined chlorine and fluorine throughout the Great Lake Tholeiite and found that fluorine content increased significantly in the late granophyres, while chlorine remained low at constant concentrations throughout the intrusion. Analyses of apatites from Shonkin Sag Laccolith and Skaergaard intrusion by Nash (1972, 1976) showed highest chlorine and fluorine concentrations in the early and late differentiates, respectively. This suggests that chlorine behaved in a different manner than that predicted by theory and experiment. This trend has also been noted in two other differentiated plutons, namely Dartmoor granite and Rhiw Complex. Fuge (1977) analysed a number of samples comprising different units in these differentiated plutons and found that chlorine is highest in the early differentiates while fluorine is highest in the latest differentiates. He linked this to the enhanced incorporation of chlorine along with water in the early formed hydroxy minerals (apatite, hornblende and biotite). Presumably, there will be least competition from fluorine at this stage of crystallizations, as it would be retained in the melt. From these observations, it can be concluded that the hydroxyl minerals occurring in early crystallized or differentiated parts of the plutonic rocks are likely to contain substantial amounts of chlorine.

In recent years, halogen-bearing minerals from high grade metamorphic terrains have been widely reported (e.g. Leelanandam, 1969a, b; Blattner, 1980; Kamineni et al., 1982). Based on these occurrences, Collerson and Fryer (1978) proposed that halogen-bearing fluids played a major role in the early evolution of the Earth's crust. The lower crustal granulites are, in general, characterized by low αH_2O and formed in presence of mantle derived exotic volatiles such as CO_2 or CO, S, SO_2, SO_3 or $SO^-{}_4$, F^- and Cl^- (see Collerson and Fryer, 1978). Because of the paucity of OH^- in this environment, any hydroxyl minerals stabilized are bound to be enriched in both flourine and chlorine. It is interesting to note that in contrast to magmatic rocks where the hydroxyl minerals concentrate only one of the two halogens (i.e. chlorine or fluorine), the hydroxyl minerals from high-grade metamorphic rocks invariably contain both chlorine and fluorine (see Leelanandam, 1969a, b; Blattner, 1980; Kamineni, 1982). From this it appears that during high grade metamorphism chlorine and fluorine are not separated (unlike during magmatic crystallization). They are incorporated together in the available hydroxyl minerals such as hornblende, biotite and apatite.

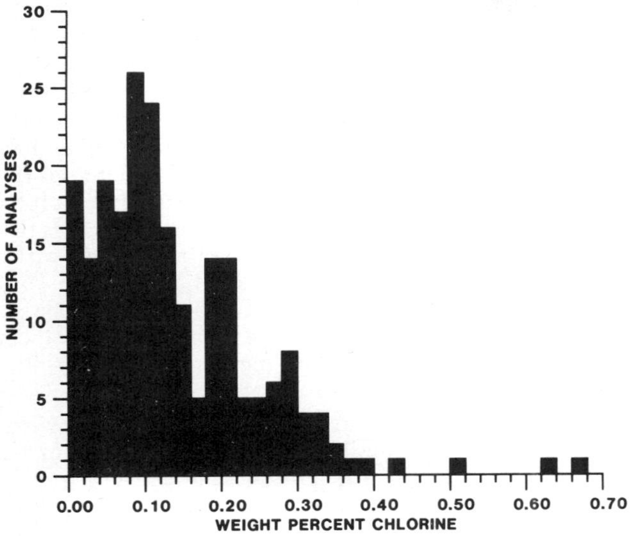

Figure 1a. *Histogram showing chlorine abundance in biotites reported from various localities in the world (219 values). Analyses included in this figure were obtained through a computer search (GEOREF) at the Library of the Geological Survey of Canada. The same applies to Figures 1b, 1c, 1d, 2a, 2b and 2c.*

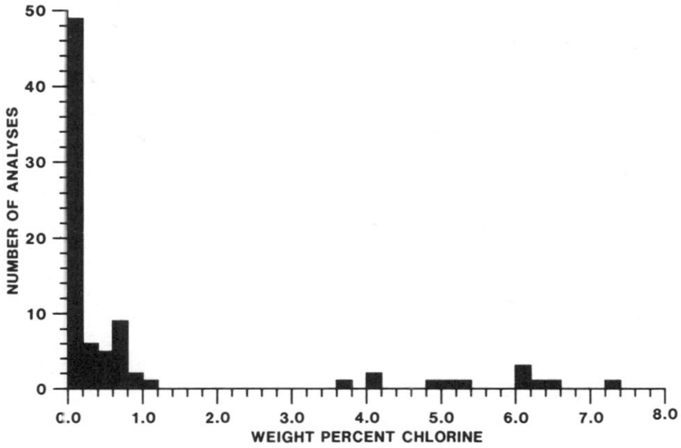

Figure 1c. *Histogram showing chlorine abundance in hornblende (84 values).*

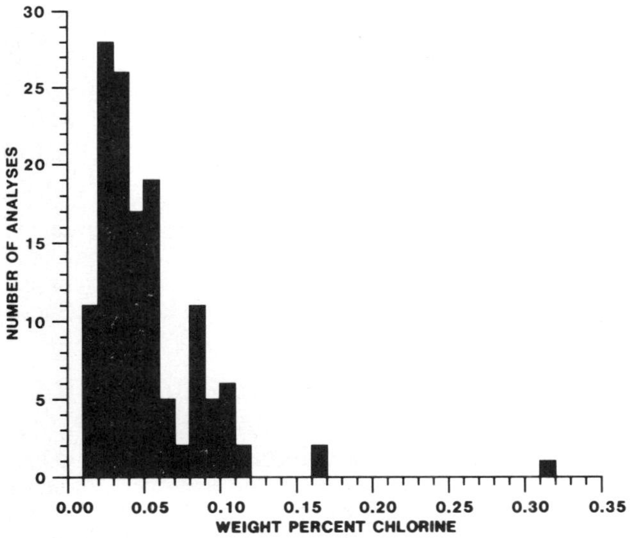

Figure 1b. *Histogram showing chlorine abundance in phlogopite (135 values).*

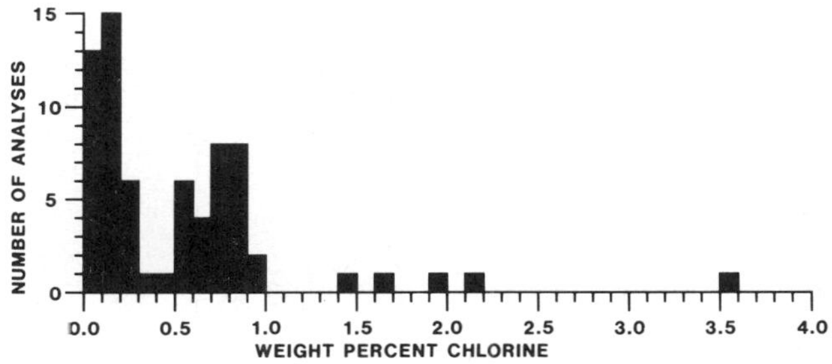

Figure 1d. *Histogram showing chlorine abundance in cpatite (69 values).*

Chlorine-bearing minerals

Chlorine-bearing minerals in plutonic rocks can be categorized into three groups:

(1) Hydroxysilicate minerals where the hydroxyl ion is substituted by chlorine. Biotite, hornblende, apatite, tourmaline and topaz are important examples of this type. Biotites occurring in various plutonic environments, ranging from granite to gabbro and schist to charnockite are known to contain chlorine. The chlorine content of biotite and phlogopite varies from 0.002 to 0.66 wt.% (Figs. 1a and 1b). Higher concentrations of chlorine (up to 2 wt.%) are also reported by Leelanandam (1970) and Kamineni et al. (1982).

Hornblendes occurring in the same environments as biotite also contain chlorine, but commonly in larger quantities. The chlorine content of hornblendes ranges from 0.01 to 3.6 wt.%, and from 0.003 to 7.24 wt.% when other amphiboles are included (Fig. 1c).

Apatite is a widely distributed accessory mineral in terrestrial rocks and usually contains less than 1 wt.% chlorine (Kind, 1938; Behne, 1953). Chlorapatite, which is a rather rare variety of apatite can contain up to 6.8 wt.% chlorine (Morton and Catanzaro, 1964). The chlorine abundance in apatite is given in Figure 1d.

Tourmaline and topaz are other hydroxysilicates that contain chlorine. Fuge and Power (1969) reported about 0.04 wt.% chlorine in tourmalines from aplites and granites. Kuroda and Sandell (1953) reported the presence of chlorine in topaz.

(2) Rock-forming silicate minerals defined by the presence of chlorine as an essential constituent. Sodalite, eudialyte and the marialite end member of scapolite are the minerals belonging to this group. Sodalite occurs in silica under-saturated igneous rocks and has a chlorine content ranging between 5.4 and 8.01 wt.% (Fig. 2a). Eudialyte is another mineral occurring in silica under-saturated rocks and contains up to 2.2 wt.% chlorine (Kostetskaya, 1961). The chlorine abundance of eudialytes is given in Fig. 2b. Marialite (scapolite), occurring in metamorphic and pegmatitic rocks, has been reported to contain up to 2.2 wt.% chlorine (Shaw, 1960), and the abundance in reported occurrences is shown in Fig. 2c.

Although not a silicate, halite is another mineral that falls in this group as it is a chloride mineral that occurs in plutonic rocks under certain conditions. It has been found in metamorphosed sedimentary rocks (Engel and Engel, 1953) and also as crystallites within fluid inclusions in plutonic rocks (Stollery et al., 1971; Roedder, 1972). Recently, Trommsdorff et al. (1985) reported the occurrence of halite and sylvite solid inclusions in some high-grade metamorphic rocks from Swiss and Italian Alps.

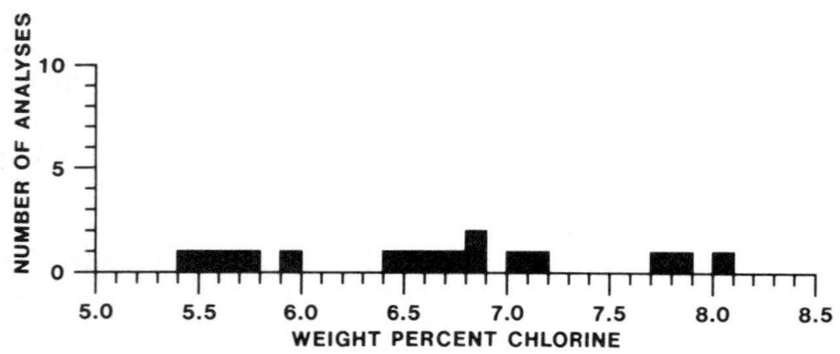

Figure 2a. *Histogram showing chlorine abundance in sodalite (16 values).*

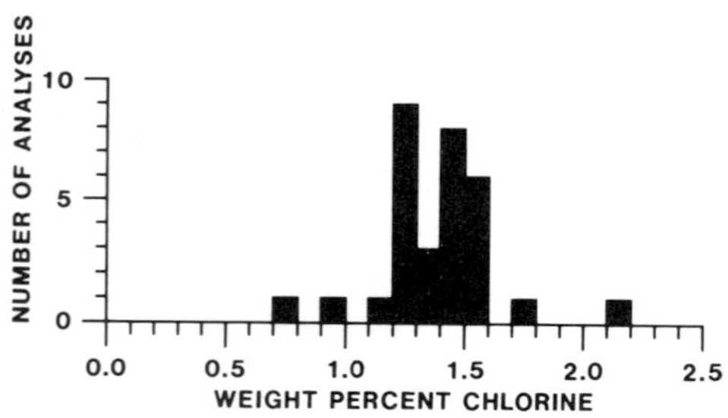

Figure 2b. *Histogram showing chlorine abundance in eudialyte (31 values).*

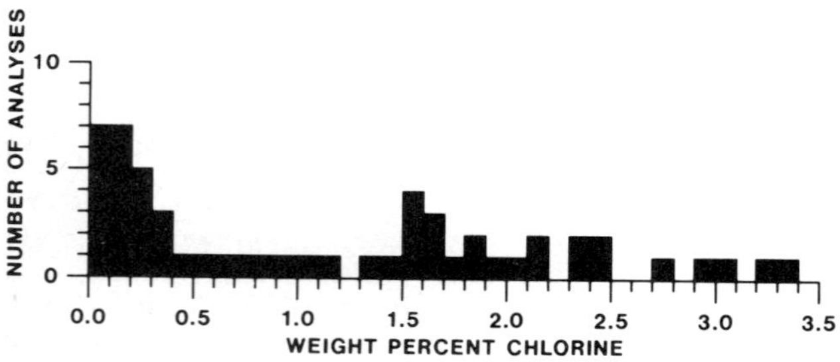

Figure 2c. *Histogram showing chlorine abundance in marialite (scapolite, 55 values).*

(3) Secondary minerals as compounds of metal hydroxides. Magnesium and ferric oxychloride in serpentinized zones of ultramafic rocks fall in this category.

Earley (1958) first recognized anomalously high chlorine values (0.1 to 0.7%) in serpentinized dunite samples from the Abitibi Region of Ontario. Later, Stueber et al. (1968) analyzed ultramafic rocks, from various localities, and noted that the mean chlorine contents of unserpentinized and serpentinized intrusions are 88 μg/g and 300 μg/g respectively. Rucklidge (1972) also reported a high chlorine concentration (0.8%) for serpentine veins in dunite from NW Quebec. A high concentration of chlorine in serpentinized ultramafic rocks is believed to be due to the presence of minerals such as Iowaite – a hydrous magnesium hydroxide-ferric hydroxide (Kohls and Rodda, 1967). More recently, Rucklidge and Patterson (1977) proposed that chlorine might have even played an active role in serpentinization. They have also suggested that compounds of $Fe_2(OH)_3Cl$ precipitate in the reaction zone during serpentinization.

In view of the data and discussion presented here, it appears that it is essential to investigate the minerals in plutonic rocks of Precambrian terrains such as the Canadian Shield in terms of their chlorine content. This would probably be an important aspect to understand the genesis and origin of saline groundwater in the shield.

CHLORINE IN THE EAST BULL LAKE ANORTHOSITE-GABBRO COMPLEX

The East Bull Lake pluton, which is a anorthosite-gabbro layered complex, is investigated with reference to the occurrence of chlorine-bearing minerals and their possible contributions towards salinity of groundwater. This is one of the plutons selected, by Atomic Energy of Canada Limited, as a research area for field studies to assess the concept of deep geologic disposal of high-level nuclear fuel waste. The chlorine content of groundwater encountered in this pluton shows a general increase with depth (Table 1, Fig. 3).

General geology

The East Bull Lake a anorthosite-gabbro layered complex, occurs within the Superior Structural Province of the Canadian Shield, about 30 km north of Massey, Ontario. The pluton intrudes Archean plutonic and supracrustal rocks (McCrank et al., 1983). An isochron based on samarium-neodymium method gave an age of 2570 ± 120 Ma for this pluton (Kamineni and Brooks, in preparation).

This preliminary age determination suggests that the pluton emplaced in the early Proterozoic. Numerous mafic dykes that give bulk rock K-Ar age ranging between 1.65 to 1.9 Ma cut the East Bull Lake pluton (Kamineni et al., 1984).

The East Bull Lake pluton is subdivided into various map units (Kamineni et al., 1984). Four boreholes, ranging in depths from 400-800 m, were drilled to investigate the shape of the pluton, fracture characteristics, fracture fillings hydrogeology and correlate map units in sub-surface. The geology of the central plateau of the pluton along with locations and orientations of boreholes are given in figure 4. The pluton has a lopolithic shape with a maximum thickness of 780 m, with depth to width to length proportions of 1:4:20.

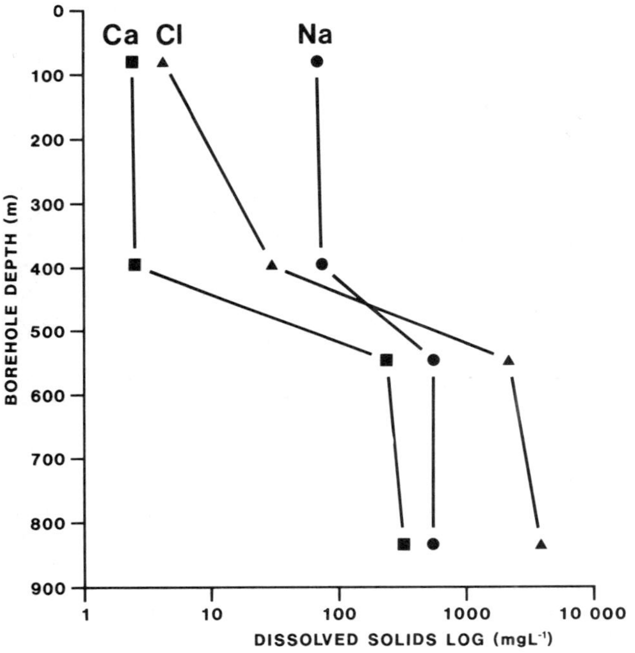

Figure 3. *Plot of calcium, sodium and chlorine concentration in EBL-2 borehole.*

The base of the pluton is composed of anorthosite, which is approximately 400 m thick that grades into a rhythmic layered anorthosite-gabbro unit. A troctolite unit separates the rhythmic layered unit from a series of gabbro units mapped as layered gabbro, crescummulate gabbro and massive gabbro.

All the map units listed above are characterized by fractures filled with a variety of minerals (Kamineni et al., 1985). Based on mineral stability (pressure-temperature conditions), the fracture-filling minerals can be related to four metamorphic facies: (1) epidote-amphibolite facies, (2) prehnite-pumpellyite facies, (3) zeolite facies, and (4) low-temperature facies due to rock-water interaction. These metamorphic facies represent progressively decreasing pressure-temperature conditions. As the pressure and temperature decreased, the spheres of influence of the three types of alteration were also reduced. For example, the early (epidote-amphibolite) metamorphism is pervasive, and minerals formed during this event are ubiquitous throughout the pluton. Amphiboles, epidote and biotite replace and form rims around pyroxenes and in microfractures that transect plagioclase grains. The calcic amphiboles contain three species: (1) actinolite-tremolite, (2) actinolitic hornblende, and (3) hornblende. Among the various minerals identified in the epidote-amphibolite facies, hornblende and biotite contain chlorine. Assemblages belonging to the prehnite-pumpellyite and zeolite facies are confined to steep and shallow-dipping fracture zones. Prehnite and associated minerals are concentrated in major fracture zones, whereas laumontite is restricted to low dip ($\leq 30°$) fractures. Higher grade minerals such as amphiboles in this zone are either absent or depleted significantly. This suggests that the low grade metamorphism is fracture-controlled and retrogressive type. The gypsum, iron-oxide-hydroxide and clay-bearing fractures represent alteration products of recent to present day low-temperature water-rock interaction. The retrogressive nature of metamorphic alteration is also supported by cross-cutting relations as prehnite- and zeolite-bearing fractures cross-cut amphibole-bearing fractures but not vice versa.

Table 1

Groundwater composition from each test zone of EBL boreholes.*

Borehole	Depths (m)	Ca	Mg	Na	K	Alk	SO_4	Cl	Br	pH
EBL 1	148-298	2.4	0.4	83.5	0.5	197	5.5	17.9	0.1	9.4
	298-513	5.5	0.1	233	0.8	66	17.8	352	1.6	10.2
	513-650	20.5	4.2	100	1.0	119	16.0	31	<0.02	7.7
EBL 2	0-75	20.0	6.3	16	0.7	144	29	4.0	0.17	6.5
	75-248	0.4	0.6	65	0.2	157	7.2	4.3	0.04	10.4
	248-460	0.5	0.3	74.2	0.5	130	21.5	31	0.3	9.7
	460-616	228	1.6	570	1.1	38	11.4	1237	3.3	8.0
	616-836	293	1.7	584	2.1	29	20.0	1350	2.1	7.4
EBL 4	0-120	13.3	4.8	15	0.9	96	9.0	0.8	0.05	6.7
	120-198	1.6	0.6	78	0.4	213	4.4	1.2	0.03	10.0
	198-396	3.6	0.9	72	0.4	186	2.0	31	0.04	10.2
	396-489	279	2.0	845	2.8	71	77	1781	5.7	7.7

*All expressed in $mg.L^{-1}$

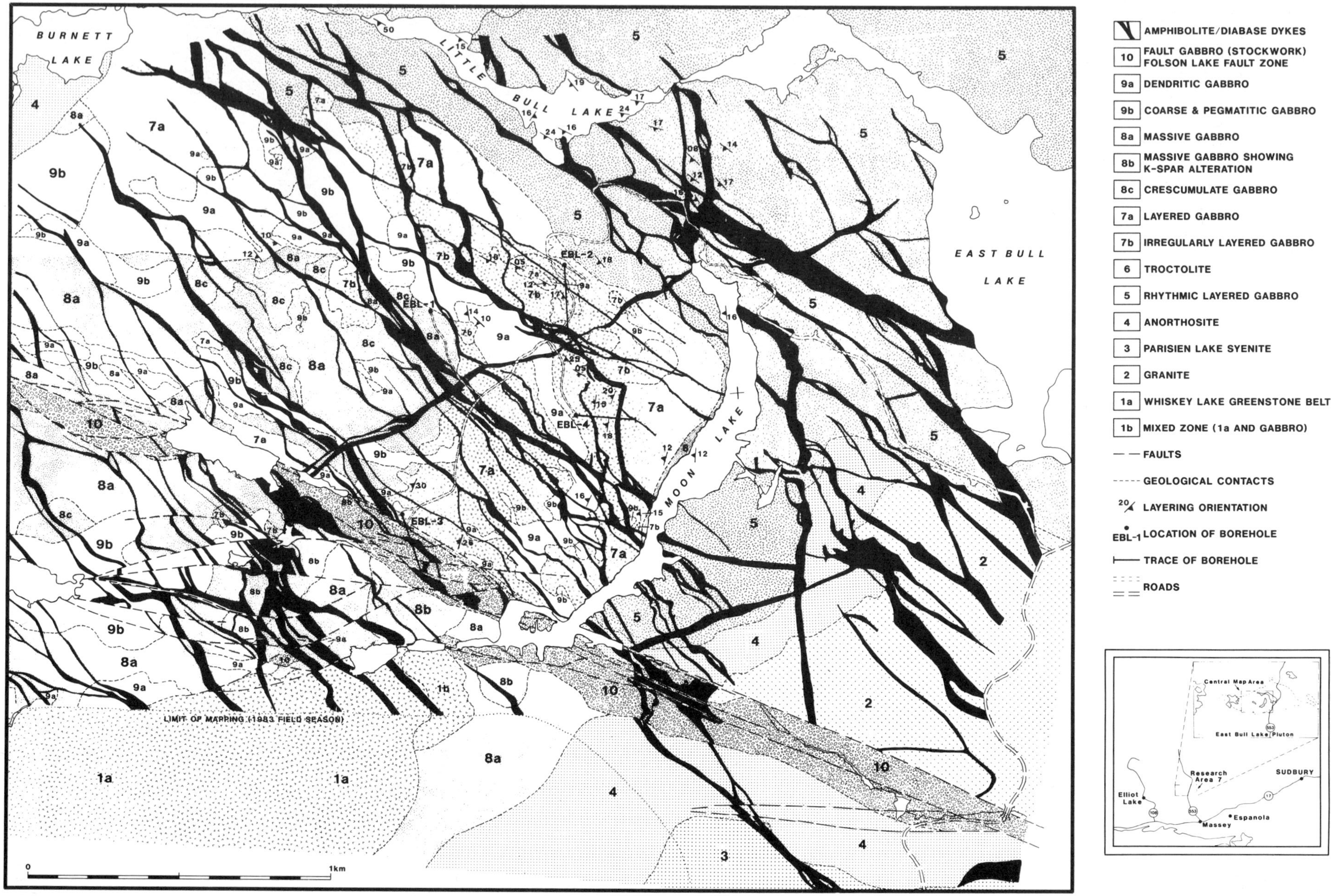

Figure 4. *Geological map of the central plateau of the East Bull Lake pluton showing various map units, dykes and location of boreholes.*

SAMPLES AND ANALYSIS

Two main groups of samples: bulk rock and silicate minerals were analyzed for chlorine. The rock samples comprise subgroups, according to the degree of alteration. Three subgroups representing samples from (1) epidote-amphibolite facies, (2) prehnite and zeolite facies, and (3) low-temperature facies were chosen for bulk rock analysis. Hornblende and biotite were the minerals analyzed.

Bulk rock powders were fused with lithium metaborate, and dissolved in 2% nitric acid. Chlorine was then determined colorimetrically on a spectrophotometer. The accuracy of the method was checked using the USGS standard GSP-1. Duplicate results for chlorine content of 0.032 and 0.030 wt.% are comparable to the values reported by Flanagan (1973).

Carbon-coated polished sections were used for the analysis of chlorine bearing silicates. The sections were first scanned using a MAC electron microprobe to identify chlorine-bearing grains with the help of an energy dispersive spectrometer. The chlorine content of these grains was then determined on a CAMECA probe, using a sodalite standard.

Table 2

Means, Standard Deviations and order statistics for chlorine from the three groups of gabbro samples (All expressed in μ g g^{-1})

	Amphibole Gabbro	Prehnite Laumontite Gabbro	Clay Gypsom Gabbro
No. of Samples	14	16	8
Minimum	189	88	29
Median	342	144	48
Maximum	588	233	88
Mean	363	150	51
Standard Deviation	108	45	19

RESULTS

Chlorine concentrations in the materials analyzed are given in Tables 2, 3, and 4.

The various groups of bulk rock samples show distinct chlorine contents (Table 2), with a progressive decrease from groups 1 to 3 (Fig. 5). The data in Table 2 indicate further that within each group (1 in particular) the chlorine content has high variability as reflected in large standard deviation. This can be attributed to the presence of variable amounts of chlorine-bearing hornblende and biotite in the amphibole gabbro (i.e. group 1) samples analysed. These minerals are generally concentrated in the top stratigraphic units of the East Bull Lake pluton. Hence, the samples from layered gabbro, crescummulate gabbro and massive gabbro units consistently gave greater chlorine values, relative to samples from the rhythmic layered and anorthosite units.

Between the two chlorine-bearing minerals, hornblende and biotite, the former contains more chlorine. However, both show variable concentrations of chlorine, which appear to be controlled by crystal structure and composition (Volfinger et al., 1985). According to these authors, the OH-sites in iron-rich hornblendes and biotites are least distorted, and thus can easily accommodate chlorine. This appears to be the case at least for hornblendes from the East Bull Lake pluton. For example, there is a strong positive correlation ($r = +0.91$) in this mineral between per cent chlorine and X_{Fe}, which is the concentration of Fe^{2+} in M_1, M_2 and M_3 sites (Fig. 6). Biotites, on the other hand, do not show significant correlation between chlorine and iron content. From the analyses (Table 4), it can be stated that

Table 3

Microprobe analyses of chlorine-bearing amphiboles

Depth (m)	10.4	11.8	12.1	39.0	60.1	65.3	87.1	103.0	182.5	254.1	351.2	414.9	702.5	767.3
SiO_2	43.13	40.81	44.19	44.12	48.10	40.86	45.51	43.73	38.75	41.60	46.76	43.28	45.54	47.09
TiO_2	0.19	0.15	0.06	0.03	0.04	0.04	0.11	0.19	0.13	0.21	0.21	0.20	0.07	0.25
Al_2O_3	12.44	14.22	9.81	10.18	7.43	17.34	10.23	12.44	17.90	11.36	10.14	13.01	8.85	9.16
Cr_2O_3	0.12	0.08	0.07	0.08	0.05	0.05	0.06	0.12	0.10	0.09	0.15	0.06	0.06	0.13
Fe_2O_3	4.30	5.12	5.08	5.12	3.96	4.28	3.99	4.30	5.10	5.27	3.73	4.00	4.93	3.64
FeO	15.86	19.13	18.92	19.13	14.68	15.84	14.82	15.86	19.05	19.20	13.77	14.91	18.31	13.46
MnO	0.36	0.30	0.27	0.26	0.28	0.24	0.22	0.36	0.30	0.30	0.24	0.25	0.25	0.31
MgO	8.28	5.80	7.08	6.73	11.17	7.80	9.77	8.28	3.98	6.58	11.12	8.53	7.72	11.32
CaO	11.46	11.50	11.73	11.70	11.94	10.69	11.64	11.46	12.01	11.86	12.08	11.55	11.64	11.75
Na_2O	1.58	1.44	0.95	0.64	0.55	1.03	1.36	1.58	0.58	0.90	0.96	1.60	0.63	1.43
K_2O	0.43	0.59	0.58	0.50	0.25	1.52	0.34	0.43	0.81	0.68	0.32	0.45	0.57	0.28
Cl	0.90	1.38	1.22	1.55	0.68	0.80	1.11	0.86	1.31	2.08	0.72	1.15	1.66	0.62
Total	99.05	100.52	99.96	100.15	99.13	100.49	99.16	99.61	100.32	100.13	100.13	99.99	100.23	99.44
Volume % Hornblende	8.0	12.0	11.0	10.0	14.0	15.0	18.0	7.0	3.0	10.0	9.0	8.0	5.0	5.0
Free Cl by alteration	0.072	0.134	0.134	0.155	0.095	0.012	0.21	0.060	0.04	0.212	0.065	0.092	0.100	0.030

Table 4

Microprobe analyses of chlorine-bearing biotites

Depth (m)	11.8	60.2	65.4	87.2	182.6(1)	182.6(2)	182.6(3)	190.31	Surface
SiO_2	35.44	34.56	34.70	35.24	39.38	39.70	37.22	39.10	34.89
TiO_2	1.72	1.51	1.57	1.88	1.13	0.05	4.70	2.09	1.82
Al_2O_3	15.82	15.23	15.80	14.21	15.95	13.64	13.02	14.19	15.64
Cr_2O_3	0.04	0.04	0.08	0.02	0.01	0.00	0.02	0.07	0.06
FeO	24.62	23.59	27.38	26.67	11.18	12.13	12.88	10.64	27.01
MnO	0.15	0.10	0.11	0.11	0.00	0.00	0.04	0.06	0.13
MgO	9.31	12.69	8.15	7.49	20.30	20.76	18.14	21.80	7.91
CaO	0.06	0.18	0.00	0.00	0.08	0.06	0.00	0.00	0.00
Na_2O	0.00	0.00	0.36	0.00	0.33	0.00	0.36	0.63	0.09
K_2O	8.51	7.77	8.31	7.80	7.20	7.73	8.38	7.23	8.90
Cl	0.28	0.35	0.33	0.38	0.20	0.12	0.18	0.16	0.58
Total	95.95	96.02	96.89	93.80	95.76	94.09	94.94	95.97	97.03
Volume % Biotite	3.0	2.0	4.0	3.0	2.0	2.0	2.0	3.0	4.0
Free Cl by alteration	0.008	0.007	0.013	0.011	0.004	0.002	0.004	0.005	0.023

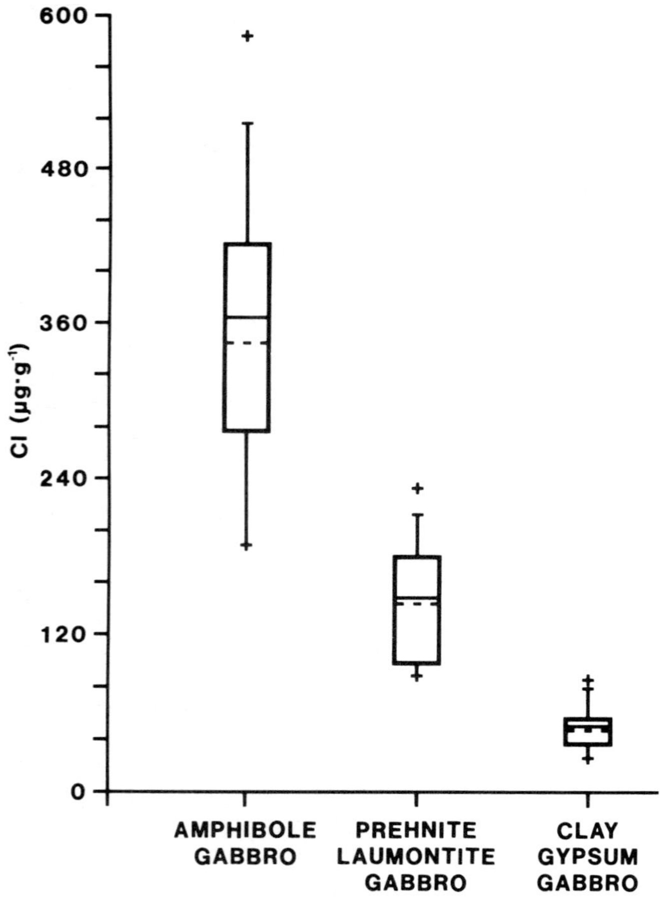

Figure 5. *Box and whisker plot for chlorine in the various groups of gabbro samples analyzed.*

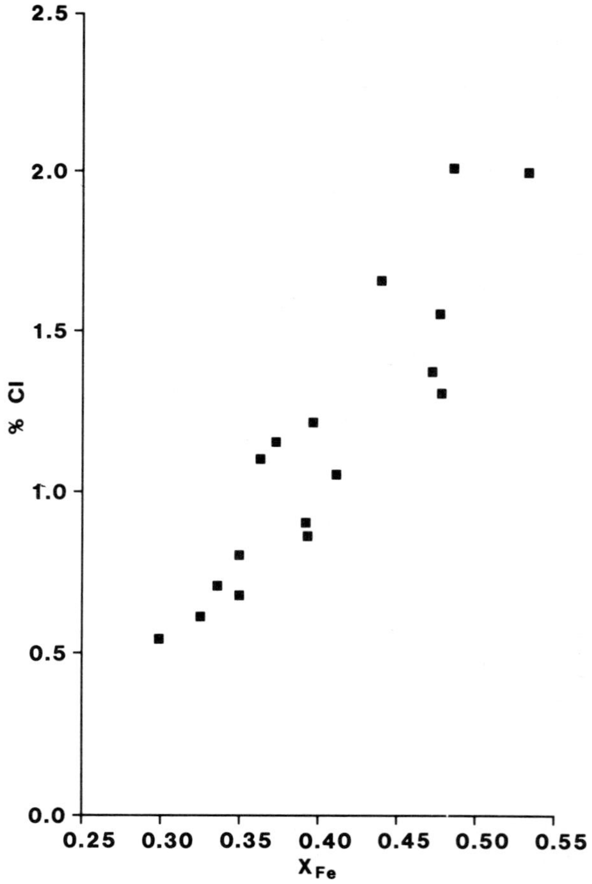

Figure 6. *Plot showing positive correlation between chlorine and $X_{Fe} = Fe^{2+}/(Al^{vi} + Mg + Fe^{2+} + Ti + Mn)_{M1-3}$ in hornblendes from East Bull Lake pluton, r, the correlation coefficient = +0.91.*

Table 5

Microprobe analyses of chlorine-free low-grade minerals

	Chlorite	Epidote	Prehnite	Actinolite	Albite	Adularia	Laumontite
SiO_2	26.18	37.55	42.85	52.48	67.40	64.58	50.80
TiO_2	0.11	0.03	0.04	0.00	0.00	0.00	0.00
Al_2O_3	20.29	24.28	23.59	2.39	19.79	18.24	21.56
Fe as FeO	26.14	12.09	1.08	15.46	0.07	0.18	0.20
MnO	0.34	0.05	0.00	0.29	0.00	0.00	0.00
MgO	17.35	0.67	0.70	14.27	0.00	0.00	0.05
CaO	0.02	23.70	27.31	12.73	0.37	0.00	11.94
Na_2O	0.00	0.00	0.00	0.00	11.54	0.71	0.18
K_2O	0.05	0.11	0.11	0.09	0.42	15.82	0.10
Total	90.48	98.49	95.68	97.71	99.59	99 53	84.83

biotites with large concentrations of iron also contain greater amounts of chlorine and vice versa. The magnesium-rich biotites are all low in chlorine and come from the troctolite unit, which is about 200 m deep in the EBL-1 borehole.

INTERPRETATION

The general increase of chlorine content in groundwater with depth can be related to indigenous sources such as the host rock. To evaluate this possibility, a study of geological factors such as chlorine-bearing minerals and their interactions with fluid in fractures in the East Bull Lake gabbro-anorthosite was undertaken.

The chlorine-bearing minerals, both hornblende and biotite, were formed under epidote-amphibolite conditions and are ubiquitous throughout the pluton. The minerals not only replace and rim pyroxene and magnetite grains, but also fill intragranular microfractures. In some specific fracture zones, which are relatively young on the basis of cross cutting relations, chlorine-bearing hornblende and biotite were transformed into low grade minerals consisting of prehnite, actinolite, chlorite and epidote. Laumontite is a dominant mineral in some fracture zones. All the low grade alteration minerals are chlorine-free (see Table 5). The alteration due to retrogressive metamorphism is reflected in rock properties also – particularly density values. For example, the low-grade rocks have an average density of 2.85 g/cm³ (σ = 0.22), whereas the amphibolite gabbro yield an average density of 2.93 g/cm³ (σ = 0.30). This suggests that the low grade retrogressive metamorphic reactions involving hydration of amphibole gabbro resulted in a decrease of density and a possible increase in volume.

The mineralogical changes involving breakdown of high grade chlorine-bearing minerals to low grade chlorine-free minerals can be documented from field and petrographic studies. A detailed study in this line is in progress.

Based on a number of thin section examinations, the following reactions are suggested for the alterations of chlorine-bearing hornblende:

(1) $25Ca_2(Mg,Fe)_3Al_4Si_6O_{22}(Cl)_2 + 69H_2O =$ (tschermakite)

$Ca_2(Mg,Fe)_5Si_8)_{22}(OH)_2 + 24Ca_2Al_3Si_3O_{12}(OH) +$ (tremolite-actinolite; epidote)

$14(Mg,Fe)_5Al_2Si_3O_{10}(OH)_8 + 28SiO_2 + 25H^+ + 25Cl^-$ (chlorite; quartz)

(2) $10Ca_2(Mg,Fe)_3Al_4Si_6O_{22}(Cl)_2 + 4CaAl_2Si_2O_8 + 40H_2O =$ (tschermakite; anorthite)

$6(Mg,Fe)_5Al_2Si_3O_{10}(OH)_8 + 12Ca_2Al_3Si_3O_{12}(OH) +$ (chlorite; epidote)

$4SiO_2 + 20H^+ + 20Cl^-$ (quartz)

(3) $NaCa_2(Mg,Fe)_5AlSi_6O_{22}(Cl)_2 + 4SiO_2 + 2H_2O =$ (edenite; quartz)

$NaAlSi_3O_8 + Ca_2(Mg,Fe)_5Si_8O_{22}(OH)_2 + 2H^+ + 2Cl^-$ (albite; tremolite-actinolite)

(4) $2.5K(Mg,Fe)_3AlSi_3O_{10}(Cl)_2 + 6CaAl_2Si_2O_8 + 1.5SiO_2$ (biotite; anorthite; quartz)

$+ 10H_2O = 1.5(Mg,Fe)_5Al_2Si_3O_{10}(OH)_8 +$ (chlorite)

$3Ca_2Al_3Si_3O_{12}(OH) + 2.5KAlSi_3O_8 + 5H^+ + 5Cl^-$ (epidote; adularia)

A number of other reactions for the breakdown of Cl-bearing hornblende and biotite to produce minerals such as prehnite, laumontite, gypsum and clays can be written. Since none of these low-grade minerals showed the presence of chlorine in microprobe analysis (see Table 5) implying that the retrograde metamorphic alteration of hornblende and biotite released chlorine to fluid phase.

The actual amount of chlorine released due to alteration of chlorine-bearing minerals can be calculated by taking into account: (1) volume per cent of chlorine-bearing minerals in rocks and (2) chlorine content in minerals. The volume per cent estimates of chlorine-bearing hornblende and biotite are given in Tables 2 and 3, respectively. The calculated amounts of free chlorine released by 1 gm of rock, when the minerals alter, are also listed (see Tables 2 and 3). From these calculations, it can be concluded that in the samples investigated from the East Bull Lake anorthosite-gabbro complex, an average 9.71 volume per cent hornblende and an average 2.78 volume percent biotite can release (by alteration) 1110 ppm and 90 ppm chlorine, respectively.

SUMMARY AND DISCUSSION

An examination of published reports on plutonic rocks from various localities throughout the world revealed that they contain chlorine-bearing minerals. These are hornblende, biotite, apatite, sodalite, scapolite and secondary minerals, including chlorine-bearing magnesium and iron oxides. Occurrence of these minerals and the ubiquitous presence of saline groundwater and brines at deeper levels of the Canadian Shield suggest that there may be a genetic relation between the two. The possibility of such a relationship has been investigated by studying some specific geological settings in the East Bull Lake anorthosite-gabbro complex. In this pluton, the chlorine-bearing minerals, hornblende and biotite, breaks down in and in the vicinity of fracture zones producing a variety of chlorine-free minerals such as actinolite, epidote, chlorite, prehnite, laumontite, gypsum and clay. These mineral products are stable in the temperature range of <100°C-300°C (Kamineni et al., 1985). This implies that the breakdown of chlorine-bearing minerals occurred through a combination of tectonism and high-temperature rock-water interaction along fractures. This type of genesis, where minerals release chlorine to fluid phase at high temperatures supports the view that the chlorine in saline ground waters could be very old. For example, the major fault (Folson Lake Fault) in the East Bull Lake pluton, where complete disappearance of chlorine-bearing hornblende and biotite was noted, appears to have had its most recent movement around late Grenville Time (≈ 1000 Ma). Argon-39/Argon-40 dating of adularia, occurring as filling in extensional fractures, gave an age of 940 ± 3 Ma (Kamineni, unpublished work).

ACKNOWLEDGMENTS

It is my pleasure to express gratitude to Robert Buchanan for cheerfully carrying out the monumental task of literature search. I am grateful to Dennis Bottomly and Mel Gascoyne for providing groundwater analyses from East Bull Lake boreholes. I am grateful to Shaun Frape and Peter Fritz for providing an opportunity to participate in the workshop on the geochemistry of brines and the associated gases. Reviews by Dennis Shaw and Ted Appleyard benefitted this paper.

REFERENCES

Behne, W., 1953. Untersuchungen zur Geochemie des Chlor und Brom: Geochimica et Cosmochimica Acta, v. 3, p. 186-215.

Blattner, P., 1980. Chlorine and fluorine in apatite, biotite and hornblende of basic charnockites from Kondapalli, India: Neues Jahrbuch fur Mineralogie, Monatschafte 6, p. 283-288.

Carmichael, I.S.E., Turner, F.J. and Verhoogen, J., 1974. Igneous petrology, McGraw Hill, New York, 733 p.

Collerson, K.D. and Fryer, B.J., 1978. The role of fluids in the formation and subsequent development of early continental crust: Contributions to Mineralogy and Petrology, v. 67, p. 151-167.

Correns, C.W., 1956. The geochemistry of halogens; *in* Ahrens, L.H., Rankama, K. and Runcorn, S.K. (editors): Physics and Chemistry of the Earth, v. 1, p. 181-233.

Earley, J.W., 1958. On chlorine in serpentinized dunite: American Mineralogist, v. 43, p. 148-155.

Edmunds, W.M., Andrews, J.N., Burgess, W.G., Kay, R.L.F. and Lee, D.J., 1984. The evolution of saline and thermal groundwaters in the Carnmenellis granite: Mineralogical Magazine, v. 48, p. 407-424.

Edmunds, W.M., Kay, R.L.F. and McCartney, 1985. Origin of saline groundwaters in the Carnmenellis granite (Cornwall, England): Natural processes and reaction during hot dry rock reservoir circulation: Chemical Geology, v. 49, p. 287-301.

Engel, E.J. and Engel, C.G., 1953. Grenville series in the northwest Adirondack Mountains, New York. Part I: General features of the Grenville Series: Bulletin of the Geological Society of America, v. 64, p. 1013-1047.

Flanagan, F.J., 1973. 1972 values for international reference samples; Geochimica et Cosmochimica Acta, v. 37, p. 1189-1200.

Frape, S.K. and Fritz, P., 1982. Saline groundwaters on the Canadian Shield – a first review; Chemical Geology, v. 36, p. 179-190.

Frape, S.K., Fritz, P. and McNutt, R.H., 1984. Water-rock interaction of groundwaters from the Canadian Shield: Geochimica et Cosmochimica Acta, v. 48, p. 1617-1627.

Fuge, R., 1977. On the behaviour of fluorine and chlorine during magmatic differentiation: Contributions to Mineralogy and Petrology, v. 61, p. 245-249.

Fuge, R. and Power, G.M., 1969. Chlorine in tourmalines from S.W. England: Mineralogical Magazine, v. 37, p. 293-294.

Greenland, L. and Lovering, J.F., 1966. Fractionation of fluorine, chlorine and other trace elements during differentiation of tholeiitic magma: Geochimica et Cosmochimica Acta, v. 30, p. 963-982.

Johns, W.D. and Huang, W.H., 1967. Distribution of chlorine in terrestrial rocks: Geochimica et Cosmochimica Acta, v. 31, p. 35-49.

Kamineni, D.C. and Dugal, J.J.B., 1982. A study of rock alteration in the Eye-Dashwa lakes pluton, Atikokan, Northwestern Ontario, Canada: Chemical Geology, v. 36, p. 35-58.

Kamineni, D.C., Bonardi, M. and Rao, A.T., 1982. Halogen-bearing minerals from Airport Hill, Visakhapatnam, India: American Mineralogist, v. 67, p. 1001-1004.

Kamineni, D.C., McCrank, G.F., Stone, D., Ejeckam, R.J., Flindall, R. and Sikorsky, R., 1984. Geology of the central plateau of the East Bull Lake pluton, Northeastern Ontario: Geological Survey of Canada, Paper 84-1B, p. 75-83.

Kamineni, D.C., McCrank, G.F., Stone, D., Ejeckam, R.J. and Sikorsky, R., 1985. A preliminary report of alteration and fracture – filling mineralogy in the East Bull Lake pluton, District of Algoma, Ontario; Geological Survey of Canada, Paper 85-1B, p. 81-88.

Kamineni, D.C. and Brooks, C., (in preparation). The Sm-Nd age of the East Bull Lake anorthosite-gabbro complex, Northeastern Ontario, Canada.

Kind, A., 1938. Der magmatische Apatit, seine chemische Zusammensetzung und sein physikalischen Eigenschaften: Chemie der Erde, v. 12, p. 50-81.

Kohls, D.W. and Rodda, J.L., 1967. Iowaite – a new hydrous magnesium hydroxide-ferric oxychloride from the Precambrian of Iowa: American Mineralogist, v. 52, p. 1261-1271.

Koster van Groos, A.F. and Wyllie, P.J., 1968. Melting relationships in the system $NaAlSi_3O_8 - NaF - H_2O$ to 4 kilobar pressure with petrological applications: Journal of Geology, v. 76, p. 50-70.

______, 1969. Melting relationships in the system $NaAlSi_3O_8 - NaCl - H_2O$ at 1 kilobar pressure with petrological applications; Journal of Geology, v. 77, p. 581-605.

Kostetskaya, E.V., 1961. Chlorine content in the eudialytes of the Lovozero Massif: Geochemistry, v. 11, p. 1099-1106.

Kuroda, P.K. and Sandell, E.B., 1953. Chlorine in igneous rocks; Bulletin of Geological Society of America: v. 64, p. 879-896.

Leelanandam, C., 1969a. Fluorine and chlorine in the charnockitic hornblendes from Kondapalli, India: Neues Jahrbuch fur Mineralogie, Monatshafte 9, p. 379-383.

______, 1969b: H_2O, F and Cl in the charnockitic biotites from Kondapalli, India: Neues Jahrbuch fur Mineralogie, Monatshafte 10, p. 461-468.

______, 1970. A chlorine rich biotite from Kondapalli, Andhra Paradesh, India: American Mineralogist, v. 55, p. 1353-1358.

Lindblom, S., 1984. Hydrogeological and hydrogeochemical investigations in boreholes – Fluid inclusions studies in Stripa Granite: SKB Stripa Project Report 84-07, Stockholm.

McCrank, G.F., Stone, D., Kamineni, D.C., Zayachkivsky, B. and Vincent, G., 1983. Regional geology of the East Bull Lake area, Ontario: Geological Survey of Canada, Paper 83-1A, p. 457-464.

McNutt, R.H., Frape, S.K. and Fritz, P. 1984. Strontium isotopic composition of some brines from the Precambrian Shield of Canada: Isotope Geoscience, v. 2, p. 205-215.

Morton, R.D. and Catanzaro, E.J., 1964. Stable chlorine isotope abundances in apatites from Ødengardens Verk, Norway; Norsk Geologisk Tidsskrift, v. 44, p. 307-314.

Nesbitt, H.W., 1985. A chemical equilibrium model for the Illinois basin formation waters: American Journal of Science, v. 285, p. 436-458.

Nash, W.P., 1972. Apatite chemistry and phosphorous fugacity in a differentiated igneous intrusion: American Journal of Science, v. 57, p. 877-886.

——, 1976. Fluorine, chlorine and OH-bearing minerals in the Skaergaard intrusion: American Journal of Science, v. 276, p. 546-557.

Nordstrom, K., 1983. Conceptual framework for the chemical processes in the Stripa groundwaters: SKB Stripa Project Report 83-01 (Chapter 5), Stockholm.

Roedder, E., 1972. Composition of fluid inclusions: *in* Data of Geochemistry: U.S. Geological Survey, Professional Paper, 440-JJ, 164 p.

Roedder, E. and Coombs, D.S., 1967. Immiscibility in granitic melts, indicated by fluid inclusions in ejected granitic blocks from Ascension Island: Journal of Petrology, v. 8, p. 417-451.

Rucklidge, J.C., 1972. Chlorine in partially serpentinized dunite: Economic Geology, v. 67, p. 38-40.

Rucklidge, J.C. and Patterson, G.C., 1977. The role of chlorine in serpentinization: Contributions to Mineralogy and Petrology, v. 64, p. 39-44.

Shaw, D.M., 1960. The geochemistry of scapolite: Journal of Petrology, v. 1, p. 218-260.

Stueber, A.M., Huang, W.H. and Johns, W.D., 1968. Chlorine and fluorine abundances in ultramafic rock; Geochimica et Cosmochimica Acta, v. 32, p. 353-358.

Stollery, G., Borcsik, M. and Holland, H.D., 1971. Chlorine in intrusives: a possible prospecting tool; Economic Geology, v. 66, p. 361-3678.

Trommsdorff, A.V., Skippen, G. and Ulmer, P., 1985. Halite and Sylvite as solid inclusions in high-grade metamorphic rocks; Contributions to Mineralogy and Petrology, v. 89, p. 24-29.

Volfinger, M., Robert J.L., Vielzeuf, D. and Neiva, A.M.R., 1985. Structural control of the chlorine content of OH-bearing silicates (micas and amphiboles); Geochimica et Cosmochimica Acta, v. 49, p. 37-48.

$^{87}Sr/^{86}Sr$ Ratios as Indicators of Water/Rock Interactions: Application to Brines found in Precambrian Age Rocks from Canada

Robert H. McNutt

Department of Geology, McMaster University, Hamilton, Ontario, Canada L8S 4M1

McNutt, R.H., $^{87}Sr/^{86}Sr$ ratios as indicators of water/rock interactions: application to brines found in precambrian age rocks from Canada; in Saline water and gases in crystalline rocks, Editors: Fritz, P., and Frape, S.K.; Geological Association of Canada Special Paper 33, 1987. p. 81–88.

Abstract

A review of $^{87}Sr/^{86}Sr$ data on saline waters and associated rocks and mineral phases, from deep bore-hole and surface samples on the Precambrian Shield of Canada, show that the brines have a similar ratio to the host rock. This is especially true for the most concentrated brines (>100,000 mg L^{-1} TDS). This indicates equilibrium between water and rock. The "age" of these waters is unknown. A comparison with oilfield waters show that brines found in sedimentary rocks up to 300Ma old can have very different $^{87}Sr/^{86}Sr$ values than their host rock. If the water is the same age as the rock, it indicates that disequilibrium can last for ~300Ma at low temperature. Experimental data on feldspar hydrolysis (a common mineral phase in Shield rocks and one that contains abundant Sr) suggests an initial low temperature stage of very rapid exchange of cation species with the water suggesting rapid equilibration of Sr isotopes. The implication of these very different rates of equilibration are discussed as it applies to the Shield brines.

When two or more waters mix, $^{87}Sr/^{86}Sr$ ratios are very sensitive indicators of this process.

The $^{87}Sr/^{86}Sr$ values of secondary mineral phases that infill fractures in Precambrian age rock are potentially a very useful application of Sr in water/rock studies. The available data on such phases indicate that in the areas studied, the fracture filling is a recent event.

Résumé

Une étude, effectuée à partir de forages profonds et d'échantillons de surface, sur le rapport $^{87}Sr/^{86}Sr$ des eaux salées, des phases minérales et des roches précambriennes du bouclier canadien révèle que ce rapport est sensiblement le même dans les eaux sursalées et la roche hôte. Ce résultat est particulièrement évident dans le cas des solutions les plus concentrées (> 100 000 mg L^{-1} pour l'ensemble des solides dissous) et indique un équilibre entre l'eau et la roche. "L'âge" de ces eaux n'est pas connu. La comparaison avec les eaux des champs pétrolifères révèle que dans les roches sédimentaires pouvant atteindre 300 Ma, le rapport $^{87}Sr/^{86}Sr$ des eaux sursalées peut différer grandement de celui de l'encaissant. Cela indique, si l'eau et la roche sont du même âge, que le déséquilibre pourrait persister pendant environ 300 Ma, à basse température. Des valeurs expérimentales sur l'hydrolyse de feldspath, minéral riche en Sr et répandu dans les roches du bouclier, suggèrent l'existence d'une phase initiale d'échange très rapide, à faible température, entre les différents cations et l'eau, entraînant vraisemblablement un équilibre rapide des isotopes de Sr. Nous traiterons de l'effet de ces taux d'équilibre très différents en relation avec les eaux sursalées du bouclier.

Le rapport $^{87}Sr/^{86}Sr$ est un excellent indicateur de mélange entre deux eaux différentes ou plus. Dans le cas des minéraux secondaires remplissant les fractures des roches précambriennes, il peut éventuellement constituer une application très pratique du Sr dans le cadre d'études d'eau et de roche. Dans les régions étudiées, les données se rapportant à ces minéraux indiquent que le remplissage des fractures est un événement récent.

INTRODUCTION

This paper reviews the use of Sr isotopes as a tracer to study the origin of Precambrian Shield brines and examines the extent of interaction and hence equilibrium, between water and host rock. Isotopes are a more sensitive indicator than elemental ratios, certainly in studying ion exchange reactions. In addition, because Sr behaves in a fashion similar to Ca in geochemical processes, it "traces" Ca, a major constituent in brines and in most rocks.

Strontium potentially is such a good indicator because different minerals in a given rock can have very different isotopic signatures depending on the mineral's Rb/Sr ratio. Sr consists of four isotopes in nature – ^{84}Sr(.56%), ^{86}Sr(9.9%), ^{87}Sr, and ^{88}Sr(83.8%). The isotopes ^{84}Sr, ^{86}Sr and ^{88}Sr do not change their relative abundance over time. In contrast, ^{87}Sr increases with time as a result of the radioactive decay of ^{87}Rb ($T_{1/2}$ = 4.88 × 10^{10} yrs.):

$$^{87}Rb \xrightarrow{\beta} {}^{87}Sr.$$

Therefore ^{87}Sr consists of two parts – the non-radiogenic component formed in the cosmos at the same time as ^{84}Sr, ^{86}Sr and ^{88}Sr, and that formed as a result of the decay of ^{87}Rb. The latter is located in a Rb site in the mineral lattice while the former resides in a Ca site. The loss of one or both components can vary depending on the relative rates of loss from the two sites within one mineral and more importantly, from different minerals.

It is analytically easier to measure ratios in a mass spectrometer than absolute amounts, so the practice is to report all data as ratios normalized to ^{86}Sr, hence $^{87}Sr/^{86}Sr$ ratios. The experimentally measured ratios are further normalized to a $^{86}Sr/^{88}Sr$ value of 0.1194 to account for isotopic fractionation in the mass spectrometer during analysis of the sample. Finally, it is assumed that the amount of isotopic fractionation caused by physical and chemical processes in nature is too small to be measured, or if present, is insignificant. Note that this lack of fractionation is in sharp contrast to the low atomic number stable isotope systems (C,O,S) where measurable fraction in nature is the basis of their utility in studying geochemical processes.

Figure 1 shows the evolution of $^{87}Sr/^{86}Sr$ values in a rock and in its constituent minerals over time, assuming a closed system with respect to gain or loss of Rb and Sr. All phases have the same initial $^{87}Sr/^{86}Sr$ at the time of formation. After a certain time interval, the ratio in a Rb rich, Sr poor mineral (mica in Figure 1), is much higher than in a Sr rich, Rb poor mineral (plagioclase in Figure 1). The rock value at any time is the weighted sum of all the minerals in the rock.

If water has reached equilibrium with its host rock, then it is argued that the water should have the same $^{87}Sr/^{86}Sr$ value as the host rock. In contrast, if equilibrium has not been achieved or only partially achieved (e.g. equilibrium with mineral A but not with mineral B), then there is no reason to expect the two phases to have the same values for their Sr ratios. If there has been no reaction at all between water and rock, then the $^{87}Sr/^{86}Sr$ ratios will reflect the isotopic character of the source, be it oceanic, hydrothermal, meteoric etc. In the rest of this paper, published data will be reviewed and reference made to data to be published in detail elsewhere, to examine how useful Sr is as a tracer in water/rock studies for saline waters in Precambrian rocks.

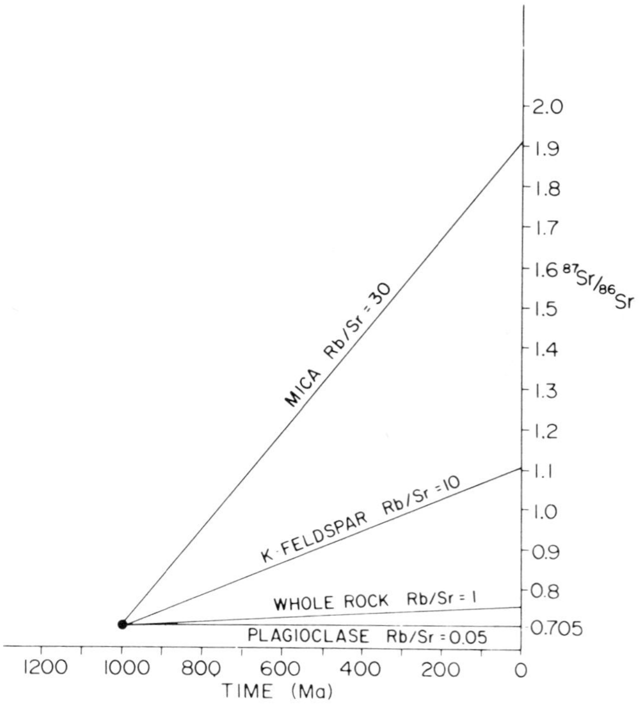

Figure 1. *Sr evolution diagram. A rock formed 1000Ma ago with an initial $^{87}Sr/^{86}Sr$ value of 0.705, evolves to a present day $^{87}Sr/^{86}Sr$ value of 0.748. The constituent mineral phases each evolve in proportion to their Rb/Sr ratio, and can have very different $^{87}Sr/^{86}Sr$ values after 1000Ma.*

DATA REVIEW

McNutt *et al.* (1984) analysed Precambrian Shield waters that vary from brines (>100,000 mg/L TDS) to brackish water (1,000-10,000 mg/L TDS) sampled in mines from Sudbury (Ontario), Thompson (Manitoba), and Yellowknife (NWT). Sample salinites ranged between 1900 and 240,000 mg/L TDS. The following points can be made (see Figure 2):

1. The brines at Sudbury give similar $^{87}Sr/^{86}Sr$ values for samples from a given borehole. As the waters become saline to brackish, the ratios start to increase with decreasing Sr (and TDS). The three sample sites each have their own distinct $^{87}Sr/^{86}Sr$ pattern.
2. A gypsum sample from one of the Sudbury boreholes gave the same Sr isotopic value as the brine.
3. The brines at Yellowknife have $^{87}Sr/^{86}Sr$ ratios that increase with decreasing Sr and TDS content. This is more apparent than real, because the data came from a comparison of samples taken from the same borehole nine months apart. They show a decrease in Sr content and in TDS, and an increase in $^{87}Sr/^{86}Sr$. The more recent samples also contained ^{3}H (Frape and Fritz, 1982), indicating mixing of dilute meteoric water with the brine.
4. The data from Thompson show some scatter in their $^{87}Sr/^{86}Sr$ ratios, but they do suggest relatively constant values over a wide range of concentration with no apparent trends. More data are needed from the dilute waters to see how Sr behaves with decreasing Sr and TDS.

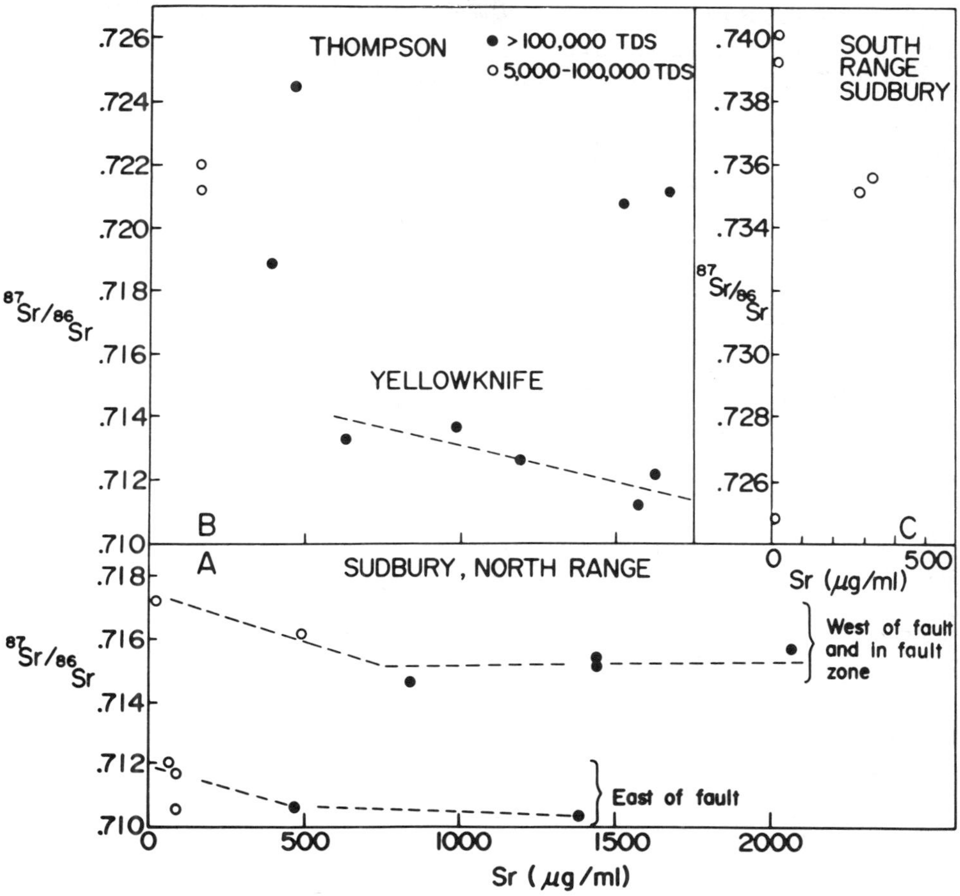

Figure 2. *$^{87}Sr/^{86}Sr$ values vs. Sr content for brines from deep boreholes at Sudbury, Yellowknife and Thompson. See text for discussion. Modified from McNutt et al. (1984) with additional date points for Thompson. The dashed lines are fitted by eye to observable trends for samples from specific areas.*

The relatively uniform values for the brines have been interpreted as reflecting equilibrium between water and rock (McNutt *et al.*, 1984). The results of brine, mineral and rock $^{87}Sr/^{86}Sr$ analyses for Sudbury samples are summarized in Figure 3. Small differences in modal mineralogy could account for the differences seen in the two North Range sites (see Figure 2). Table 1 shows how subtle this change can be. The values for $^{87}Sr/^{86}Sr$ ratios and for Rb and Sr contents are very close to the actual values found by Hurst and Farhat (1977) and by Gibbins and McNutt (1975a,b) for Sudbury felsic norite samples from the North Range. They have been altered slightly for east of calculation. As Table 1 shows, a change in the rock $^{87}Sr/^{86}Sr$ from 0.712 to 0.715, requires only minor changes to the rock mode (Compare 1 and 2, Table 1), i.e., they are both "felsic norites". The very different values for the South Range samples reflect the significantly different rock type there, i.e., a quartz diorite, containing considerably more hornblende and biotite (the latter a mineral with high $^{87}Sr/^{86}Sr$ ratios).

5. When more dilute waters are examined, a similar picture emerges. Figure 4 (from Frape et al. in prep) shows data from a salt lick in the Nipigon region of NW Ontario where the water seeps to the surface. The nearest outcrops are Archaen in age and granitic in character and probably represent the subsurface bedrock in the area. The TDS in

Table 1

Typical chemical concentrations and isotopic ratios for minerals of the felsic norite, North Range, Sudbury

Mineral	Mode(%)		Sr(ppm)	Rb(ppm)	$^{87}Rb/^{86}Sr$	$^{87}Sr/^{86}Sr$
	1	2				
Plagioclase, An_{50}	60	50	790	25	.089	.7105
Pyroxene	15	20	12	3	.705	.7240
Hornblende	15	20	40	15	1.05	.7325
Biotite	3	5	30	400	37.60	1.870
Alkali Feldspar/Quartz	7	5	400	120	0.845	.728
*Whole Rock (Mode 1)			500	35	.200	.7120
*Whole Rock (Mode 2)			425	45	.300	.7150

*The values are the weighted averages based on the given modes.

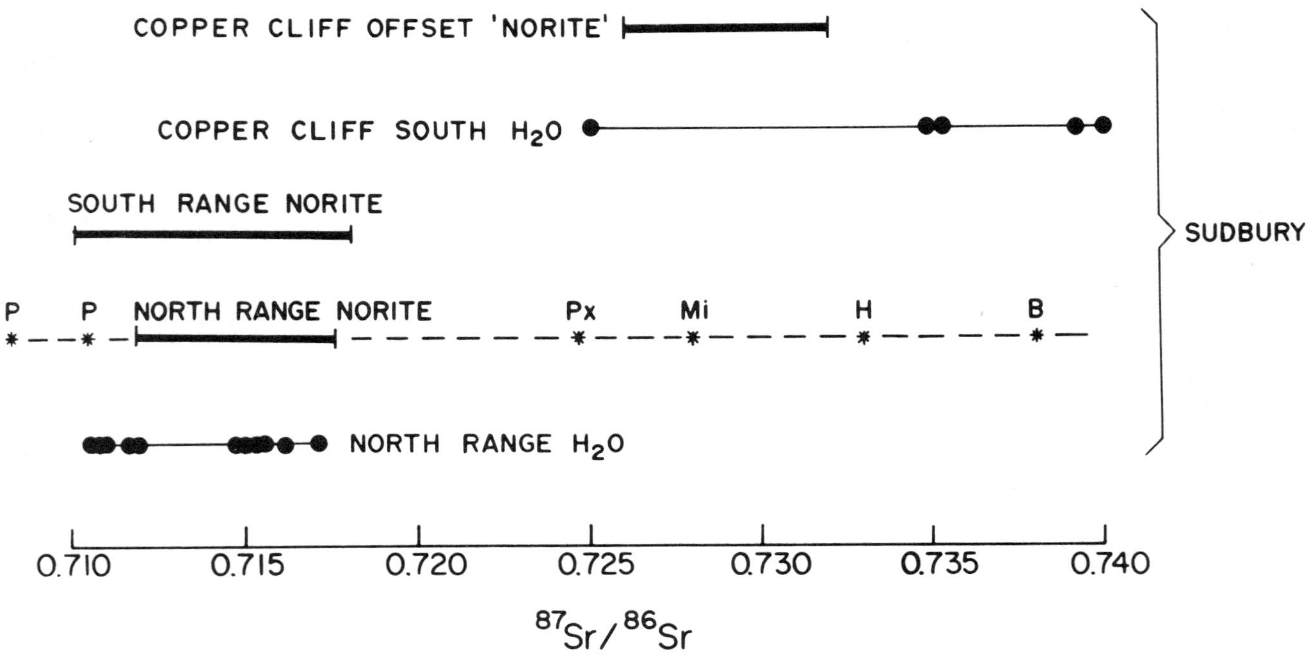

Figure 3. *Comparison of rock, mineral and brine $^{87}Sr/^{86}Sr$ date for Sudbury. The solid bars represent the range in present day $^{87}Sr/^{86}Sr$ whole rock values, the solid circles are brines and the asterisks are mineral phases as follows:*

P = Plagioclase
Px = Pyroxene
MI = micrographic intergrowth of quartz and alkali feldspar
H = hornblende
B = biotite
Modified from McNutt et al. (1984).

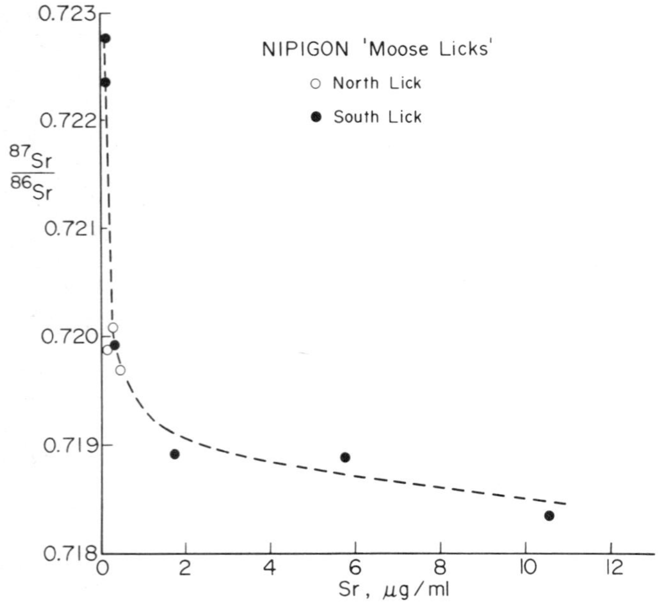

Figure 4. *$^{87}Sr/^{86}Sr$ values vs Sr content for salt licks from the Nipigon area, N-W Ontario. Note the very pronounced increase is isotopic ratios as the Sr content decreases, suggesting the mixing of (at least) two waters.*

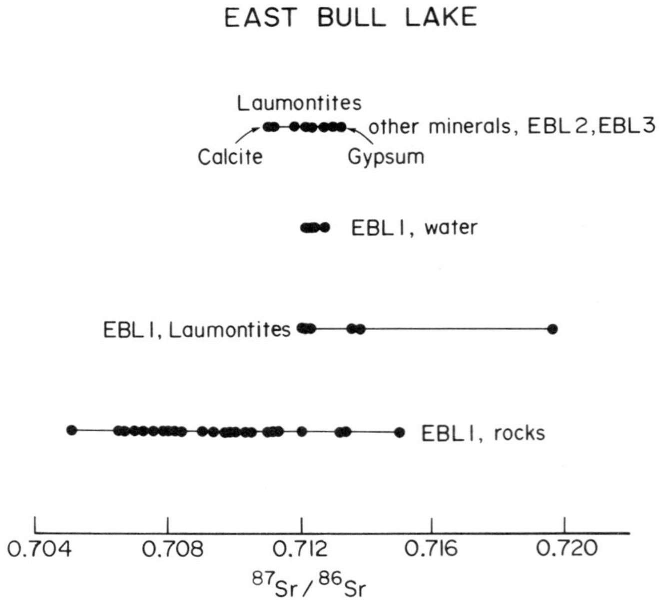

Figure 5. *$^{87}Sr/^{86}Sr$ value of whole rocks, minerals and water samples from the East Bull Lake, Ontario, gabbroic complex. See text for discussion.*

these waters ranges from 120-9,300 mg/L, and 0.1-5.7 mg/L Sr. At higher concentrations of Sr the $^{87}Sr/^{86}Sr$ ratios approach a constant value and show a pronounced increase with decreasing Sr content. This pattern, and similar ones at Sudbury and at Thompson, are here interpreted as the mixing of (at least) two waters – a dilute meteoric component from the surface and a brackish water/brine from depth.

6. At East Bull Lake, Ontario, water samples were taken from a borehole (EBL1 in Figure 5) drilled to the base of a gabbroic complex. The rock types include gabbro, troctolite and anorthosite. The Sr content of most of the waters is very low, in the range of 20-300 μg/L and only two have any significant Sr content, of 1600 and 2900 μg/L. The TDS values are correspondingly low, in the range of fresh to brackish. The host rocks have Sr contents in the range of 50-400 ppm (McNutt, Gascoyne, Kamineni, in prep). Figure 5 shows that these very dilute waters have a narrow range in isotopic composition and it overlaps with the higher values for the rocks. Such rocks are enriched in the Rb-bearing minerals such as hornblende, biotite etc. The agreement between water and the secondary mineral, laumontite (formed by the alteration of plagioclase) is pronounced. Analyses of secondary mineral phases from two other deep boreholes near to EBL1, and intersecting the same rock types, have Sr ratios that overlap those found for EBL1 waters.

In summary, the $^{87}Sr/^{86}Sr$ data show that: (a) brines from a given locality have uniform isotopic value; (b) brine and saline water values are the same as values for host rocks; (c) as waters at a given locality become more dilute the isotope values increase; and (d) secondary minerals, including silicates, carbonates and sulphates, have the same isotopic values as the associated waters.

DISCUSSION

The similarity between the Sr isotopic composition of water and rock strongly suggests that "equilibrium" between these two phases has been achieved. This is most evident in the case of the brines. Furthermore, the equilibrium applies to the *present day* Sr isotopic values, indicating that equilibration is relatively fast and therefore recent. The meaning of the term "fast" and "recent" however, depends on the host rock. For a felsic norite, as at Sudbury, the $^{87}Sr/^{86}Sr$ whole-rock value increases at a rate of 0.0006/100Ma. Therefore a range of ∓ 0.001 for measured $^{87}Sr/^{86}Sr$ values allows a latitude of 150Ma i.e. the water could have acquired its isotopic signature as much as 150Ma ago. On the other hand the $^{87}Sr/^{86}Sr$ for a granite, with a high Rb/Sr ratio, would increase at a faster rate, decreasing the time interval of uncertainty (a Rb/Sr ratio of 1.00 would give an increase in $^{87}Sr/^{86}Sr$ at a rate of 0.004/100Ma). In any event, the calcite and gypsum found in the fracture zones of the boreholes at Sudbury and at East Bull Lake, are not "old". Therefore, if these minerals are among those that seal fractures, then the sealing is a relatively recent geological event, in that it occurred long after the rock's formation in Precambrian time.

If the Sr data indicate the possible maintenance of extensive equilibration up to recent times then the chemistry of the water must also have been extensively modified over this same period. (see Frape et al., 1984 for a discussion on the extent of the chemical changes to Shield brines) and it is therefore not possible to determine the origin of the water. In fact the chemical and isotopic evolution of the water over time may be indeterminate because equilibration could have been too rapid.

In the remainder of this discussion, some of the literature on reaction rates in low temperature (<100°C) water/rock systems will be reviewed to examine this question of the rate of equilibration. Fritz and Frape (1982) have demonstrated that the distinctive D/H and $^{18}O/^{16}O$ of the Shield brines are a result of reaction at low temperature. In addition, a brief review of oilfield brines also presumably formed at low temperatures, will be given.

Comparison of sedimentary basin and oilfield brines with shield brines

In a review of the data on oilfield brines Collins (1976) finds that in rocks as young as the Tertiary, the brines can have similar TDS and Sr contents to the Precambrian Shield brines. There are some important geochemical differences however between the two types – oilfield brines are Na-Ca-Cl brines with weight ratios of Na/Ca>1 and with significant amounts of Mg, K, SO_4 and HCO_3, while the Shield waters are Ca-Na-Cl brines (Na/Ca<1) with very low values of Mg, K, SO_4 and HCO_3 (Thompson is an exception in that the brines have a significant Mg content, Frape *et al.*, 1984). These differences may simply reflect equilibration with different rock types i.e., sedimentary vs igneous/metamorphic; or it may reflect the greater age of the Shield brines, especially since Frape et al. (1984) showed that, as the brines from the Shield rocks became more concentrated, they approached the characteristic Ca-dominent chemistry. Other possibilities include a different temperature-pressure history and mixing history (e.g. the Shield brines may have mixed with hydrothermal solutions as all the brines studied to date have come from mines where one would expect to find such solutions). However, further speculation is not warranted at this point.

Nesbitt (1985) proposed a model for the composition of the brines found in Carboniferous strata in the Illinois basin. He argues that the chemical trends are the result of the equilibration of trapped and concentrated seawater (connate water) with clay minerals (Na-beidellite, kaolinite and illite) and calcite. Both the proportion of the minerals and the brine change with time. He argues that minerals such as feldspar and mica are too "unreactive" to have control over the composition of the brines. While ion exchange reactions between mineral and water can play a role, he believes it is not a dominant one. He uses published data on oxygen isotopes to show that the combination of water ultrafiltration through a shale membrane and mineral equilibration, can produce a brine with "old" solute constituents and "young" water molecules. Certainly a similar process is possible in a Shield brine if it comes in contact with meteoric water from the surface. Using the Cl content of the brines as a measure of the degree of increase in concentration, Nesbitt explains the decreasing SO_4+HCO_3/Cl and Na/Cl vs. Cl, and the increasing Ca/Cl trends in terms of such equilibration. In comparison, the Shield brines show the same behaviour for SO_4,HCO_3 and Na but not for Ca and Sr. This suggests that an equilibration model such as Nesbitt's may be applicable, but with different minerals producing the different behaviour of Ca and Sr.

Table 2

Chemical characteristics of the potassium feldspar used in the dissolution calculations

Element	Concentration
K	13%
Al	10%
Si	30%
O	46%
Rb	650ppm (assumed a K/Rb = 200)
Sr	200ppm (invited $^{87}Sr/^{86}Sr$ = 0.705)
Age	1.85 Ga.
$^{87}Sr/^{86}Sr$) today	0.945.
Radiogenic ^{87}Sr formed in last 1.85 Ga.~4 ppm	

The $^{87}Sr/^{86}Sr$ data from oilfield brines give conflicting results if one is to interpret them in terms of an equilibrium model between water and rock. Chauduri *et al.* (1985) found that brines from a Mississippan carbonate in Kansas have $^{87}Sr/^{86}Sr$ values in a narrow range of 0.722-0.723 while the carbonate host rock has 0.7086. The clay mineral fraction, which contains the Rb in the rock, and hence any radiogenic Sr, gives a value of 0.714. The authors conclude that the Sr isotopic composition is externally controlled and they propose that the water was in contact with the Precambrian basement at some stage in its migration to its present site in the Mississippan rocks.

Steuber *et al.* (1984) found that brines in the Upper Jurassic Smackover carbonate formation have Sr values from 0.7071 to 0.710, while the host carbonate gave 0.707-0.708 in agreement with the values for seawater at that time (Burke *et al.*, 1982). They feel the brine migrated from an underlying salt formation, through a shale unit where it reacted with the shale to modify its chemistry, and subsequently reacted with connate seawater in the Smackover Formation. However, Pushkar *et al.* (1985) find that brines in Silurian and Devonian pinnacle reefs have values in agreement with the rock which, in turn are equivalent to the expected value for seawater at that time.

The above examples give evidence both for and against isotopic equilibrium between water and rock. If the water is as old as the rock that hosts it, then it is possible that water as old as 300Ma (e.g. the Kansas field) can be isotopically out of equilibrium with its host.

Experimental reaction rates

As feldspar is a major phase of the host rocks for the Shield brines, some experimental data on feldspar alteration is reviewed as a means of estimating the possible behaviour of Sr in the history of brines.

Helgeson *et al.* (1984), Helgeson and Murphy (1983) and Aagaard and Helgeson (1982) have developed a model of feldspar hydrolysis called the transition state theory. This involves two stages. The first, an incongruent step, is the development of an activated complex on the surface of the mineral, caused by the rapid exchange of H_3O^+ for Na^+, K^+ or Ca^+. The complex has the composition $(H_3O)AlSi_3O_8{}^+$,(H_3O) $AlSi_3O_8(H_2O)_n$ and $NaAl(OH)_{0.4}$ $Si_3O_8{}^{-0.4}$ at low, intermediate and high pH respectively.

The second stage is the decomposition of the activated complex and it is this decomposition that is the rate controlling step. The authors argue that this decomposition is not diffusion controlled (see below), and that it is pH dependent at low and high pH. The process is strictly a surface reaction, estimated by the authors to occur to an average depth of 50 μm if the entire surface takes part in the alteration process. However it is generally agreed that mineral surfaces have reactive sites where reaction proceeds such as dislocations, site vacancies, kinks, cleavage fractures, etc. Feldspars and other minerals occuring in fracture zones of plutonic rocks, where brines are concentrated, invariably contain these characteristics (C. Kamineni, Pers. Comm.). Estimates of the ratio of reactive sites to total surface area vary from less than 0.01 to 0.1. If a value of 0.01 is used then one makes a corresponding increase in the depth estimates of the reaction (e.g., to 50/0.01 = 5000 μm or 5.10^{-4} mm).

Helgeson *et al.* (1984) estimate the rate of feldspar dissolution at approximately 10^{-16} moles feldspar cm^{-2} sec^{-1} at 25°C and 1 atm over a pH range of 2-8. This increases to 10^{-14} at 100°C and 10^{-13} at 200°C (see their figure 24 for a compilation). Using the value supplied by Helgeson and Murphy (1983), a calculation can be made to give an estimate of the approximate amount of feldspar that will be dissolved in a given time interval. Table 2 gives the chemical characteristics of the K feldspar used in the calculation. The Rb and Sr values are based on actual values found for Sudbury feldspar together with its age of 1.85Ga. It is assumed that Rb and Sr (both the non-radiogenic and the radiogenic ^{87}Sr) show the same kinetic behaviour (to a first approximation) as K, Na or Ca. Considering the geochemical similarity and the "ball-park" nature of these calculations, this is a reasonable assumption. Following Helgeson and Murphy, the following values are used:

(a) The total surface area of the feldspar is $10^{4.6}$ cm^2 $Kg(H_2O)^{-1}$. This is equivalent to a network of cracks approximately 0.5 mm in width; or to the total surface area of spherical grains with a radius of 0.1 mm in a 11.6% suspension; or to the total surface area of a mineral aggregate of spherical grains with a radius of 1.75 mm and with 30% porosity.

(b) The depth of the altered zone is 50 μm assuming that the entire grain surface takes part in the reaction.

Using the above figures, one can calculate the volume of mineral reacted per Kg H_2O to be $10^{4.6}$ $10^{-6.3}$ = $10^{-1.7}$ cc $Kg(H_2O)^{-1}$. Assuming a density of 2.6 gms cc^{-1}, one can calculate the mass of dissolved feldspar as $2.6 \cdot 10^{-1.7} = 5.2$ 10^{-2} gms $Kg(H_2O)^{-1}$. Using the assigned value of Sr of 200 ppm (table 2), gives $5.2\ 10^{-2} \cdot 2\ 10^{-4} = 1.04\ 10^{-5}$ gms Sr $Kg(H_2O)^{-1}$. Or in other terms, approximately 10 ppb of Sr with a $^{87}Sr/^{86}Sr$ ratio of 0.945.

Helgeson and Murphy (1983) calculated that hydrolysis of microcline in a dilute HCL solution (pH = 4) at 25°C, 1 bar, under the total surface conditions in (a) above, will, after 800 years, destroy approximately 0.04 gms of feldspar to produce 0.015 gms of muscovite and lesser amounts of kaolinite. Using the data of table 2 would give 8 ppb Sr in solution. An important feature of their calculations is that after 800 yrs, the reaction has essentially ended (see their figure 2).

Admittedly, calculations given above are the slowest rate that could be considered because the rate of reaction will increase with increasing temperature. On the other hand, a system with maximum porosity has been chosen. Shield igneous and metamorphic rocks will have much lower porosities, on the order of 1% or less. Also the calculations applied to dilute solutions and it becomes a question of how applicable they are to the more concentrated brines. For example, Wollast and Chou (1985) maintain that in the presence of seawater, the surface sites of the feldspar will be occupied by Na rather than H+. How this affects the rate of Sr exchange is unknown. It might even work to the favour of Sr being exchanged into the solution as it is replaced by Na in the lattice site.

Wollast and Chou (1985) present a model based largely on one developed in a quantitative way by Pačes (1973). Their experimental data on Amelia albite shows that the first stage is a very rapid exchange of Na for H+. At 25°C and 1 atm, such an exchange between DD H_2O and albite has a rate constant of $10^{5.76}$ i.e., equilibration within minutes. The second stage is the build up of a residual layer of (Al,Si)O,(OH), with continual preferential leaching of Na (and possibly Al). This is followed by the rate controlling step which is the surface reaction of the residual layer with the solution. This reaction is diffusion controlled and at steady state therefore, according to Wollast and Chou, there is a more or a less linear increase of all species in solution. If the water remains in contact with the rock for a long enough time, then saturation can occur with subsequent precipitation. The rate of dissolution is pH dependant.

Regardless of the mode of dissolution chosen, it seems that there is an initial release of alkalis and alkaline earths to the solution followed by a slow rate controlling step of additional release of material to solution. In an open system, a given volume of water passes through successive volumes of rock. Each volume of rock will have the initial rapid reaction with the water, and as the water passes through the rock it continues to increase in solute components. Theoretically this can continue until the saturation is reached and there is subsequent precipitation of one or more mineral phases. Such a process will proceed with the most efficiency with respect to time, if the water/rock ratio is small, such as in intergranular flow. If on the other hand, there is a closed system, with a constant water/rock ratio, then the buildup of solute constituents will be much slower and, on the basis of the experimental data reviewed above, the solution would not achieve the concentrations observed in the Shield brines. As pointed out by Helgeson and Murphy, equilibrium among reactant minerals and solution, is achieved more rapidly in an open system. Admittedly, the time differences found in the experiments (200 vs. 800 years) is very small in a geological context, but the principle is the same. In fact "the length of time required for a given mineral to reach a quasistatic state in a geochemical process is a function not only of the magnitude of its rate constant, effective surface area, initial solution composition, fluid flow rate, pressure, temperature, and local equilibrium constraints encountered along the reaction path, but it also depends on the compositions, rate constants, and effective surface areas of all other reactant minerals involved in the irreversible process" (Helgeson and Murphy, 1983, page 122).

CONCLUSIONS

The Sr isotopic data on saline waters suggest that both the dilute and concentrated waters have "equilibrated".

1. If it is assumed that the brines are much older than the brackish waters, it would appear that there are two stages of "equilibration". The dilute waters, such as at East Bull Lake, with their low TDS and Sr content in the ng/ml range are "young" and the Sr isotope composition is the result of early, rapid, incongruent reactions of the type proposed by Helgeson et al. (1984) on the surface sites of the more reactive minerals in a relatively open system, with a low water/rock ratio.

2. The Shield brines have been sealed in fracture zones until intersected by drilling. The Sudbury data shows that two brine systems, very close in location, have not communicated with each other. Their isotopic equilibration with the rock must be a long process, because of an essentially fixed water/rock ratio. The brine probably achieved the bulk of its solute constituents (including Sr) before being isolated (a long history of flow in a more open rock system, or an origin involving seawater or hydrothermal solutions), followed by a slow equilibration by the rate controlling process discussed by Helgeson et al. (1984) or the diffusion controlled models of Paćes (1973) or Wollast and Chow (1985)). The oilfield data suggest this could be as long as 300Ma.

3. When two or more waters mix, one concentrated and one dilute, Sr isotopes seem to be a very sensitive indicator of this process. The salt-lick data show this especially well.

4. The Sr isotopic analyses of secondary mineral phases that infill fractures in the rock, are potentially the most useful application of Sr in water/rock studies. The available data on calcites and gypsum from Sudbury and East Bull Lake indicate that in both areas, the fracture-filling is a recent event.

ACKNOWLEDGEMENTS

The research on which this discussion is based was funded by contracts with the AECL and by NSERC grants. The author was introduced to the brine problem by Drs. P. Fritz and S.K. Frape, University of Waterloo and he gladly acknowledges their input. The East Bull Lake samples are from a joint study with Drs. M. Gascoyne and C. Kamineni of AECL, and I appreciate their thoughtful comments on this work and on the paper. The technical support of J. McAndrew, D. Jones and C. Jager is gratefully acknowledged.

REFERENCES

Aagaard, P. and Helgeson, H.C., 1982. Thermodynamic and kinetic constraints on reaction rates among minerals and aqueous solution. I. theoretical considerations: American Journal of Science, v. 282, p. 237-285.

Burke, W.H., Denison, R.E., Hetherington, E.A., Koepnick, R.B., Nelson, H.F. and Otto, J.B., 1982. Variation of seawater $^{87}Sr/^{86}Sr$ throughout Phanerozoic time: Geology, v. 10, p. 516-519.

Chaudhuri, S., Clauer, N. and Broedeh, V., 1985. Strontium isotopes of oil-field water as a key to understanding of their chemical evolution and of diagenetic recrystallization of carbonate rocks: Abstract, AAPG Research Conference, New Orleans, 1985.

Collins, A.G., 1975. Geochemistry of oilfield waters: Elsevier, 496 p.

Frape, S.K., Fritz, P. and McNutt, R.H., 1984. Water-rock interaction and chemistry of groundwaters from the Canadian Shield: Geochimica et Cosmochimica Acta, v. 48, p. 1617-1627.

Frape, S.K. and Fritz, P., 1982. The chemistry and isotopic composition of saline groundwaters from the Sudbury basin, Ontario: Canadian Journal of Earth Science, v. 19, p. 645-661.

Fritz, P. and Frape, S.K., 1982. Saline groundwaters on the Canadian Shield – a first overview: Chemical Geology, v. 36, p. 179-190.

Gibbins, W.A. and McNutt, R.H., 1975a. The age of the Sudbury Nickel Irruptive and the Murray Granite: Canadian Journal of Earth Science, v. 12, p. 1970-1989.

______, 1975b. Rubidium-strontium Mineral ages and polymetamorphism at Sudbury, Ontario: Canadian Journal of Earth Science, v. 12, p. 1990-2003.

Helgeson, H.C., Murphy, W.M. and Aagaard, P., 1984. Thermodynamic and kinetic constraints on reaction rates among minerals and aqueous solution. II. Rate constants, effective surface area, and the hydrolysis of feldspar: Geochimica et Cosmochimica Acta, v. 48, p. 2405-2432.

Helgeson, H.C. and Murphy, W.M., 1983. Calculation of mass transfer among minerals and aqueous solutions as a function of time and surface area in geochemical processes. I. computational approach: Mathematical Geology, v. 15, p. 109-130.

Hurst, R.W. and Farhat, J., 1977. Geochronologic investigation of the Sudbury Nickel Irruptive and the Superior Province granites north of Sudbury: Geochimica et Cosmochimica Acta, v. 41, p. 1803-1816.

McNutt, R.H., Frape, S.K. and Fritz, P., 1984. Strontium isotopic composition of some brines from the Precambrian Shield of Canada: isotope Geoscience, v. 2, p. 205-215.

Nesbitt, H.W., 1985. A chemical equilibrium model for the Illinois basin formation waters: American Journal of Science, v. 285, p. 436-458.

Pačes, T., 1973. Steady-state kinetics and equilibrium between ground water and granitic rock: Geochimica et Cosmochimica Acta, v. 37, p. 2641-2663.

Pushkar, P., Stueber, A.M. and Hetherington, E.A., 1985. Application of variation in $^{87}Sr/^{86}Sr$ ratios to studies of carbonate paragenesis and the origin of associated brine: Abstract, AAPG Research Conference, New Orleans, 1985.

Stueber, A.M., Pushkar, P. and Hetherington, E.A., 1984. A strontium isotopic study of Smackover brines and associated solids, southern Arkansas: Geochimica et Cosmochimica Acta, v. 48, p. 1637-1649.

Wollast, R. and Chou, L., 1985. Kinetic study of the dissolution of albite with a continuous flow-through fluidized bed reactor: *In* The Chemistry of Weathering J.I. Dreven (ed.), Reidel Pub. Co., p. 75-96.

Chlorine stable isotope composition of Canadian Shield brines

Ronald Kaufmann*
Laboratory of Isotope Geochemistry, University of Arizona, Tucson, Arizona 85721, U.S.A.

Shaun Frape, Peter Fritz
Department of Earth Sciences, University of Waterloo, Waterloo, Ontario, Canada N2L 3G1

Harold Bentley
Hydro Geo Chem, Inc., Tucson, Arizona 85705, U.S.A.

Kaufmann, R., Frape, S., Fritz, P. and Bentley, H., Chlorine stable isotope composition of Canadian Shield brines; in Saline water and gases in crystalline rocks, Editors: Fritz, P., and Frape, S.K.; Geological Association of Canada Special Paper 33, 1987. p. 89–93.

Abstract

The Canadian Shield contains Ca-Na-Cl brines at depths in excess of 1000 metres. The potential source of these brines may be post diagenetic penetration of concentrated ocean water, residual rock-forming fluids, or migration from surrounding rocks and sedimentary basins. The spatial distribution of the chemistry of brines sampled indicates mixing between brines and fresh water, as well as rock-water interactions. We attempted to investigate the origin of these fluids by measuring the $^{37}Cl/^{35}Cl$ in eleven brine samples from the Sudbury, Thompson and Yellowknife mining districts. Chloride is a conservative aqueous component and, excluding diffusion related processes, its isotopic ratio is unaffected by low temperature processes. Thus the ratio may provide evidence for the genesis of the brines.

The chlorine isotope ratios of eight samples were enriched in ^{37}Cl with respect to ocean chloride (SMOC), and ranged from +0.38 ‰ to + 0.77 ‰ (precision = 0.12 ‰, SMOC = 0.00 ‰). Two samples, one from Thompson and one from Yellowknife, had compositions similar to SMOC (+0.01 ‰), and one sample from Sudbury was depleted by −0.45 ‰ with respect to SMOC. Ratio variations suggest that the brines did not result from concentrated ocean water, and that the ratios were fractionated in situ, possibly by volatile segregation, diffusion, or ion filtration.

Résumé

On trouve dans le bouclier canadien des eaux sursalées riches en Ca-Na-Cl à des profondeurs de plus de 1000 mètres. Elles pourraient provenir d'une pénétration post-diagénétique d'eau marine concentrée, de fluides pétrogénétiques résiduels ou d'une migration à partir de roches et de bassins sédimentaires environnants. La répartition spatiale de la composition chimique des eaux sursalées échantillonnées indique un mélange entre celles-ci et l'eau douce de même que des interactions entre l'eau et la roche. Nous avons étudié l'origine de ces fluides d'après le rapport $^{37}Cl/^{35}Cl$ de onze échantillons provenant des régions minières de Sudbury, Thompson et Yellowknife. Le chlore est un composant dissous stable et, si l'on néglige l'influence de la diffusion, son rapport isotopique n'est pas modifié par les processus de faible température. Par conséquent, ce rapport peut fournir des renseignements sur la genèse des eaux sursalées.

Huit échantillons sont enrichis en ^{37}Cl comparativement au chlore océanique (SMOC). Les valeurs du rapport isotopique varient entre +0,38‰ et + 0,77‰ (précision: 0,12‰; SMOC: 0,00‰). Un échantillon provenant de Thompson et un autre, de Yellowknife ont une composition semblabe à celle du chlore océanique (+0,01‰) et un échantillon de Sudbury est appauvri en ^{37}Cl comparativement au chlore océanique (−0,45‰). Les variations du rapport isotopique indiquent que les eaux sursalées ne résultent pas d'une concentration d'eau marine et que le fractionnement s'est effectué "in situ" soit par ségrégation volatile, diffusion ou filtration ionique.

**Now: Dept. of Geology, Northern Illinois University, Delcalb, Il. 60115*

INTRODUCTION

The Canadian Shield has been a stable craton for 1.3 b.y. Shield rocks are primarily granitic (granite, felsite, diabase, gabbro), although mafic rocks are present (basalt, amphibolite) and metamorphic alteration is common in some areas. Mining of metal deposits and associated exploration drilling encountered substantial volumes of chloride brine deep in the Shield. Fritz and Frape (1982) discussed the possible origin of these brines, including post diagenetic penetration of concentrated ocean water, residual rock-forming fluids, or modification of formation waters by rock-water interactions. Although, no specific mode of formation was proposed, Frape et al. (1984) suggested that highly saline, Ca-Na-Cl type fluids had to be expected in any rock with a long "diagenetic" history and that it may not be possible or even relevant to search for a specific source.

This reasoning was based on chemical and isotopic components of the Shield ground waters which indicate a complex history of development. Shallow ground waters in the Shield are fresh whereas waters below 1 km depth usually exceed 200 g/L Total Dissolved Solids (TDS). Distribution of most major ions and stable isotopes of oxygen and hydrogen in the brines suggests that mixing with fresh water occurs, at least in mining environments. Other geochemical indicators point to substantial rock-water interaction which must have occurred so that mixing and rock-water interactions have modified most of the dissolved compositions, often to a point where the geochemical nature of the parent brine(s) is lost.

Thus, only very conservative constituents might have preserved the primary characteristics of the present brine's precursors and isotope analyses of the chloride component in the brines appeared particularly promising. The chlorine stable isotope ratio, $^{37}Cl/^{35}Cl$, may be one of the few brine characteristics to be preserved in the chemically dynamic Shield environment sampled because aqueous chloride has few sources and sinks in igneous rock environments (Wedepohl et al., 1969), and only diffusion significantly affects the ratio at low temperatures.

This paper reports the results of $^{37}Cl/^{35}Cl$ measurements on samples of deep brines from three Canadian Shield mining districts (Figure 1). Knowing the chlorine isotope compositions of the brines should improve our understanding of brine origin.

CANADIAN SHEILD BRINES

Ground-water samples from the Sudbury, Thompson and Yellowknife mining districts (Fritz and Frape, 1982) indicate that shallow groundwater, above 1 km depth, is fresh, with $^{2}H/H$ and $^{18}O/^{16}O$ compositions that plot on the local Meteoric Water Line. Deep groundwaters occur in fractures and pockets within shear zones and are often under high pressure, having salt contents exceeding 200 g/L TDS and they are classified as Ca-Na-Cl brines. Hydrogen and oxygen isotope compositions increase linearly above and to the left of the GMWL and D and ^{18}O enrichment correlate with salinity and depth. Fritz and Frape (1982) suggested that the shallow water is primarily influenced by "modern" recharge and the deep brines are isolated at depth. They interpreted the linear increases in isotope and TDS compositions as indications of mixing between a deep brine of unknown origin and shallow water, most probably as a consequence of man's activities in the areas where it was possible to collect samples.

Many of the major ions (Ca, Na, Sr, Cl and Br) found in the groundwaters show substantial increases in concentration with increasing depth. However, unlike other major dissolved components and isotopes analyzed in the brine samples, Mg and $^{87}Sr/^{86}Sr$ show little or no correlation with depth and salinity. Frape et al. (1984) suggested that these two parameters (among others) reflect rock-water interactions of varying intensity. High Mg concentrations in brines from the Thompson district, for example, seem to relate to the magnesium-rich ultramafic rocks in the district. In other areas, formation of chlorite and other secondary minerals may have removed Mg from solution. Variations of $^{87}Sr/^{86}Sr$ among and within districts also result from rock-water interaction (McNutt et al., 1984) because they closely reflect the strontium isotopic composition of the host rock. Different rocks in the same mining district can thus retain brines with quite different strontium isotope ratios, but otherwise similar geochemical characteristics.

As noted in Frape et al. (1984) and Frape and Fritz (this volume), in the most concentrated brines many of the parameters measured, such as Ca, Sr, Br and several of the heavy metals have very uniform Cl ratios from site to site. Therefore, it is suggested that strong solubility controls in Ca-Na-Cl brines exists for most elements, which governs their absolute concentrations. However, isotopic compositions are controlled by isotope exchange reactions during intense rock-water interactions and "equilibration" with the host rock.

CHLORINE ISOTOPE GEOCHEMISTRY

Naturally occurring chlorine has two stable isotopes in near constant ratio of 0.324 ($^{37}Cl/^{35}Cl$). Chlorine is found predominantly in a single oxidation state as (−1) chloride. Chloride is conservative in aqueous solutions exhibiting high solubility and low sorption. Only at high concentrations does chlorine interact with other species as complexes and precipitates. Excluding evaporites, chloride is a minor component in most rocks.

Kaufmann et al. (1984) report that $^{37}Cl/^{35}Cl$ is invariant in modern oceans, but does differ among halites, hydrothermal waters and groundwaters. Desaulniers et al. (Geoch. Cosmochim. Acta, 1986) indicate that chlorine fractionation does occur during molecular diffusion in glaciolacustrine clays in Ontario. They show that ^{35}Cl diffuses

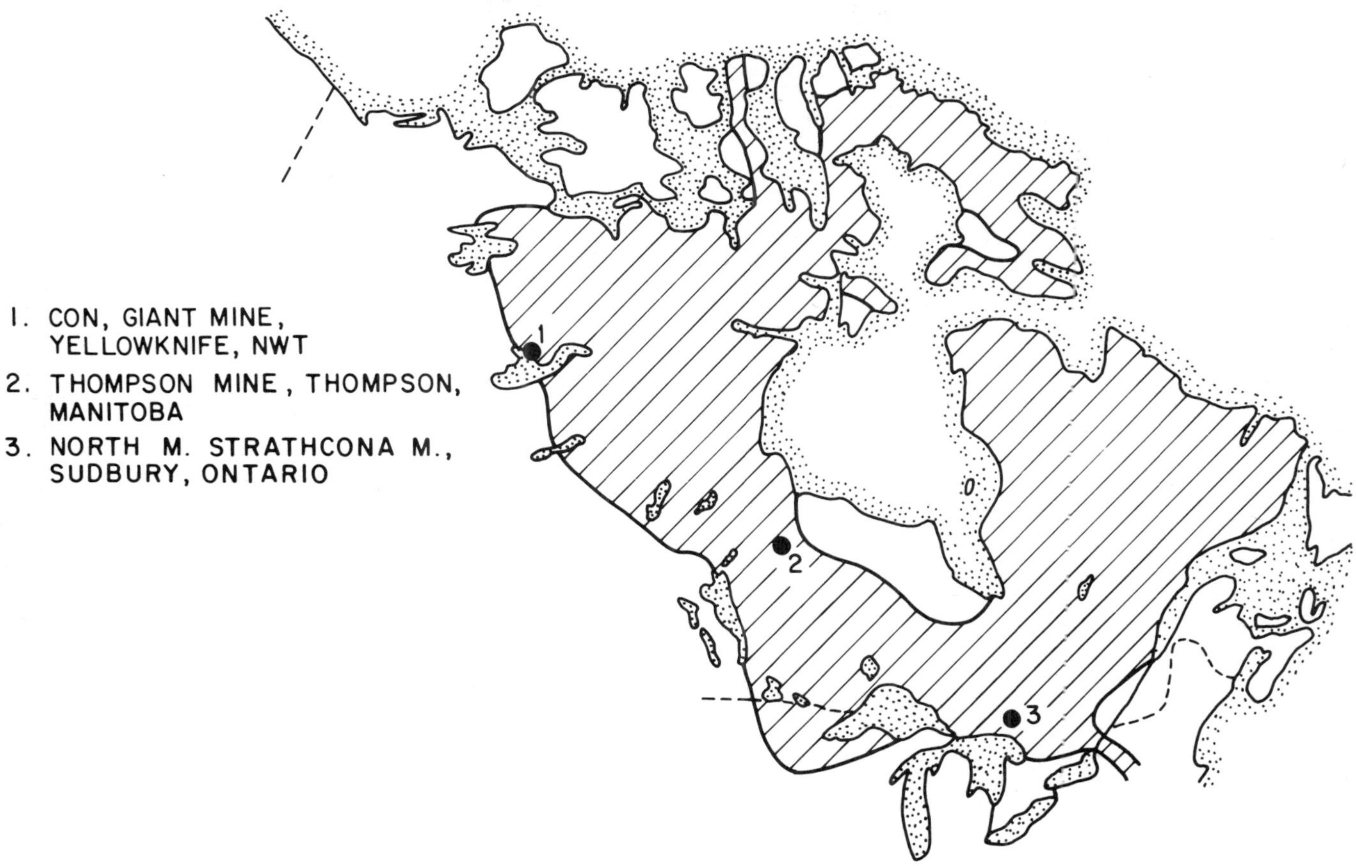

Figure 1. *Location of Sudbury, Thompson and Yellowknife mine districts.*

more rapidly that ^{37}Cl. Based on a study of Texas-Louisiana oil field brines, Kaufmann et al. (submitted to AAPG Bulletin, 1986), suggested that where diffusion occurs in low advective regional flow systems, fractionation may also result.

Campbell (1985) demonstrated that chlorine isotopes fractionate during laboratory ion filtration experiments. Effluent derived by forcing a saline solution through a clay plug was enriched by +1.7 ‰ (precision 0.12 ‰) with respect to the influent. This experiment suggests that effluent from ion filtration in nature will produce enrichment in ^{37}Cl unlike diffusion which enriches ^{35}Cl in effluents.

Howald (1960) showed that chlorine isotopes will exchange between HCl gas and liquid. Kaufmann and Long (in preparation) suggest that mechanisms such as phase equilibrium exchange may fractionate chlorine in hydrothermal waters.

SAMPLE SELECTION AND ANALYSIS

Samples for this study were deep ground-water brines from the Sudbury, Thompson, and Yellowknife mining districts (Figure 1). They represent brines with a range of dissolved solids from each of the three regions; Table 1 shows the chemical compositions.

Chlorine isotope ratios were measured by the Taylor and Grimsrud (1969) method modified by Kaufmann (1984). In this procedure chloride was quantitatively precipitated from solution as AgCl and reacted with methyl iodide in a sealed tube for two days at 70°C-100°C to produce methyl chloride (CH_3Cl). A gas chromatograph purified the CH_3Cl, separating it from excess methyl iodide. Isotope ratios were measured with gas mass spectrometer (a modified MM 602) as $^{+}CH_3{}^{37}Cl/{}^{+}CH_3{}^{35}Cl$ with collector slit widths masking all but m/e 50 + 51 for the major beam and m/e of 52 for the minor. Because the 51 peak is only about 3% of the 50 peak, 52/50 + 51 approximately equals 52/50. The mass spectrometer measures the peak intensities to within a precision of 0.03 ‰. The overall precision of the method, based on the average difference between replicate measurements is 0.12 ‰.

Table 1 shows the chlorine isotope ratios of the Canadian Shield brines. With a total range of 1.22 ‰ (−0.45 to +0.77 ‰), the samples vary significantly. STR31-178 (−0.45 ‰) is lighter than SMOC (Standard Mean Ocean Chloride) and the other samples from Sudbury. BL4400 and 4500 lc are both similar to SMOC and are lighter than other samples from Thomspon and Yellowknife. The remaining eight samples are heavier than 0.37 ‰.

DISCUSSION

The question of chloride origin within the brines is complicated by the high mobility of chloride in the crust. Most

Table 1

Chemical and isotopic analysis from Sudbury, Thompson and Yellowknife mine districts.

	Ca	Mg	Na	mg/1 Cl	HCO_3	SO_4	TDS	δ‰ SMOW ^{18}O	^2H	δ‰ SMOC ^{37}Cl
Sudbury:										
N3645	31100	92	7800	62700	<1	155	102,400	−11.5	−68	+0.77
N3640 b	6400	29	21000	15300	<1	335	240,000	−10.3	−42	+0.65
STR31-178	41300	11	8200	67700	<1	665	118,440	−10.5	−35	−0.45
Thompson:										
T3-344	31800	1310	20500	74400	24	20	129,600	−15.3	−107	+0.44
T3-330	14500	790	6000	36100	13	10	58,200	−16.6	−118	+0.62
BL4400	44300	1960	28000	123000	5	10	200,000	−12.9	−76	+0.00
Yellowknife:										
YK 2387	4430	23	3040	12400	11	590	20,500	−22.2	−162	+0.56
YK 3464	26400	406	19880	82700	14	428	130,290	−19.3	−126	+0.42
YK 3457	10500	109	8050	29400	34	715	49,538	−22.4	−162	+0.38
4500 1b	37000	525	23700	90500	18	346	153,073	−18.7	−114	+0.74
4500 1c	49400	712	27900	132800	16	8	213,938	−15.7	−84	+0.56

$\delta^{37}Cl$ expressed as per mil difference from SMOC.
$\delta^{37}Cl$ precision is 0.12 ‰.

crustal chloride has been circulated through the oceans: chloride in rainfall is primarily from marine aerosols; most sediment chloride was deposited in marine-related environments; and volcanic chloride can be derived from marine sediments or recirculated from tectonic plate subduction zones (Wedepohl et al., 1969). High chloride ground water in the Canadian Shield may have evolved directly from ocean water; been buried with sediments (formation water); be associated with igneous and metamorphic processes (magmatic water, and hydrothermal water); or diffuses into fractures and fissures from surrounding rocks. Thus, the origin question becomes one of mode of emplacement. Brine chlorine isotope ratios reflect both mode of emplacement and in situ fractionation processes.

Post diagenetic penetration of concentrated ocean chloride is the least tenable of emplacement modes proposed because only two of the 11 brines are similar to SMOC. Neither brine production by subaerial evaporation of ocean water nor advective flow through fractures are likely to change the chlorine ratios of the brine (Kaufmann, 1984). Chlorine composition of the ocean brine might be influenced by mixing; however, the concentration of chloride in water available for mixing is insufficient to affect the ratios of the brine without substantially diluting the brine.

Either residual ore and hydrothermal fluids, or Shield rocks (contributing chloride by solid state diffusion) could be the immediate source of brine chloride. The two sources may be distinguished by examining chloride in the rocks. Diffusion depends on the presence of a chemical potential. If the chemical potential of chloride in the brines is higher than that in the surrounding rocks, residual rock-forming fluids would be the probable source of the brines.

The broad range of chlorine isotope results (−0.45 ‰ to +0.77 ‰) suggests the influences of in situ fractionation processes; among these are volatile segregation, diffusion, and ion filtration. Volatile segregation is the loss from magmatic melts and metasomatic fluids of volatile components including chlorine. The isotopic signature of brines remaining after volatile segregation will reflect any phase isotopic differences. No data are available yet to examine this fractionation process.

Liquid phase diffusional effects on chlorine ratios have been identified in both field and laboratory studies (Desaulniers et al., 1986; Madorsky and Straus, 1947) and could have affected Shield chlorine ratios. Diffusion would have to be identified in the Shield through detailed sampling and measurements in a limited geographic area, this has not yet been possible. Solid phase diffusional effects should be evaluated through studies of chlorine ratios in rocks related to brines.

Ion filtration is yet another process to consider which has been observed to alter chlorine isotope ratios in laboratory experiments (Campbell, 1985). Again, it may be identified through detailed studies in the Shield; however, minerals that could produce the membrane effects must also be identified before accepting this mechanism.

The chlorine isotope results may also support the mixing hypothesis of Fritz and Frape (1982) and the uniform brine hypothesis (see J.F. Pearson, this volume). In each mine district the lightest samples are also the most enriched in deuterium. Although the data base is very limited, this could suggest that "mixing lines" between ^{37}Cl enriched non saline waters and ^{37}Cl depleted brines do exist. Should further studies verify this then the chlorine isotope data would support the existence of a unique brine across the Canadian Shield, characterized by $\delta^{37}Cl$ values between (extrapolation) − 0.5 and 1 ‰ SMOC.

SUMMARY AND CONCLUSIONS

At this stage insufficient chlorine stable isotope data are available to allow us to draw firm conclusions about the origin and/or geochemical history of the Canadian Shield brines. However, it is possible to formulate hypotheses on which to base future sampling programs.

Our measurements show a broad range of $^{37}Cl/^{35}Cl$ ratios from −0.45 ‰ to +0.74 ‰ SMOC. The relatively wide range of chlorine ratios makes oceans as a direct chloride source of the brines unlikely. Thus, infiltration of (evaporated) seawater or sedimentary basin brines cannot be seen as brine precursors, unless in situ fractionation has altered the ratios. Volatile segregation, diffusion and ion filtration are possible mechanisms of in situ fractionation. This would support the interpretation of Frape et al. (1984) who consider the brines as fluids of possibly (but not necessarily) variable origins which have been modified to a more or less unique composition through intense rock water interactions in a crystalline rock regime. An interesting aspect is the observation that possibly all brines across the Shield have similar, negative $S^{37}Cl$ ratios. Since the extrapolated S^2H values (see Pearson, this volume) are also rather uniform across the Shield and since only a few processes can significantly modify the hydrogen isotopic compositions of fluids participating in rock-water interactions, one could speculate that the brines have undergone not only a similar geochemical evolution but have also a unique origin.

REFERENCES

Campbell, D.J. 1985. Fractionation of stable chlorine isotopes during transport through semipermeable membranes: Masters Thesis, University of Arizona, May 1985, 103 p.

Desaulniers, D.E., Kaufmann, R.S., Cherry, J.A. and Bentley, H.W. 1986. ^{37}Cl-^{35}Cl variations in a diffusion-controlled groundwater system. Geochimica et Cosmochimica Acta, v. 50, p. 1757-1764.

Frape, S.K., Fritz, P. and McNutt, R.H. 1984. Water rock interaction and chemistry of groundwaters form the Canadian Shield: Geochimica et Cosmochimica Acta, v. 48, p. 1617-1627.

Fritz, P. and Frape, S.K. 1982. Saline groundwaters in the Canadian Shield – A first overview: Chemical Geology, v. 36, p. 179-190.

Howald, R.A. 1960. Ion pairs. I. Isotope effects shown by chloride solutions in glacial acetic acid: Journal of the American Chemical Society, v. 82, p. 20-24.

Kaufmann, R.S. 1984. Chlorine in ground water: Stable isotope distribution: Dissertation, University of Arizona, May 1984, 137 p.

Kaufmann, R.S., Long, A., Bentley, H. and Davis, S. 1984. Natural chlorine variations: Nature, v. 309, p. 338-340.

Madorsky, S.L. and Straus, S. 1947. Concentration of isotopes of chlorine by counter-current electromigration method. Journal of Research of the National Bureau of Standards, v. 38, p. 185-189.

McNutt, R.H., Frape, S.K. and Fritz, P. 1984. Strontium isotopic composition of some brines from the Precambrian Shield of Canada: Isotope Geology, v. 2, p. 205-215.

Taylor, J.W. and Grimsrud, E.P. 1969. Chlorine isotopic ratios by negative ion mass spectrometry. Analytical Chemistry, v. 41, p. 805-810.

Wedepohl, K., Correns, C., Shaw, D., Turekian, K. and Zemann, J. 1969. Handbook of Geochemistry, Chapter 17.

Vug Brines-Fluid Inclusions: A Key to the Understanding of Secondary Gold Enrichment Processes and the Evolution of Deep Brines in the Canadian Shield

Jayanta Guha
Sciences de la Terre, Université du Québec à Chicoutimi, Chicoutimi, Québec, Canada G7H 2B1

Ram Kanwar
Northgate Patino Mines, Box 8000, Chibougamau, Québec, Canada G8P 2L1

Jayanta, G. and Kanwar, R., Vug brines-fluid inclusions: a key to the understanding of secondary gold enrichment processes and the evolution of deep brines in the Canadian Shield; in Saline water and gases in crystalline rocks, Editors: Fritz, P., and Frape, S.K.; Geological Association of Canada Special Paper 33, 1987. p. 95–101.

Abstract

A systematic evaluation of $CaCl_2$ – NaCl – CH_4 fluid-bearing vugs in the Copper Rand Mine in the Chibougamau mining district indicates that zones of anomalously high gold concentration are spatially associated with these vugs. They occur in late structures and the gold mineralization in them has a completely different character from that found in the majority of the deposit. The latter is in the form of finely dispersed native gold grains associated with pyrite and chalcopyrite whereas the former contains nuggets of varying dimensions associated with quartz and carbonate fillings.

Based on structure, lithology, fluid inclusion and isotope data of these deposits it is postulated that the formation of nugget type gold occurred after the initial mineralizing period. The fluids in the vugs, which are comparble to the "deep brines" in their chemical and isotopic composition, are modified formation waters that infiltrated from the Proterozoic sedimentary basin during the stages of shearing.

This study also suggests the possibility that some of the other "deep brines" reported in the Canadian Shield may have similar origins and that the correlation between these brines and gold enrichment needs more investigation.

Résumé

L'évaluation systématique des poches de fluides renfermant du $CaCl_2$, du NaCl et du CH_4 compris dans les amas minéralisés de la mine Copper Rand, dans la région de Chibougamau, indique que la localisation des concentrations d'or anormalement élevées est reliée à ces poches. Ces dernieres se trouvent dans des structures tardives et la minéralisation aurifère qu'elles renferment diffère totalement du type le plus répandu de la mine. Celui-ci se présente sous la forme de grains d'or natif finement disséminés associés à la pyrite et à la chalcopyrite tandis que le premier type renferment des pépites de dimension variable associées aux quartz et de carbonate.

En se basant sur la structure, la lithologie, les inclusions fluides et les données isotopiques de ces gîtes, il ressort que la formation de pépites serait postérieure au stade initial de minéralisation. Les fluides de ces poches, dont la composition chimique et isotopique s'apparente à celle des saumures de profondeur, sont des eaux de formation modifiées qui se sont échappées du bassin sédimentaire protérozoïque durant les étapes de cisaillement.

Cette étude soulève la possibilité d'une origine semblable pour d'autres saumures profondes du bouclier canadien et la pertinence d'une étude plus détaillée de la corrélation entre ces saumures et l'enrichissement en or.

INTRODUCTION

Interest in the calcium- and sodium-rich brines in the Chibougamau mining district (Fig. 1) dates back to the early sixties when brines were intercepted while drilling in the shear zone controlled Cu-Au orebodies hosted by the Dore Lake Complex. No record of analysis could be traced, but the fluids were remembered to be "salty". In the course of a long term fluid inclusion study undertaken by the first author since 1977, it became evident that these brines could be of some significance as the ore fluids were identified as coexisting sodium and calcium chloride and methane rich fluids. Attempts were made, therefore, to sample the brine, but not until 1979, could the first sample be collected. The sample was taken from a vug, exposed during mining operations. Subsequently more vugs were intercepted either during drilling or drifting and were carefully sampled under the supervision of the second author. It is noteworthy that all the "strikes" not only showed the presence of methane but also that the "vein structures" associated with these vugs showed consistently high gold values. In the Copper Rand Mine two areas have been discovered so far where the gold is in the form of "nuggets", in close association with vugs containing brine and methane. This paper not only sums up preliminary ideas on the relation between high gold bearing areas and the presence of the brines but also underlines the fact that fluid inclusion data from the deposit in which the brines are located may be an important key to the understanding of the evolution of similar brines within the Canadian Shield. More detail on the characteristics and geological implications of the brine, including that from sites in other mines, will be given separately (Guha et al. in preparation).

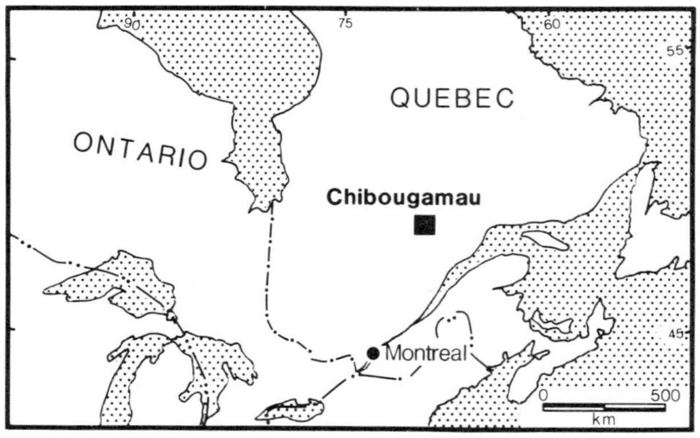

Figure 1. *Location of the Chibougamau mining district.*

REGIONAL SETTING OF THE COPPER RAND TYPE CU-AU OREBODIES

The rocks of the area belong to the Chibougamau-Matagami Archean greenstone belt and comprise volcanic and volcano-sedimentary sequences with mafic to ultramafic sills and felsic intrusions (Fig. 2). The Archean rocks are overlain by "patches" of sedimentary units of the Proterzoic Chibougamau Formation and are intruded by dykes of Proterozoic age (Chown, 1984).

The tectonic evolution of the area can be summed up as follows: at the beginning of the Kenoran Orogeny an incipient north-south folding was followed by the major east-west folding and east-west trending regional thrust faults. This was followed by major NE trending shear belts exemplified by the Gwillim Shear Belt which is late Kenoran in age (Dimroth et al., 1984). The shear zones within the Dore Lake Complex now appear to have been active in Proterozoic time. That is the shear zones can be related to the fault-shear structures affecting the Chibougamau Formation of Proterozoic age (Archambault et al. in preparation).

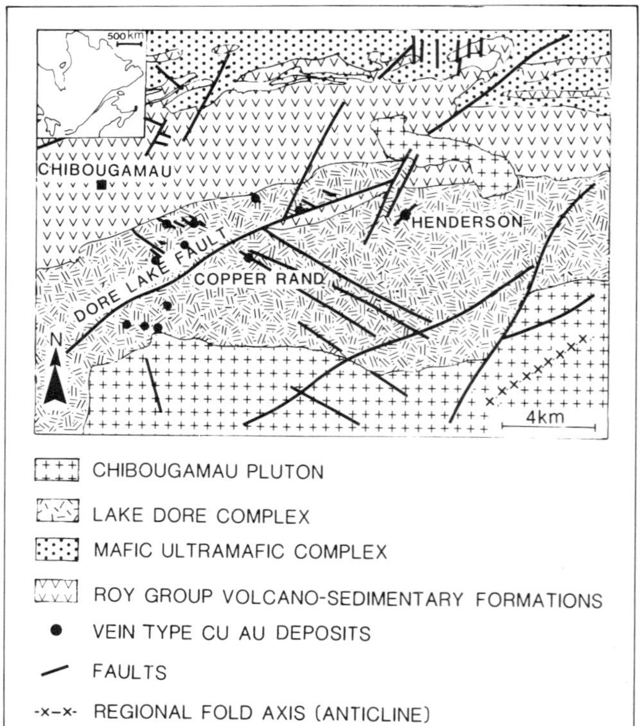

Figure 2. *A simplified geological map showing the vein type Cu-Au deposits in the Dore Lake complex. (Modified from Guha et al. (1983)).*

LOCAL SETTING AND CHARACTERISTICS OF THE COPPER RAND TYPE CU-AU OREBODIES

The host rock of these deposits is the meta-anorthosite unit of the Dore Lake Complex, a layered sill of mafic composition (Allard, 1973). The original composition of the anorthositic rock has been transformed into an albite-zoisite-chlorite assemblage by regional metamorphism.

The orebodies, which are lensoid in nature, are contained in shear zones striking NW or NE with variable dips. The immediate host rocks are quartz-sericite-chlorite-carbonate schists. The principal sulphide minerals are pyrite, pyrrhotite, chalcopyrite and sphalerite with a multitude of accessory minerals. Gold is found as discrete grains either associated with pyrite or chalcopyrite.

The ore fluids were introduced during the shearing (Guha et al. 1979) and the geometry of the orebody and the physical nature of the ore has been influenced by the evolution of the shear zones. The pattern is set by the relative movements of the R, P and D shears the overriding of which created the dilation and compression zones (Guha et al., 1983; Archambault et al., 1984). The ore types occupying these zones have been identified as the sulphide schist (Guha and Koo, 1975) and dilation type (Guha, 1984).

The age of the ore formation has been shown to be Proterozoic by Pb isotope modelling (Thorpe et al. 1984). Guha (1984) proposed that if the modelling of the Pb isotope data is geologically verifiable, then the ore fluids could be derived from the formation waters of Proterozoic age. The recent interpretation that the shear zones were produced during the Proterozoic, and the modelling of oxygen and hydrogen isotope data of fluids from these deposits (Tremblay, M.Sc. thesis in preparation; Tremblay et al., 1985) suggests formation waters as the source of the fluid.

"Nugget type" and high anomalous gold zones

Anomalous high gold zones were found in close proximity to the vugs containing brine and methane in several locations. These zones are in structures which can be shown to be late in the evolution of the shearing. The characteristic feature of these high gold zones is the phenomenal concentration of visible gold which occurs along with quartz, carbonate and chlorite in very limited areas. This is contrary to the normal gold occurrence in the ore lenses where the gold is rarely visible. To differentiate from the normal gold occurrence the term "nugget" gold has been used even though the masses of visible gold do not show the rounded

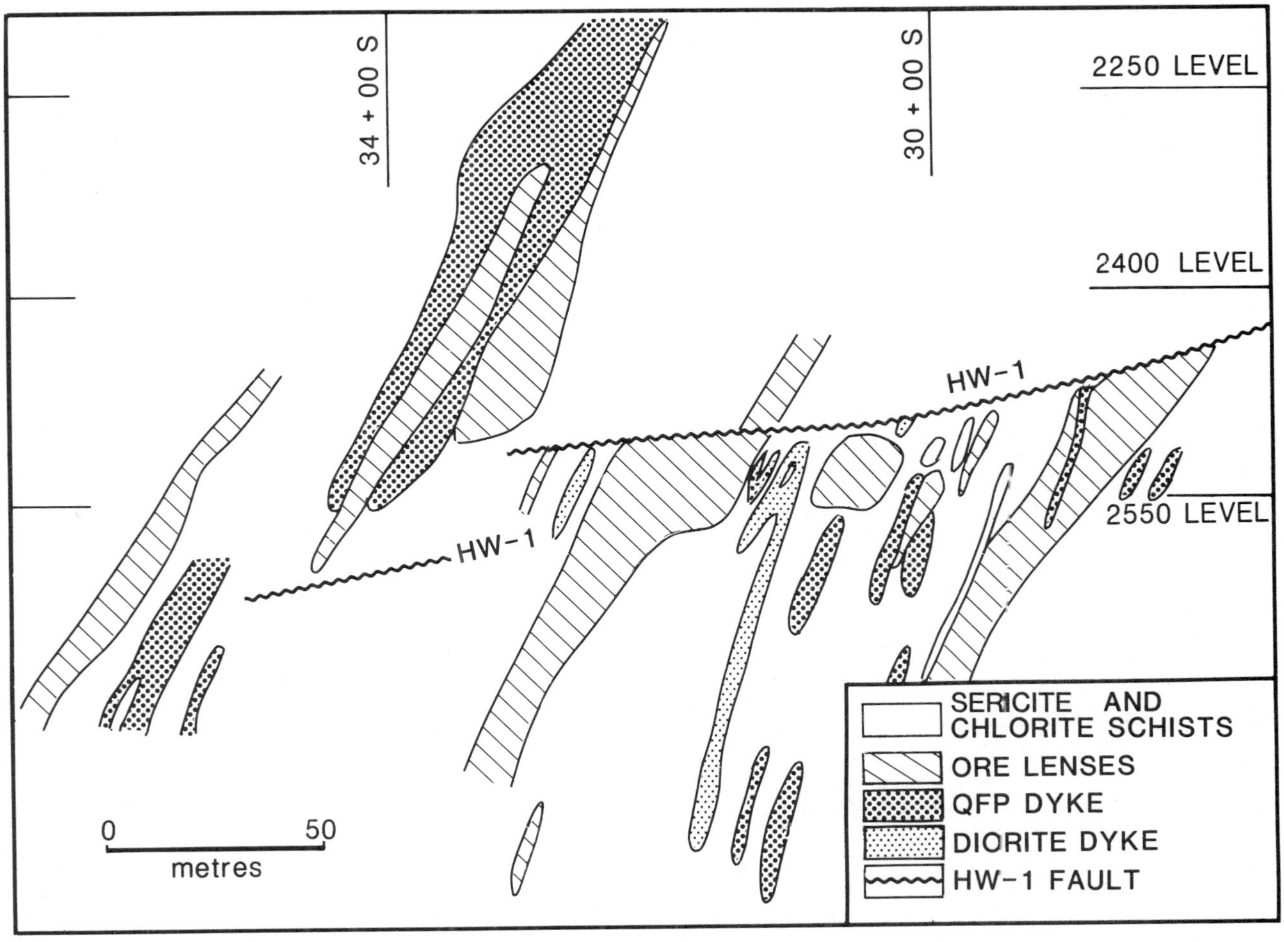

Figure 3. *A simplifed section (52 + 00E) showing the HW-1 Fault, Copper Rand Mine.*

Table 1

Compositions of vug brines from the Copper Rand Mine (CRV) and brines from other localities. Michigan (White et al. 1963), Sudbury and Yellowknife (Frape et al. 1984). Concentrations in mg l^{-1}.

	Na^+	K^+	Ca^{++}	Mg^{++}	Cl^-	SO_4^-	TDS	$\delta^{18}O\%$ (SMOW)	$\delta D\%$ (SMOW)
Michigan	11900	38	62900	179	128000	88(?)	203105		
Sudbury	16880	122	65000	12	156000	138(?)	238152	−11.3	−33
Yellowknife	326000	495	57300	920	142000		233315	−14.4	−71
CRV 1	51045	219	67576	2267	204776	170	326053	−16.9	−50
CRV 2	49170	219	65405	4818	206547	135	326294	−17.0	−48
CRV 3	16563	76	127382	1653	260857	19	406550	−14.6	−43

appearance of a typical nugget. Newberry (1881 in Liversidge, 1893) cites the occurrence of a mass of vein gold weighing ninety-five and a half pounds (43 kilos), in the form of nugget from the Monumental Mine, Sierra Buttes, California. So the use of the term is not without precedent. One nugget in the Copper Rand Mine yielded 5406 gms of gold.

One of the late structures that can be clearly identified contained the most important strike of nugget gold. It is the HW-1 fault (Fig. 3). This fault has been defined by underground openings and diamond drilling and has been traced between the 1910 and 2700 levels for at least 2000 metres down dip and approximately 1000 metres along strike. The fault strikes roughly in a north-south direction and dips 25-35° towards the west. This general attitude of the fault does not vary greatly except west of section 51 + 00 E where it appears to flatten to a near horizontal position. Several ore lenses (53-3, 54-3, 56-3, 56-4, 51-5, 52-5) are definitely truncated by the fault. However, the true displacement along the HW-1 fault is not known. The HW-1 fault in turn is displaced by steeply dipping cross faults. A contour of the HW-1 fault surface indicate a gently undulating plane having a series of synclinal and anticlinal axes oriented roughly east-west. This "folding" of the fault plane may be due to the offsetting of the HW-1 fault by the east-west strike faults.

In underground workings the HW-1 fault can be easily recognized from its attitude and the crenulations on its fault plane. The crenulation is also displayed by minerals formed in the fault zone. The width of the fault zone may vary from a few centimetres to three quarters of a metre. The fault zone is characterized by tightly folded bands of fine grained blackish brown material consisting of chlorite and tourmaline, bands and stringers of quartz-carbonate and also bands of chlorite. Halite has also been observed within the fault zone occupying spaces between the other minerals.

Besides the nugget gold found in the HW-1 structure, high anomalous gold zones are localized in the 53-3, 54-3, and 52-5 lenses in the vicinity of the fault. Concentrations of gold five times the average lens value are not uncommon. However, this increase in concentration has not been found in other lenses cut by the HW-1 fault.

Table 2

Atomic ratios of vug brine and brines from other localities.

	K/Na	Ca/Na	Ca/Mg
Michigan	0.0019	3.04	213.23
Sudbury	$.4 \times 10^{-5}$	2.20	3285.5
Yellowknife	$.8 \times 10^{-6}$	1.00	37.78
Copper Rand Vugs Mean	0.0026	1.93	23.76

Vug-brines

Brine and methane filled vugs were intersected in the Hanging Wall Zone of the Copper Rand Mine at three places to date. Two vugs were intersected in the 21-41-6 stope between sections 45 + 00 E to 46 + 00 E and 28 + 80 S, approximately 10 metres above the 2100 level. The vugs were located in cross fractures within the vein system (striking N55W and dipping S35W) dipping 5-10° towards N70W. The third vug was located on 2850 level, near sections 57 + 00 E and 35 + 00 S. The vug structure dips 10-15° towards N70W and cuts across the vein. The dimensions of the vugs are variable. For instance the third vug measured 10 metres in length, 4-6 metres in width and 0.4 metres in height. When intercepted the discharge of the fluid continued for a few days and methane was detected by methanometer. The pH measured at the time of sampling was between 3 and 4.

The chemical composition of the vug fluids is given in Table 1 which shows that these fluids are effectively sodium- and calcium-rich brines. A quick comparison with the saline groundwater samples from Yellowknife and Sudbury (Frape et al., 1984) as well as the brine sample from the Centennial Mine seepage, Michigan (White et al., 1963) in Table 1 show an overall similarity. However, comparing some of the ratios of cations (Table 2), it can be seen that there are differences on an individual element basis.

The oxygen and hydrogen isotope data (analyses by S.M.F. Sheppard, C.R.P.G., Nancy, France) of the brines from the vugs plotted on Figure 4 also show a remarkable similarity with the saline groundwaters of Frape et al. (op. cit.). That is the δD values plot above the global meteoritic water line (GMWL) (Craig, 1961) and the $\delta^{18}O$ values plot to the left of the GMWL. The values are very similar to those of the Sudbury (North Mine).

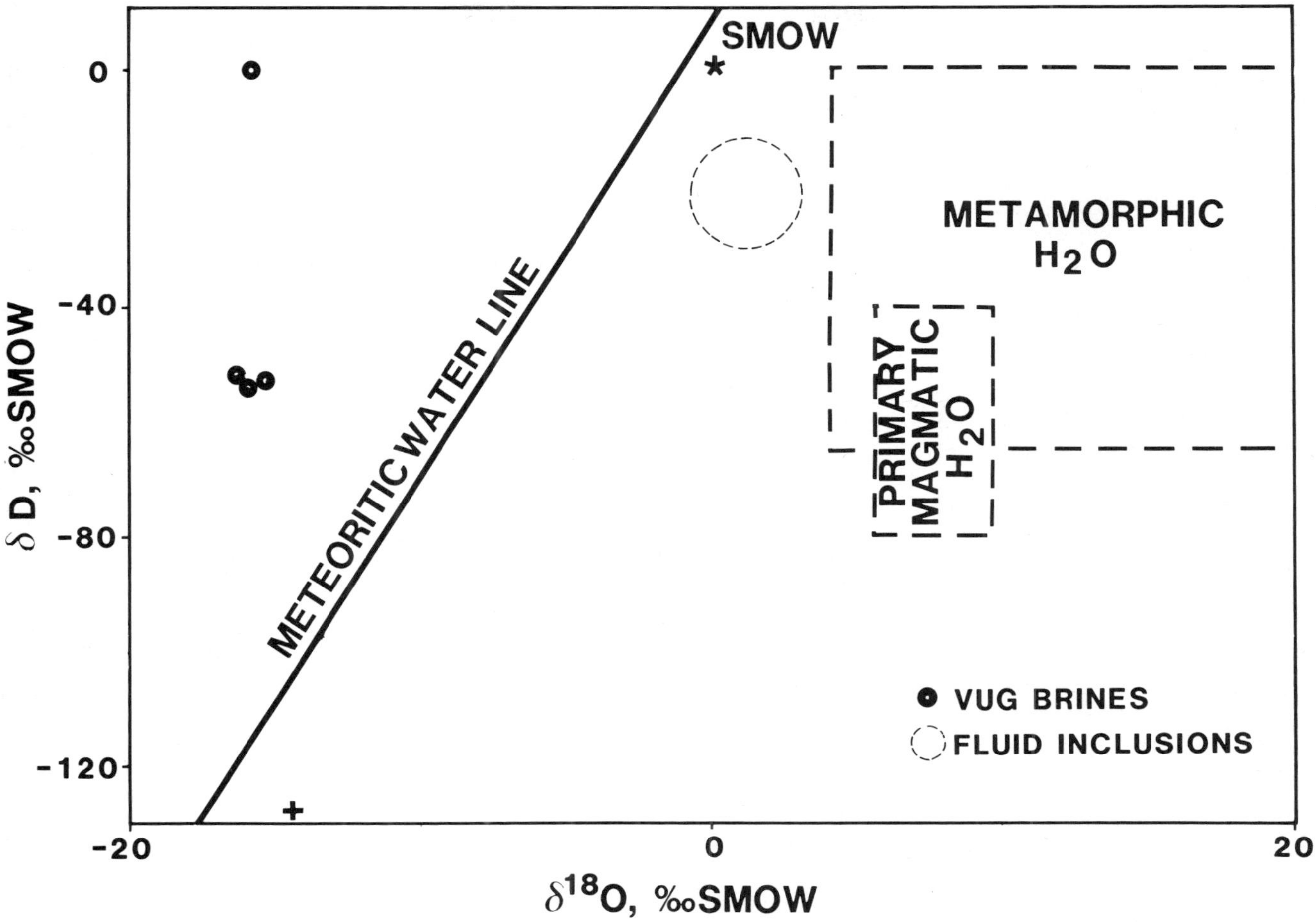

Figure 4. *A δD/δ¹⁸O plot showing the difference between the vug brines and the fluid inclusions.*

Fluid inclusion data

Fluid inclusion study (Guha et al., 1979, 1983; Guha, 1984) of the shear zone controlled Cu-Au orebodies in which the vugs occur indicates that the solutions from which the sulfide precipitated were aqueous chloride solutions highly charged with salts (essentially NaCl and $CaCl_2$) at the limit of NaCl saturation or oversaturated, and a coexisting methane rich fluid. The fluid inclusions occurring in quartz are of secondary type and they have been shown to be related to the Cu-Au mineralization. Analyses of dissolved ions indicate that the Na^+ and Ca^{+2} represent nearly the total quantity of dissolved cations in the fluid.

A calcite crystal within the vug showed the presence of similar fluid inclusions as in the orebodies (Guha et al. in preparation).

DISCUSSION

The basic assumption that the abnormal gold concentration is related to the vug brines can be demonstrated by the evidence presented earlier. The nugget and high anomalous gold zones have always been found in close association with vugs or related to structures that formed late in the history of shearing. This brings up the matter of the timing of formation of the vugs. As stated earlier the mechanism of evolution of the shear zone calls for a progressive overrriding of successive shear systems and conjugate faults. Therefore it is difficult to envisage sealing and conservation of the vugs through the period of major deformation. It is more plausible that the vugs that are now preserved formed late in the evolution of the shear zone. This is corroborated by the location of the vugs in cross structures within the vein.

The HW-1 fault has been shown to be later than the ore lenses which have been truncated by it. Also the HW-1 fault has been shown to be generated from D type shears which had repeated movements and displacements giving rise to the subhorizontal and undulating nature of its plane. The relative tightness of this fault makes it difficult for the formation and preservation of vugs inside the fault zone. However, the presence of halite within the fault zone indicates that the brine must have been present in the fault zone. The occurrence of nugget and high anomalous gold zones in some of the ore lenses associated with the HW-1 fault is therefore not a fortuitous relationship and that effectively the HW-1 fault plane was responsible for channelling the gold bearing fluid, which was highly saline.

From the above considerations it is evident that saline fluids were present during the later stages of the evolution of the shear zone. Traces of their existence either as halite in the fault planes or as trapped fluids in the vugs coincide with areas of high gold concentrations. The next question that arises is how the brines evolved.

What is known is that the vug brines were trapped during the stages of evolution of the shear zone which also localized the Cu-Au lenses. The ore fluids responsible for their emplacement were sodium- and calcium-chloride-rich fluid coexisting with methane introduced during the shearing (Guha et al., 1979, 1983). It is known from the age of formation of the ores, the timing of the shearing and the isotopic signatures of the ore fluids that the provenance of the fluid is the formation waters related to the Proterozoic sedimentary basin that covered the Archean rocks (Guha, 1984; Tremblay, M.Sc. thesis in preparation; Tremblay et al., 1985).

Since both the ores and the vugs are structurally controlled by the same shear system and both contain sodium- and calcium-chloride-rich fluids plus methane, it is logical to compare their chemistry. Direct comparison of absolute values is not possible since concentrations in fluid inclusions can only be examined in terms of cation ratios, therefore ratios of cations of the vug brines and fluid inclusions are shown in Table 3. From this table it can be seen that despite the resemblance of the overall signature (the ubiquitous presence of calcium chloride, sodium chloride and methane), the ratios of the cations of the vug brines and the ore fluids vary indicating that the vug brines have undergone changes. This variation between the ore fluids and the vug brines as well as between samples of the vug brine can be explained as the result of reactions between the fluid and the minerals with which they were reacting or forming since the shear zone was constantly evolving. It would be impossible to have identical cation ratios. However, the difference in the cation ratios between the two fluids necessitate that the vug fluid evolved chemically and it cannot be simply trapped "residual ore fluid". This is further evident from Figure 4 which shows a marked difference in the isotopic composition between these two fluids. The vug brines lie above and to the left of the global meteoritic water line as do the brines of Frape et al. (1984, Fig. 1), showing a specific isotopic evolution of these brine samples.

Table 3

Atomic ratios of fluid inclusions (Copper Rand-mean) and vug brines.

Samples	K/Na	Ca/Na	Ca/Mg	Σ Cations / Cl
Copper Rand	0.016	0.68	42.5	0.90
CRV 1	0.0025	0.740	17.64	0.96
CRV 2	0.0026	0.744	8.03	0.980
CRV 3	0.0026	4.304	45.6	0.99

As stated earlier, the ore fluids formed from the formation waters that were introduced syntectonically during Proterozoic time. Also the vugs therefore formed during the late stages of the evolution of the shear zone. This opens a few possibilities regarding the source of the vug brines. The vugs contain trapped residual ore fluids which evolved through exchange, or by the addition of unmodified formation waters (recharge), or a mixture of both processes.

A reconstruction of the scenario is as follows: during the shearing and emplacement of the orebody the formation waters entering the shear zone formed alteration minerals such as sericite and chlorite under relatively high water/rock ratios and a temperature of 250°C, as determined by fluid inclusion study. The alteration minerals will be slightly enriched in $\delta^{18}O$. With time, sealing of the system will begin due to the deposition of the gangue minerals such as quartz, thereby greatly reducing the intake of the brine and thus lowering the water/rock ratio. The temperature of the system decreases at the same time. The late stage movements reactivate and control the movement of this fluid which will be much more restricted than at the beginning. This fluid will form and react with alteration minerals along the fault planes through which they move. At lower temperatures and low water rock ratio the fluid would progressively be depleted in $\delta^{18}O$ with a possible enrichment of δD. This is reflected in the isotopic composition of the vug brines, making it plot to the left and above the GMWL. It is possible that the activation of these faults will draw in new basinal waters depending on the sedimentary cover remaining, but the entry of the water will be restricted because of prior sealing and only a limited flow can be evisaged. In any case, the low water/rock ratio and lower temperatures would control the isotopic exchange in the same manner as above. Recently Kelly et al. (1986) have proposed a similar brine evolution although they support a Paleozoic formational source for the Michigan and other shield areas.

Work is underway to study the precise controls and hydrodynamic pattern of fluid flow enabling the enrichment of gold. It is hoped that the key to the control of deposition will be extremely helpful in locating other pockets in the HW-1 fault plane and other later structures.

Brines and methane gas have been found in the gold mines of the Yellowknife area (Frape et al., 1984). It would be interesting to check whether higher gold values are associated with these finds. However, in order to do this, the precise location of the vugs from which the seepage occurs must be determined. It would also be interesting to consider whether these brines could be modified basinal waters.

CONCLUSIONS

Though the exact nature of the process of transport and deposition of the gold in the gold enriched zones is yet to be worked out, the present study demonstrates the following.

1. The evidences point to the fact that the brine which circulated through late structures such as the HW-1 fault and trapped in the vugs is responsible for the gold enrichment.
2. The brine in the vugs originated from chemically and isotopically evolved formation waters.
3. The brine shows a characteristic chemical and isotopic signature, very similar to the deep saline groundwaters described by Frape et al. (1984) from the Canadian Shield. This leads us to believe that evolved formation waters may explain the origin of some of the other occurrences.
4. And lastly it would be interesting to investigate whether brines found in the shear zone like Yellowknife could be responsible for similar gold enrichment.

ACKNOWLEDGEMENTS

The authors gratefully acknowledge the continuing cooperation of the management of the Northgate Patino Mines. The authors also acknowledge the analytical support of C.R.P.G. Nancy France and S.M.F. Sheppard who is collaborating on the overall project. This study was financed by the Natural Sciences and Engineering Research Council of Canada (Grant A 8374 to J. Guha), for which we are grateful.

REFERENCES

Allard, G.O., 1973. Some speculations regarding the lower hidder zone of the Dore Lake complex and its potential mineral resources: Geological Society of America Bulletin, v. 84, p. 717-724.

Archambault, G., Guha, J., Tremblay, A. and Kanwar, R., 1984. Implications of the geomechanical interpretation of the copper rand deposit on the Dore Lake shear belt: in Guha, J. and Chown, E.H. (eds.) Chibougamau-Stratigraphy and Mineralization, Canadian Institute of Mining and Metallurgy, Special Volume 34, p. 300-318.

Chown, E.H., 1984. Mineralization controls in the aphebian formations, Chibougamau, Mistassini and Otish areas: in Guha, J. and Chown. E.H. (eds.) Chibougamau-Stratigraphy and mineralization, Canadian Institute of Mining and Metallurgy, Special Volume 34, p. 229-243.

Craig, H., 1961. Isotopic variations in meteoric waters: Science, v. 133, p. 1702-1703.

Dimroth, E., Archambault, G., Goulet, N., Guha, J. and Mueller, W., 1984. A mechanical analysis of the late Archean Gwillim Lake shear belt, Chibougamau area, Quebec: Canadian Journal of Earth Sciences, v. 21, p. 963-968.

Frape, S.K., Fritz, P. and McNutt, R.H., 1984. Water-rock interaction and chemistry of groundwaters from the Canadian shield: Geochimica et Cosmochimica Acta, v. 48, p. 1617-1627.

Guha, J., 1984. Hydrothermal systems and correlations of mineral deposits in the Chibougamau mining district – an overview: In Guha, J. and Chown, E.H. (eds.) Chibougamau-Stratigraphy and Mineralization, Canadian Institute of Mining and Metallurgy, Special Volume 34, p. 517-534.

Guha, J., Archambault, G. and Leroy, J., 1983. A correlation between the evolution of mineralizing fluids and the geomechanical development of a shear zone as illustrated by the Henderson 2 Mine, Quebec: Economic Geology, v. 78, p. 1605-1618.

Guha, J. and Koo, J., 1975. Role of fluid state mobilization during metamorphism of the Henderson orebodies, Chibougamau, Quebec, Canada: Canadian Journal of Earth Sciences, v. 12, p. 1516-1523.

Guha, J., Leroy, J. and Guha, D., 1979. Significance of fluid phases associated with shear zone Cu-Au mineralization in the Dore Lake complex, Chibougamau, Quebec: Bulletin Minéralogie, v. 102, p. 569-576.

Kelly, W.C., Rye, R.O. and Livnat, A. (1986). Saline minewaters of the Keweenaw Peninsula, Northern Michigan: Their nature, origin and relation to similar deep waters in Precambrian crystalline rocks of the Canadian Shield: American Journal of Science, v. 286, p. 281-308.

Liversidge, A., 1983. On the origin of gold nuggets: Royal Society of New South Wales Journal and Proceedings, v. 27, p. 303-343.

Thorpe, R.I., Guha, J., Franklin, J.M. and Loveridge, W.D., 1984. Use of a superior province lead Isotope framework in interpreting mineralization stages in the Chibougamau district: In Guha, J. and Chown, E.H. (eds.), Chibougamau-Stratigraphy and Mineralization. Canadian Institute of Mining and Metallurgy, Special Volume 34, p. 496-516.

Tremblay, F., Guha, J., Sheppard, S.M.F., 1985. Reconnaissance en géochimie isotopique de l'oxygène et de l'hydrogène sur des gisements d'âge précambrien de la région de Chibougamau (résumé). L'Association Canadienne Française pour l'Avancement des Sciences, Annales, v. 52-53, p. 336.

White, D.E., Hem, J.D. and Waring, G.A., 1963. Geochemical compositions of subsurface waters *in* Data of Geochemistry: U.S. Geological Survey Professional Paper 440-F, 65 p.

Brackish and Saline Groundwaters in Finland

P.W. Lahermo,
Geological Survey of Finland, Kivimiehentie 1, 02150 Espoo, Finland

P.H. Lampén,
Imatran Voima Oy (Imatra Power Company Ltd), P.O. Box 138, 00101 Helsinki, Finland

Lahermo, P.W. and Lampén, P.H., Brackish and saline groundwaters in Finland; in Saline water and gases in crystalline rocks, Editors: Fritz, P., and Frape, S.K.; Geological Association of Canada Special Paper 33, 1987. p. 103–109.

Abstract

Brackish and saline groundwater occurrences in Finland are discussed on the basis of their geographical distribution and the results of chemical, gas and isotopic analyses. Most of the wells with brackish or saline water are situated in the coastal belt below the highest shoreline of the postglacial sea. The salinity is attributed primarily to relict salts of the more saline Litorina sea stage (c. 7000 years ago) of the Baltic Sea. New observations are reported on the geochemistry of deep saline inland groundwaters, suggesting that the character of these old groundwaters was affected mainly by long-lasting water-rock interactions. Some results of radio-isotope analyses show that in coastal areas too, deep groundwaters are old and were modified by a long history of geochemical evolution.

Résumé

Les eaux souterraines saumâtres et salées trouvées en Finlande sont discutées d'après leur distribution géographique et les résultats des analyses chimiques et isotopiques. La plupart des souraces d'eau saumâtre ou d'eau salée sont situées dans la zone littorale au-dessous du niveau le plus éléve (il y a à peu près 7 000 ans) de l'ancienne mer Litorina. On suppose que la période de la mer Litorina, ayant été une période plus salée de la mer Baltique, soit la raison principale de la salinité. Il existe quelques observations sur la géochimie des eaux souterraines salées et profondes situées a l'interieur du pays finlandais, qui indiquent que surtout les interactions de longue durée entre l'eau et la roche eu un effet sur le caractère de ces eaux souterraines anciennes. Quelques résultats des analyses isotopiques indiquent que aussi dans les zones littorales les eaux souterraines profondes sont très ansiennes et à part l'influence ansienne et récente des mers, elles ont été modifiées par une evolution géochimique de longue durée.

INTRODUCTION

The purpose of this paper is to give a short review of the research made into the occurrence of brackish and, in a few cases, saline groundwaters and their geochemistry in Finland. A few studies on fresh and brackish groundwaters in the overburden and Precambrian bedrock in Finland have been published by Salmi (1963), Hyyppä (1973), Lahermo (1971), Rönkä et al. (1980) and Rönkä (1983). The radioactivity of groundwater and especially its radon content have been discussed by Kahlos et al. (1973), Asikainen et al. (1979) and Asikainen (1981a and 1981b). The most recent studies concerned investigations dealing with the disposal of nuclear waste into bedrock Hyyppä (1984a and 1984b), Kankainen (1981 and 1982), and Lampén (1985). Much of the information given in this paper has been gathered from the papers referred to above and from unpublished material of the Geological Survey of Finland.

Table 1.

Analyses data on different kinds of groundwater in Finland.

TYPE OF WATER	Number of analyses	pH	Electrical conductivity mS/m, 25°C	Cl mg/l	Br mg/l	HCO_3 mg/l	SO_4 mg/l	SiO_2 mg/l	Na mg/l	K mg/l	Ca mg/l	Mg mg/l	Na/Cl	K/Cl	Ca/Cl	Mg/Cl	Ca/Mg	Ca/HCO_3	Cl/SO_4	Ref. no.
PRECIPITATION																				
Jokioinen, 1971-84	150-168	4.8	3.5	0.8	n.d.	n.d.	4.4	n.d.	0.5	0.4	0.8	0.1	0.57	0.45	0.93	0.17	5.6	n.d.	0.2	1
Sipoo, 1971-84	152-168	5.0	3.5	1.0	n.d.	n.d.	5.2	n.d.	0.5	0.6	1.1	0.2	0.56	0.60	1.15	0.20	5.8	n.d.	0.2	1
FRESH GROUNDWATER																				
Quaternary deposits:																				
Natural springs	420-570	6.6	3.7	2	n.d.	24	3.6	11	1.5	0.5	5	1.5	0.75	0.25	2.50	0.75	3.3	0.21	0.6	2
Domestic wells	3550	6.4	19	9	n.d.	46	13	11	7	5	17	4	0.78	0.56	1.89	0.44	4.3	0.4	0.7	2
Bedrock																				
Drilled wells	1021	6.7	24	13	n.d.	74	11	13	9	4	20	6	0.69	0.31	1.54	0.46	3.3	0.3	1.1	3
Lavia, 907 m, 1984	1	8.9	17	14	0.1	76	0.08	5	23	2.7	8	1.3	1.64	0.19	0.57	0.09	6.4	0.1	175	4
BRACKISH/SALINE GROUNDWATER																				
Coastal regions:																				
Jurmo (island)	1	7.0	n.d.	880	n.d.	201	150	12	420	18	180	51	0.48	0.02	0.20	0.06	3.5	0.9	5.9	5
Hiittinen (island)	1	7.1	840	3250	n.d.	37	171	12	n.d.	n.d.	887	130	—	—	0.27	0.04	6.8	24	19	6
Olkiluoto, YD1, 188 m, 1981	1	8.3	340	980	n.d.	18	130	21	510	14	110	30	0.52	0.01	0.11	0.03	3.7	6.1	7.5	7
Hästholmen, Y1, 193 m, 1983	1	7.7	1500	5040	18	92	647	9	2350	25	370	307	0.47	0.01	0.07	0.06	1.2	4.0	7.8	8
Mapping of brackish/saline wells																				
Well group III (Cl <500 mg/l)	43-91	7.6	144	297	n.d.	218	65	19	192	8.5	61	18	0.65	0.03	0.21	0.06	3.4	0.3	4.6	3
Well group II (Cl 500-1000 mg/l)	46-68	7.5	273	736	n.d.	187	94	14	344	8.6	150	31	0.47	0.01	0.20	0.04	4.8	0.8	7.8	3
Well group I (Cl >1000 mg/l)	56-90	7.3	710	2570	n.d.	157	149	12	920	18	544	156	0.36	0.01	0.21	0.06	3.5	3.5	17.3	3
Ylistaro, 600 m	1	n.d.	2018	8940	n.d.	114	15	n.d.	1900	11	350	2600	0.21	0.001	0.04	0.29	0.1	3.1	596	3
Pori, 600 m	1	8.2	323	1078	n.d.	87	47	n.d.	540	1.1	180	8.8	0.50	0.001	0.17	0.01	20.5	2.1	22.9	9,11
Eurajoki, flowing well, 32 m	1	6.9	1346	5390	n.d.	64	79	n.d.	1500	20	1800	120	0.28	0.004	0.33	0.02	15.0	28.1	68	9,11
Eura, 232 m	1	6.8	n.d.	3124	n.d.	104	85	n.d.	1010	35	784	80	0.32	0.011	0.25	0.03	9.8	7.6	37	9,10
Inland:																				
Urjala, 76 m	1	n.d.	614	2120	n.d.	n.d.	1.4	n.d.	450	n.d.	625	80	0.21	n.d.	0.29	0.04	7.8	n.d.	1514	11
Polvijärvi	1	6.8	n.d.	1875	n.d.	35	17	7	500	10	650	25	0.27	0.01	0.35	0.01	26	18.6	110	1
OKU 741 1250 m (vertical depth c. 1050 m)	1	8.6	3300	16800	n.d.	35	2.0	2.5	3200	32	5700	1100	0.19	0.19	0.34	0.07	5.2	163	8400	3,16
Kotalahti, mine, 700 + 136 m	1	7.8	3832	18000	n.d.	91	500	n.d.	4000	17	7500	22	0.22	0.001	0.42	0.01	341	82	36	5,9
Reference areas:																				
Nipigon, Ontario, Canada, seeps.	1	6.7	n.d.	5680	14.9	206	0.5	n.d.	2440	9.6	890	18	0.43	0.002	0.16	0.01	49	50	11360	12
Sudbury, Ontario	1	6.8	n.d.	156000	n.d.	31	138	n.d.	16880	122	65000	12	0.11	0.001	0.42	0.01	5417	2097	1130	13
Stripa, Sweden, V1 410-506 m	1	9.3	134	700	6.2	13	93	14	304	2.5	170	0.3	0.43	0.004	0.24	0.01	531	13.1	7.5	14
Stripa, V2, 492 m	1	9.9	122	460	5.1	18	55	15	224	0.7	89	0.4	0.49	0.002	0.19	0.01	228	4.9	8.3	14
BALTIC SEA																				
Hirvensalo, west coast	1	8.4	800	2944	n.d.	n.d.	100	1.5	1600	59	86	209	0.54	0.02	0.03	0.07	0.4	n.d.	29.4	3
Hästholmen, south coast, 19 m	1	7.5	770	2709	16	75	450	2.2	1440	73	58	172	0.53	0.03	0.02	0.06	0.3	0.8	6.02	8
OCEAN		n.d.	n.d.	19000	67	140	2650	7	10600	380	400	1270	0.56	0.02	0.02	0.07	0.3	2.9	7.2	15

n.d. = not determined

Sources of data: 1. Järvinen, et al. 1986; 2. Lahermo, 1970; 3. Geological Survey of Finland, unpublished material; 4. Lampén, 1985; 5. Hyyppä, 1985b; 6. Vesto Oy archives; 7. Suunnittelukeskus Oy, 1984; 8. Hyyppä 1984a; 9. Hyyppä, 1981; 10. Hyyppä, 1963; 11. Perttunen et al., 1984; 12. Frape et al. 1984b; 13. Frape et al., 1984; 14. Nordstrom et al., 1985, 15. Rankama and Sahama, 1950; 16. Nurmi and Kukkonen, 1986.

EXPERIMENTAL

The methods used for sampling groundwater have varied from simple scooping to advanced equipment consisting of inflatable double-straddle packers and a membrane pump. In simple technique for deep water and gas sampling recently developed at the Geological Survey of Finland (Nurmi et al., 1985b), a continuous column of water is raised inside a thin polyamide tube with a back-pressure valve at the lower end and with shut-off valves every 50 metres. The wide variety of sampling and preservation methods combined with the shortage of special chemical analyses or the fact that they have been performed by several different laboratories impedes interpretation of the information presented to some extent.

The analytical data collected from the sources listed in Table I are grouped by type of aquifer, salinity of the water and geographical location of the sampling sites. Most of the figures are medians of several samples or, when referring to one sample only, the actual analysis results. Because of insufficient data only conductivity, pH, and main cations and anions are given together with ion ratios for specific parameters.

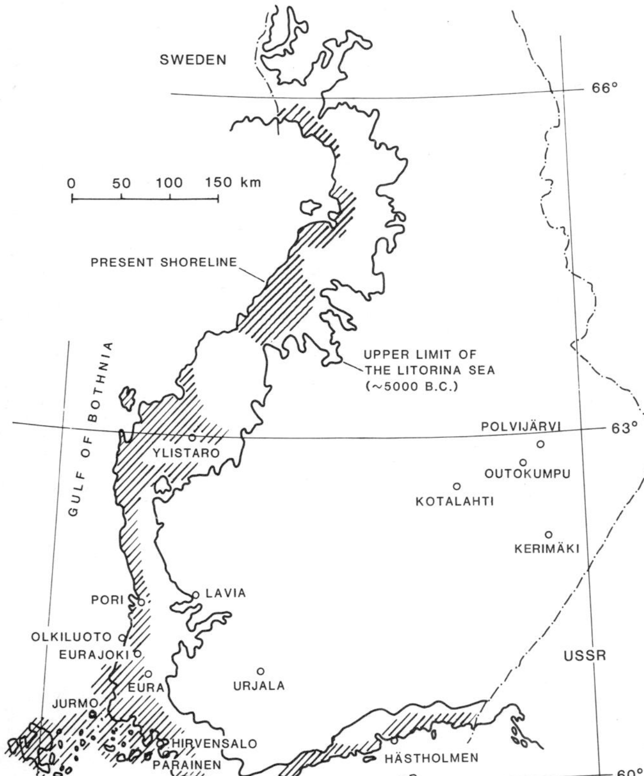

Figure 1. *The highest shoreline of the Litorina Sea some 7000 years ago. The shaded areas show the highest frequency of the occurrence of domestic brackish or saline wells drilled into bedrock. Some of the sampled well sites listed in Table I are presented on the map.*

BRACKISH AND SALINE GROUNDWATERS IN THE COASTAL BELT

The Litorina sea stage of the Baltic Sea covered a considerable part of the Finnish coastal region about 7500-5000 years ago (Figure 1). It has been estimated that the salinity of Litorina sea water was 5-10‰. (Eronen et al. 1979). This is considerably higher than that of the present-day Baltic Sea, which ranges from 4 to 6‰. (Perttilä et al. 1980, Pietikäinen et al. 1980). In the Helsinki region the highest Litorina shoreline is about 30 m above the present sea level, while in Ostrobothnia (Pohjanmaa), on the west coast of Finland, it is some 100 m above the level of the Gulf of Bothnia (Eronen 1983).

The shaded areas in Figure 1 represent the coastal areas where brackish or saline groundwaters have most often been observed in domestic drilled wells. Most of the saline wells are situated below the highest Litorina shoreline. Low-lying depressions filled with late or postglacial clay deposits have trapped the sea water into the underlying bedrock. The sluggish exchange of water has diminished the effect of dilution by a later, less saline, stage of the present Baltic Sea and by slowly circulating, fresh groundwater.

The thickness of the fresh water layer varies from some metres to several hundreds of metres the average thickness being from 20 to 60 metres. The change in salinity is generally abrupt (cf. Fig. 3). Although the probability of encountering highly saline water near the surface is greater on the coast than inland, there does not seem to be a clear-cut dependence between the depth of the saline water and distance from the sea. This is seen in several experimental holes drilled on the sites of the nuclear power plants of Hästholmen on the southeast coast and of Olkiluoto on the west coast, where layers of brackish or saline water are met with at quite different levels (40 to 120 m below sea level) within a small area (cf. Hyyppä 1984a, Novakowski 1984).

BRACKISH AND SALINE GROUNDWATERS IN INLAND AREAS

To date, only a few brackish or saline wells have been found inland. The following is a description of saline wells or diamond-drill holes in Urjala, Polvijärvi, Outokumpu and Kotalahti.

Two wells drilled into bedrock for domestic purposes at Urjala, far inland from the highest Litorina limit, have been reported to yield brackish water (Fig. 1, Perttunen et al. 1984). The results of chemical analyses of one of them are given in Table I. The well, which is 76 m deep, is situated in the lowest part of the clay-covered depression and surrounded by glacial till deposits and hillocks. The bedrock is composed of granodiorites, mica schists and amphibolites. The low topographic location of the mineralized water occurrence in relation to the environment and consequent stagnation has favoured its preservation and geochemical evolution for long periods.

The Quaternary deposits in the Kotalahti, Outokumpu and Polvijärvi areas, where deep drill holes were sampled (Fig. 1, Table I), comprise mainly hilly ground and hummocky moraine, whereas the depressions are filled with peat beds and water courses. The area belongs to the Outokumpu sulphide-ore rich belt composed of e.g. mica shists, skarns and quartz rocks, associated with serpentinites. The

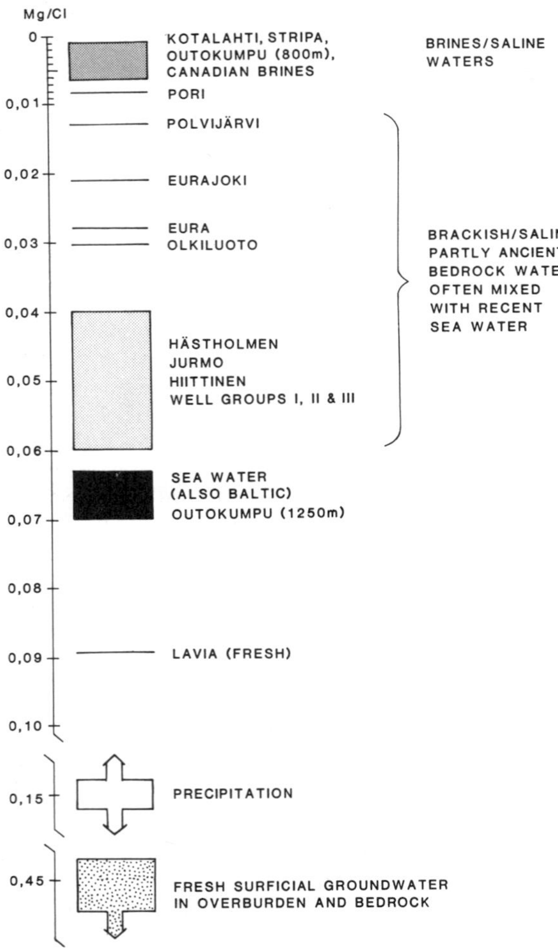

Figure 2. *Mg/Cl ion ratios in different types of groundwater and surface water in Finland. For sample descriptions see Table I.*

main ore minerals are pentlandite, pyrrhotite and chalcopyrite. The water samples discussed in the present context were collected from exploratory diamond-drill holes. At Outokumpu the saline groundwater begins at a depth of 300 metres below sea level (Fig. 3). The increased Mg contents below 800 metres is inconsistent with the occurrence of serpentinites at that level.

In an inland drill hole at Kerimäki, not listed in the table (see Nurmi et al. 1985), salinity increased sharply at a depth of more than 320 metres below the ground surface. At Lavia, (Fig. 1, Table I), however, no mineralized groundwater at all was encountered in a 1001-m-deep experimental hole drilled by the Industrial Power Company Ltd (TVO). For comparison, in a diamond-drill hole in the southwestern archipelago of Parainen (Fig. 1), salinity increased at a depth of 490 metres below the ground surface even though the site is close to the sea shore.

Data on a sample containing 18,000 mg/l of chloride and collected from a flowing drill hole in the Kotalahti mine are presented together with those on the samples from Urjala. Outokumpu and Polvijärvi and the reference material: saline seeps and shield brines from Ontario, Canada (Frape et al. 1984a, 1984b), and brackish water from the Stripa mine, Sweden (Nordstrom et al. 1985) (Table I).

According to the general concept of brines (>100 g/l TDS) no Finnish groundwater belongs to this group.

ORIGIN OF BRACKISH AND SALINE GROUNDWATER

Ion ratios

The chemical analysis of water can be interpreted in many ways, such as ion ratios, molar proportions and correlative interrelations of main components (e.g. Cl, SO_4, Na, Ca, Mg) or minor elements (e.g. Br, J, B, Sr) in evaluating the origin of water. In this brief account, only the Mg/Cl ratio is interpreted as one of the most useful parameters (Fig. 2) although some other most widely used ion ratios are presented in Table I.

The brackish and saline wells in the coastal belt, which are grouped into three categories according to the concentration of chloride (cf. Table I), form a separate group close to the water of the present-day Baltic in terms of their Mg/Cl ratios. The groundwater samples taken from the open sea islands of Jurmo and Hiittinen, together with those taken from Hästholmen island near the coast, also belong to this group. The Mg/Cl ratios of the groundwater samples from Urjala are in the same range, even though the Urjala well is lying above the ancient Litorina shoreline. Other inland drill holes yield water with a proportional composition similar to that of brines (Kotalahti, Polvijärvi) and some deep coastal wells (Pori). The ancient marine component is very low in the deep-seated fresh water from Lavia.

Gas and isotope analyses

The gases evacuated from Finnish groundwater have been sampled and analysed in only a few cases. Hyyppä (1981) has reported some results from drilled wells in Pori, Eurajoki and Eura on the west coast and from the Kotalahti mine, but he has not described the sampling methods. Some very recent results of gas analyses are available from the deep test hole at Lavia and the diamond drill hole at Outokumpu (cf. Nurmi et al. 1986). Table II gives the results of the gas analyses.

The sample from the Kotalahti mine and that from the Outokumpu drill hole contain a large amount of methane.

Table 2.

Chloride contents and some gases evacuated from medium or deep-seated groundwater in Finland.

sampling site	Cl mg/l	N_2	O_2	Ar	He vol %	CO_2	CH_4	C_2H_x
Pori	1078	96.0	1.7	1.5	0.7	0	0	0
Eurajoki	5390	92.7	2.0	1.1	3.7	0.7	0.4	0
Eura	3124	90.0	7.6	—	2.2	0.2	—	—
Kotalahti	18000	43.0	1.0	0.7	4.5	0	49.9	0.9
Outokumpu Perttilahti (OKU 741)	15000	14.7	2.7	0.2	n.d.	0.1	81.6	n.d.
Lavia	14	82.5	3.0		5.1	1.4	2.2	0

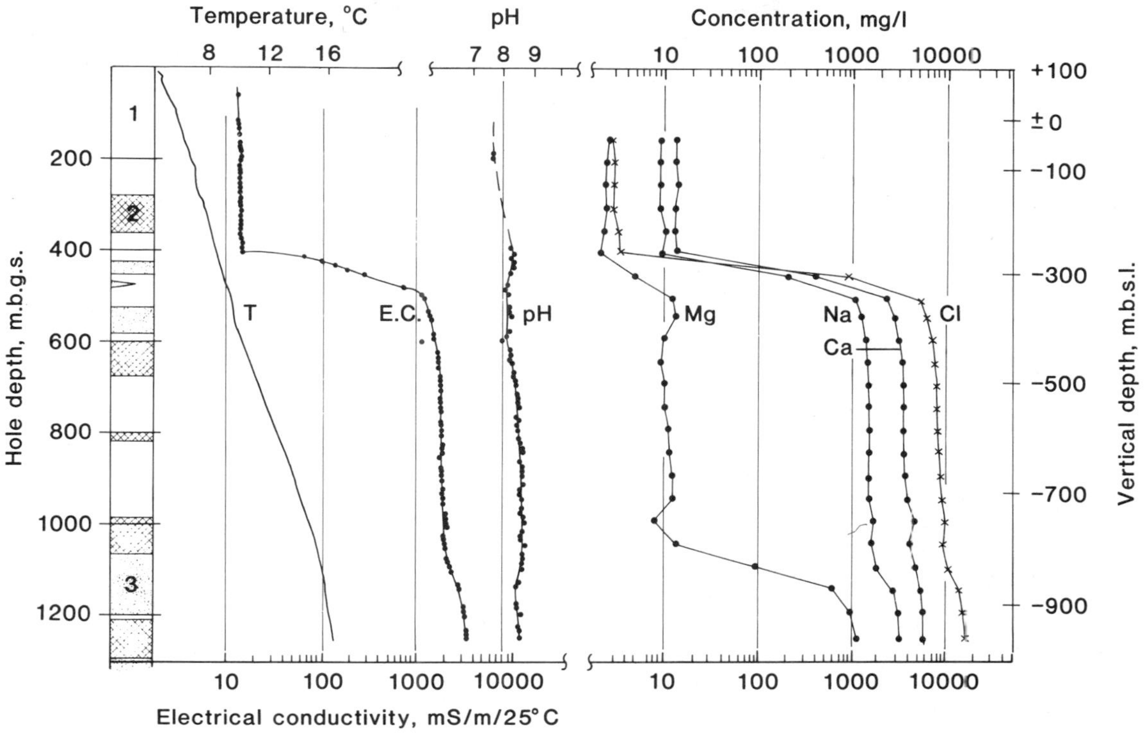

Figure 3. *Vertical distribution of temperature, electric conductance as well as some main components in deep diamond drill hole at Sukkulansalo, Outokumpu (OKU-741). Modified from Nurmi et al. (1986). Legend: 1. Mica gneiss; 2. Black schist; 3. Serpentinite, skarn, quartzite, dolomite.*

The helium contents must be considered with caution because of the ready leakage of helium from the sampling bags. For the Lavia test hole, noble gases dissolved in water were also analysed, but these results are not yet available.

Some groundwater samples have been analysed for various environmental isotopes, especially ^{14}C, ^{13}C, ^{18}O, ^{2}H and ^{3}H. Table III lists some isotope results from brackish and saline groundwaters (Hyyppä 1981, Kankainen 1981, 1983).

The amount of total uranium and the ratio of the ^{234}U and ^{238}U isotopes have been determined from the saline groundwater of Hästholmen (Hyyppä 1984a). The results are shown in Table IV.

Table 3.

Tritium and ^{14}C analyses from brackish and saline groundwaters.

sampling site	Cl mg/l	^{3}H TU	$\delta^{13}C_{PDB}$ %	mean residence time, years (according to Pearson's model)
Pori	1,078	<2.3	−18.5	18,900 – 20,800
Eura	3,124	40	n.d.	n.d.
Eurajoki	5,390	24	−14.1	10,100 – 11,500
Kotalahti	18,000	n.d.	(−25.0)	(gas) > 37,000
Hästholmen	4,759	<1.2	−14.3	10,420 – 11,650
Olkiluoto	980	47	−14.7	1,150 – 1,770

Most of the few $\delta^{18}O$ and $\delta^{2}H$ determinations fall near the meteoric water line except Urjala (−14.4‰, −90.4‰), Kotalahti (−12.9‰, −84.4‰) and Outokumpu (−13.0‰, −73.4‰), which fall to the left hand side (Nurmi et al. 1985).

The isotopic data obtained so far is insufficient for any regional evaluations to be made. It is evident, however, that some deep inland waters are very old (Kotalahti) as are also deep coastal groundwaters (Pori). The high ^{3}H values of most brackish saline groundwaters imply that they are mixed with recent fresh water and recent or relict sea water.

THE DEEP TEST HOLE AT LAVIA

A special test hole with a length of 1001 m was drilled at Lavia, western Finland, in the spring of 1984. This hole is used as a site for testing geological, geophysical and hydrogeochemical research methods as a part of the investigations to find a site for nuclear waste disposal.

Table 4.

Uranium analyses from Hästholmen, Loviisa.

sampling site	sampling depth, m	U (tot) μg/l	^{234}U / ^{238}U
Y 1	191.8 – 195.3	151	5.29 ± 0.17
Y 5	190.0 – 193.5	4.8	2.50 ± 0.25
Baltic Sea		1.2	1.30 ± 0.20
Ocean		3.3	1.13-1.15 ± 0.02

New groundwater sampling equipment comprising a membrane pump and an inflatable double packer system was developed for Lavia and similar deep drill holes (Rouhiainen et al. 1985). The equipment enables field measurements of conductivity, pH, Eh, pS, dissolved oxygen and temperature to be performed simultaneously. Groundwater samples have been taken in the autumn of 1984 and in the spring of 1985.

The results obtained so far indicate that the groundwater is very diluted for the whole length of the Lavia test hole (Table I). This hole is very important for future hydrogeochemical research into deep-seated bedrock groundwater in Finland by promoting the development of testing methods and the equipment required.

CONCLUSIONS

Complete chemical, gas and isotopic data on coastal and especially on inland brackish or saline groundwater are still scarce. However, present knowledge indicates that the samples of brackish or saline groundwater taken from areas below the ancient Litorina shoreline are characterized by higher ratios of Mg/Cl and lower ratios of Ca/Mg and Cl/SO_4 than those taken from inland. The isotope analyses show that in most cases brackish and saline groundwaters are the mixtures of different saline components of various ages and of fresh groundwaters. The samples taken from the Kotalahti mine and from the Outokumpu drill hole show the effects of the special environment of the bedrock geology.

In summary, most coastal brackish and saline groundwaters still bear the overprint of ancient sea-water components whereas deep-seated coastal groundwaters and inland groundwaters more clearly reflect long-lasting water-rock interaction. Therefore there are inevitably no striking differences between the chemical character of deep groundwater in coastal regions and that inland.

A research project into the geochemistry of deep-seated bedrock groundwater has just started, and a new, simple sampling method has been developed (cf. Nurmi and Kukkonen, 1986) that enables scores of exploratory diamond drill holes more than 1,000 m deep all over Finland to be sampled. In the addition the research into and development of methods being carried out by the Industrial Power Company Ltd and the Imatra Power Company Ltd will soon help us to piece together a more comprehensive picture of the nature of the geochemistry of deep groundwater and their genesis.

REFERENCES

Asikainen, M. and Kahlos, H., 1979. Anomalously high concentrations of uranium, radium and radon in water from drilled wells in the Helsinki region: Geochimica et Cosmochimica Acta, v. 43, p. 1681-1686.

Asikainen, M., 1981a. State of disequilibrium between ^{238}U, ^{234}U, ^{226}Ra and ^{222}Ra in groundwater from bedrock: Geochimica et Cosmochimica Acta, v. 45, p. 201-206.

Asikainen, M. 1981b. Radium content and $^{226}Rn/^{228}R$ activity ratio in groundwater from bedrock: Geochimica et Cosmochimica Acta, v. 45, p. 1375-1381.

Eronen, M., Uusinoka, R. and Jungner, H., 1979. ^{14}C-ajoitet tu sinisimpukkalöytö Seinäjoelta ja tietoja muista kuoriesiintymistä Itämeren piirissä (English summary: ^{14}C dated subfossil find of Mytilus edulis L. in Seinäjoki, western Finland, and discussion on some other shell deposits in the Baltic area): Terra 91, v. 4, p. 209-218.

Eronen, M., 1983. Late Weichselian and Holocene shore displacement in Finland: *in* Smith, D.E. and Dawson, A.G., ed., Shorelines and Isostacy: Institute of British Geographers. Special Publication 16. Academic Press.

Frape, S.K., Fritz, P. and McNutt, R.H., 1984a. Water-rock interaction and chemistry of groundwaters from the Canadian Shield: Geochimica et Cosmochimica Acta, v. 48, p. 1617-1627.

Frape, S.K., Fritz, P., Blackmer, A.J., 1984b. Saline groundwater discharges from crystalline rocks near Thunder Bay, Ontario, Canada: *in* Eriksson, E., ed., Hydrochemical Balances of Freshwater Systems: IAHS Publication No. 150, p. 369-379.

Grasshoff, K. and Voipio, A., 1981. Chemical oceanography: *in* Voipio, A., ed., The Baltic Sea. Elsevier Oceanography Series, 30. Amsterdam, p. 184-218.

Hyyppä, J., 1963. Eurassa kallioperään poratun syväkaivon suolaisen veden koostumuksesta: Geologi, v. 6, p. 61-63.

Hyyppä, J., 1973. Maamme pohjavesien geokemiasta. English summary: On Finnish groundwater geochemistry: Papers of the Engineering-Geological Society of Finland, v. 8, 60, p. 1-3.

Hyyppä, J., 1981. Geologisten erityisolojen vaikutus pohjaveden laatuun. Vesihallituksen ja geologisen tutkimuslaitoksen pohjavesisymposiumi 15.-16.10. 1981: Vesihallituksen monistesarja 91, p. 117-130.

Hyyppä, J., 1984a. Loviisan Hästholmenin kallioperän pohjavesien geokemiasta. (English summary: Geochemistry of the groundwaters of the bedrock on Hästholmen, Loviisa): Report YJT-84-05, 93 p., appendices.

Hyyppä, J., 1984b. Pohjaveden kemiallinen koostumus Suomen kallioperässä. (English summary: Chemical composition of groundwater in the bedrock of Finland: Report YJT-84-10, 69 p, appendices.

Järvinen, O. and Haapala, K. 1980. Sadeveden laatu Suomessa 1971-1977. (English summary: The quality of wet and dry deposition in Finland according to observations made from 1971 to 1977): National Board of Waters, Finland, Report 198.

Järvinen, O., 1986. Unpublished material.

Kahlos, H. and Asikainen, M., 1973. Natural radioactivity of groundwater in the Helsinki area: Institute of Radiation Physics, Report SFLA-19.

Kankainen, T., 1981. Eurajoen Olkiuodon Ulkopäänniemen kairausreikien YD1 ja YD2 pohjaveden iänmääritykset. Geologinen tutkimuslaitos.

Kankainen, T., 1983. Väliraportti Loviisan Hästholmenin pohjaveden isotooppianalyyseistä (unpublished report).

Lahermo, P., 1970. Chemical geology of ground- and surface waters in Finnish Lapland: Bulletin de la Commission Géologique de Finlande, v. 242, 106 p.

Lahermo, P., 1971. On the hydrogeology of the coastal region of south-eastern Finland: Geological Survey of Finland, Bulletin 252, 44 p.

Lampén, P., 1985. Koereiän pohjavesinäytteiden kenttämittaukset ja fysikaalis-kemiallisten analyysien tulokset syksyllä 1984. TVO/koereikä, työraportti 85-03.

Lindeval, H., 1985. Salt grundvatten i Sverige: Sveriges Geologiska Undersökning, Raportten och meddelanden 39, 69 p, appendices.

Nordstrom, D.K., Andrews, J.N., Carlsson, L., Fontes, J-C., Fritz, P., Moser, H. and Olsson, T., 1985. Hydrogeological and hydrogeochemical investigations in boreholes – Final report of the phase 1, geochemical investigations of the Stripa groundwaters. SKB Stripa Project Report 85-06, 250 p, Stockholm.

Novakowski, C. and Lindberg, A., 1984. Hydrogeological conditions in the coastal zone of southwestern Finland with respect to the disposal of radioactive waste: Geological Survey of Finland: Report of Investigation 66, 21 p.

Nurmi, P., Kankainen, T. and Kukkonen, O., (1985). Lavian koereijän pohjavesitutkimukset syksyllä 1984 – Isotooppi-ja kaasuanalyysit sekä geokemiallisten, termisten ja hydraulisten tulosten alustava arviointi: Geologian tutkimuskeskus, Ydinjätteiden sijoitustutkimukset. Työraportti 85-07, 26 p, appendices.

Nurmi, P., Kukkonen, I., Lahermo, P., Salmi, M. and Rahkola, P. 1985b. Kalliopohjavesien geokemia: Alustavat tulokset syvistä kairareí istä Outokummun, Kerimäen, Paraisten ja Limingan alueelta. Abstract: Geochemistry of groundwater in bedrock: Preliminary results of groundwater sampling in deep drill holes at Outokumpu, Kerimäki, Parainen and Liminka. Geological Survey of Finland, Nuclear Waste Disposal Research. Report YST-50. 25 p, appendices.

Nurmi, P. and Kukkonen, I., 1986. A new technique for sampling water and gas from deep drill holes: Canadian Journal of Earth Sciences, v. 23, in press.

Perttilä, M., Tulkki, P. and Pietikäinen, S., 1980. Mean values and trends of hydrographical and chemical properties in the Gulf of Finland 1962-1978: Finnish Marine Research, v. 247, p. 38-50.

Perttunen, M., Sten, C-G., Hyyppä, J. and Grönlund, T., 1984. Toijalan kartta-alueen maaperä. (English summary: Quaternary deposits in the Toijala map-sheet area. Geological map of Finland 1:100 000, Explanation to the Maps of Quaternary Deposits, Sheet 2114, 61 p.

Pietikäinen, S., Niemi, Å., Tulkki, P. and Aurimaa, K., 1978. Mean values and trends of physical and chemical properties in the Gulf of Bothnia 1962-1975: Finnish Marine Research No. 244, p. 64-75.

Rankama, K. and Sahama, T.S., 1950. Geochemistry. University of Chicago Press, 911 p.

Rouhiainen, P. and Lampén, P., 1985. Vesinäytelaite. Tekninen kuvaus ja käyttöohje. TVO/koereikä, työraportti 85-01.

Rönkä, E., Uusinoka, R. and Vuorinen, A., 1980. Geochemistry of groundwater in the Precambrian crystalline bedrock of Finland in relation to the chemical composition of the reservoir rocks: Publications of the Water Research Institute 38, p. 41-53.

Rönkä, E., 1983. Drilled wells and groundwater in the Precambrian crystalline bedrock of Finland: Publications of the Water Research Institute 52, 57 p.

Salmi, M., 1963. Suolaista vettä kallio-kaivosta Seinäjoella. Summary: Saline water from a well drilled into bedrock at Seinäjoki. Geologi, v. 6, p. 56-60.

Suunnittelukeskus, Oy. 1984. Teollisuuden Voima Oy, Olkiluodon pohjavesitutkimusten vuosien 1980-1982 analyysitulosten käsittely. Työ 1621/2883. Unpublished report.

Fluid Inclusions as a Source of Dissolved Salts in Deep Granitic Groundwaters

Darrell Kirk Nordstrom
U.S. Geological Survey, 345 Middlefield Road, Menlo Park, California 94025, U.S.A.

Tommy Olsson
Geosystem AB, Box 8020, S-75008 Uppsala, Sweden

Nordstrom, D.K. and Olsson, T., Fluid inclusions as a source of dissolved salts in deep granitic groundwaters; in Saline water and gases in crystalline rocks, Editors: Fritz, P., and Frape, S.K.; Geological Association of Canada Special Paper 33, 1987. p. 111–119.

Abstract

The feasibility of fluid inclusions leaking into local groundwater systems from granitic rocks is examined from several points of view: (1) Is there enough salt present? (2) Does this salt contain a chemical signature that can help to distinguish it from other sources? (3) How much of this salt is water soluble? (4) What is the maximum amount of chloride that might be contributed from this source under optimal conditions? and (5) What mechanisms might exist for fluid inclusion leakage? The answers to (1) and (2) are affirmative, with approximately 50% of the total chloride being water-soluble (on the average). A maximum possible chloride concentration of 20,000 mg/L can be attained in a groundwater associated with minimum porosities, maximum soluble rock chloride and static groundwater conditions. Leakage can probably occur through microfracturing of any mineral in the rock under sufficient stress, but microfracturing of calcite, zeolites and other easily broken minerals in fracture-fillings or veins may be highly significant. Stress changes leading to fluid-inclusion leakage could occur in deep mines and radioactive waste repositories as a result of blasting, drilling and dewatering large volumes of rock.

Résumé

Dans cet article, la possibilité d'un drainage des inclusions fluides des roches granitiques vers des régimes locaux d'eau souterraine sera étudiée sous divers aspects: (1) La quantité de sel présent est-elle suffisante? (2) Ce sel possède-t-il des caractéristiques chimiques distinctes de celles d'un sel d'origine différente? (3) Quelle proportion de ce sel est soluble dans l'eau? (4) Quelle quantité maximale de chlore peut être tirée de cette source dans des conditions optimales? (5) Quels sont les mécanismes qui pourraient provoquer le drainage des inclusions fluides? La réponse aux questions 1 et 2 est affirmative, quant à la question 3, environ 50% du chlore total (en moyenne) est soluble dans l'eau. Une eau souterraine peut contenir une concentration maximale de 20 000 mg/l de chlore si les conditions suivantes sont réunies: porosité minimale, proportion maximale de chlore soluble dans la roche et régime hydraulique souterrain statique. Le drainage peut probablement survenir quand une contrainte suffisante provoque la microfracturation d'un minéral quelconque, mais la microfracturation de la calcite, des zéolites et d'autres minéraux de faible dureté contenus dans les fractures ou les filons semblerait très propice. Les modifications de contraintes entraînant le drainage des inclusions fluides peuvent se rencontrer dans les mines profondes ou les sites d'enfouissement de déchets radioactifs et résulteraient du dynamitage, des forages et de la déshydratation d'importants volumes rocheux.

INTRODUCTION

Recent geochemical and hydrological investigations of crystalline bedrock, under the auspices and sponsorship of radioactive waste management agencies, have made major contributions to our knowledge of the deep (>500 m) groundwater chemistry in Precambrian shield areas. Groundwaters of unexpected chemical and isotopic compositions have been reported from Canada (Frape, *et al.*, 1984), the Soviet Union (Vovk, 1982) and Sweden (Nordstrom, *et al.*, 1985). Highly saline groundwaters, up to about 325 g/L total dissolved solids, have been found from many localities throughout the Canadian Shield at depths below 650 m (Frape, *et al.*, 1984). Their isotopic compositions are anomalously enriched in deuterium, such that they plot well to the left of the global meteoric water line, and their water chemistry is dominated by Ca-Na-Cl. Brines with very similar chemistry have been found in the Kola Peninsula of the Baltic Shield and in the Ukrainian Shield (see Vovk, this volume). In addition to shield areas, salty groundwaters have been found in granitic intrusions, such as the occurrence of chloride concentrations up to 1500 mg/L in a granodiorite at North Saanich, British Columbia (Fritz, *et al.*, 1983), and chloride concentrations up to 12,000 mg/L in the Carnmenellis granite, southwestern England (Edmunds, *et al.*, 1984).

Several different hypotheses have been proposed for these occurrences of salty groundwaters. In parts of the world where most of the recharge water is lost by evapotranspiration, increased salinities would be expected, such as in groundwaters of the Brazilian Shield, where up to 25,000 mg/L have been observed (Fritz, *et al.*, 1983), and in the unsaturated zone of weathered materials overlying Precambrian basement in western Australia, where up to 9,000 mg/L chloride occur (Turner, 1984). In shield areas where evapotranspiration cannot account for the high salinities, other sources must be considered.

Precambrian shields may contain ancient brines because the lack of hydraulic gradients, combined with low permeabilities at considerable depths, has prevented recharge and dilution from meteoric infiltration. Assuming that these brines were originally seawater and must be at least several millions of years old, two general sources have been suggested: Precambrian seawater or Paleozoic seawater. Kelly, *et al.* (1986) have discussed the brines found in the deep mines of the Keweenaw Peninsula, Michigan, and suggested that they (along with the Canadian Shield brines) may be Paleozoic formation waters that infiltrated the Precambrian basement and were subjected to chemical and isotopic modifications. These modifications were caused by moderate temperature increases (up to 250°C) at high water-rock ratios and low-temperature re-equilibration at low water-rock ratios. An alternate hypothesis is that they are Precambrian formation waters that might better be called residual metamorphic fluids since they must have undergone extensive modification during metamorphism after their original burial. Attempting to distinguish between a Precambrian and a Paleozoic seawater, both modified by metamorphic conditions and by low-temperature re-equilibration over many millions of years will be exceedingly difficult. As pointed out by Frape, *et al.* (1984) and McNutt, *et al.* (1984), there is good evidence for extensive water-rock interaction of the Canadian Shield brines at low temperature and at a local scale. With these strong overprints, the original chemical and isotopic signatures may be effectively disguised. Vovk (this volume) has presented a good argument for shield brines as being residual metamorphic fluids from Precambrian times modified by radiolytic reactions that can account for the anomalous levels of hydrogen and oxygen that have been found in them. The radiolytic model was first proposed by Gutsalo (1971) when substantial quantities of oxygen were discovered at more than 2 km depth in the Dnieper-Donets depression.

The hypothesis offered for the saline groundwaters occurring in the Carnmenellis granite, England, is that dissolution of biotite provides the major source of halogens and of certain phyllo-silicate cations, such as Mg, Li, and K that reflect the stoichiometry of the Carnmenellis biotite (Edmunds, *et al.*, 1984). Isotopic data and hot, dry rock-injection test data are given for support of this hypothesis (Edmunds, *et al.*, 1985).

The Swedish investigations, sponsored by the International Stripa Project, includes hydrologic studies, geochemical studies, tracer migration experiments, geophysical studies of fractures and studies of engineered barriers in a small Proterozoic granitic intrusion in the Baltic Shield. The geochemical studies have been focused on interpreting the origin and evolution of the groundwater through several chemical and isotopic methods. One of the hypotheses that developed from the Stripa investigations is that fluid inclusions, from either the granite or the surrounding metamorphics, may be the source of several of the major salt components in the deep groundwater. Other hypotheses, such as Holocene or older seawater intrusion or infiltration of dissolved Paleozoic evaporate salts, do not seem to be consistent with the available data. The purpose of this paper is to examine the general feasibility of fluid-inclusion leakage in granitic rocks and to identify the limits of fluid-inclusion influence on granitic groundwater chemistry. This examination also provides an interesting comparison to Precambrian-hosted brines in shield areas that could not have been derived from fluid inclusions, but may have a similar evolutionary history.

BACKGROUND

It is difficult to determine when it was first proposed that fluid inclusions may contribute to the solute composition of groundwaters, but it has been well known for over a hundred years that leaching of pulverized granite samples will produce a solution enriched in alkalies, calcium and chloride (see reviews by Faber, 1941, and Roedder, 1972, 1984). Rankama and Sahama (1950) mention that "During weathering the cavity solutions [i.e., fluid inclusions] are liberated, except from the most resistant minerals, and are subsequently incorporated into the weathering solutions." One of the conclusions reached by Gambell and Fisher (1966) during a study of solute mass balances from the weathering of igneous bedrock was that a substantial input of chloride is likely to be derived from the subsurface, i.e., rock weathering. Garrels (1967) made perhaps the first clear proposition that dissolved chloride in groundwaters may be derived from fluid inclusions in feldspathic igneous rocks. Jacks (1978) and Fritz, *et al.* (1979) mentioned the possibility that the increased chloride concentrations found at depths of 800 m in groundwaters of the Stripa granite, Sweden, could be caused by fluid-inclusion leakage, although there were no data to support it. The idea was given support by Nordstrom (1983), Nordstrom, *et al.* (1985) and Nordstrom and Andrews (1985) for the Stripa groundwaters when Br/Cl and I/Cl ratios of the fluid-inclusion leachates were found to match those of the groundwater.

The feasibility of this hypothesis will now be explored in light of our general knowledge of the geochemistry and structure of granitic rocks. The properties of fluid inclusions in granites need to be reviewed first, and then the possible mechanisms of leakage are considered.

THE OCCURRENCE AND PROPERTIES OF FLUID INCLUSIONS, MICROCAVITIES AND SALT IN GRANITIC ROCKS

a) *The occurrence of halogens in granites.* One of the earliest compilations of chlorine analyses for igneous rocks is found in Clarke's Data of Geochemistry published in five editions from 1908-1924. Values range from below detection up to 0.8 wt % Cl. This wide range of values has been substantiated in several papers and reviews (Tröger, 1934; Selivanov, 1940; Kuroda and Sandell, 1953; Correns, 1956), although the Cl content is rarely higher than 0.1%. The most recent estimates are provided by Fuge (1974a), who has compiled much of the earlier data and contributed new data of his own. His frequency distribution diagram indicates an average value of about 200 ppm Cl for granites and granodiorites, a number that is consistent with other estimates (Turekian and Wedepohl, 1961; Rankama and Sahama, 1950; Vinogradov, 1962). His most recent estimate is unchanged (Fuge, *et al.*, 1986).

Fewer analyses are available for Br and I, and there is usually a larger uncertainty associated with the results. Fuge (1974b) has reviewed earlier data and suggests an upper limit of 1 ppm for Br in igneous rocks. The distribution of values given by Fuge (1974b) would suggest that an average value of 0.5 ppm for granites would be a reasonable estimate. This concentration is lower than the estimates of Vinogradov (1962), Turekian and Wedepohl (1961) and Rankama and Sahama (1950), but the earlier data could be biased by lower sensitivity of the analytical method. More Br analyses are needed to support Fuge's contention.

New data by Fuge, *et al.* (1978) on iodine in granites supports his earlier contention (Fuge, 1974c) that the concentrations of 0.4 and 0.5 ppm cited by Vinogradov (1962) and Turekian and Wedepohl (1961), respectively, are too high for granitic rocks. The mean value for 24 British granites is 0.27 ppm (Fuge, *et al.*, 1978). Fuge (1974c) suggested that the mean value for iodine in igneous rocks lies in the range of 0.08-0.15 ppm, but his recent study suggests 0.24 ppm for granites (Fuge, *et al.*, 1986), and we shall adopt this value.

No matter whose values are adopted, there is a clear trend for Br/Cl and especially for I/Cl ratios in granites to be higher than for seawater, except for one estimate based on Fuge's data, which we suspect may be biased by an estimate for Br that is slightly too low. This trend can be seen by comparing values in Table 1. The low Br/Cl adopted from Fuge's work is not from direct measurements on a set of granites, but rather from the attempt to recommend a mean Br concentration for granites and a mean Cl concentration for granites and then taking the ratio of the two means. Hence, this particular ratio carries a higher degree of uncertainty. If the estimate for the mean concentration of Br in granites were higher by a factor of 2 (a reasonable error estimate), then the Br/Cl ratio becomes higher than that of seawater.

Correns (1956) pointed out that halogens can occur in at least three forms in igneous rocks: (1) as an essential part of the lattice network, i.e., halogen minerals, such as chloroapatite, $Ca_5(PO_4)_3Cl$, sodalite, $Na_8Al_6Si_6O_{24}Cl_2$, and scapolite, var. mariolite, $Na_4Al_3Si_9O_{24}Cl$; (2) as a substitional replacement for OH^- in micas and amphiboles (primarily biotite and hornblende, respectively); and (3) as inclusions in other minerals, either solid or liquid inclusions. Halogen minerals can contain up to 7% chlorine, micas and amphiboles can contain from 0-1% chlorine (Fuge, 1974a), and inclusions can contain up to 70% NaCl or about 42% chlorine (Roedder, 1984). If one considers the contribution of each of these sources to the total rock chloride, there is considerable overlap that can only be resolved by direct analyses of the rock and each phase containing chloride. We do not know of one study where this has been done.

Another approach is to distinguish between the water-soluble and water-insoluble halogens. Roedder (1958) suggested that as much as 100% of the chlorine in some rocks could be from fluid inclusions, and, therefore, would be water soluble. Behne (1953) found that 100% of the chlorine in granitic feldspars was water soluble. Fuge (1979) showed that 20-100% of the whole-rock chlorine was in a water-soluble state, and 60-100% of the whole-rock iodine was water soluble (Fuge, *et al.*, 1978) for 48 samples of granites

Table 1.

Halogen weight ratios for granitic bedrock and seawater

	Br/Cl	I/Cl	Reference
SEAWATER	0.00347	0.0000030	Morris and Riley (1966), Drever (1982)
GRANITES	0.0065	0.0025	Turekian and Wedepohl (1961)
GRANITES	0.0071	0.0017	Vinogradov (1962)
GRANITES	0.0036		Selivanov (1940)
GRANITES	0.0044-0.37		Behne (1953)
GRANITES	0.0025	0.0012	Fuge, *et al.* (1986)
HYDROTHERMAL QUARTZ	0.015		Kozlowski and Karwowski (1974)

from the British Isles. Unfortunately, there is no guarantee that the procedures used were 100% successful in leaching all of the fluid inclusions. Such inclusions are commonly less than 2 micrometers in size, and nearly all crushing and milling methods are unable to open all the inclusions. Nonetheless, the available data indicate that nearly all the Br and I should be in a water-soluble form, and at least 50% of the Cl is likely to be in a water-soluble form because of the great abundance of fluid inclusions in most high-grade crystalline silicate rocks, and because Br and I do not occur in OH^- or other lattice sites to any significant extent (Correns, 1956; Fuge, 1974b, c). Consequently, the mean Br/Cl and I/Cl for the water-soluble (or fluid-inclusion) fraction of granites is likely to be as high as those ratios for total halogens, and they could easily be higher if significant quantities of scapolite, sodalite or Cl-enriched apatite, biotite or amphiboles occurred in the rock. Fuge (1979) has pointed out that there is a positive correlation between the total Cl content of igneous rocks and the proportion of water-insoluble or lattice-bound Cl. The alteration of feldspars to scapolite, and OH-rich to Cl-rich micas, must certainly be induced by the presence of high-Cl fluids. Therefore, one might expect the post-reaction fluid to be enriched in Br and I and the halogen ratios in the residual fluid to be positively correlated with the total halogen content of the rock.

b) *The occurrence of fluid inclusions and microcavities in granites.* Fluid inclusions are microscopic inclusions (seldom larger than 1 mm) that were trapped either within mineral grains, i.e., intracrystalline inclusions, or between crystal grain boundaries, i.e., intercrystalline inclusions (Roedder, 1984). Even though these inclusions may be totally gaseous or solid in their present state as observed now, the term refers to their condition at the time of trapping. For example, salt crystals found along grain boundaries would also be called fluid inclusions. Hence, references to "grain boundary salts" or "interstitial pore fluids" or "micropores," etc. are all likely to fall under the heading of fluid inclusion. Then the question arises, how does one distinguish between a fluid inclusion and the bulk groundwater that often occurs in micrometer to millimeter-sized fracture openings in crystalline bedrock? It would appear that there is no clear distinction. End members are easy to distinguish, but to define an unequivocal division between the two doesn't seem possible. There may be a continuum all the way from a water molecule up to brine pools covering several km^3 of rock. In our attempts to classify properties of natural systems, we are often confronted with continuums that defy our conceptual efforts at taxonomy. The most important point, in the context of this paper, is to recognize that small voids in crystalline rocks are generally classified as fluid inclusions. This label holds regardless of whether these "voids" contain liquid, gas or salt crystals, and regardless of whether they occur within mineral grains or between mineral grains.

Fluid inclusions are a common feature of granitic rocks. Felsic magmas at high pressures and temperatures can dissolve as much as 5% H_2O, most of which will be released during intrusion and cooling at lower pressures and temperatures. Large quantities of fluid will separate from the magma with the remainder entrapped as minute inclusions. Considerable hydrofracting can also occur both in the country rock and within the plutonic stock itself (i.e., Burnham, 1979). Fluid inclusions within granitic stocks will occur in two distinct parts of the granite: (1) the bulk rock composed of feldspars, quartz, micas and amphiboles, and (2) the fracture-fill minerals, usually composed of quartz, calcite, chlorite, epidote, hornblende with minor fluorite, pyrite, laumontite, prehnite and feldspar. The size, composition, population density and phase assemblage of the inclusions may vary significantly, depending on whether they are intercrystalline, intracrystalline, trapped in the primary minerals or trapped in vein minerals, and depending on the geological history of the rock mass. Intercrystalline inclusions would be most susceptible to microfracturing and leakage; hence, they are rarely studied. Although relatively clear minerals like quartz, fluorite, calcite and sphalerite are most frequently used for inclusion measurements, water-soluble salts and fluid inclusions are also found in feldspars, micas and amphiboles (Faber, 1941; Roedder, 1984).

Although fluid inclusions range in size from <1 micrometer to several millimeters in maximum dimension, they frequently fall in the range of 1-10 micrometers. They are found to be as small as 0.02 micrometers and, as Roedder (1984) mentions, there is probably a continuum from single water molecules ($\sim 2 \times 10^{-8}$ cm) up to the largest inclusions of hundreds to thousands of milliliters (Roedder, 1984; Guha and Kanwar, 1985). There usually exists an inverse relationship between the size and the number of inclusions in any given sample so that very small inclusions are much more abundant than large ones. For inclusions in the range of 1-100 micrometers, a population density of 10^8-10^{10} inclusions/cm^3 would be fairly common for white quartz and calcite (Roedder, 1984).

The composition of the aqueous fluid in inclusions can range anywhere from 0% up to 70 wt. % NaCl when trapped, and many inclusions contain halite and sylvite daughter minerals (Roedder, 1984). Although it is very risky to estimate a typical or average salinity of fluid inclusions in granitic rocks, it would not be unreasonable to choose a value of 10 wt. % NaCl for the purposes of the calculations in this paper. Granites are frequently subject to repeated tectonism, causing fracturing, introduction of new fluids and rehealing to form new inclusions in planar structures. Consequently, a wide range of inclusion compositions can be found in the same rock and even the same mineral grain.

Assuming a 2 micrometer diameter inclusion with a uniform population density of 10^9 inclusions/cm^3 and average salinity of 10 wt. % NaCl, the total fluid-inclusion chloride concentration will be 95 ppm in a granite. Hence, for the average granite containing 200 ppm total chloride, about 50% can occur as fluid inclusions. Put another way, in a cubic meter of "average" granite there can be 253 g of fluid-inclusion chloride (assuming a rock density of 2.65). If 1% of the inclusion fluid is mixed with fresh groundwater containing negligible chloride for a porosity of 1%, then 253 mg/L of chloride could be achieved in a static groundwater system. This amount is only a small contribution to the groundwater chemistry, but if we ask the question: "What would be the maximum contribution of chloride from fluid inclusions?" we would have to know (a) the upper range on all of these numerical estimates, and (b) the mechanism and rate of fluid-inclusion leakage. These are difficult to assess, but the data described above can provide some limits. One important limit is a reasonable maximum to the chloride content of granites. Based on all the available data, this value should be about 0.1 wt. %. Using the correlation between total and water-soluble chlorine depicted by Fuge (1979), the maximum fluid-inclusion chloride would then be about 0.03 wt. %. For a minimum porosity of 0.4% (Norton and Knapp, 1977) and a leakage of about 10% of all fluid inclusions, then the maximum chloride concentration would be about 20,000 mg/L (rounded to the nearest significant figure). A more typical situation might be 0.01 wt. % fluid-inclusion chloride in the whole rock at 1% porosity and 5% leakage, giving about 1,000 mg/L chloride concentration in the groundwater. At Stripa, where the total porosity at depths >350 m is typically 0.5% and fluid-inclusion salinities average 3% NaCl, only 1-2% of the fluid inclusions need to leak to obtain the groundwater concentrations of 600-700 mg/L of chloride (Nordstrom, *et al.*, 1985). As a qualitative guide then, granitic groundwater concentrations of Cl in the range of 50 mg/L-20,000 mg/L could be derived from fluid inclusions. Concentrations of approximately 50 mg/L or less are usually derived from atmospheric or surface sources. However, we would maintain that concentrations greater than about 5,000 mg/L Cl are likely to be derived from other sources because in the typical situation only a few thousand mg/L can be achieved from fluid-inclusion leakage in a granitic groundwater. If fluid-inclusion leakage occurs by mixing of groundwater with the inclusion fluid along fractured surfaces, then there will be some type of relationship between the porosity (or number of fractures) and the percent leakage. The physical properties of the fractures and their evolutionary history may very well govern the groundwater chemistry, as well as the groundwater flow conditions and mineral solubilities.

HALOGEN RATIOS FOR GROUNDWATERS, GRANITES AND GRANITE LEACHATES

It is a striking observation that halogen ratios (Br/Cl and I/Cl) are very similar for granitic rocks, granite leachates and deep brines and saline groundwaters from Precambrian shield areas. As shown earlier in Table 1, the Br/Cl and I/Cl ratios for granitic rocks indicate a trend towards significant enrichment in Br and I relative to seawater. This same enrichment has been found in the brines and saline groundwaters from the Canadian Shield (Frape, *et al.*, 1984), from the Baltic Shield in Finland (Nurmi, *et al.*, 1985), and from the Keweenaw Peninsula (Kelly, *et al.*, 1986). The ranges of the Br/Cl and I/Cl ratios for these groundwaters are shown in Table 2, along with the range of ratios for granites. In addition, the mean ratios from granite-leaching experiments by Nordstrom, *et al.* (1985) for the Stripa granite, and by Bourg, *et al.* (1985) for the Fanay granite are compared. All of these ratios show enrichment relative to seawater, and in some instances, as in the Fanay granite, an extreme enrichment.

An unresolved question is whether (1) this halogen enrichment is a signature related to metamorphic or igneous processes, or (2) it is a signature remaining from pre-metamorphic times and suggests a residual bittern origin, indicating seawater evaporation, or (3) it is a signature of Precambrian seawater of different composition than modern seawater, or (4) it is indicative of halogen chemistry from a deep crustal or mantle origin, so-called juvenile volatiles from degassing of the earth. As Kelly, *et al.* (1986) point out, not enough is known about halogen partitioning during high-temperature events to better discriminate between some of these possibilities. However, since the whole-rock halogen ratios are enriched, it seems likely that the original fluid was also enriched before metamorphism. Hence, possibility (1) is least likely. Since the world's oceans seem to be of fairly constant composition throughout the Phanerozoic and probably earlier (Holser, 1979), then possibilities (2) or (4) seem most likely. Perhaps, the use of Ca/Mg ratios, along with Br/Cl ratios, might also be used as a discriminating parameter, as shown by Frape, *et al.* (1984) and Nordstrom, *et al.* (1985).

POSSIBLE MECHANISMS FOR FLUID INCLUSION LEAKAGE

We have shown that granitic rocks have (a) anomalous Br/Cl and I/Cl ratios that tend to be higher than the seawater ratios (whether leached or whole-rock ratios), (b) halogen ratios comparable to those found for deep-seated brines in

Table 2.

Halogen Weight Ratios for Granites, Granite Leachates and Shield Groundwaters

	Cl(mg/L)	Br/Cl	I/Cl	Reference
SEAWATER	19,800	0.00347	0.0000030	see Table 1
GRANITES		0.0036-0.37	0.0017-0.0025	see Table 1
SHIELD GROUNDWATERS				
Canadian Shield	325,000(max.)	0.0060-0.012	0.000017-0.000035	Frape, *et al.*, 1984
Baltic Shield	16,800(max.)	0.0073-0.011	0.000039-0.00029	Nurmi, *et al.*, 1985
Stripa	700(max.)	0.0070-0.012	0.01-0.08	Nordstrom, *et al.*, 1985
Keweenaw	156,500(max.)	0.0080-0.010	—	Kelly, *et al.*, 1986
GRANITE LEACHATES				
Stripa	10(max.)	0.0080-0.013	0.022-0.081	Nordstrom, *et al.*, 1985
Fanay	2.5(max.)	0.040-0.30	0.30-1.0	Bourg, *et al.*, 1985

crystalline bedrock, and (c) sufficient soluble salt content to contribute a few thousands of mg/L chloride to the groundwater. Now we shall consider the mixing process of inclusion fluids and their daughter minerals with the groundwater.

For leakage and mixing to occur, there must be a legitimate pathway and an adequate gradient. The existence of pathways is well documented. Fracturing of the granite can easily lead to exposure of large quantities of fluid inclusions, both directly and by slow leakage, from microfractures. Granites commonly have many macro- and microfractures. Leakage may occur either by diffusion or by flow through these microfractures. Diffusion through solid mineral grains is too slow to be important. Microfractures are likely to be abundant near major fractured zones and may be present in large enough quantities for significant fluid flow. If a small volume of groundwater flows by a large surface area of rock (a typical situation in deep crystalline rocks) containing abundant inclusions, salinity of the groundwater is likely to increase.

Another important observation is the common occurrence of fluid inclusions along planar structures in crystalline rock. These fluid-inclusion planes are usually secondary features which reflect early fracturing and rehealing of the rock around these fractures. Many fluid-inclusion planes cut across mineral grain boundaries, and are parallel to and coincident with microjoints. Wise (1964) found evidence for younger planes of fluid inclusions reopening older fluid-inclusion planes. These older planes represent Precambrian joint systems, and they controlled the orientation of subsequent fracturing during the Laramide Orogeny (Late Cretaceous to early Tertiary). Dale (1923) found numerous examples of fluid-inclusion planes that were parallel and coincident to microfractures in granites and gneisses from New England. He provides some striking examples of fluid-inclusion planes becoming microfractures with the same strike and dip. Large numbers of very small inclusions have been found on mineral grain boundaries of fractured mineral surfaces by electron microscopy (Sella and Deicha, 1962a, b, 1963). Hence, refractured secondary fluid-inclusion planes can provide both the salt and the flow path for infiltrating groundwater.

Microcracks and micropores are common in igneous rocks and have been investigated for their effect on the physical properties of the rock, especially electrical resistivity, compressibility, and permeability (Brace *et al.*, 1972). These microcavities have been found both along grain boundaries and within mineral grains (Sprunt and Brace, 1974). They have been found to connect and to form larger microcracks when placed under subcritical stress (Tapponier and Brace, 1976). They are more common in rocks with a higher water content than those that are "drier" (Montgomery and Brace, 1975), and they are more abundant in sodic feldspar than in quartz or potassium feldspar (Sprunt and Brace, 1974; Montgomery and Brace, 1975). This last observation is particularly important because many of these microcavities are identical in description to fluid inclusions and are thought to be fluid inclusions by Sprunt and Brace (1974) and Montgomery and Brace (1975). Feldspars are notoriously difficult to study for fluid inclusions because of their opacity, and little is really known about the distribution of inclusions between quartz and feldspars in a granite. If the results of Montgomery and Brace (1975) are valid for most granites, then the density of fluid inclusions in sodic feldspar must be considered greater than in coexisting quartz grains, and the volume contained by fluid inclusions could take up to 2% of the feldspar. Fluid inclusions may be, in fact, the reason for the turbid, opaque appearance of feldspars (Folk, 1955) and the milky white appearance of quartz (Roedder, 1972, 1984).

Thus, fluid-inclusion planes can be zones of weakness where additional fracturing has a higher probability of occurring, and groundwater would have direct access to the highest density of fluid inclusions. This close relationship between fluid-inclusion planes and microfractures/microjoints could make the accessibility of fluid inclusions to the groundwater nearly instantaneous in response to tectonically- or anthropogenically-derived changes in rock stress. It may not be necessary for fluid inclusions to be transported any significant distances in order for them to mix with the groundwaters; only micro-fracturing needs to occur.

The various types of gradients that could influence the movement of inclusion fluid into the groundwater include: (a) diffusion due to a chemical gradient, (b) pressure gradient due to differences in fluid inclusion and hydrostatic pressures, including expansion of CO_2 from liquid to gas from the inclusion, (c) pressure gradient due to differences in lithostatic and hydrostatic pressures, (d) pressure gradient reversals between microcavities and fracture flow zones caused by seasonal variations in the groundwater head, and (e) thermal gradients and thermomechanical stresses that can fracture fluid inclusions and/or close up fracture apertures that would squeeze out fluid inclusions into the major flow zones. If fluid flow occurs in a system, then it will dominate over diffusion, and for the three categories of pressure gradients, (b) and (c) seem most likely. However, phenomena (a) and (b) require active microfracturing of a granite that would have to be caused by additional forces, such as active tectonism. Phenomenon (c) might also be caused by tectonic forces, such as extension, compression, or isostatic rebound, but it could also result from an accumulation of stress in a deep mine where slight closure of fracture apertures might occur from dewatering the subsurface excavations. Local stresses would also be exerted in a mine from blasting and excavation. Since calcite is a common fracture-fill mineral that both easily fractures and dissolves more readily under slight increases in stress, these stresses could cause numerous calcite fluid inclusions to be released into the groundwater system. It is likely that local stress release after drilling of a borehole could cause some release of inclusion fluids in addition to those released by breakage and microfracturing during drilling. These scenarios probably apply to the Stripa groundwaters, where mine excavations exist to depths of 400 m and drillholes to depths of 1200 m. Thermal cracking is evident in geothermally active systems, and thermomechanical stress may have opened fluid-inclusions, causing an increase in chloride concentrations at the Stripa site (Nordstrom, *et al.*, 1985).

The discussions in this paper have some important implications regarding the nature of porosity in crystalline bedrock. This topic has rarely been examined, especially with respect to the known properties of granitic rocks and the occurrence of fluid inclusions. Furthermore, much confusion exists in the literature with regard to the various types of measured porosites. The only careful and comprehensive review is by Norton and Knapp (1977), who point out that there are three types of porosity that contribute to the total porosity in fractured bedrock: flow porosity, n_F, diffusion porosity, n_D, and residual porosity, n_R, for which

$$n_R \gg n_D \simeq n_F$$

From their experimental data, total porosities were in the range of 0.1-7% for granitic rocks, and the residual porosities contributed over 90% of the total. The flow porosity constitutes the dominant mode of fluid transport through main fracture systems, but only accounts for a small fraction of the total porosity. The diffusion porosity represents the interconnected microcavities that may include microfractured and interconnected fluid inclusions. Hence, the fluid in the microfractures could be from opened fluid inclusions or from a pre-existing groundwater, depending on the timing and evolutionary history of the fracturing. Since many deep groundwaters in crystalline bedrock are brines or saline waters similar in composition to fluid inclusions, it may be impossible to distinguish between them. In fact, they could be of the same origin, and have a similar history. Kelley, *et al.* (1986) used this type of comparison as a basis for suggesting that the deep Precambrian-hosted brines in the Keweenaw Peninsula are Paleozoic formation waters because the fluid-inclusion brines are isotopically dissimilar to the groundwater brines.

The residual porosity, or closed micropores, should be dominated by fluid inclusions. The residual porosity is always determined by difference. However, both the residual porosity and the total porosity are subject to significant errors not usually considered in hydrologic investigations. The best estimate of total porosity was described by Norton and Knapp (1977), who used the ratio of the dry rock bulk density to the grain density "determined on a finely ground sample with a grain size less than the minimum pore diameter." The error introduced with this approach is caused by the difficulty in getting the sample ground fine enough. Fluid inclusions are commonly less than 2 micrometers in diameter, and if a continuum in size exists from the size of water molecules on up, then it would be practically impossible to grind less than the smallest pore size. Consequently, the total porosity will be biased towards a value that may be too low. A better approach might be to do direct counting of fluid inclusions per volume of rock as an estimate of residual porosity, and then compare the sum of the component porosities (flow + diffusion + residual) with the total porosity measurement.

At Stripa, estimates have been made of the flow porosity, the diffusion porosity and the residual porosity. The results show that the diffusion porosity is one to two orders of magnitude greater than the fracture flow porosity. For example, at depths below 300 m at Stripa, the "total" porosity from resistivity logs averages about 0.5% (Olsson and Jämtlid, 1984) of which 0.005% is the flow porosity (Nordstrom, *et al.*, 1985) and 0.15% would be the diffusion porosity (Skagius and Neretnieks, 1985). Although the remaining 0.345% has been assumed to be residual porosity, this is incorrect. Only the effective porosity can be deduced from the electrical resistivity log (Keys and MacCary, 1971), and the spaces occupied by intracrystalline fluid inclusions are not measured by this method. At Stripa, the quartz fluid inclusions average $10^8/cm^3$ with a mean diameter of 6 micrometers and a mean volume of about 10-20 microliters per cm^3 or 1-2% by volume of the rock (Lindblom, 1984). Hence, the true residual porosity is 2-4 times greater than the effective porosity. Since the effective porosities are undoubtedly higher than *in situ* values due to the release of stress upon removal of a core from depth, the 0.345% "residual" porosity may be largely induced by removing core (Skagius and Neretnieks, 1985) and the effective porosity should be even less. If we assume an effective diffusion porosity of 0.15% based on Skagius and Neretniek's (1985) data, and 1% for the residual porosity, then the total porosity becomes 1.155%, of which the flow porosity and microporosity come to 13% of the total, and 87% are due to fluid inclusions.

Rock-matrix-to-fracture crossflow may take place provided that there exists a potential gradient from microfractures out to the fractures. If a near-static groundwater system is disturbed by an introduced sink, such as a deep mine, the pressure in the fractures carrying the main flow will decrease rapidly, while the flow resistance in the microfractures will cause a higher pressure to prevail for a longer period of time within the diffusion porosity. Hence, the salinity should decrease with time from an on-going micropore discharge.

An estimate of matrix-to-fracture crossflow can be made at Stripa with the available hydraulic data and a few reasonable assumptions. The Stripa mine contains tunnels down to 410 m. Drill-holes act as additional sinks for groundwater, extending the influence of the mine to over a kilometer depth. The outflow of fluid inclusions and saline fluids in connected micropores should have been at its highest immediately after drilling. In addition, the drainage from intercepted fractures introduces a significant stress on the rock that could open up some of the residual porosity to the outflow of diffusion porosity. Simple calculations for the Vl borehole at Stripa with respect to the pressure drop immediately after drilling gives a basis for estimating the volumetric mixing between diffusional flow and fracture flow. Assuming that the overall mean value for the hydraulic conductivity of 5×10^{-11} m/s is valid, and that the conductivity for the diffusion porosity is in the range of 10^{-14}-10^{-16} m/s with an average porosity of 0.1%, then it is possible to estimate the amount of recharge from matrix-to-fracture. With an assumed head drop to the fractures of 50 m, and an assumed spacing of water-bearing fractures of 2 m, the mixing ratio of micropore fluid to fracture fluid is 0.01 to 1, depending on the matrix conductivity. Under these conditions it would take roughly 100-10,000 years to deplete the saline source, provided that the matrix flow decreases linearly with time. Thus, it is possible to assign the source of salt in the Stripa groundwaters to matrix-to-fracture crossflow. Furthermore, the duration of crossflow influence may be long enough to make it impossible to observe any decrease in the salinity without monitoring the groundwater chemistry for several generations.

CONCLUSIONS

1) Br/Cl and I/Cl ratios of granites and their leachates show enrichment relative to seawater. Leachate ratios are likely to be more enriched than the whole rock because of the preferential substitution of Cl over Br and I into hydroxyl lattice sites. These ratios may provide a clue to the early history of the fluids incorporated by granites.

2) Br/Cl and I/Cl ratios of deep-seated brines from crystalline shields are similar to those ratios for granites and their leachates. Hence, their histories could be linked or, at

least, the history of the original halogenic fluid could be the same.

3) Likely origins for a Br- and I-enriched fluid in shield areas include an early Paleozoic or Precambrian formation water (or residual bittern) or juvenile fluids from depth during degassing of the earth or fluid inclusions.

4) Granites contain sufficient salt content in the form of soluble fluid inclusions to contribute a few thousand mg/L chloride to the groundwater under conditions of low permeability.

5) Microfractures are a common feature of granites, and they may be formed along fluid-inclusion planes that were formed from the healing of earlier fractures. These microfractures provide a flow path for mixing between fluid-inclusion salts and infiltrating groundwaters.

6) Natural or anthropogenic disturbances may provide the mechanism for fluid-inclusion leakage into groundwaters *via* microfracturing from stress changes.

7) Dewatering of deep mines in crystalline bedrock may induce sufficient stress changes to cause fluid-inclusion leakage. Such stress changes would tend to increase the pressure on fracture-fill minerals like calcite and zeolites, thereby releasing numerous fluid inclusions. The dewatering would also tend to produce a pressure gradient that would cause the microfracture fluid to flow into the main water-bearing fractures.

8) "Residual" or fluid-inclusion porosities are best measured by making measurements of fluid-inclusion population densities. Such porosities may greatly overshadow the effective porosity, the diffusion porosity or the flow porosity.

9) Fluid-inclusion leakage must be considered as one of the possible processes affecting deep groundwater chemistry in crystalline bedrock, and the results from Stripa show that it most likely does occur in crystalline bedrock at low permeability for chloride concentrations less than 20,000 mg/L, and where significant drawdown of the water table occurs as in deep mines.

ACKNOWLEDGEMENTS

The researchers and administrators associated with the International Stripa Project especially the project manger, Hans Carlsson, are gratefully acknowledged for their support of these investigations. The support of the U.S. Geological Survey is greatly appreciated, especially the many friends and colleagues who have contributed numerous comments during the evolution of our thinking. Thanks are due to Peter Fritz, who invited one of us (D.K.N.) to join the Stripa Project and who has continued to be an excellent source of sound advice. We are also grateful to Mike Edmunds, Bob Fournier, Paul Hsieh, Dave John and Ed Roedder for their many helpful comments and careful reviews of our manuscript.

REFERENCES

Behne, W., 1953. Untersuchungen zur Geochemie des Chlor und Brom: Geochimica et Cosmochimica Acta, v. 3, p. 186-214.

Bourg, A.C.M., Oustriere, P. and Sureau, J.F., 1985. Experimental investigation of clogging of fissures in granite: Mineralogical Magazine, v. 49, p. 223-231.

Brace, W.F., Silver, E., Hadley, K. and Goetze, C., 1972. Cracks and pores: A closer look: Science, v. 178, p. 162-164.

Burnham, C.W., 1979. Magmas and hydrothermal fluids: *in* Barnes, H.L., ed., Geochemistry of Hydrothermal Ore Deposits: Wiley-Interscience, New York, p. 71-136.

Correns, C.W., 1956. The geochemistry of the halogens: *in* Ahrens, L.F., Rankama, K., and Runcorn, S.K., eds., Physics and Chemistry of the Earth, v. 1, p. 181-233.

Dale, T.N., 1923. The commercial granites of New England: United States Geological Survey Bulletin 738, p. 14-26.

Drever, J.I., 1982. The Geochemistry of Natural Waters: Prentice-Hall, Englewood Cliffs, 338 p.

Edmunds, W.M., Andres, J.N., Burgess, W.G., Kay, R.L.F. and Lee, D.J., 1984. The evolution of saline and thermal groundwaters in the Carnmenellis granite: Mineralogical Magazine, v. 48, p. 407-424.

Edmunds, W.M., Kay, R.L.F. and McCartney, R.A., 1985. Origin of saline groundwaters in the Carnmenellis granite (Cornwall, England): Natural processes and reaction during hot dry rock reservoir circulation: Chemical Geology, v. 49, p. 287-301.

Faber, H., 1941. On the salt-solutions in microscopic cavities in granites: Danmarks Geologiske Undersögelse II, no. 67, 45 p.

Folk, R.L., 1955. Note on the significance of "turbid" feldspars: American Mineralogist, v. 40, p. 356-357.

Frape, S.K., Fritz, P. and McNutt, R.H., 1984. Water-rock interaction and chemistry of groundwaters from the Canadian Shield: Geochimica et Cosmochimica Acta, v. 48, p. 1617-1627.

Fritz, P., Barker, J.F., and Gale, J.E., 1979. Geochemistry and isotope hydrology of groundwaters in the Stripa Granite: Swedish-American Cooperative Program on Radioactive Waste Storage in Mine Caverns in Crystalline Rock, Technical Information Report 12, 135 p.

Fritz, P., Al-Mooji, Y., Frape, S.K., and Salati, E., 1983. Saline ground waters in crystalline rocks, isotopic and geochemical consideration on their origin: International Conference on Ground Water and Man, Sydney, Australia.

Fuge, R., 1974a. Chlorine: *in* Wedepohl, K.H., ed., Handbook of Geochemistry, v. II/2, Chapter 17-B – 17-0.

Fuge, R., 1974b. Bromine: *in* Wedepohl, K.H., ed., Handbook of Geochemistry, v. II/3, Chapter 35-B – 35-G, 35-I, 35-L, 35-O.

Fuge, R., 1974c. Iodine: *in* Wedepohl, K.H., ed., Handbook of Geochemistry, v. II/4, Chapter 53-B – 53-M, 53-O.

Fuge, R., 1979. Water-soluble chlorine in granitic rocks: Chemical Geology, v. 25, p. 169-174.

Fuge, R., Johnson, C.C., and Phillips, W.J., 1978. Iodine in granitic and associated rocks: Chemical Geology, v. 22, p. 347-352.

Fuge, R., Andrews, M.J., and Johnson, C.C., 1986. Chlorine and iodine, potential pathfinder elements in exploration geochemistry: Applied Geochemistry, v. 1, p. 111-116.

Gambell, A.W., and Fisher, D.W., 1966. Chemical composition of rainfall in eastern North Carolina and southeastern Virginia: United States Geological Survey Water-Supply Paper 1535-K, 41 p.

Garrels, R.M., 1967. Genesis of some ground waters from igneous rocks: *in* Abelson, P.H., ed., Researches in Geochemistry, v. 2, p. 405-420.

Guha, J., and Kanwar, R., 1986. The role of $CaCl_2$-NaCl-CH_4 rich fluids in the secondary gold concentration process at the Copper Rand Mine, Chibougamau, P.Q., Canada: *This Volume*.

Gutsalo, L.K., 1971. Radiolysis of water as a source of free oxygen in the subsurface hydrosphere: Geokhimiya, v. 12, p. 1473-1481.

Holser, W.T., 1979. Mineralogy of evaporites, *in* Burns, R.G., ed., Marine Minerals: Mineralogical Society of America Short Course Notes, v. 6, p. 211-294.

Jacks, G., 1978. Ground water chemistry at depth in granites and gneisses: Swedish Nuclear Fuel and Waste Management Company (KBS) Technical Report, v. 88, 28 p.

Keys, W.S., and MacCary, L.M., 1971. Application of borehole geophysics to water-resources investigations: United States Geological Survey Techniques of Water Resources Investigations, Chapter E1, Book 2, 126 p.

Kelly, W.C., Rye, R.O., and Livnat, A., 1986. Saline mine waters of the Keweenaw Peninsula, northern Michigan: Their nature, origin and relation to similar deep waters in Precambrian crystalline rocks of the Canadian Shield: American Journal of Science, v. 286, p. 281-308.

Kozlowski, A., and Karwowski, L., 1974. Chlorine/bromine ratios in fluid inclusions: Economic Geology, v. 69, p. 268-271.
Kuroda, P.K., amd Sandell, E.B., 1953. Chlorine in igneous rocks: Geological Society of America Bulletin, v. 64, p. 879-896.
Lindblom, S., 1984. Fluid inclusion studies of the Stripa Granite: Stripa Project Internal Report, v. 84-07, 42 p.
McNutt, R.H., Frape, S.K., and Fritz, P., 1984. Strontium isotope composition of some brines from the Precambrian Shield of Canada: Isotope Geoscience, v. 2, p. 205-215.
Montgomery, C.W., and Brace, W.F., 1975. Micropores in plagioclase: Contributions to Mineralogy and Petrology, v. 52, p. 17-28.
Morris, A.W., and Riley, J.P., 1966. The bromide/chlorinity and sulphate/chlorinity ratio in seawater: Deep Sea Research, v. 13, p. 699-705.
Nordstrom, D.K., 1983. Preliminary data on the geochemical characteristics of groundwater at Stripa: *in* Geological Disposal of Radioactive Waste in Situ Experiments in Granite, Proceedings of the Workshop, Organization for Economic Cooperation and Development/Nuclear Energy Agency, Stockholm, 1982, p. 143-153.
Nordstrom, D.K. (ed.), Andrews, J.N., Carlsson, L., Fontes, J.-C., Fritz, P., Moser, H., and Olsson, T., 1985. Final report of the Phase I geochemical investigations of the Stripa groundwaters: SKB Stripa Project Report 85-06, 250 p, Stockholm.
Nordstrom, D.K., and Andrews, J.C., 1985. Progress on geochemical investigations at Stripa: *in* In Situ Experiments in Granite Associated with the Disposal of Radioactive Waste, Proceedings of the Workshop, Organization for Economic Cooperation and Development/Nuclear Energy Agency, Stockholm, 1985, p. 96-106.
Norton, D., and Knapp, R., 1977. Transport phenomena in hydrothermal systems: The nature of porosity: American Journal of Science, v. 277, p. 913-936.
Nurmi, P., Kukkonen, I., Lahermo, P., Salmi, M., and Rahkola, P., 1985. Geochemistry of groundwater in bedrock: Preliminary results of groundwater sampling in deep drill-holes at Outokumpu, Kerimäki, Parainen and Liminka: Geological Survey of Finaland Report YST-50, 60 p.
Olsson, O., and Jämtlid, 1984. Geophysical borehole measurements: SKB Stripa Project Report 84-03, 83 p, Stockholm.
Rankama, R., and Sahama, T.G., 1950. Geochemistry: Chicago University Press, Chicago, 912 p.
Roedder, E., 1958. Technique for the extraction and partial chemical analysis of fluid-filled inclusions from minerals: Economic Geology, v. 53, p. 235-269.
Roedder, E., 1972. Composition of fluid inclusions: United States Geological Survey Professional Paper 440-JJ, 164 p.
Roedder, E., 1984. Fluid inclusions: Reviews in Mineralogy, v. 12, Mineralogical Society of America, 644 p.
Selivanov, L.S., 1940. Chlorine and bromine in massive crystalline rocks: Doklady Akademia Nauk SSSR, Comptes rendus, v. 28, p. 809-813.
Sella, C., and Deicha, G., 1962a. Etude au microscope electronqiue des pores intergranulaires des gangues et des roches: Comptes rendus, v. 254, p. 2796-2798.
Sella, C. and Deicha, G., 1962b. Lacunes de cristallisation et pores intergranulaires du quartz: Abstract, *in* S. Breese, Jr., ed., Electron Microscopy, Academic Press, New York.
Sella, C., and Deicha, G., 1963. Importance des cavites intra et intercristallines dans l'Architecture des mineraux et des roches: Journal de Microscopie, v. 2, p. 283-296.
Skagius, K., and Neretnieks, I., 1985. Diffusivity measurements and electrical resistivity measurements in rock samples under mechanical stress: Swedish Nuclear Fuel and Waste Management Company (SKB) Technical Report, v. 85-05, 30 p.
Sprunt, E.S., and Brace, W.F., 1974. Direct observations of microcavities in crystalline rocks: International Journal of Rock Mechanics and Mining Sciences, v. 11, p. 139-150.
Tapponier, P., and Brace, W.F., 1976. Development of stress-incuded microcracks in Westerly granite: International Journal of Rock Mechanics and Mining Sciences, v. 13, p. 103-112.
Tröger, E., 1934. Der Gehalt an Selteneren Elementen Bei Eruptivegesteinen: Chemisches Erde, v. 9, p. 286-294.
Turekian, K.K., and Wedepohl, K.H., 1961. Distribution of the elements in some major units of the earth's crust: Geological Society of America Bulletin, v. 72, p. 175-192.
Turner, J.V., 1984. Hydrogeochemical evolution and variability of groundwater in experimental basins of S.W. Western Australia: *in* Eriksson, E., ed., Hydrochemical Balances of Freshwater Systems: IAHS Publication No. 150, p. 41-54.
Vinogradov, A.P., 1962. Average contents of chemical elements in the principal types of igneous rocks of the earth's crust: Geokhimiya, no. 7, p. 555-571.
Vovk, I.F., 1982. Brines in the crystalline basement of the shields: Ukrainian Akademy of Science, Kiev, 163 p.
Vovk, I.F., 1986. Saline waters and brines in the crystalline basement of the east European platform: *This Volume.*
Wise, D.U., 1964. Microjointing in basement, middle Rocky Mountains of Montana and Wyoming: Geological Society of America Bulletin, v. 75, p. 287-306.

Strontium Isotopic Data and Geochemical Calculations as Indicators for the Origin of Saline Waters in Crystalline Rocks

Bertrand Fritz, Norbert Clauer and Marie Kam
Centre de Sédimentologie et de Géochimie de la Surface (CNRS), 1, rue Blessig, 67084 Strasbourg-Cedex, France

Fritz, B., Clauer, N., and Kam, M., Strontium isotopic data and geochemical calculations as indicators for the origin of saline waters in crystalline rocks; in Saline water and gases in crystalline rocks, Editors: Fritz, P., and Frape, S.K.; Geological Association of Canada Special Paper 33, 1987. p. 121–126.

Abstract

Groundwaters collected at deep levels in the granitic body of Stripa in central Sweden, are of a saline type. Their origin is still matter of debate with two major theories: the primary waters might either be of marine origin, or they might have been very diluted and could have leached progressively the fluid inclusions of some minerals of the plutonic massif.

Rb-Sr data on a rock and minerals from granitic body, on fracture coatings and on waters, as well as geochemical calculations on these waters allow a somewhat different hypothesis for the origin of their salinity, which is based on a long-term rock-water interaction in the fractures of the granite. Our results clearly discard a marine origin for the primary waters, but they do not definitely adresse the alternative possibility which is a leaching of fluid inclusions.

Résumé

Des eaux souterraines salées ont été recueillies à de grandes profondeurs dans le massif granitique de Stripa, en Suède centrale. Deux théories sont en concurrence pour expliquer leur origine. Les eaux initiales peuvent être des eaux marines ou des eaux très peu concentrées qui auraient progressivement lessivé les inclusions fluides de certains minéraux du pluton. Cependant, le rapport Rb-Sr du matériel à proximité des fractures, de l'eau, d'un échantillon et de minéraux provenant du massif granitique, ainsi que des calculs géochimiques effectués sur ces eaux permettent d'avancer une hypothèse quelque peu différente quant à l'origine de la salinité, qui proviendrait d'une interaction prolongée entre l'eau et la roche dans les fractures du granite. Nos résultats s'opposent clairement à l'origine marine des eaux initiales, mais ne permettent pas de réfuter la possibilité d'un lessivage des inclusions fluides.

INTRODUCTION

An extensive geochemical prospect of the deep groundwaters from Stripa granite located in central Sweden at 170 km to the WNW of Stockholm (KBS, 1978), has pointed to a surprisingly saline type for these waters. Several reports including analytical data for the major and trace elements, as well as preliminary discussions and interpretations for the possible origin of this salinity, have been already published. Nordstrom et al., (1985) summarized all results as following:

(1) the deep groundwaters in the granitic body show abnormally high salinities: up to 700 mg/l of chloride at the Stripa site and even several g/l at other sites;

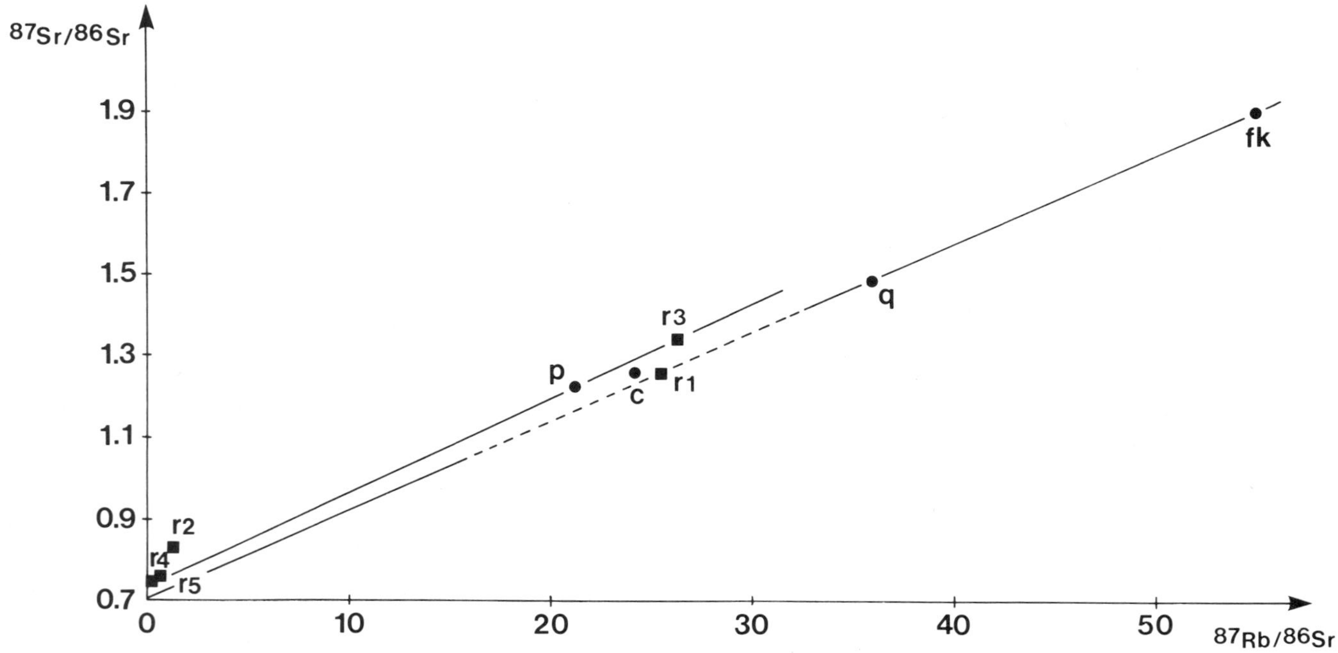

Figure 1. *Rb-Sr isochron-diagram for the rock of the granitic body, some of its components and the four fracture coatings (p = plagioclase; c = mixture biotite-chlorite; r = whole rock; fk = k-feldspar; m = muscovite).*

(2) the saline waters contain Na-Ca-Cl and SO_4 as major components and therefore differ significantly from the shallow groundwaters which generally are of a Ca-Mg-Na-HCO_3 type;

(3) the saline waters from Stripa site have a very high pH which increases with depth up to values between 9 and 10, the alkalinity decreasing concomitantly;

(4) the Ca/Mg ratios of the waters are exceptionally high for low temperature waters: the Ca/Mg ratio is 100 to 200 times higher in the most saline waters than in any known seawater, which is apparently due to a low Mg amount, unusual for cold waters;

(5) anomalies were also detected in the ratios of halogenes, when compared to seawater or related evaporitic waters: higher Br/Cl and I/Cl ratios for instance.

These different anomalies in the chemical composition of the groundwaters of the Stripa site were confirmed by a further study using stable isotope geochemistry (Nordstrom et al., 1985; Fritz et al., 1985).

The aim of the present study is to provide new information and arguments to the discussion and debate on the origin of the unexpected salinity found in deep groundwaters of granitic environments, in focusing on two complementary methods: (1) a geochemical and geochronological Rb-Sr approach on solutions and rocks from the Stripa site and (2) a geochemical modelling (speciation calculations) for the Ca-Mg-HCO_3-CO_2 values of the waters with respect to some components of the rocks. The study is still in progress, but it already might provide some new insights to the above summarized results.

THE RB-SR GEOCHRONOLOGICAL AND GEOCHEMICAL STUDY

The Rb-Sr method was applied on two kinds of material from the Stripa site: (1) on some rocks and minerals from granitic body and from fracture coatings, and (2) on some solutions collected in the fractures and fissures of the main body. The chemical and analytical procedures are similar to those published by Clauer (1976). The usual constants were used, especially $\lambda^{87}Rb = 1.42 \times 10^{-11}\ a^{-1}$ for the age calculations.

Rb-Sr data on the rocks

Five samples were taken from cores of the V1 drilling in the Stripa mine (Table 1). One of these cores is representative of the granitic body; it was a cylinder with a radius and a length of 6 and 20 centimeters, respectively. The four other samples are fracture coatings which were scraped off the fracture surfaces cutting the granite. The thickness of these coatings were of 1 to 2 millimeters.

The major components from the sample of the granitic body were separated, purified and analyzed apart: muscovite (subscript m), microcline including a few (less than 3%) plagioclase impurities detected by XRD (fk), plagioclases with a little microcline and quartz impurities (p), a mixture of 85% chlorite and 15% biotite (c), and a quartz fraction (q) were obtained. The analytical data of all the samples are scattered in a $^{87}Sr/^{86}Sr = f\,(^{87}Rb/^{86}Sr)$ diagram (Figure 1). This suggests a complex history for the granitic body, as the mineral separates especially do not fit on an isochron with their whole rock. The plagioclase fraction gives the highest apparent age together with its whole rock (r3) at about 1.63 Ga with an initial $^{87}Sr/^{86}Sr$ ratio at

Table 1.

Description of the rocks and mineral separates studied by the Rb-Sr method.

Sample numbers	Location in the drill-holes (1)	Sample description
r1	− 42.9	Fracture coatings with quartz, microcline, plagioclases and traces of chlorite, sericite and calcite
r2	− 47.5	—″—
r3	−318.5	Whole rock of the granitic body and following separates:
m		muscovite
fk		microcline and traces of plagioclases
p		plagioclases and traces of microcline and quartz
q		quartz with microcline intergrowths
c		85% chlorite mixed up with 15% biotite
r4	−496.7	Fracture coatings with quartz, microcline, plagioclases and chlorite
r5	−502.2	—″—

(1) locations are depths in meters underneath the Stripa mine level

Table 2.

Rb-Sr data for the rocks and mineral separates of the Stripa site.

Sample numbers (1)	Rb (μg/g)	Sr (μg/g)	$^{87}Rb/^{86}Sr$	$^{87}Sr/^{86}Sr \pm 2\,\sigma / \sqrt{N}$ (2)
r1	559.	67.5	25.23	1.25195 ± 14
r2	197.6	469.	1.234	0.82981 ± 6
r3	237.	27.9	26.14	1.33803 ± 4
m	921.	5.99	908.2	11.3690 ± 15
fk	497.	29.3	54.8	1.89580 ± 10
p	243.	35.0	21.12	1.22160 ± 12
q	39.8	3.46	35.81	1.48111 ± 13
c	314.	39.7	24.09	1.25525 ± 11
r4	112.7	619.	0.529	0.75078 ± 7
r5	100.5	934.	0.313	0.74529 ± 4

(1) the signs for the different minerals are given in the text
(2) the errors are given in 10^{-5} except for the muscovite m for which it is in 10^{-4}

Table 3.

Rb-Sr data for the solutions collected in three drill-holes at Stripa.

Sample numbers (1)	Location in the drill-holes (2)	Rb (mg/l)	Sr (mg/l)	Rb/Sr	$^{87}Sr/^{86}Sr \pm 2\,\sigma / \sqrt{N}$ (3)
M_3	− 5 m	0.001	0.13	0.008	0.74056 ± 9
V_1	−505 m	0.010	1.51	0.007	0.75360 ± 6
V_2	−822 m	0.005	0.76	0.007	0.74748 ± 5
Seawater (Faure 1982)	—	0.110	7.60	0.0145	0.70906 ± 3

(1) sample numbers are drill-hole references
(2) locations are depths underneath surface
(3) errors given en 10^{-5}

about 0.748. A value of 0.748 for the Sr initially present is too high to represent the time of rock formation. Such values are indeed rare for Precambrian plutonic rocks which rather contain initial Sr with $^{87}Sr/^{86}Sr$ ratios around 0.705. A reasonable minimum age for the granitic rock studied here can therefore be assumed at 1.71 Ga with an initial $^{87}Sr/^{86}Sr$ of 0.705. This estimate is close to a K-Ar age determination of 1691 ± 16 Ma reported by Wollenberg et al. (1980). The Rb-Sr apparent age for the microcline and the whole rock is significantly lower at 1.39 Ga and that by combining the muscovite with the whole rock is even lower at 0.80 Ga suggesting a very late crystallization. The position of the chlorite-biotite mixture close but below the whole rock could indicate a chloritization of the primary biotite in an open system with a preferential loss of radiogenic ^{87}Sr in relation to ^{87}Rb and ^{86}Sr.

The fracture coatings scraped from the cores collected at great depth (r4 and r5 in Table 1) contain mineral parageneses similar to that of the granitic body, with quartz, microcline, plagioclases and chlorite which is sometimes present in large amounts. These mineral assemblages are typical for high temperature conditions as no low temperature weathering can be suggested to explain them. The two samples plot on the line drawn through the granitic whole rock and its plagioclase fraction. This relationship suggests that these coatings could have formed at the same time than the plagioclases, perhaps during early hydrothermal activities in the granitic body, which could have been contemporaneous to fracturing during cooling of the pluton.

The fracture coatings collected near the surface contain a mineral paragenesis similar to that of the deeper levels (r1 and r2 in Table 1). However these shallow samples contain also calcite and sample r1 also has a large amount of sericite. Sample r2 plots away above the plagioclase-whole rock r3 line, while sample r1 plots below. The near surface coatings probably underwent a complicated post-formational history, which might include weathering processes and which cannot be clearly deciphered by the methods utilized here.

Rb-Sr data on the solutions

Three waters were analyzed by the Rb-Sr method. Two samples were taken from deep levels in drill holes V1 and V2 and one was collected at a shallow depth (Table 3). All three samples are strongly depleted in Rb and Sr if compared to seawater; 10 to 100 times for Rb and 5 to 60 times for Sr. Their $^{87}Sr/^{86}Sr$ ratios, however, are significantly more radiogenic, 0.741 to 0.754 while the mean seawater value is 0.709 (Faure, 1982). The very low Rb/Sr ratio of about 0.007 excludes the possibility of an evaporitic concentration of marine origin, by either an in situ enrichment of radiogenic ^{87}Sr from decay of ^{87}Rb, or by leaching of alkali-rich salts. Carnallite, for instance, has a Rb/Sr ratio greater than 400 (Oesterle, 1974) and neither the Rb and Sr contents, nor the $^{87}Sr/^{86}Sr$ ratio can be explained by simple dilution of seawater by groundwaters. The Rb/Sr ratios of the Stripa waters are 0.007 to 0.008 and that of the seawater is 0.015. No mixture of these two members will produce the desired solutions. To generate such solutions, the dilute end member must be very depleted in Rb and Sr and have $^{87}Sr/^{86}Sr$ ratios as high as 0.800. This is an abnormally high value for waters, but it is still lower than the $^{87}Sr/^{86}Sr$ ratios of all major components of the rocks (Figure 2).

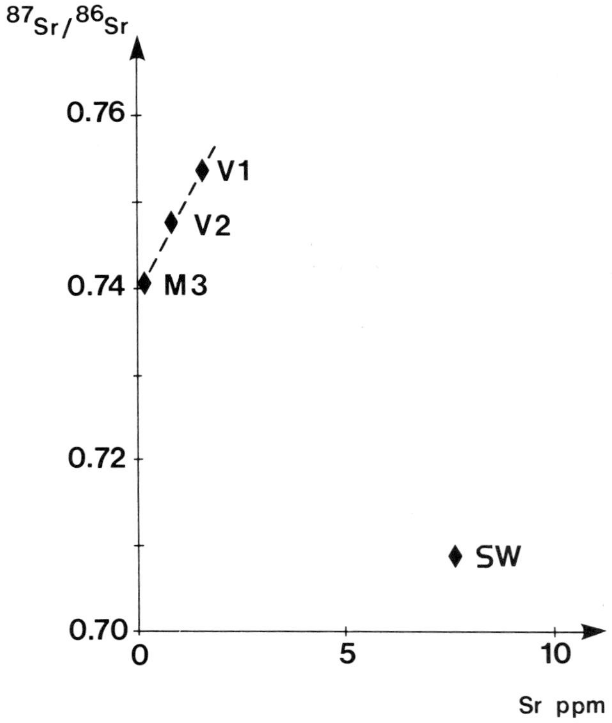

Figure 2. *$^{87}Sr/^{86}Sr$ ratio vs Sr content of the Stripa waters compared to the standard seawater values.*

The $^{87}Sr/^{86}Sr$ ratios of the deep groundwaters (V1 and V2) are very close to those of the two deep fracture coatings. These coatings are assumed to have been developed closely after intrusion of the granitic body. It cannot be excluded, however, that the whole rock-plagioclase apparent age, as well as those obtained on the other mineral fractions result partly from an open system behaviour at a small scale due to the hand specimen size of the whole rock and that subsequently all the ages have no real geological meaning. We, nevertheless, believe that the age of the waters is much younger than that of the granite itself, and therefore propose that the waters were initially dilute waters which subsequently leached the fracture zones of the granite.

CALCIUM, MAGNESIUM AND ALKALINITY CONTROLS OF THE STRIPA WATERS AND THERMODYNAMIC CONSTRAINTS

The origin of the salinity measured in the Stripa waters cannot be determined directly, it can be estimated through a geochemical modeling. The first step is the identification of the source for salinity. Since analytical results have excluded simple mixing of aqueous sources, such as seawater and groundwaters, the origin of salinity has to be related to the rock-body itself. This means a long term solution-rock interaction involving either a leaching of fluid inclusions and/or a reaction with minerals in the "porosity" (fractures and fissures) of the rock. Such a long term interaction should have given specific signature to water chemistry. Specifically, this could explain the particularity of the Ca-Mg-pH-pCO_2 system.

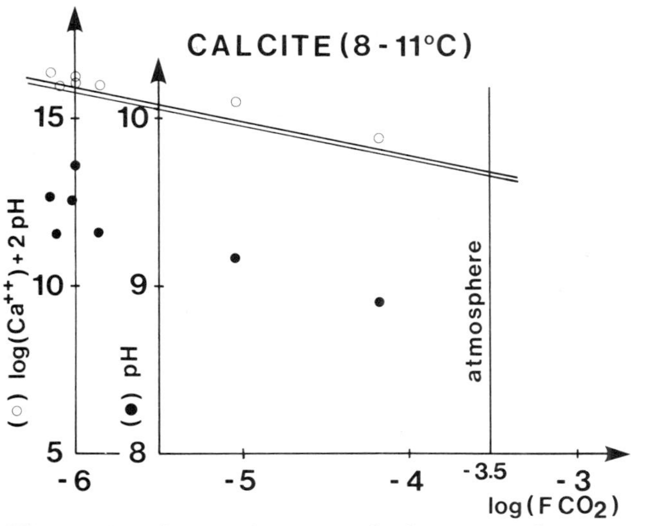

Figure 3. *Saturation state of solutions with respect to calcite and pH in a [Ca + +], [H+], FCO_2, activity diagram at field temperatures (8 to 11°C).*

Calcite control, pH and pCO_2

A proportion of salinity in the Stripa waters is due to the Ca content. In the most concentrated solution published by Nordstrom et al. (1985), that is the sample 81 AW 208, the results are:

Na	= 304 mg/l	= 13.2 mMoles/l
Ca	= 170 mg/l	= 4.24 mMoles/l
K	= 2.5 mg/l	= 0.064 mMoles/l
Mg	= 0.32 mg/l	= 0.013 mMoles/l
Cl	= 700 mg/l	= 19.74 mMoles/l
SO_4	= 95 mg/l	= 0.99 mMoles/l
Alkalinity (total)		= 0.21 meq./l
pH	= 9.27	

The pH above is high enough to induce chemical speciation with a very high $[Ca^{2+}]/[H^+]^2$ activity ratio. This favours saturation with respect to calcite (Figure 3) even when the alkalinity is very low as is the case here. It also induces a very low equilibrium value for fCO_2. This can be seen in figure 3, where different waters from the Stripa site are plotted. The activity ratio $[Ca^{2+}]/[H^+]^2$ and the carbon dioxide fugacity were calculated using the speciation calculation model EQUIL(T) at 8 to 11°C (Fritz, 1981). The waters are slightly oversaturated with respect to calcite over a very wide range of fCO_2 values, 2.7 orders of magnitude below atmospheric levels. This is an important argument for discussion, because it underscores that high pH values and high Ca are consistent with low carbon dioxide values at saturation with respect to calcite, the most common carbonate mineral. The high Ca/alkanity ratio (10 to 20) allows this saturation without the presence of abundant calcite in the fractures. In addition, the very low calculated carbon dioxide values show also that sampling and pH measurements have not been significantly disturbed by the atmosphere and that such values are consistent with a long-term reaction for a water-rock system closed to the atmosphere and without sources of inorganic carbon.

Magnesium control

The Ca content and the alkalinity of the Stripa waters are clearly related through an obvious calcite control. The Mg, however, is too low to be involved in a carbonate phase, and the solutions are significantly undersaturated with respect to magnesite and dolomite. The very low Mg concentration cannot be explained only by the occurrence of a trace mineral, which cannot produce such a depletion of Mg concentration. The pH values are, however, high enough to produce high $[Mg^{2+}]/[H^+]^2$ activity ratios, which have been considered in the chlorite-kaolinite diagram (Figure 4) between 8 and 11°C. It can be noticed that the Stripa waters with the highest pH values, in other words those which are not contaminated by atmospheric CO_2 plot around the saturation state for the Mg-chlorite or clinochlore. If this is true, it could be assumed that the low Mg content could be related here to a low temperature chlorite control on the chemistry of the solution. This kind of control is more usual in hydrothermal waters, at higher temperature, higher silica content but lower pH value. This conclusion represents, in fact, a second argument in favour of relations between the chemistry of the waters and the mineral components of the fractures in the granitic body.

DISCUSSION

The results obtained by a combined isotopic and thermodynamic study suggest that any seawater origin, directly through mixing or indirectly through evaporitic concentration, can be disregarded for the deep groundwaters sampled in the Stripa granite. As previously for the Br/Cl ratio for example, which has similar values in the solutions and rocks of Stripa, but different from seawater, the Sr isotopic data cannot be explained by a mixing of a marine endmember diluted by groundwater.

The waters in the granitic body of Stripa more likely result from a rock-water interaction and from a "fracture leaching", which might include a possible leaching of fluid inclusions such as those of the quartz grains. This hypothesis is, furthermore, consistent with results obtained by hydrological studies which show a very low regional hydraulic flow (Carlsson and Winberg, 1983).

The groundwaters are probably not responsible for the fracture coatings. More likely they may have resulted from long term interactions with their minerals. As they have low Rb/Sr ratios compared to the fracture minerals, it could be that they preferentially leached the minerals having low Rb/Sr ratios, such as chlorite and/or epidote, taking in account their saturation with respect to quartz. This kind of interaction would then explain the calculated saturation state and their low Sr and Mg concentrations.

A further support of the water-rock interaction hypothesis would be to determine if leaching of fluid inclusions could have occurred in the microfractures of quartz grains, even though quartz is stable in the solutions presently in the pluton. To test that possibility, a quartz fraction was separated and its purity was checked by XRD. The Rb and Sr amounts and the $^{87}Sr/^{86}Sr$ ratio of this fraction are much too high to be assigned to quartz only, even if the XRD control did not show any contamination (Table 2). Further binocular observation showed tiny intergrowths with microcline. Nevertheless, the analytical result allows some crude mass balance calculations for the repartition of Rb and Sr between quartz and microcline as their contents are

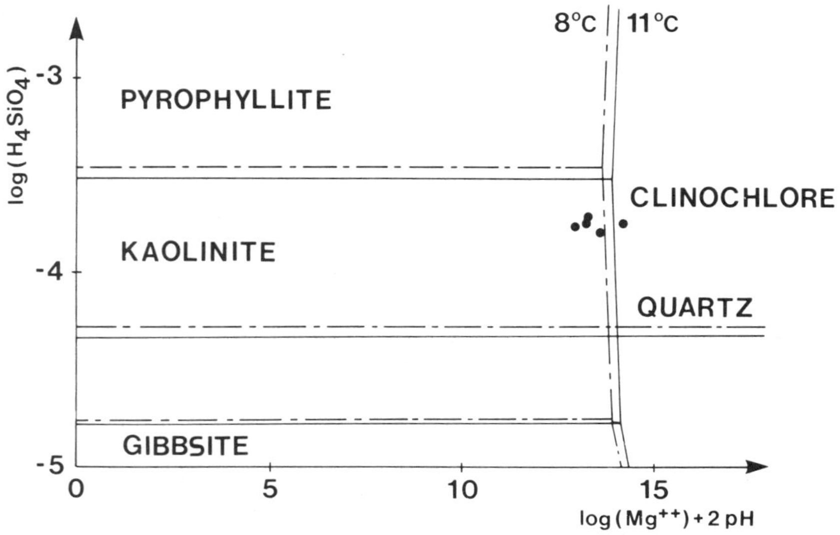

Figure 4. *Saturation state of solutions with respect to gibbsite, kaolinite, pyrophyllite and clinochlore in a [Mg^{++}], [H^{+}], [H_4SiO_4], Khorzinsky's diagram at field temperatures (8 to 11°C).*

known in the microcline end-member. If the entire Rb amount measured in the mixture is assumed to be in the microcline, the mixture would contain 8% microcline at the most as some Rb is probably also present in the fluid inclusions of the quartz grains. This percentage is, in fact, unrealistically high, as it would result in an $^{87}Sr/^{86}Sr$ ratio of 0.603 for quartz inclusions, which is impossible since the $^{86}Sr/^{86}Sr$ ratio of terrestrial material was never lower than 0.699. Considering 7.5% microcline contamination allows calculation of a $^{87}Sr/^{86}Sr$ ratio of 0.757 for the quartz fluid inclusions, which is fairly close to the values in the waters. If this assumption is correct, the Rb/Sr ratio of the quartz fluid inclusions would be close to .05; subsequently if they contributed to the chemistry of the groundwaters, the Rb/Sr of the latter would have to be about 0.5 and not the measured 0.008, a difference of two orders of magnitude. Furthermore, the line drawn through the microcline fraction and the quartz microcline mixture has an initial $^{87}Sr/^{86}Sr$ ratio significantly lower than the value of 0.757 obtained by the above mass balance calculation (Figure 1). Close to the intercept of this line is located the second end-member (e.g. the quartz fluid inclusions) of the quartz-microcline mixture. In summary, the available results suggest that if fluid inclusions contributed at all to the salinity of the Stripa groundwaters, they could not have originated solely from quartz grains of the granitic body.

CONCLUSIONS

The Rb-Sr geochemical and geochronological study on rocks, minerals and groundwaters from Stripa granite in Sweden, combined with thermodynamic considerations of alkalinity with Ca and Mg controls in the waters, point to the following scenario for the origin of water salinity.

1. The magmatic intrusion of the Stripa body occurred at least 1.71 Ga ago.
2. A later hydrothermal activity which may occurred 1.63 Ga ago and which may be related to post-emplacement cooling, produced fracture coatings. These coatings are characterized by mineral parageneses similar to that of the rock matrix. This similarity argues for a high temperature origin of the deep coating minerals, while some in the shallower underwent more complicated processes.
3. The granitic body could have underwent further events.
4. Recently (hundred, thousand, million years ago), dilute waters entered the granitic body, predominantly along drains, such as the previously formed fractures. These waters interacted with minerals having low Rb/Sr in the fractures (e.g. chlorites) and leached their Sr.

The hypothesis proposed here for the origin of salinity in the Stripa groundwaters is based on a long-term rock-water interaction. It clearly discards a marine origin for the primary waters, but does not give a definite answer as to the necessity of salinity contribution from quartz fluid inclusions. Further investigations are in progress for resolution of this ambiguity.

ACKNOWLEDGEMENT

We would like to thank B. Kiefel, D. Tisserant, R.A. Wendling and R. Winkler from Centre de Sédimentologie et Géochimie de la Surface (CNRS) for technical assistance, two anonymous reviewers for improving discussion and suggestions, and the SKB Company (Stripa Project) in Stockholm for partial financial support of this research.

REFERENCES

Carlsson, L. and Winberg, A. 1983. Model calculations of the groundwater flow at Finnsjön, Gideo and Kamlunge: SKB Stripa Project Report 83-45, 208 p., Stockholm.

Clauer, N. 1976. Géochimie du strontium des milieux sédimentaires. Application à la géochronologie de la couverture du craton ouest africain. Sciences Géologiques Mémoire, Strasbourg, 45, 256 p.

Faure, G. 1982. The marine-strontium geochronometer. In: Numerical Dating in Stratigraphy, G.S. Odin (Ed.), J. Wiley and Sons Ltd, p. 73-79.

Fritz, B. 1981. Etude thermodynamique et modélisation des réactions hydrothermales et diagénétiques: Sciences Géologiques Mémoire, Strasbourg, 65, 197 p.

Fritz, P., Frape, S.K., and Miles, M. 1985. Methane in the crystalline rocks of the Canadian Shield: *This volume.*

K.B.S. Report 1978. Handling of spent nuclear fuel and final storage of vitrified high level reprocessing waste. II. Geology, 161 p.

Nordstrom, D.K., Andrews, J.N., Carlsson, L., Fontes, J.C., Fritz, P., Moser, H. and Olsson, T. 1985. Hydrogeological and hydrogeochemical investigations in boreholes: Final report of the phase I geochemical investigations of the Stripa groundwaters. SKB Stripa Project Report 85-06, 250 p., Stockholm.

Oesterle, F.P. 1974. Radiogenes Strontium in Saltzlagerstätten am Beispiel des Oberrheinischen Lagerstätte und der von Neuhof – Ellers: Unpublished PhD Thesis, Univ. of Heidelberg, 174 p.

Wollenberg, H., Flexser, S. and Andersson 1982. Petrology and Radiogeology of the Stripa pluton: Lawrence Berkeley Laboratories Report LBL 11654, SAC-36.

The Origin of Saline Groundwaters in the Carnmenellis Granite, Cornwall (U.K.): Further Evidence from Minor and Trace Elements

W.M. Edmunds, R.L.F. Kay, D.L. Miles and J.M. Cook
British Geological Survey, Wallingford, OX10 8BB U.K.

Edmunds, W.M., Kay, R.L.F., Miles, D.L. and Cook, J.M., The origin of saline groundwaters in the Carnmenellis Granite, Cornwall (U.K.): further evidence from minor and trace elements; in Saline water and gases in crystalline rocks, Editors: Fritz, P., and Frape, S.K.; Geological Association of Canada Special Paper 33, 1987. p. 127–143.

Abstract

Thermal groundwaters (up to 55°C), some of them saline (up to 19 300 mg l^{-1} total mineralisation), occur as inflows in tin mines in the Hercynian Carnmenellis Granite at depths between 200 and 700 m below ground level. These waters are recognised as meteoric and have circulated to depths of c.1200 m. The saline waters are mixtures, in varying proportions, of a recent, low salinity, tritium-containing component and an older, saline component, possibly with total mineralisation as high as 30 000 mg l^{-1}, which is of the order 10^4-10^6 years old. In this paper the chemistry of waters occurring naturally in the granite is compared with that of waters which have circulated to a depth of over 2 km in the UK Hot Dry Rock experiment.

Acid hydrolysis of plagioclase and biotite are proposed as the principal origins of the groundwater salinity and the distinctive ionic compositions. Enriched Ca/Na ratios compared with bulk rock compositions are achieved in the groundwaters by selective reaction of the more calcic centres of zoned plagioclase. The groundwaters contain high lithium concentrations (up to 125 mg l^{-1} Li^+) which can be related to stoichiometric biotite reaction; chloride, which is enriched by a factor of about 20 over Li^+, is considered to be derived from hydroxyl exchange for Cl^- in the biotite interlayer, without structural breakdown. This reaction also generates H^+ which may further drive the hydrolysis reactions.

The dissolved concentration of several minor and trace elements (Li, Sr, Ba, F, B, Br, Rb, Cs) can be related to reaction of aluminosilicate minerals. Other elements, such as Mn, As, Al, Cd, Cu, Co, Be, Ge, Ni and Zn, are detected in acidic and shallow minewaters. These elements may either be derived from oxidation of sulphide minerals or released from oxides, the latter possibly being an important adsorption sink for some of the trace elements originally derived from silicate breakdown reactions.

Résumé

On trouve aux Cornouailles dans le sud-ouest du Royaume Uni, des eaux souterraines thermales (jusqu'à 55°C) et parfois salées (jusqu'à 19 300 mgl^{-1} de minéralisation totale) qui affluent dans les mines d'étain du granite hercynien de Carnmenellis à des profondeurs variant entre 200 et 700 m. L'origine météorique de ces eaux, qui ont circulé à des profondeurs d'environ 1200 m, a déjà été établie. Les eaux salées sont un mélange comprenant, dans des proportions variables, une composante récente à faible salinité renfermant du tritium et une composante salée plus ancienne pouvant atteindre 30 000 mg l^{-1} de minéralisation total et âgée de 10^4 à 10^6 ans. Nous effectuons dans cet article une comparaison entre la composition chimique de ces eaux, qui se présentent à l'état naturel dans le granite, et celle des eaux qui ont circulé à des profondeurs de plus de 2 km et ont fait l'objet de l'expérience britannique "Hot Dry Rock".

L'hydrolyse acide du plagioclase et de la biotite est, dans une large mesure, à l'origine de la salinité de l'eau souterraine et des compositions ioniques caractéristiques. Le rapport Ca/Na des eaux souterraines, plus élevé que celui des échantillons mégascopiques du massif résulte d'une réaction sélective avec le coeur de plagioclases zonés, de composition plus calcique. Les eaux souterraines renferment également des teneurs élevées en lithium (jusqu'à 125 mg l^{-1} de Li^+) pouvant être associées à une réaction stoechiométrique de la biotite. L'enrichissement en chlore, environ 20 fois plus élevé que celui en Li^+, pourrait provenir d'un échange entre les hydroxyles et le Cl^- des biotites intercalaires, effectué sans effondrement structural. Cette réaction génère aussi du H^+ pouvant à son tour alimenter la réaction d'hydrolyse.

La concentration de quelques autres éléments traces et éléments mineurs dissous (Li, Sr, Ba, F, B, Br, Rb, Cs) peut être associée à la réaction des aluminosilicates. D'autres éléments, tels que Mn, As, Al, Cd, Co, Cu, Be, Ge, Ni et Zn sont présents dans les eaux acides et peu profondes des mines. Ils peuvent résulter de l'oxydation des sulfures ou des oxydes, ces derniers pouvant constituer un puits d'absorption important pour les éléments issus de la dégradation des silicates.

INTRODUCTION

Saline groundwaters (mineralisation >10 000 mg l^{-1} TDS) have been described from a number of altered granites and related igneous rocks around the world and some controversy exists concerning their mode or modes of origin. These include: 1) marine transgressions and subsequent concentration (Frape and Fritz, 1982), 2) migration of sedimentary formation brines (Fritz and Frape, 1982), 3) residual hydrothermal fluids (Alderton and Sheppard, 1977), 4) dissolution of grain boundary salts (Grigsby *et al.*, 1983), 5) silicate mineral hydrolysis and related water-rock interaction (Edmunds *et al.*, 1984; 1985; Frape *et al.*, 1984), 6) breakdown of fluid inclusions in quartz and other minerals (Nordstrom, 1983), 7) radiolytic decomposition of water during α-series decay (Vovk, 1981). Some of these authors also discuss the possibility that several processes may have been involved in producing the observed chemistries, especially where extended timescales, possibly exceeding 10^9 years, may have been involved.

Most of the best described occurrences are from Precambrian granitic rocks. These include highly saline waters with total mineralisations up to 325 g l^{-1} from the Canadian Shield (Fritz and Frape, 1982) and calcium chloride brines up to 550 g l^{-1} TDS from the Russian platform (Vovk, 1981). Enrichment of calcium over sodium is a common feature of such brines.

The saline waters of the Carnmenellis Granite may provide important evidence for the origin of groundwater mineralisation. This is because they occur not in ancient crystalline rocks but near the roof of a much younger (Hercynian) granite which has had a relatively simple geological history. Although Carnmenellis groundwater salinities are lower than those in the Shield areas, they are considerably higher than those in other granitic plutons. For example, at Los Alamos, New Mexico (Grigsby *et al.*, 1983) and at Stripa, Sweden (Nordstrom, 1983), the maximum chloride concentration is only about 600 mg l^{-1}. No other enhanced salinities from high-level granites of Hercynian or more recent age have yet been recorded. Low chloride concentrations have been recorded from granitic terrains, for example, in the Bohemian Massif (Paces, 1972), the Sierra Nevada, USA (Feth *et al.*, 1964), and the Pyrenees (Fouillac and Michard, 1981).

Cornish tin mines have yielded saline and thermal groundwaters for more than 100 years (Miller, 1865; Phillips, 1873). Maximum salinities of 19 300 mg l^{-1} TDS (11 890 Cl^-) and temperatures of up to 55°C have been recorded. These are the only saline groundwaters known from igneous rocks in Britain and their origin has been the subject of debate since the metalliferous mines were first developed. The fluids contain very high concentrations of lithium, up to 125 mg l^{-1} (Beer *et al.*, 1978), as well as enhanced levels of other minor elements.

This paper reviews the geochemical evidence for the origin of the salinity of Carnmenellis groundwaters. New data for minor and trace elements in the saline and related groundwaters from the granite are presented to improve the overall model of the origin of the salinity.

THE CARNMENELLIS GRANITE

The Carnmenellis Granite is one of a series of granite bosses exposed in SW England (Figure 1). These were believed to be formed by denudation of a large batholith but recent evidence suggests that they might not be connected at depth (Brooks *et al.*, 1984). The Carnmenellis Granite was intruded about 285 Ma ago into Devonian argillaceous sediments which had been regionally metamorphosed to greenschist facies, producing a contact aureole including cordierite and andalusite (Figure 1). The granite is a micaceous adamellite with primary quartz, plagioclase (zoned oligoclase with An-rich cores), orthoclase (generally perthitic), biotite and/or muscovite (Exley and Stone, 1964). Together with the other SW England granites, it is notably enriched in volatile elements (B, Cl, F, Li) compared with the average granite composition (Fuge and Power, 1969). Boron and fluorine metasomatism has given rise to extensive tourmalinisation and greisenisation.

There is extensive hydrothermal mineralisation, especially on the northern flank of the granite, which has produced economic Sn-W-Cu vein deposits, dated at about 270 Ma (Jackson *et al.*, 1982) and slightly later Pb-Zn mineralisation, which is generally further away from the granite contact. Later processes such as uranium mineralisation and kaolinisation are also recorded. The latter may represent the final stages of hydrothermal activity, involving significant amounts of meteoric water, and may still be active today (Durrance *et al.*, 1982).

Fluid inclusions are common in quartz and, to a lesser extent, in other minerals in the SW England granites; they are most widespread in the north of the Carnmenellis Granite (Rankin *et al.*, 1982), where there is more mineralisation. Rare primary inclusions must represent late-stage consolidation fluids that have passed through the granite, but several generations and types of fluid inclusions have been recognised and monophase inclusions of variable salinity are particularly common. Clearly, lower-salinity, moderate temperature inclusions are associated with the extensive Sn-W-Cu mineralisation; the high-salinity, halite-bearing inclusions show no clear relationship to mineralisation.

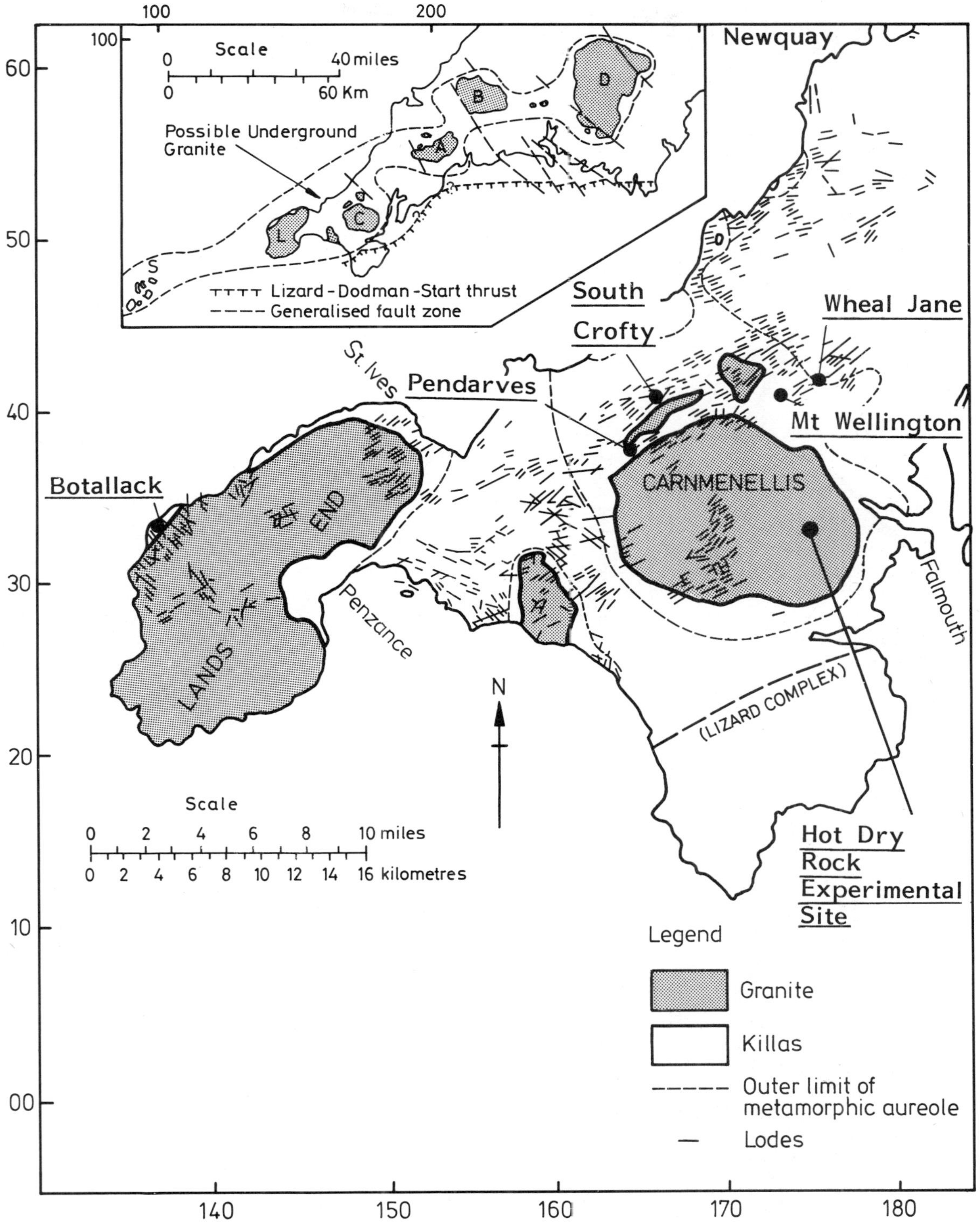

Figure 1. *Location map of the Carnmenellis Granite relative to other granites of the Cornubian Batholith (S – Scilly Isles, L – Land's End, A – St. Austell, B – Bodmin, C – Carnmenellis, D – Dartmoor). The location of active tin mines and the site of the UK Hot Dry Rock Experiment, from which water samples have been taken, are also shown.*

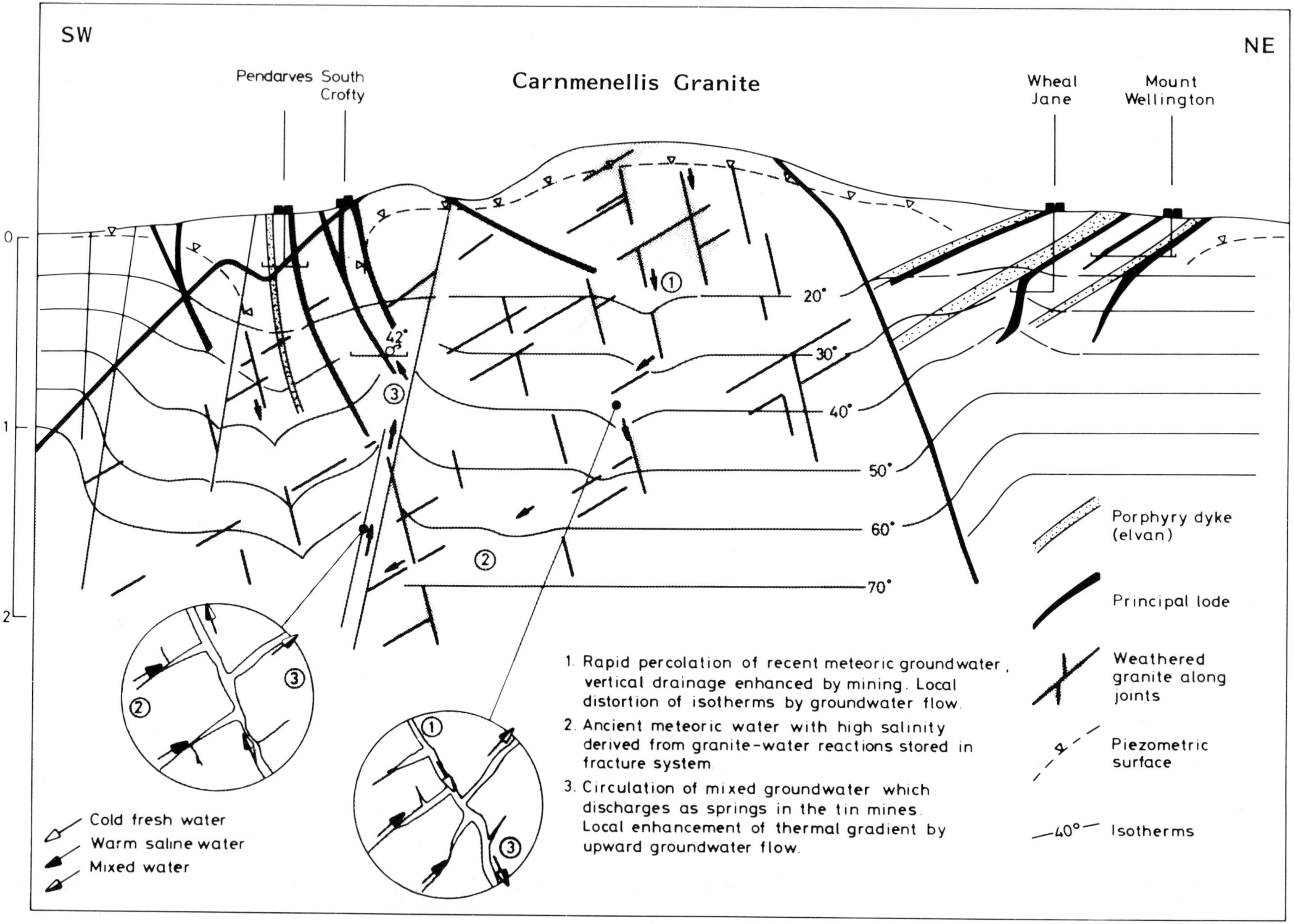

Figure 2. *Conceptual cross section through the Carnmenellis Granite showing the location of the thermal groundwaters and active tin mines. Isotherms are drawn in the granite and its aureole showing local convective distortion due to groundwater circulation and possible flow routes.*

SALINE GROUNDWATERS IN THE CARNMENELLIS GRANITE

The occurrence of thermal and saline groundwaters has been described by Edmunds *et al.* (1984). The main features are:

Location

Saline groundwaters have been encountered during mining almost exclusively at the northern margin of the granite and its aureole and are now found in two active tin mines: near 600 metres below sea level in the South Crofty Mine and in working levels below 150 metres below sea level in the Wheal Jane Mine (Figures 1 and 2). Typical water-bearing structures are cross-courses running NW-SE and thus normal to the mineral lodes. There is no record of any surface discharge of saline or thermal water.

Temperature

The average heat flow for the granite is about 120 mWm^{-2}, over twice the average for the UK (Wheildon *et al.*, 1977). The average thermal gradient is 29.8°C km^{-1} across the granite but rises to 50°C km^{-1} in the metapelitic aureole (Figure 2). Higher temperatures are therefore likely to occur at shallower depths in mines in the aureole. Discharge temperatures up to 55°C have been measured at 400 metres bsl at Wheal Jane during recent deepening of the mine. The most saline water at South Crofty has a nearly constant temperature of 41.5°C. Groundwater temperature is generally higher than that expected from the thermal gradient and implies upward convection of thermal water. Although it is recognised that they may be unreliable at low temperatures, SiO_2 and Na/K geothermometers suggest a maximum equilibration temperature of 54°C, which would imply circulation to a depth of at least 1.1 km.

Origin of the water

The thermal groundwaters are almost identical in isotopic composition to shallow non-thermal waters ($\delta^{18}O = -5.5 \pm 0.5‰$; $\delta^2H = -34 \pm 5‰$), indicating that they are most likely to be of local meteoric origin. The groundwater itself cannot therefore be derived from seawater; neither can there have been any significant oxygen and hydrogen isotopic interaction between water and rock. It further seems improbable that the salinity has a marine origin.

Mixing of recent and older groundwater

The range of salinities observed in the thermal groundwater is principally the result of mixing between the saline component and recent groundwaters. Rapid downflow of recent recharge is induced by the hydraulic sink created by the mining activity. All groundwaters contain tritium of thermo-nuclear origin (Andrews and Kay, 1982) and resampling in the mines during the present study (Table 1) yielded very similar tritium concentrations. The most saline water (the Hot Drive Source from South Crofty Mine) has a ^3H content of about 6 T.R. Tritium input to the complex hydraulic system is estimated to have been in the range 20-80 T.R. There is thus a large uncertainty as to the proportion of the saline end-member in the mixture. This could be as low as 70%, leading to a salinity near 30000 mg l^{-1} TDS for the saline component, although it is probably nearer the higher estimate of 94% giving a maximum salinity c. 21 000 mg l^{-1} TDS. A hundred years of saline water discharge equivalent to the South Crofty Hot Drive source represents (very approximately) 1% of the total water stored in the upper kilometre of the granite.

Age of the saline component

The estimation of groundwater ages is extremely difficult because all samples are probably mixtures of at least two components. Although the waters contain ^{14}C (42-72% modern carbon), no reliable age can be derived due to the complexity of the carbon geochemistry and the uncertainty of mixing proportions. The enhanced ^4He contents and the uranium series geochemistry suggest that, allowing for mixing, the saline component has evolved over a time-scale greater than 5×10^4 and probably of the order of 10^6 years.

Chemistry of the saline groundwaters

The major element characteristics of saline and dilute groundwaters are summarised in Figure 3. In addition to the enhanced salinity, the most important features are the depletion of Na^+ relative to Cl^- and the enhancement of Ca^{2+}. These are attributed to plagioclase hydrolysis which is discussed more fully below. In addition, it is important to recognise that the sources of Na^+ and Cl^- are different, an unusual hydrogeochemical feature. The magnesium to chloride relationship is not as simple as those for Na^+ and Ca^{2+}. The saline groundwaters are relatively depleted in Mg^{2+} when compared with the dilute waters. The low Mg^{2+}/Ca^{2+} ratios (about 0.1) imply that Mg^{2+} loss has occurred due to formation of chlorite or other hydroxyl silicates, a reaction which would be enhanced at higher temperatures and which may have affected the saline waters but not the dilute waters. The latter may owe their relative Mg^{2+} enrichment to congruent reaction of biotite. Potassium is also relatively enriched in the dilute groundwaters and, although some may be related to biotite and/or K-feldspar dissolution, there is also evidence of anthropogenic pollution in the form of enhanced K^+ (and NO_3).

Salinity generation during HDR circulation

The Carnmenellis Granite provides the site for the first European 'Hot Dry Rock' experiment (Batchelor, 1984). Three boreholes have now been drilled to a maximum depth of 2.8 km. Comprehensive chemical sampling has been undertaken during circulation experiments, the results of which have been presented elsewhere (McCartney, 1984; Edmunds *et al.*, 1985).

The mean residence time of water during the circulation experiments, determined using tracers, was about 770 hours. During this period, a net gain in solutes in the return water over the injection waters was observed for many elements including (mg l^{-1}): Na (40), Cl (12), Li (0.16), Si (9.0) and B (0.09). The enhanced and almost constant levels of these elements under steady-state conditions (reached after 600 hr) demonstrate significant water-rock interaction; the major ionic ratios are comparable to those observed in the natural system. Thus, it has been concluded that artificial cirulation is promoting hydrolytic processes similar to those which have produced the natural saline system. Progressive dissolution, leading to salinity increase, and dilution/mixing, resulting in lower salinities in thermal groundwaters, are therefore both occurring at the present time.

Experimental studies with biotite at 80°C and 1 atm have also demonstrated that Li and Cl, as well as other elements, are enhanced in the reaction fluids over successive five-day runs and provide further confirmation of the mineral hydrolysis model (McCartney, 1984; Edmunds *et al.*, 1985).

TRACE ELEMENTS – SAMPLING AND ANALYSIS

All accessible minewaters were resampled in 1984. Together with two samples from the HDR circulation experiment, these were analysed to provide the best state-of-the-art data set for minor and trace elements, and a double-

Table 1

Chemical analyses of selected Cornish minewaters and HDR circulation return waters.

		SOUTH CROFTY MINE (SW 670 398)			WHEAL JANE (SW 775 426)			WHEAL PENDARVES (SW 647 382)			HDR PUMP TEST RT2B01 (SW 735 345)	
SAMPLE		SC1	SC2	SC3	WJ1	WJ2	WJ3	P1	P2	P3	HDR1	HDR2
Depth		580	580	580	150	280	400	130	40	130	2400	2400
Date	mbsl	15/03/84	15/03/84	15/03/84	13/03/84	13/03/84	26/10/84	14/03/84	14/03/84	14/03/84	04/06/84	23/05/84
Temp.	°C	40	44.0	—	21.6	—	47.8	17.8	15.0	18.9	37	37
pH		6.65	8.0	3.9	3.5	6.3	8.3	6.35	6.25	6.35	8.3	8.35
Eh	mV	—	—	—	—	—	—	—	—	—	100	100
HCO_3	mgl^{-1}	60	61	18	12.7	24	54	73	58	56	88*	94*
Cl	"	9280	1900	2910	179	4460	6090	29	32.2	32	70*	71*
F	"	3.0	3.1	8.3	44	3.5	3.0	1.5	2.75	—	6.0*	5.7*
I	"	0.091	0.025	0.024	0.035	0.041	0.24	0.0027	0.0016	0.0029	0.028	0.032
Br	"	35	8.02	11.4	0.63	17.2	18.6	0.134	0.156	—	0.27	0.31
SO_4	"	129	80	260	1390	189	114	33	62	74	8.2	6.9
NO_3	"	<1.0	<1.0	<1.0	<1.0	<1.0	—	<1.0	<1.0	<1.0	—	—
HS	"	0.024	0.020	0.014	0.014	0.016	—	—	—	—	—	—
Ca	"	1840	573	675	191	995	1300	15.3	24	44	17.6	19.2
Mg	"	55	40	47	43	23	13	10.5	11.8	10	0.48	0.6
Na	"	3520	660	1020	93	1620	2090	26	25	26	72	67.2
K	"	153	29	49	12.0	100	132	4.4	2.9	2.7	2.8	2.7
Al	"	<0.1	<0.1	0.48	25	<0.1	<0.1	<0.005	0.013	<0.02	0.015	<0.006
Si	"	16.0	14.6	9.8	11.9	16	16	17.6	8.0	10.1	13.1	13.1
Sr	"	30	4.6	9.7	1.87	16.8	23	0.02	0.096	0.109	0.13	0.10
Ba	"	0.94	0.24	2.8	0.052	0.62	0.76	0.005	0.030	0.018	0.006	0.005
Li	"	107	17.7	31	2.7	49	64	0.048	0.025	0.028	0.209	0.16
B	"	13.9	2.1	3.9	<0.5	6.1	7.9	0.031	0.034	0.043	0.22	0.24
Fe	"	4.2	0.6	19.3	346	20.2	0.028	9.7	0.15	0.016	0.064	0.11
Mn	"	3.5	1.67	4.0	19.7	3.5	0.62	0.93	0.61	1.01	0.025	0.034
Ag	"	<0.01	<0.01	<0.01	<0.01	<0.01	<0.01	<0.0005	<0.001	<0.002	0.010	0.002
As	"	0.033	0.056	0.13	2.1	0.12	0.065	0.013	0.005	0.008	1.6	1.4
Be	"	<0.006	0.002	0.068	0.15	0.007	<0.001	0.0006	<0.0045	0.0007	<0.0001	<0.0001
Cd	"	<0.025	<0.025	<0.025	<0.025	<0.025	<0.025	<0.002	0.011	<0.004	<0.001	<0.001
Co	"	<0.02	<0.02	0.041	0.27	<0.02	<0.02	<0.001	0.011	0.16	<0.0008	<0.0008
Cs	"	4.28	0.41	0.68	3.9	3.25	4.58	0.0034	0.0043	—	0.031	0.031
Cr	"	<0.03	<0.03	<0.03	<0.03	<0.03	<0.03	0.006	0.006	<0.004	<0.001	<0.001
Cu	"	<0.008	<0.008	<0.008	<0.008	<0.008	<0.007	0.0016	0.32	0.047	<0.0004	<0.0003
Ge	"	<0.3	<0.3	<0.3	<0.3	<0.3	<0.3	0.018	<0.015	<0.04	0.025	<0.015
Mo	"	<0.1	<0.1	<0.1	<0.1	<0.1	<0.1	<0.005	<0.005	<0.01	0.061	0.061
Ni	"	0.075	0.075	0.13	0.86	0.084	0.075	0.018	0.022	0.03	<0.003	<0.003
Rb	"	2.5	1.7	0.47	0.21	1.51	2.2	<0.02	<0.02	—	<0.02	<0.02
U	"	0.000081	0.000067	—	0.000071	0.000019	0.044	0.000075	0.016	—	0.0003	—
W	"	<0.15	<0.15	<0.15	<0.15	<0.15	<0.15	<0.008	<0.008	<0.02	0.035	0.031
Zn	"	<0.02	<0.02	0.12	125	0.38	<0.02	<0.003	1.78	0.64	0.004	0.008
TDS	"	15250	3400	5080	2470	7550	9930	222	230	257	278	280
^3H	T.R.	6.1	13	23	17	7.0	3.0	21	21	—	—	—
δ^2H	‰ SMOW	−28	−38	−32	−35	−34	−37	−34	−35	−36	—	−31
$\delta^{18}O$	"	−5.3	−6.0	−5.9	−5.4	−5.8	−5.6	−5.6	−5.3	−5.7	—	−5.4

*indicates analyses performed by staff of the Camborne School of Mines Geothermal Energy Project check on tritium contents. Thermal and saline waters encountered during mining have been analysed over many decades and some of these historical data are also used in this paper.

Water samples were filtered *in-situ* through 0.45 μm membrane filters and collected in 500 ml polyethylene bottles. The latter had been initially washed in 50% v/v reagent grade HCl and rinsed with distilled water. Subsequently they were soaked in 10% (v/v) AristaR grade HCl, soaked and rinsed in deionised distilled water. Samples which were to be analysed for Br, I, F (as well as Cl, NO_3, HCO_3) received no further treatment. Samples to be analysed for trace elements (and also SO_4 and major cations) were acidified by the addition of 5 ml AristaR grade HCl. Iodide, bromide and nitrate were determined by automated colorimetry and fluoride was determined by ion selective electrode. Major cations and some minor elements were determined by aspirating samples directly into an ARL 34000C inductively-coupled plasma optical emission spectrometer (ICP-OES). Emission at the 180.73 nm sulphur line was monitored as an indicator of the sulphate concentration (Miles and Cook, 1982). Rubidium was determined by flame atomic absorption spectrometry and caesium by inductively-coupled plasma mass spectrometry (Gray and Date, 1983).

Four samples were preconcentrated and reanalysed by ICP-OES using both on-peak intensity measurements and wavelength scanning to establish trace element concentrations. Aliquots (400 ml) of the filtered and acidified samples were transferred to PTFE beakers and evaporated to 5 ml under infra-red lamps in a laminar-flow hood. The concentrated residues were washed into 20 ml volumetric flasks

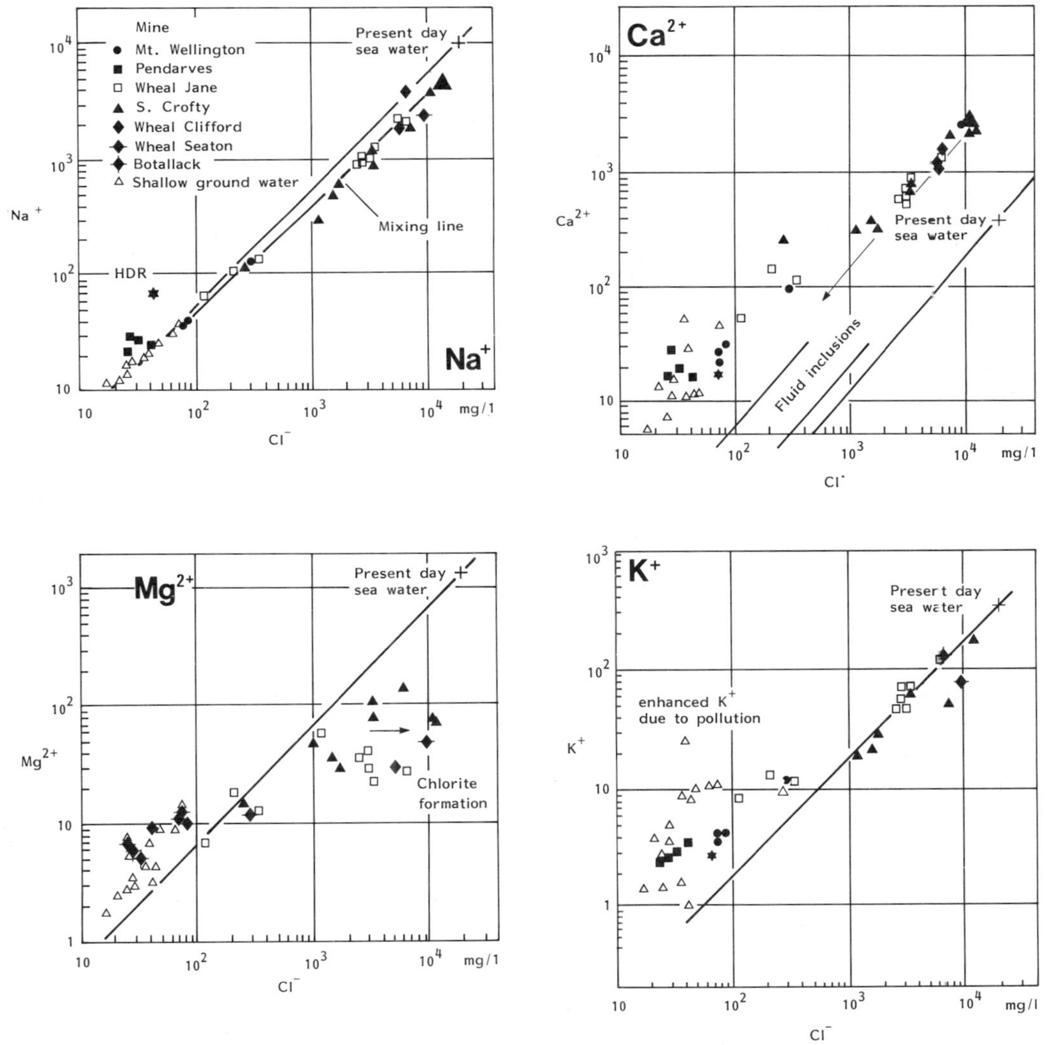

Figure 3. *Major element (Na^+, Ca^{2+}, Mg^{2+} and K^+) variations relative to chloride. Fluid inclusion ratios from quartz samples in the Carnmenellis Granite (Rankin et al., 1983) have been calculated on the basis of $Cl^- \equiv Na^+$. Trends for reaction and mixing are shown relative to the dilution line for present-day seawater.*

and made up to volume. Blank solutions, consisting of 400 ml of 1% (v/v) AristaR grade HCl in deionised/distilled water, were carried through the preconcentration procedure at the same time as the samples (Trafford, 1982 unpublished report). Precipitation of solids during evaporation, and spectral interferences caused by high concentrations of alkaline earth elements in the resulting solutions precluded the preconcentration of the more saline samples.

RESULTS

Analyses of selected groundwaters from the three mines are presented in Table 1 together with two analyses from HDR circulation fluids. Some 50 species were determined and data are compared in Figures 4 and 5. Positive values were reported for approximately 80% of the determinands; detection limits, appropriate to the analytical procedures employed, are quoted to provide reliable values for the upper limit of concentration for the remainder. For the dilute waters, which were preconcentrated by evaporation, the detection limits for many of the cations are correspondingly lower, generally by a factor of 20.

The following elements were not detected in any of the samples, and their detection limits in saline and dilute waters respectively (in mg l^{-1}) are: Bi 0.4, 0.02; Ce 0.05, 0.0025; Eu 0.008, 0.001; Ga 0.3, 0.015; Hg 0.3, 0.015; La 0.01, 0.0005; Pb 0.2, 0.01; Sc 0.0001, 0.00005; Se 0.8, 0.04; Sn 0.1, 0.005; Te 0.6, 0.03; Ti 0.5, 0.025; V 0.02, 0.0007; Y 0.001, 0.00005; Zr 0.02, 0.001.

The principal characteristics of each site may be summarised with reference to Figures 4 and 5.

South Crofty Mine

The relative abundance of the elements in the most saline water source from South Crofty is shown in Figure 4 and

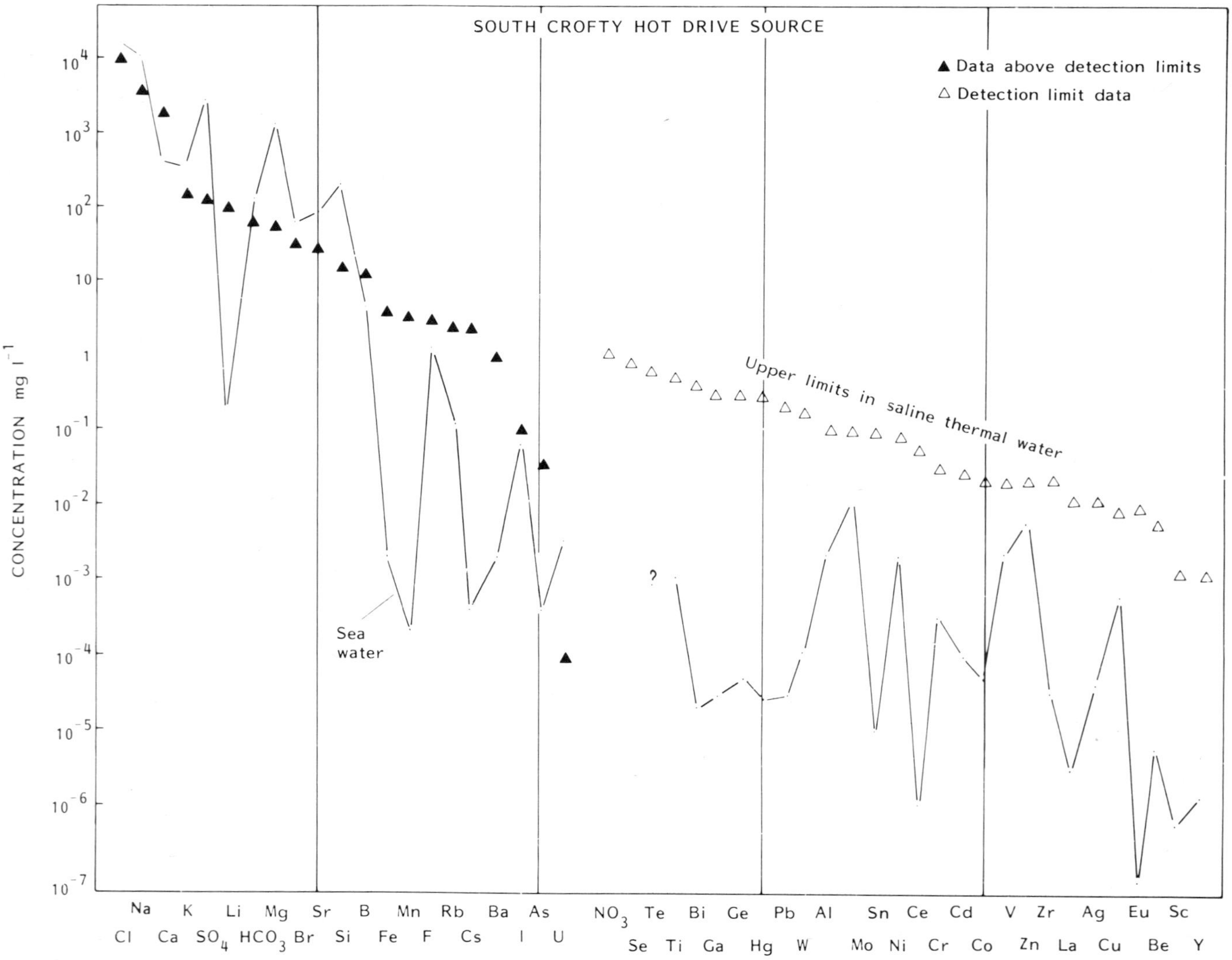

Figure 4. Multielement profile of the saline groundwater end member from South Crofty Mine. Detection limit data (△) indicate the reliable upper limits of occurence for many trace elements. Seawater composition (Bruland, 1983) is shown for reference.

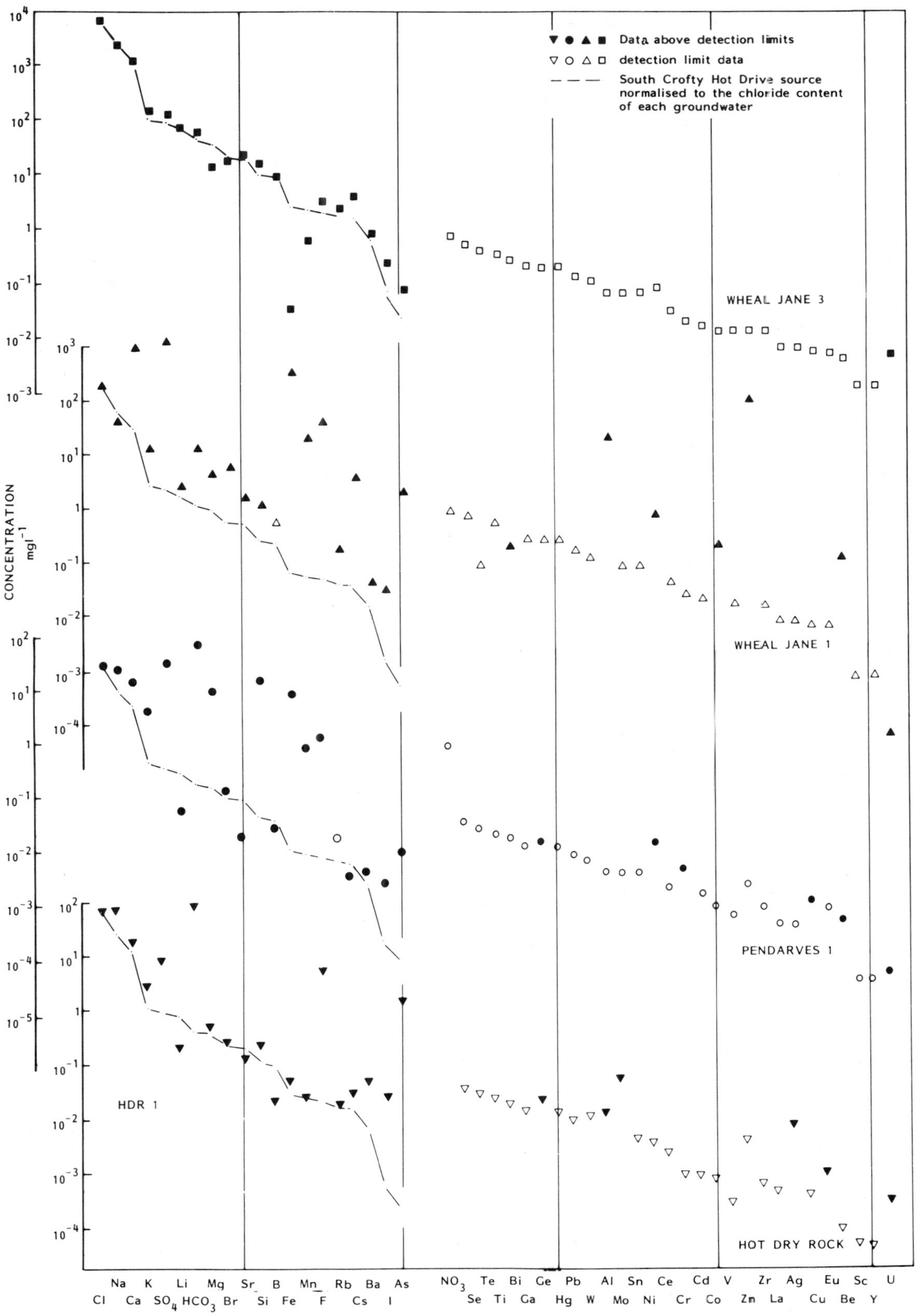

Figure 5. *Multielement profiles of waters from Wheal Jane, Wheal Pendarves and return-water from the steady-state conditions in the Hot Dry Rock circulation system. Shown for comparison is the saline water end-member at South Crofty normalised to the chloride content of each groundwater.*

compared with values for seawater (Bruland, 1983). This analysis, made in 1984, has a slightly lower salinity than that quoted in Edmunds *et al.* (1984), probably because of the effect of mining on the flow system. Two analyses of less saline waters (incorporating a greater proportion of the descending dilute component) from the same depth (580 m) as the saline inflow are listed for comparison in Table 1. Most minor and detected trace elements are considerably enhanced above equivalent seawater compositions, whilst magnesium and sulphate are the only exceptions among major species. Lithium is enriched by almost three orders of magnitude. One of the dilute minewaters is an acid sulphate water (pH 3.9) which contains enhanced levels of As, Mn, Fe, Ba, Al, F, Co, Cs, Be, Zn compared with the saline end member. Other elements in this and other waters are diluted approximately in proportion to chloride.

Pendarves

Wheal Pendarves contains low-salinity groundwaters, with pH in the range 6.0-6.5 and a maximum temperature of 18.9°C, which have evolved almost exclusively within the granite. These exemplify the short-residence-time waters, which have been almost neutralised during percolation through the rock mass. Concentrations are generally lower than those in the deeper waters, but elements such as Li and B are relatively much less enriched than in the long-residence-time groundwaters. The elements As, Co, Cr, Cu, Ge, Ni and Zn are present in detectable quantities at or above 0.001 mg l^{-1}.

Wheal Jane

Wheal Jane, developed in the metamorphic aureole, contains the warmest recorded groundwaters (55°C)* with salinities up to 9930 mg l^{-1} (6090 Cl^-). The element distribution in the most saline groundwater closely mirrors that in saline water from South Crofty (Figure 5), which implies a common origin and evolution within the granite. It is inferred that migration from the granite into the aureole has been of relatively short duration and has not modified the saline waters to any significant extent. Acidic, sulphate-rich waters (pH 3.5) also occur at Wheal Jane but these are at a relatively low temperature (22°C) and represent a short residence component of the flow system. The acidic waters are enriched in Fe, Mn, As, Ni, Zn and Be compared with the saline waters.

'Hot Dry Rock' reservoir

The return water from the HDR reservoir has had a short residence time, yet has reached temperatures up to or exceeding those of the mines; the surface discharge temperature is 36°C. The relative enrichment or depletion of elements, normalised to chloride in South Crofty, is shown in Figure 5. Sodium is enriched relative to Cl. Other elements – B and Li, for example – which characterise the long-residence saline waters are relatively depleted, although at higher levels than in the injection waters (McCartney personal communications, 1984). Bicarbonate, Si, F, and Ba are relatively high in the HDR circulation and As, Ag, Al, and Mo occur in detectable quantities. Iodine is also enriched.

*The highest temperature was recorded two months before the present sampling in a cross-course during mine development.

MINOR AND TRACE ELEMENTS AND THE EVOLUTION OF SALINITY

The minor element behaviour, as with the major elements (Edmunds *et al.*, 1984), can be described with reference to chloride. It is important to note that, contrary to most hydrogeochemical systems, chloride does not behave conservatively but is added to the water during its evolution. Although chloride may not be the best indicator of congruent mineral dissolution, once in solution it is unlikely to be incorporated in secondary minerals and is thus an index of the overall extent of reaction or mixing. Saturation indices quoted were calculated by WATEQF (Plummer *et al.*, 1976).

Bromide

Bromide, like chloride, is unlikely to be incorporated in new low-temperature minerals, and the Br/Cl ratio is therefore a useful indicator of such hydrogeochemical processes as mixing with water affected by evaporites, contact with organic sediments and the presence of seawater. In the present case, the Br/Cl ratio may also be derived from mineral weathering. The most saline water at South Crofty has a Br/Cl ratio virtually identical to that of seawater (Figure 6a). Before stable isotope analyses were available this was taken as evidence for a seawater origin. Since seawater can now be ruled out, it is likely that the bromide source is the same as that of the chloride.

At present there are insufficient data on bromide in biotite or other candidate minerals, from the Carnmenellis Granite or elsewhere, to determine its source. Fuge (1974) has shown, however, that igneous biotites have Br/Cl ratios similar to seawater. Recent data presented here, albeit close to the detection limit, suggest that mineral Br/Cl ratios are similar to, or slightly depleted, relative to that of seawater, but within the range of experimental error of those in the saline waters. Bromide, therefore, is likely to be present in biotites at a Br/Cl ratio similar to that of seawater and to be released on reaction in the same manner as the chloride.

Iodine

The iodine concentration in South Crofty saline water is 0.091 mg l^{-1} and the other saline groundwaters in the granite are related to it by a mixing line (Figure 6b). These values represent an enrichment above an equivalent seawater composition. In the shallow groundwaters from the granite and in waters slightly enriched in chloride, including the HDR circulation fluid, iodine appears to be highly enriched, implying an addition during weathering reactions. This may be due to mobilisation as IO_3^- which will be the dominant and more mobile species in shallow oxidising waters. At present there is little evidence to indicate the source of iodine in granite but the water analyses available highlight the difference between saline waters evolved by mixing and those deriving their chloride enhancement from present-day processes (Figure 6b). It is also implied that iodine might be removed during long residence in the granite.

Fluoride

Apart from one anomalous acid-sulphate sample from Wheal Jane, fluoride concentrations are less than 8.5 mg

l^{-1}. The principal control on aqueous F^- levels is fluorite saturation which is exceeded in all the saline waters. The very high F^- in sample WJ1 (44 mg l^{-1}) occurs in an acidic water in which the high Al content (25 mg l^{-1}) allows formation of aluminium fluoride complexes; this water is actually undersaturated with respect to fluorite (SI= −0.51). Fluoride is released together with chloride (and probably bromide) either during hydrolysis of biotite and/or by interlayer exchange. Fluoride may also be released to solution by reaction of muscovite or possibly tourmaline. It might be expected that compared with chloride, F^- would be more strongly bonded to the biotite lattice and that exchange with hydroxyl ion might not be as freely achieved. Nevertheless, fluoride ion is likely to be readily available during any period of groundwater circulation and weathering, and fluorite is found quite widely as a mineral on joint surfaces. Experimental studies with Carnmenellis Granite show that under static conditions, fluorite and/or topaz dissolution, rather than biotite alteration, controls the concentration of dissolved F^- (Savage *et al.*, 1985).

It is interesting that the HDR fluid is marginally undersaturated with respect to fluorite (SI= −0.08) at the pre-circulation bottom-hole temperature of 80°C, but over saturated (SI= +0.28) at the surface discharge temperature of 36°C. Equilibrium (SI = 0) would occur at 68°C which may be indicative of the maximum temperature reached during circulation.

Lithium

The high concentration of lithium (125 mg l^{-1}) in the saline water at South Crofty represents an extreme enrichment relative to most groundwaters and to seawater. All the available data from mines and near-surface groundwaters are plotted in Figure 6c. Saline groundwaters and mixed waters lie on a line representing Li:Cl = 1:100. One exception is the saline water from the Botallack Mine in the Land's End Granite which, from other evidence, must contain seawater. Nevertheless, the lithium content (12 mg l^{-1}) implies that mixing with evolved, saline groundwater (more than 10%) has occurred and that similar saline waters must also occur in the Land's End Granite – this so far being the only recorded occurrence in Southwest England outside the Carnmenellis Granite.

Shallow groundwaters from the area have lithium contents between 0.001 and 0.01 mg l^{-1}, but lithium has increased significantly in shallow minewaters (Pendarves) and also in return-water from the HDR circulation system. It is clear that lithium enrichment is taking place over a relatively short time scale, and that its release is accompanied by an increase in chloride. The Li^+ increase has been interpreted (Edmunds *et al.*, 1985) as evidence of strong congruent reaction of biotite, but the greater chloride increase suggests that chloride is also being added by $Cl^- \rightleftharpoons OH^-$ exchange.

The enrichment of lithium in the Southwest England granites is well known and data for the Carnmenellis Granite have been given by Exley and Stone (1964), Fuge and Power (1969) and Beer *et al.* (1978). The last authors quote a mean Li content for the Carnmenellis Granite of 278 ppm (range 40-473); biotites are reported with Li contents ranging from 650-6500 ppm. Wilson and Long (1983) analysed granite minerals by ion microprobe and found that biotite is the mineral most enriched in Li, with concentrations in the range 4000 to >10 000 ppm. Muscovites also have high lithium concentrations whilst tourmalines and other primary and alteration minerals have concentrations typically two orders of magnitude lower. Biotite is therefore the prime candidate for the source of lithium in the saline waters and, compared with most other elements released from the biotite structure, lithium is likely to be strongly partitioned in the water rather than in secondary minerals; further analysis of alteration minerals is required to support this.

Boron

Boron occurs in the most saline groundwaters at 11 mg l^{-1}, which is significantly enhanced above an equivalent seawater composition; other minewater samples are diluted by mixing (Figure 6d). Although there are no data for shallow groundwaters, those for shallow minewaters and HDR return-water also suggest that boron is being added to the water (up to 0.2 mg l^{-1}) by water-rock interaction at the present day. Boron occurs in the main granite at around 400 ppm (Exley and Stone, 1964) but concentrations may be considerably higher in marginal facies. Tourmalinisation is a widespread phenomenon in the SW England granites, originating at high temperatures at a late stage in the granite emplacement; it is most abundant in marginal areas of the granites (Exley and Stone, 1964; Charoy, 1982). Although tourmaline contains about 5% B, it is usually considered to be very stable at low temperatures. However, Savage *et al.* (1985) noted etching and hence considerable reaction of tourmaline in autoclave experiments using Carnmenellis Granite and meteoric water in the temperature range 80°-250°C. Boron is likely to be present in equal amounts in the main rock-forming minerals of granite (Harder, 1975) and the quantities released could also be derived from biotite or plagioclase where boron is substituted for aluminium.

Strontium

Strontium is present at 41 mg l^{-1} in the most saline groundwater. Although strontianite saturation is indicated, this mineral may not control Sr^{2+} concentration in view of the low dissolved Sr/Ca ratios (Helz and Holland, 1965). However, low sulphate concentrations also keep the waters below celestite saturation. It is also possible that strontium is buffered by calcite reactions; calcite saturation is reached in the saline groundwaters and in the HDR circulation fluid. The other saline groundwaters are related to the South Crofty saline source by the mixing line but in the shallower groundwaters (Figure 6e) strontium increases with chloride in a similar fashion to lithium, boron and iodine. In the HDR circulation return-fluid, however, strontium is considerably enhanced relative to other groundwaters of low salinity.

Strontium is present in plagioclase at about 100 ppm but occurs at less than a tenth of this value in biotite; either mineral could thus be a source, as could calcite or siderite vein material. Strontium isotope systematics are being studied to resolve these possibilities. Preliminary data suggest that the more saline waters and HDR fluids approach isotopic equilibrium with plagioclase feldspar which, at 0.735, has the lowest $^{87}Sr/^{86}Sr$ ratio of the rock forming minerals (Darbyshire and Shepherd, 1985). More dilute

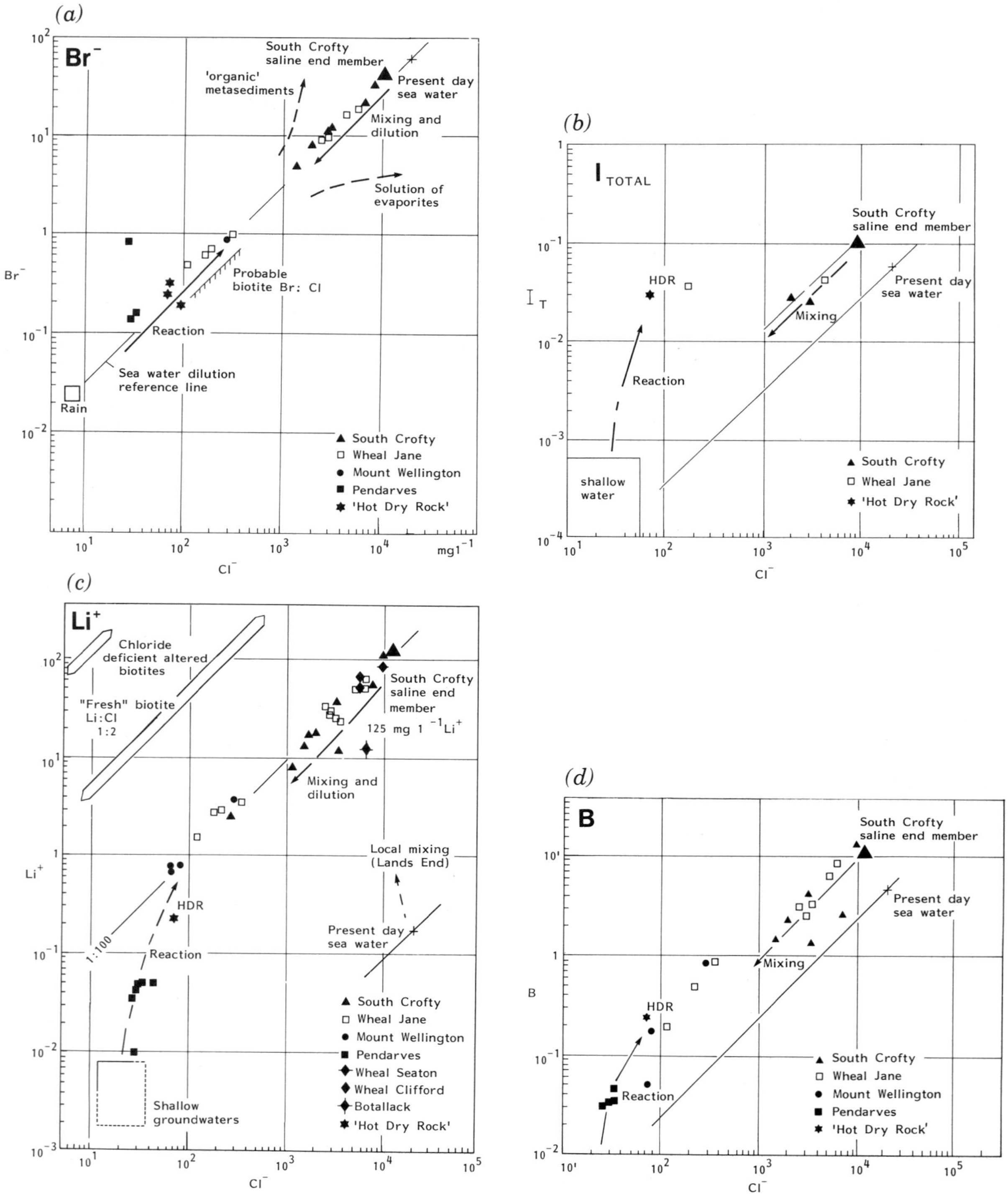

Figure 6. *Minor and trace element relationships with chloride. a) bromide, b) iodine, c) lithium. d) boron, e) strontium. f) barium, g) rubidium, h) caesium. Trends for reaction and mixing are shown relative to the dilution line for present-day seawater.*

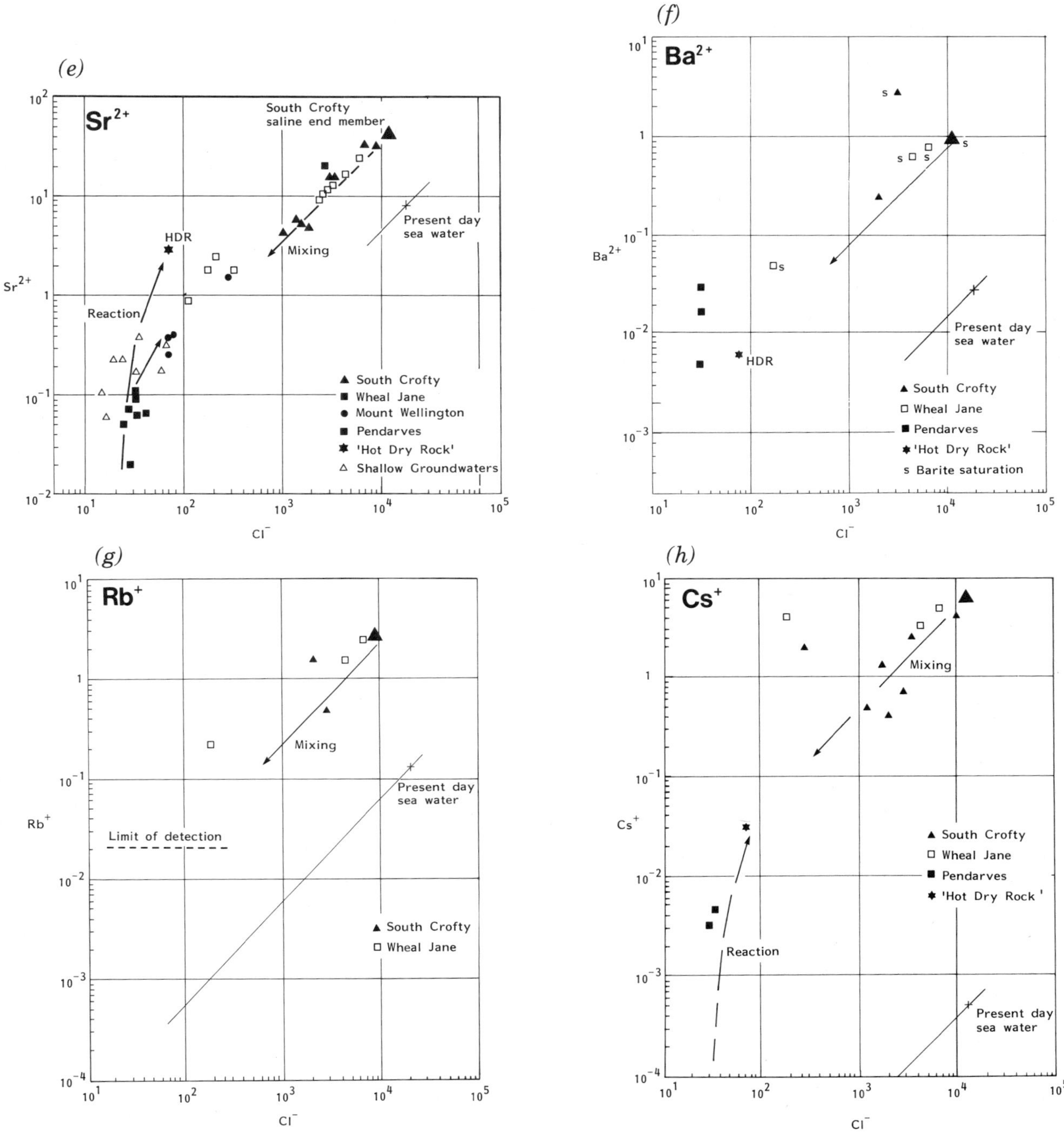

minewaters have lower $^{87}Sr/^{86}Sr$ ratios (0.714-0.715) suggesting that dissolved strontium derives from a strontium containing mineral even lower in rubidium than plagioclase. Carbonate or phosphate minerals (apatite, monazite) are thus likely initial sources for dissolved Sr, with feldspar dissolution becoming more important with increasing residence times.

Barium

The saline water at South Crofty is barite-saturated (0.94 mg l^{-1} Ba^{2+}, 129 mg l^{-1} SO_4^{2-}) and there is a good mixing relationship between the saturated saline water and the less mature and more dilute waters (Figure 6f). Other less saline waters (except acid minewaters which are supersaturated) are undersaturated with respect to barite. The very low Ba^{2+} concentrations (0.006 mg l^{-1}) in the HDR circulation system and in other shallow minewaters must reflect the low geochemical abundance in the granite and in secondary minerals. The mean barium content of the Carnmenellis Granite is 192 ppm (Darbyshire and Shepherd, 1985) which is very low compared with an average for granites of around 800 ppm (Puchelt, 1972); barium is likely to be rather evenly distributed between the two feldspars and biotite or muscovite. Considerable residence times are thus probably required to achieve barite saturation, while lower barium concentrations may indicate the extent of rock-water interaction.

Rubidium and caesium

Rubidium and caesium concentrations in the granite are 448 and 41 ppm respectively (Darbyshire and Shepherd, 1985). Rubidium levels in plagioclase, orthoclase and biotite are respectively up to 100, 600-700 and up to 3500 ppm. The caesium content of biotite from Carnmenellis is reported to be 1350 ppm (Butler, 1953). In relation to the other minerals the abundance of Cs is likely to be muscovite > biotite > feldspar > plagioclase (Heier and Billings, 1970). Despite the ten-fold excess of Rb over Cs in the rock, the Cs (4.3 mg l^{-1}) is nearly double that of Rb in the saline water at South Crofty. Preferential reaction/leaching of caesium from biotite or other primary minerals must be occurring to account for the ratios observed. There is also evidence that Rb and K behave similarly, suggesting incorporation of both into smectite and/or mica. This is apparent both from K loss during HDR circulation and the steady states observed in autoclave experiments (Savage *et al.*, 1985). It is also clear from the element/chloride relationships that caesium is enriched relative to Rb in some dilute waters, especially acidic samples. This might also suggest that under acidic conditions Cs is being desorbed from, or dissolved with, ferric hydroxide.

Other elements

Considerable enhancement of several metals can be observed in the acidic sulphate waters. Iron, Al, Mn, As, Be, Co, Cs, Ni and Zn are particularly enriched, and this must relate mainly to the reactions of sulphide and oxide minerals. The saline water at South Crofty does not contain abnormally high metal levels, although diluted saline waters, e.g. from Wheal Jane, show some of the characteristics of both types of water. The HDR return-water contains detectable Ag, Mo, and W but it is not clear whether these relate to alteration of primary silicate or secondary minerals. These results suggest that although beryllium must be derived initially from primary silicates, it is being mobilised largely from secondary phases, in particular iron oxide minerals. Germanium is present only in Wheal Pendarves, in dilute, low temperature and near neutral-waters. It is most likely to be derived from sulphide minerals where germanium substitutes for iron.

DISCUSSION

The saline groundwaters in the Carnmenellis Granite have an unusual chemistry that is dominated by sodium depletion and calcium enrichment relative to chloride. In particular, the low Na/Cl ratios contrast with those found in groundwaters from most sedimentary environments where values close to unity are usual. The Na/Cl ratios in groundwaters from sedimentary basins are derived either from marine sources, solution of halite or the release of interstitial formation waters. In fact, UK formation brines tend to have Na/Cl ratios slightly greater than one (Edmunds, 1985). In addition to eliminating present-day seawater as a source of the salinity, it is also likely that migration of formation brines from any source can be ruled out.

Earlier work has suggested that the salinity and characteristic hydrogeochemistry can be accounted for by hydrolytic and related reactions of rock-forming minerals, especially plagioclase and biotite (Edmunds *et al.*, 1984; 1985). These and other rock-forming minerals owe their compositions to high temperature fluid-rock equilibria and are mostly unstable under present, low temperature conditions. There are several possible reaction products of the acid hydrolysis of plagioclase but the two most likely from petrographic and field observations are kaolinite and laumontite. The laumontite reactions were proposed by McCartney (1984) who also suggested that phillipsite and other zeolites might be stable in the granite. The following reactions are considered probable:

$$\underset{\text{Oligoclase}}{5\,Na_{0.8}Ca_{0.2}\,Al_{1.2}Si_{2.8}O_8} + 6H^+ + 19H_2O \rightarrow \underset{\text{Kaolinite}}{3Al_2Si_2O_5(OH)_4} + 4Na^+ + Ca^{2+} + 4H_4SiO_4$$

$$\underset{\text{Oligoclase}}{5\,Na_{0.8}Ca_{0.2}\,Al_{1.2}Si_{2.8}O_8} + 16H_2O + 2Ca^{2+} \rightarrow \underset{\text{Laumontite}}{3CaAl_2Si_4O_{12}.4H_2O} + 4Na^+ + 2H_4SiO_4$$

The molar Na/Ca ratio of the groundwaters vary between 1.3 and 4.0. These are generally low compared with the bulk composition of feldspars. The difference in chemistry may be accounted for either by considering the preferential solution of Ca-rich zoned centres or by incongruent dissolution of oligoclase; 'late' albite is also found in the granite. Reaction kinetics for anorthite are faster by some two orders of magnitude compared with albite (Lasaga, 1984) and this could partially explain the observed instability of zoned feldspar centres and the present water compositions. The production of laumontite requires more Ca^{2+} than can be obtained from dissolution of feldspar, but the imbalance can be made up if a more Ca-rich feldspar is substituted in the reaction. Calcite is also found in the laumontite assemblage, and if congruent-only reaction is assumed and a water Na/Ca ratio below that of plagioclase is to be accounted for, there is clearly a calcium 'supply' problem. Laumontite has also been experimentally produced at 250°C by reaction of granite with 'local' groundwater (water-rock ratio 1:10) by Savage *et al.* (1985). These authors also found that plagioclase and orthoclase are apparently much more reactive than biotite under the relatively static conditions of the experiments.

Congruent and incongruent reactions of biotite, and possibly also to a lesser extent of muscovite (there is no amphibole, although apatite and tourmaline may be involved), are considered to be the main processes leading to the build-up of Cl, Li and B and other solutes in the groundwater:

$$\underset{\text{Biotite}}{K_2(Mg,Fe)_4\,(Fe,Al,Li)_2\,[Si_6Al_2O_{20}]\,(OH)_2(F,Cl)_2} + 3H_2O + 12H^+ \rightarrow$$

$$\underset{\text{Kaolinite}}{Al_2\,Si_2O_5(OH)_4} + 4H_4SiO_4 + 2K^+ + 4(Mg^{2+},Fe^{2+}) +$$

$$2(Fe^{3+},Al^{3+},Li^+) + 2(F^-,Cl^-)$$

Chloride and lithium may be considered conservative products of these reactions whereas elements such as K, Si, Mg, Fe and F are more likely to be assimilated into secondary minerals. The observed aqueous Li/Cl ratios of 1:100, described earlier, must indicate significant chloride release from the biotite interlayer as well as roughly equimolar release during congruent hydrolysis reactions. From structural considerations, muscovite is likely to contain considerably less chloride substitution than coexisting biotites (Volfinger *et al.*, 1985).

Reactions of both plagioclase and biotite as well as those of other primary silicates lead to the production of silica in excess of that required for most secondary silicate minerals. Ultimately the dissolved silica should precipitate as chalcedony or quartz. The supply of H^+ is critical to silicate hydrolysis. The principal initial source of H^+ must be atmospheric precipitation, enhanced by CO_2 dissolution during percolation through the soil. Oxidation of sulphide minerals may provide an important additional source. Groundwater circulation is therefore essential to promote rapid influx of reactants and removal of products. On a longer time-scale and in deeper, an aerobic and more alkaline systems the exchange of Cl^- and OH^- provides a possible additional internal source of H^+: increased electrostatic interactions resulting from higher chloride and/or fluoride in solution would lead to further dissociation of H_2O. This is seen as being important over geological timescales with reaction rates controlled by diffusion processes. This latter process is considered to be a dominant control over the salinity of the groundwater.

The range of salinity in fluid inclusions in quartz from the Carnmenellis Granite varies widely – averaging 10-20 wt % NaCl (60-120 g l^{-1} Cl^-). Fluid inclusions probably contain 35% of the total chlorine in the granite but this chlorine probably does not contribute significantly, if at all, to chloride in the groundwater. Rock-forming quartz is almost inert under the field conditions. If mineral hydrolysis is taking place, excess silica is being generated and silica minerals must be precipitating. Thus, new fluid inclusions with relatively low salinities would be predicted from the water-rock interaction model and this fits in with the available field evidence of Alderton and Rankin (1983) who record a large proportion of monophase inclusions indicating temperatures of origin from 170°C down to 70°C or less.

The groundwater chemistry is also inconsistent with an origin from fluid inclusions. The depleted Na/Cl ratios do not fit such an origin, especially since plagioclase dissolution, in addition to fluid inclusions, would tend to increase the ratio above 1. The average Ca/Na ratio (weight units) of fluid inclusions from twelve unaltered samples of SW England granites (Rankin *et al.*, 1982) is 1:21. This is in contrast to 1:1.74 for the saline end-member at South Crofty Mine and it seems that Ca enrichment does not relate to fluid inclusion chemistry.

Some contribution to salinity by fluid inclusions due to decrepitation and weathering cannot be ruled out but this influence must be negligible when compared with mineral hydrolysis reactions. The contribution from weathering of feldspar fluid inclusions has yet to be quantified, but the abundance of inclusions in feldspars is small when compared with that in quartz.

Minor- and trace-element compositions of the groundwaters generally support the water-rock interaction model. The elements Sr, Ba, F, B, Rb and Cs are derived mainly from reaction of the silicate minerals and lithium, in particular, provides a relatively conservative record of the extent of water-rock interactions. Many other elements may be released during reaction of primary silicate minerals but are not enhanced in the saline fluids. Elements such as Mn, As, Al, Cd, Co, Cu, Be, Ge, Ni, Zn are mobilised under acidic conditions and may have been incorporated in sulphide minerals or oxides, the latter providing an important sink for many of the trace elements derived originally from the silicates.

CONCLUSIONS

The available evidence points strongly to the prime source of groundwater salinity being the reaction of circulating groundwaters with original, high-temperature rock-forming minerals both in the short-term in the circulation system in the hot dry rock reservoir and over geological time-intervals. This explanation of salinity development in Cornwall is, therefore, relatively straightforward when compared with other granite areas. The Carnmenellis Granite may be regarded as a model for this mode of origin since the unusual composition of the granite enhances the evidence of water-rock interaction and other modes of salinity generation can be largely discounted. It would be valuable to compare Cornwall with other high-level Hercynian and later granites around the world. However, in other granites, as in the Pyrenees and Carpathians, salinities as high as those in Cornish groundwaters have not been recorded. In Shield areas, the timescales for the evolution of both the granitic terrains and their groundwater is likely to be considerable and groundwater circulation has probably been restricted.

In Cornwall, the hydrodynamic conditions are likely to have changed frequently over geological time as a result of tectonic activity, changes in geothermal gradient and weathering. Eustatic and sea-level changes during the Pleistocene glaciation would also have influenced the hydraulic régime. It is therefore remarkable that salinities as high as 20 000 mg l^{-1} TDS appear to have evolved over the order of one million years and yet have been retained in granite. In mountainous terrains such as the Alps, Pyrenees and Bohemian Massif, solute release due to water-rock interaction is also likely to be occurring, but only low salinities are recorded because of short residence times and low temperatures.

The saline waters found in Cornwall and those in granites in quite different geological settings may therefore share a similar origin, although salinity derived from other sources may be superimposed in some areas. The characteristic calcium enrichment, leading eventually in some Shield areas to calcium-chloride brines with near-saturated composition, may be explained by prolonged congruent and especially incongruent reaction of plagioclase feldspar and other calcium-rich primary minerals. This process may be more enhanced than in Cornwall if the granite contains feldspars more enriched in anorthite.

The water-rock interaction model proposed for the Carnmenellis Granite, however, is most notable for the demonstration it provides of the generation of Cl^- from rock-forming minerals. The reservoir of chloride derived from hydrothermal and magmatic processes, contained especially in quartz, remains inert under the present-day, low-temperature conditions which favour the formation rather than the destruction of new fluid inclusions. Notwithstanding the proximity to the sea, the absence of marine chloride can also be demonstrated. Finally, although the granite was emplaced within sedimentary country rocks, it is also unnecessary to invoke sedimentary formation waters in the origin of the salinity.

ACKNOWLEDGEMENTS

We are grateful to J.M. Trafford for assistance with the trace-element analysis and to A.R. Date and Y.Y. Cheung of BGS Analytical Chemistry Group for carrying out the caesium analyses. We particularly wish to acknowledge the continuing collaboration of the management and staff of South Crofty, Wheal Pendarves and Wheal Jane mines and of the Hot Dry Rock Project. We thank D. Savage and A.E. Milodowski for valuable discussion and for reading the manuscript. This paper has also benefitted from two helpful external reviews. This work forms part of a broader programme of geochemical studies of granite-water interaction related to Hot Dry Rock geothermal development funded by the UK Department of Energy and the Commission of the European Communities. The paper is published with the permission of the Director, British Geological Survey, Natural Environment Research Council.

REFERENCES

Alderton, D.H.M. and Rankin, A.H., 1983. The character and evolution of hydrothermal fluids associated with the kaolinised St. Austell granite, SW England: Journal Geological Society London, v. 140, p. 297-309.

Alderton, D.H.M. and Sheppard, S.M.F., 1977. Chemistry and origin of thermal waters from southwest England: Transactions Institute Mining and Metallurgy, v. (B)86, p. 191-194.

Andrews, J.N. and Kay, R.L.F. 1982. Natural production of tritium in permeable rocks: Nature, v. 298, p. 361-363.

Batchelor, A.S., 1984. Hot dry rock geothermal exploitation in the United Kingdom: Modern Geology, v. 9, p. 1-41.

Beer, K.E., Edmunds, W.M. and Hawkes, J.R., 1978. A preliminary look at lithium in the United Kingdom: Energy, v. 3, p. 281-292.

Brooks, M., Doody, J.J. and Al-Rawi, F.R.J., 1984. Major crustal reflectors beneath SW England: Journal Geological Society London, v. 141, p. 97-103.

Bruland, K.W., 1983. Trace elements in sea-water: *in* Riley, J.P., and Chester, R., ed., Chemical Oceanography, v. 8, p. 157-220.

Butler, L.R., 1953. The geochemistry and petrology of rock weathering (1) The Lizard area, Cornwall: Geochimica et Cosmochimica Acta, v. 4, p. 157-178.

Charoy, B., 1982. Tourmalinization in Cornwall, England: *in* Evans, A.M., ed., Metallization Associated with Acid Magmatism: Wiley, New York, p. 63-70.

Darbyshire, D.P.F. and Shepherd, T.J., 1985. Chronology of granite magmatism and associated mineralisation, SW England: Journal Geological Society London, in press.

Durrance, E.M., Bromley, A.G., Bristow, C.M., Heath, M.J. and Penman, J.M., 1982. Hydrothermal circulation and postmagmatic changes in granites of south-west England: Proceedings of the Ussher Society, v. 5, p. 304-320.

Edmunds, W.M., 1985. Geochemistry of geothermal waters in the UK: *in* Downing, R.A., and Gray, D.A., ed., Geothermal Energy – the Potential in the United Kingdom: Her Majesty's Stationery Office, London, Chapter 6, in press.

Edmunds, W.M., Andrews, J.N., Burgess, W.G., Kay, R.L.F. and Lee, D.J., 1984. The evolution of saline and thermal groundwaters in the Carnmenellis granite: Mineralogical Magazine, v. 48, p. 407-424.

Edmunds, W.M., Kay, R.L.F. and McCartney, R.A., 1985. Origin of saline groundwaters in the Carnmenellis granite: natural processes and reaction during Hot Dry Rock reservoir circulation: Chemical Geology, v. 49, p. 287-301.

Exley, C.S. and Stone, M., 1964. The granitic rocks of southwest England: *in* Present Views of some Aspects of the Geology of Cornwall and Devon 150th Anniversary volume: Transactions Royal Geological Society Cornwall. p. 131-184.

Feth, J.H., Robertson, C.E., and Polzer, W.L., 1964. Sources of mineral constituents in water from granitic rocks, Sierra Nevada California and Nevada: United States Geological Survey Water-Supply Paper 1535-I, 70 p.

Fouillac, C., and Michard, G., 1981. Sodium/lithium ratio in water applied to geochemistry of geothermal reservoirs: Geothermics, v. 10, p. 55-70.

Frape, S.K. and Fritz, P., 1982. The chemistry and isotopic composition of saline groundwaters from the Sudbury basin, Ontario: Canadian Journal of Earth Sciences, v. 19, p. 645-661.

Frape, S.K., Fritz, P. and McNutt, R.H., 1984. Water-rock interaction and chemistry of groundwaters from the Canadian Shield: Geochimica et Cosmochimica Acta, v. 48, p. 1617-1627.

Fritz, P. and Frape, S.K., 1982. Saline groundwaters in the Canadian Shield – a first overview: Chemical Geology, v. 36, p. 179-190.

Fuge, R. and Power, G.M., 1969. Chlorine and fluorine in granitic rocks from SW England: Geochimica et Cosmochimica Acta, v. 33, p. 888-893.

Fuge, R., 1974. Chlorine: *in* Wedepohl, K.H., ed., Handbook of Geochemistry: Springer Verlag, New York, Chapter 17.

Gray, A.L., and Date, A.R., 1983. Inductively coupled plasma source mass spectrometry using continuous flow ion extraction: The Analyst, v. 108, p. 1033-50.

Grigsby, C.O., Tester, J.W., Trujillo, P.E., Counce, D.A., Abbott, J., Holley, C.E., and Blatz, L.A., 1983. Rock-water interactions in hot dry rock geothermal systems: field investigations of *in situ* geochimcal behavior: Journal of Volcanology and Geothermal Research, v. 15, p. 101-136.

Harder, H., 1975. Boron in minerals and igneous rock: *in* Walker, C.T., ed., Geochemistry of boron: Dowden, Hutchinson and Ross Inc. Stroudsberg, PA.

Heier, K.S., and Billings, G.K., 1970. Cesium – Abundance in rock-forming minerals: *in* Wedepohl K.H., ed., Handbook of Goechemistry: Springer Verlag, New York, Section 55-D. 1 p.

Helz, G.R., and Holland, D.G., 1965. The solubility and geologic occurrence of strontianite: Geochimica et Cosmochimica Acta, v. 29, p. 1303-1315.

Jackson, N.J., Halliday, A.N., Sheppard, S.M.F., and Mitchell, J.G., 1982. Hydrothermal activity in the St. Just mining district, Cornwall, England: *in* Evans A.M., ed., Metallization Associated with Acid Magmatism: Wiley, New York, p. 137-179.

Lasaga, A.C., 1984. Chemical kinetics of water-rock interaction: Journal Geophysical Research, v. 89, p. 4009-4025.

McCartney, R.A., 1984. Geochemical investigation of two Hot Dry Rock reservoirs in Cornwall, UK: Unpublished PhD thesis: Camborne School of Mines.

Miles, D.L., and Cook, J.M., 1982. The determination of sulphate in natural waters by inductively coupled plasma emission spectrometry: Analytica Chimica Acta, v. 141, p. 207-212.

Miller, W.A., 1865. Chemical examination of a hot spring containing caesium and lithium in Wheal Clifford, Cornwall: Report of the British Association, 34th Meeting, p. 35-36.

Nordstrom, D.K., 1983. Preliminary data on the geochemial characteristics of groundwater at Stripa: Proceedings NEA Workshop Geological Disposal of Radioactive Waste: *in situ* experiments in granite. Stockholm, Sweden 25-27 Oct. 1982, Nuclear Energy Agency, Paris, p. 143-153.

Paces, T., 1972. Chemical characteristics and equilibrium in natural water-felsic rock-CO2 systems: Geochimica et Cosmochimica Acta, v. 36, p. 217-240.

Phillips, J.A., 1873, The rocks of the mining districts of Cornwall and their relation to metalliferous deposits: Quarterly Journal Geological Society London, v. 31, p. 319-345.

Plummer, N., Jones, B.F. and Truesdell, A.H., 1976. WATEQF – A Fortran-IV version of WATEQ, a computer program for calculating chemical equilibrium of natural waters: United States Geological Survey Water Resources Investigations, 76-13, 61 p.

Puchelt, H., 1972. Barium – Abundance in common igneous rock types: *in* Wedepohl, K.H., ed., Handbook of Geochemistry: Springer Verlag, New York, Section 56E, 9 p.

Rankin, A.H., Alderton, D.H.M. and Thompson, M. and Goulter, J.E., 1982. Determination of uranium:carbon ratios in fluid inclusion decrepitates by inductively coupled plasma emission spectroscopy: Mineralogical Magazine, v. 46, p. 179-186.

Savage, D., Cave, M.R. and Milodowski, A.E., 1985. The origin of saline groundwaters in granitic rocks: evidence from hydrothermal experiments: Proceedings of the 9th International Symposium: Scientific Basis for Nuclear Waste Management, Sept. 1985, Stockholm, Sweden, in press.

Trafford, J.M., 1982. The analysis of 13 trace elements by inductively coupled plasma atomic emission spectrometry preceded by an evaporative method of preconcentration: Unpublished Report Hydrogeology Unit Institute of Geological Sciences, London.

Volfinger, M., Robert, J-L., Vielzeuf, D. and Neiva, A.M.R., 1985. Structural controls of the chlorine content of OH-bearing silicates (micas and amphiboles): Geochimica et Cosmochimica Acta, v. 49, p. 37-48.

Vovk, I.F., 1981. Radiolytic model of formation of brine compositions in the crystalline basement of Shields: Geokhimiya No. 4, p. 467-480 (Translation: Geochemistry International 1981, p. 80-93).

Wheildon, J., Francis, M.F. and Thomas-Bettes, A., 1977. Investigation of the SW England thermal anomaly zone: Seminar on Geothermal Energy, Brussels, EUR 5920, v. 1, p. 175-188.

Wilson, G.C. and Long, J.V.P., 1983. The distribution of lithium in some Cornish minerals: ion microprobe measurements: Mineralogical Magazine, v. 47, p. 191-199.

Hydrochemical evolution of saline waters from crystalline rocks of the Bohemian Massif (Czechoslovakia)

Tomáš Pačes[1]

Department of Land Improvement and Drainage, Royal Technical Institute, Stockholm, Sweden

[1]Present address: Geological Survey (ÚÚG), Malostranské nám. 19, 118 21 Praha 1, Czechoslovakia

Pačes, Tomáš, Hydrochemical evolution of saline waters from crystalline rocks of the Bohemian Massif (Czechoslovakia); in Saline water and gases in crystalline rocks, Editors: Fritz, P., and Frape, S.K.; Geological Association of Canada Special Paper 33, 1987. p. 145–156.

Abstract

The composition of saline waters in the Bohemian Massif is highly variable. The variation probably reflects the complex geological history of the region. The waters can be represented by a hydrochemical-field diagram reflecting water-rock interaction and input of some components from surrounding hydrogeological structures. The waters contain C + Ca and C + Na components (C represents carbon in all dissolved carbonate species), resulting from hydrolysis of alumosilicate rocks, S + Ca component (S represents all dissolved species containing sulphur) dissolved after oxidation of sulphide minerals, haline components Cl + Na and Cl + Ca, and a volcanic component S + Na of ambiguous origin. Most of the saline waters are different from oil field waters or other brines in crystalline rocks because they are enriched in sulphates and sometimes in carbon dioxide of a deep origin. Part of the dissolved content may be a product of Tertiary rift activity, and some of the haline components may be preserved from the Upper Cretaceous shallow sea. The high concentrations of the S + Na component may have accumulated in a closed Tertiary rift lake that received a high input of volcanic sulphur. The isotopic composition of sulphur was $\delta^{34}S = +5.2‰$*. Reduced sulphur species were oxidized in the lake and the solution was concentrated by evaporation. The resulting brine or just the salt content of the brine migrated into bottom fractures by convective flow and by molecular diffusion. The Tertiary volcanic component S + Na has been preserved in the fractures till today. The origin of Cl + Ca saline water in a granodioritic massif has yet to be explained. The different isotopic composition of sulphur in the brine indicates that the origin of the solution is different from that of the saline waters associated with the Tertiary rifts. The isotopic composition of the water is usually similar or identical to that of local meteoric water. This implies that the water molecules and the salt content may have different origins. Water in the brine containing the S + Na volcanic component deviates in its isotopic composition from local meteoric waters. The isotopic fractionation has shifted D and* ^{18}O *above the global meteoric line in a similar manner as observed in the Canadian Shield. The saline waters in the Bohemian Massif suggest a diverse involvement of water in geological processes. Details of the involvement are still not fully understood.*

Résumé

La composition des eaux salées du massif bohémien révèle d'importantes variations reflétant probablement l'histoire géologique complexe de la région. Ces eaux peuvent être portées sur un diagramme hydrochimique qui tient compte de l'interaction entre l'eau et la roche et de l'apport de composantes provenant de structures hydrogéologiques environnantes. Ces eaux contiennent les composantes C + Ca et C + Na (la lettre C désignant tous les carbonates dissous issus de l'hydrolyse des aluminosilicates), la composante S + Ca (la lettre S *lettre S désignant toutes les phases contenant du soufre dissous issu de l'oxydation des sulfures), les composantes salines Cl+Na et Cl+Ca et une composante volcanique, S+Na, d'origine indéterminée. La plupart des eaux salées diffèrent de celles des champs pétrolifères ou des autres types d'eaux sursalées des roches cristallines par le fait qu'elles sont enrichies en sulfates et parfois en gaz carbonique d'origine profonde. Une partie du matériel dissous pourrait provenir d'un rift actif au Tertiaire et une certaine proportion des composantes salines a pu être préservée depuis l'époque de la mer peu profonde du Crétacé supérieur. La forte concentration de la composante S+Na peut résulter d'une accumulation dans un lac fermé limité par un rift d'âge tertiaire recevant un apport de soufre volcanique élevé dont la composition isotopique ($\delta^{34}S$) était de +5,2‰. Les différents types de soufre réduit ont été oxydés dans le lac et la solution s'est concentrée par évaporation. L'eau sursalée résultante, ou simplement son contenu salin, a migré vers les fractures sur le fond au lac par circulation convective ou diffusion moléculaire. La composante volcanique S+Na, d'âge tertiaire, aurait ainsi été préservée jusqu'à maintenant. L'origine de la composante Cl+Ca dans l'eau salée du massif granodioritique reste indéterminée. La composition isotopique particulière du soufre dans cette eau sursalée, indique que l'origine de la solution est différente de celle des eaux salées associées aux rifts tertiaires. La composition isotopique de l'eau est généralement semblable ou identique à celle de l'eau météorique locale, ce qui implique que les molécules d'eau et le contenu salin pourraient provenir de sources différentes. La composition isotopique de l'eau sursalée contenant la composante volcanique S+Na diffère de celle de l'eau météorique locale. Le fractionnement isotopique a fait passer les teneurs D et ^{18}O au delà de la limite météorique d'une façon qui s'apparente au processus ayant cours dans le bouclier canadien. Les eaux salées du massif bohémien indiquent que l'eau n'a pas participé de la même façon à tous les processus géologiques. Tous les aspects du comportement détaillé de l'eau ne sont pas encore connus.*

GEOLOGICAL INTRODUCTION

The Bohemian Massif is an important geological structure in central Europe (Buday 1969, Svoboda 1966, Suk 1984). The basement was consolidated during the Variscan orogeny. A deep sea is documented only by some Paleozoic sediments. Most of the sedimentary sequences are metamorphosed to phyllite or mica gneisses. Platform sediments of a shallow sea were deposited during the Upper Cretaceous. They form a sedimentary basin whose rocks are mostly sandstones and marlstones. The thickness of the sediments rarely exceeds 800 m. Alpine orogeny caused only rejuvenation of old block tectonic patterns without pronounced folding. The Bohemian Massif acted as a rigid block against the south-to-north thrust of Alpine nappes and southeast-to-northwest thrust of the Carpathian nappes. Two continental rifts were formed during the Miocene and Oligocene. They are the Ohře River Rift and the Elbe River Line. The Ohře River Rift is a graben which extends in SW to NE direction, 280 km long and about 25 km wide. The rift activity is manifested by a linear trend and zoning of basaltic volcanism, elevated heat flow, and the increased flux of CO_2 from below. The rift valley is filled with Tertiary lake sediments. Geological evidence suggests that the sediments were deposited in closed basins. According to the paleontological record and local pyritic cement in sandstones, water in the lakes was saline, perhaps containing sulphide ions. Observed die-backs of biological communities indicate that the chemical composition of the closed lakes could change abruptly, perhaps due to an input of chemicals from local volcanos. The Elbe Line is not a typical continental rift. However, it is a deep-reaching zone of SE to NW direction connected with Tertiary basaltic volcanism. Today, a slightly elevated heat flow and anomalous flux of deepseated CO_2 is still observed (Čermák et al., 1981). The volcanic activity along both rift-like zones ceased about one million years ago or earlier.

This complex geological history is probably reflected by the wide range of saline waters whose chemical composition cannot be explained by water-rock interaction or simple evolution of Cretaceous sea water. The saline waters are discharged in "mineral-water" springs in famous Bohemian spas at Karlovy Vary (Carlsbad), Marianské Lázně (Marienbad) and Františkovy Lázně (Francisbad). Related types of water were tapped by drill holes in the basement rocks underlying the Tertiary sediments of the Cheb basin which is a part of the Ohře River Rift, and in Permocarboniferous sandstones underlying the Cretaceous sediments which cover the Elbe Line. A different type of saline water was encountered during mining in a granodiorite of the Bor Massif. The small massif is of a prevariscan age and follows a tectonic zone transverse to the Ohře River Rift.

CHEMICAL AND ISOTOPIC COMPOSITION OF WATER SAMPLES

The characteristics of the water samples used in the following discussion are summarized in Table 1. The table contains available data on the isotopic composition of the waters, and composite hydrochemical characteristics whose meaning is explained in the next section. Chemical analyses of waters from the Bohemian Massif and the Carpathians are given in Table 2.

The chemical data include groundwaters of meteoric origin whose chemical composition can be explained by water-rock interaction (No. 1 to 6 in Table 2) and saline waters whose chemical composition cannot be derived from water-rock interaction only so that a more complex history must be assumed to explain their evolution (No. 7 to 12 in Table 2). Three standards are used in the discussion of their evolution. They are: "pp" – typical atmospheric wet precipitation in the Bohemian Massif (No. 13 in Table 2); "smr" – standard mean river water calculated from published data on 20 major world rivers; and "smow" – standard mean ocean water.

Two genetic types of brines described from other regions are assumed to be relevant to the explanation of the saline waters in the Bohemian Massif. They are oil-field brines of Cl-Na and Cl-Ca composition, and brines with very high contents of sulphate related to volcanic activity. These samples are described in Table 1. The chemical composition of the samples is presented only in the form of the composite hydrochemical characteristics. The complete chemical analyses can be found in the literature cited in Table 1.

Table 1

Symbols and composite chemical characteristics of water samples plotted in the hydrochemical-field diagrams in Figs. 1 to 6

Symbol	Description of the source	$\log_{10}\frac{C+Ca}{Cl+Na}$	$\log_{10}\frac{S}{C+Na}$	$\log_{10}$ TDS
		(molar concentrations)		(g/L)
STANDARDS				
smrw	Standard mean river water (Livingstone, 1963, Martin and Meybeck, 1978)	0.48	−1.01	−0.94
smow	Standard mean ocean water (Turekian, 1969)	−1.89	−1.22	1.55
pp	Wet precipitation in central Europe is assumed to be the starting point of the evolution of meteoric groundwaters (sample 13, Table 2)			
GROUNDWATERS AND SALINE WATERS FROM THE BOHEMIAN MASSIF AND WESTERN CARPATHIANS				
C	Groundwater from Cretaceous sandstones and marlstones overlying the crystalline basement and Permocarboniferous sediments (Sample 1, Table 2)			
J	Seepage of groundwater on the 5th level of Svornost mine, Jáchymov; water reacts with sulfide minerals and carbonate minerals in fractures in mica gneiss (Sample 2, Table 2)			
L	Groundwater tapped by a deep drillhole in granite underlying Cretaceous and Permocarboniferous sandstones, Louny; water is charged with CO_2 of a deep origin (Sample 3, Table 2)			
MP	Groundwater from limestones discharged in a natural spring in Muranská Planina Flat (Sample 4, Table 2)			
TB	Groundwater from limestones charged with CO_2 of a deep origin, Teplice nad Bečvou (Sample 5, Table 2)			
T	Saline water tapped by a drillhole in halite and gypsum layers of the Tortonian period; according to the low contents of Br (10.9 mg/L) and I (2.5 mg/L), and meteoric values of $\delta^{18}O$ and δD, it is a meteoric groundwater which has dissolved salts in Trebišovská Tabule flat (Sample 6, Table 2)			
S	Saline water from Permocarboniferous sandstones underlying Cenomanian sandstones at Stránka u Mšena; a major characteristic are anomalously high contents of SO_4 and HCO_3; they cannot be explained by a simple groundwater – rock interaction (Sample 7, Table 2)			
B	Saline water from granodiorite tapped by a shaft at Vítkov at a depth of 400 m; it seems to be a diluted calcium – chloride brine. The $\delta^{18}O$ value corresponds to local meteoric waters (−10% SMOW) and the isotopic composition of sulphur in sulphate is $\delta^{34}S = +28‰$ CDT (isotopic data by V. Šmejkal and F. Buzek in Pačes, 1983) (Sample 8, Table 2)			
K	Hot saline water discharged from granite in Karlovy Vary; it is a heterogenous mixture of water and gas containing more than 99% CO_2; the gas/water volume ratio is 10/1. The isotopic composition is: $\delta^{18}O = -10.2‰$ (F. Buzek), $\delta D = -69.1‰$ (J. O'Neil), $\delta^{34}S = +5.6‰$ (V. Šmejkal), $\delta^{13}C$ in $HCO_3 = -3.7‰$ (V. Šmejkal) (in Pačes, 1983); the isotopic composition of H_2O corresponds to local meteoric waters (Sample 9, Table 2)			
G	Saline water tapped by a drillhole in Tertiary sandstones overlying crystalline rocks at Františkovy Lázně. According to hydrological data water is connected with water in fractures of underlying gneiss and granite; it is charged with CO_2 of a deep origin. The isotopic composition of sulphur determined in similar waters from other drillholes in the vicinity varies from $\delta^{34}S = +4.2$ to $+5.5‰$ (Hladíkova and Šmejkal, 1973) (Sample, Table 2)			
F	Saline water discharged by a natural spring from amphibolite at Mariánské Lázně. It is a mixture of water and very pure CO_2 gas. According to the isotopic composition, the water is of meteoric origin $\delta^{18}O = -10.51‰$, $\delta D = -66.4‰$, (J. O'Neil in Pačes, 1974) Composition of sulphur in sulphate ions is $\delta^{34}S = +5.2‰$ (Hladíková and Šmejkal, 1973) (Sample 10, Table 2)			
O	Saline water without gaseous CO_2 tapped by a drillhole in phyllite underlying Tertiary lake sediments of the Cheb basin; it is the most saline water found in the rocks of the Bohemian Massif by A. Pazdera, Geoindustria, Praha. The isotopic composition of H_2O differs from local meteoric water: $\delta^{18}O = -11‰$ (F. Buzek and V. Šmejkal in Pačes, 1983), $\delta^{18}O = -10.4$, $-10.3‰$ (S. Nuti, personal written communication, 1982), $\delta D = -59‰$ (P. Fritz in Pačes, 1983) $\delta D = -55$, $-54‰$ (S. Nuti, personal written communication, 1982); isotopic composition of sulphur is $\delta^{34}S = +5.4‰$ (F. Buzek and V. Šmejkal, in Pačes, 1983) (Sample 12, Table 2)			
TYPICAL OIL-FIELD BRINES				
ALN	Cl-Na brine from the Angaro-Lenska basin, USSR (Krainov and Shvec, 1980, Table 32)	−2.48	−2.08	2.49
ALC	Cl-Ca brine from the Angaro-Lenska basin, USSR (Krainov and Shvec, 1980, Table 32)	−0.42	−1.98	2.71
ES	Brine from East Siberia, USSR (Krotova, 1957 in Frape and Fritz, 1982, Sample 18, Table 3)	−0.18	−2.12	2.71
NA	Cl-Na brine from Alberta basin, Canada (Hitchon et al., 1971 in Frape and Fritz, 1982, Sample 1, Table 3)	−1.43	−2.51	2.01
AC	Cl-Ca-Na brine from Alberta basin, Canada (Hitchon et al., 1971 in Frape and Fritz, 1982, Sample 3, Table 3)	−0.77	−2.73	2.38
WC	Mean composition of oil-field brines in Western Canadian Basin (Billings et al., 1969)	−1.16	−2.39	2.02
MB	Michigan Basin (Graf et al., 1966 in Frape and Fritz, 1982, Sample 5, Table 3)	−1.39	−2.92	2.47
TS	Mean composition of oil-field brines from Tertiary sandstones, Texas (Collins 1975, Table 7. xix)	−1.82	−3.18	2.02
IB	Oil-field brine from Illinois Basin (Graf et al., 1966 in Frape and Fritz, 1982, Sample 4, Table 3)	−1.42	−2.13	2.20
T93	The most concentrated brine from Canadian Shield from Thompson (Frape et al. 1984)	−0.69	−2.82	2.51
YK	Brine from Yellowknife, Canada (Fritz and Frape 1980)	−0.63	−5.13	2.41
NY	Brine from salt in New York (White et al., 1963, in Frape and Fritz, 1982, Sample 7, Table 3)	−1.32	−1.26	1.58
BRINES CONTAINING EXTREME CONCENTRATIONS OF SULPHATE IONS				
Kg	Brine from the playa lake Kitagata in a volcanic region Katwe, Uganda (Arad, 1969)	−0.93	−0.58	2.31
KI	Brine from a crater lake of the volcano Kawa Idgen, Java (Tkaczenko and Zotov, 1974)	−1.45	1.11	1.94
Sd	Brine from Salafo, New Mexico (White et al., 1963 in Frape and Fritz, 1982, Sample 8, Table 3)	−2.23	0.08	2.51

A characteristic feature of the saline waters in the Bohemian Massif is an enrichment with SO_4 + Na component. The isotopic composition of sulphur in waters from the Ohře River Rift is almost identical, with $\delta^{34}S = +5.2‰$ (Šmejkal, 1978). The isotopic composition of oxygen and hydrogen in the saline waters is the same as the composition in local meteoric water. Only in the case of the most saline water from the Cheb basin (No. 12, Table 2), hydrogen contains about 20‰ more deuterium and oxygen contains about 1‰ more ^{18}O than in the meteoric water (unpublished data by P. Fritz and S. Nuti).

HYDROCHEMICAL FIELD OF NATURAL WATERS

Hydrochemical field describes hydrochemical variations among hypothetical or real end-member solutions in a given hydrogeological structure (Pačes, 1980). The end members are not just solutions of simple salts. They are dissolved components whose origin is linked to individual hydrological, geological and geochemical processes which determine the evolution of groundwaters and brines. Some end members result from well-defined processes such as oxidation of sulphides or hydrolysis of feldspars. Other end members result from processes that may be more complex and not well understood; as in the case of the calcium-chloride component in highly concentrated brines, or the sodium-sulphate component in the saline waters of the Bohemian Massif. The value of the end members is calculated as a sum of characteristic cation and characteristic anion molar concentrations in typical water samples in a given hydrogeological structure.

Six hydrochemical end members define the field of the waters in the Bohemian Massif:

C + Ca end member (C is carbon in all dissolved carbonate species, expressed in mole/L, and Ca is dissolved calcium) represents a water component which originates from hydrolysis of aluminosilicate rocks and carbonate rocks in the

Table 2

Chemical composition of selected groundwater samples and saline waters from the Bohemian Massif and Western Carpathians (Czechoslovakia), concentrations are in mg/L; logarithmic coefficients are calculated from molar concentrations and total dissolved solids (TDS) is in g/L

No. of sample	1	2	3	4	5
Symbol in diagrams	C	J	L	MP	TB
Locality	246 wells west of Jizera river Bohemian Massif	Shaft Svornost Jáchymov, depth 340 m	Louny borehole Ln-1	Muráňská Planina, flat West Carpathians	Teplice nad Bečvou, West Carpathians
Source	Cretaceous sandstones with calcareous cement	seepage from gneiss	granite depth 1100 m	Limestones	Devonian limestones
References	Pištora, 1978	Pačes, 1969	Krajča et al. 1971	Gazda and Kullman, 1964	Franko et al. 1975
Temperature °C	11	18.3	39	5.7	22
Presence of $CO_{2(g)}$ at surface	no	no	1738 mg/L	no	5574 mL/L
pH	7.0	6.80	6.3	7.2	6.2
Na	35.6	50.0	4140	1.8	71.2
K	11	14.5	450	1.5	11.5
NH_4		0.00	0.0		0.8
Mg	27	171.5	99.3	0.3	48.6
Ca	105	531.1	53.7	61.2	490.6
Fe	2.7	trace	7.7	0.00	2.7
Mn		0.00	0.0		
Cl	25.6	17.7	710.5	2.4	39.4
F		1.3	3.3		2.5
HCO_3	363	237.9	11773	179.6	1879
NO_3	14.1	0	0.0		1.5
SO_4	137	1852.6	0.82	15.2	9.5
SiO_2	7.9	14.4	39		31.0
$\log\frac{C+Ca}{Cl+Na}$ (molar concentrations)	0.58	0.81	−1.19	1.49	1.01
$\log\frac{S}{C+Na}$ (molar concentrations)	−0.72	0.50	−4.34	−1.28	−2.53
log TDS (total dissolved solids, g/L)	−0.14	0.46	0.79	−0.58	0.41

near-surface zone enriched with biogenic CO_2.

S + Ca end member (S represents all sulphur species dissolved in water) results from the oxidation of bedrock containing sulphide minerals, and the consumption of generated protons by hydrolysis of calcium bearing minerals. Both the calcium end members characterize open hydrogeological structures with fast exchange of groundwater and input of atmospheric gases.

C + Na end member is characteristic for semi-open structures with an intensive hydrolysis of aluminosilicate rocks containing sodium feldspars (or other sodium-bearing alumosilicates). The input of CO_2 necessary to promote the hydrolysis at depth is of an endogenous origin, either metamorphic or magmatic. There is an anomalous flux of CO_2 of a deep origin along major fault zones in the Bohemian Massif (Pačes, 1972).

Cl + Na and Cl + Ca end members are dominant in closed hydrogeological structures. The high concentrations found in brines can usually not be explained by water-rock interaction unless the solutions result from simple dissolution of salt deposits by meteoric water. The chemical variations observed in the brines usually exhibit a characteristic vertical zonation independent of wall-rock composition. The component Cl + Na predominates in less deep water bodies with total dissolved solids less than 150 g/L. The component Cl + Ca predominates in the deepest aquifers where the total dissolved solids may exceed even 500 g/L (Pinneker, 1966). This zonation described from deep sedimentary basins, has not been found in irregular water bodies filling the fractures of crystalline massifs. The origin of these end members is usually a subject of disputes. Evaporation, diagenesis, membrane filtration, and magmatic processes may play different roles in different structures.

S + Na end member is an unusual component in natural brines. It is characteristic for closed systems in volcanic regions, such as playa lakes in continental rifts. The isotopic composition of sulphur in this component of the saline waters in the Bohemian Massif does not depend on location and is different from the isotopic composition of sulphur in wall-rock. Therefore, this component may be allochtonous in the sense that it was supplied from the surrounding structures prior to the formation of the water composition. The source may be the volcanic activity along the Tertiary rifts.

As groundwater evolves due to water-rock interaction and the input of allochtonous components, its position in the hydrochemical field changes and the concentration of dissolved solids increases. These variations can be illustrated

Table 2 continued

No. of sample	6	7	8	9
Symbol in diagrams	T	S	B	K
Locality	Trebišovská Tabule flat, Western Carpathians borehole Stretava-2,	Stránka u Mšena Bohemian Massif borehole MJ-2	Bor u Tachova Bohemian Massif Shaft Vítkov II	Karlovy Vary Bohemian Massif Spring Vřídlo II
Source	3000 m halite, gypsum	Permocarboniferous sandstones underlying Cretaceous sediments	Granodiorite	Granite
References	Michalíček, 1970	Jetel, 1970	Skořepa in Pačes, 1983	Macháček and Šulcek, 1964
Temperature °C		40	37	72
Presence of $CO_{2(g)}$ at surface	no			10 L of CO_2/L H_2O
pH	5.2		7.0	6.6
Na	182400	20400	1786	1713
K	43	194	32	118
NH_4	0		0.001	0.2
Mg	23.4	157	23.1	44
Ca	41.4	138	2189	129
Fe		157	3.6	1.03
Mn		0.73		0.16
Cl	280800	22700	6458	618
F		2.9	Br 23, I 0.3	F 6.2
HCO_3	878	8415	56.7	2152
NO_3			85	
SO_4	39.1	7396	338	1702
SiO_2		51	19.4	71
$\log \frac{C+Ca}{Cl+Na}$ (molar concentrations)	−3.01	−1.03	−0.67	−0.38
$\log \frac{S}{C+Na}$ (molar concentrations)	−4.29	−1.12	−1.35	−0.79
log TDS (total dissolved solids, g/L)	2.67	1.78	−1.04	0.82

on a simple hydrochemical-field diagram defined by three variables: $X=\log_{10}$ [(C+Ca) / (Cl+Na)], $Y=\log_{10}$ [S/(C+Na)], $Z=\log_{10}$ TDS where C, Cl, S, Ca and Na are concentrations of dissolved elements in mole/L and TDS are total dissolved solids in g/L. The value of X differentiates between water samples from open structures (high values) and closed structures (low values). The value of Y discriminates between oxidation (high values) and hydrolysis (low values) in open and semi-open hydrogeological structures, and between high input of S+Na component and low input of the component in closed structures and volcanic regions.

An input of Cl+Ca component moves the position of brines to the central part of the diagram. Total dissolved solids, or Z value, differentiates between waters which dissolve calcium from wall-rock (Z is low) and waters which contain the Cl+Ca component of a deep origin (Z is high). Both types of water may fall in the same spot in the X-Y diagram.

APPLICATION OF HYDROCHEMICAL-FIELD DIAGRAMS

The distribution of various genetic types of waters in the diagram is shown in Figures 1, 2 and 3. The water samples plotted in the diagrams were selected in such a way that they represent typical chemical composition in which the individual end-member components dominate. Figure 4 illustrates the position of the end members in the hydrochemical-field diagrams. The position of the end members is given by extreme compositions of actual natural water samples and not by hypothetical pure solutions of simple salts. Such pure solutions cannot be shown in this diagram because their values of X and Y are not defined.

A genetic interpretation of the hydrochemical-field diagram in Figure 5 is based on observed hydrochemical variations (Figures 1 to 3), and hydrogeological and geochemical characteristics of the water samples. Meteoric waters dominate in the upper part of the diagram. They are so-called lithogenic waters chemical composition of which is determined by kinetics of hydrolysis and oxidation of wall-rocks. Chemical equilibrium usually does not control the concentrations of major dissolved components. Fossil waters dominate the lower part of the diagram. This is, however, not a rule without exceptions. Meteoric waters which dissolve salt deposits are located in the left-lower corner of the diagram. Fossil brines containing Cl+Ca component are located in the center of the diagram, in the domain of semi-open structures, although they are typically the product of evolution in closed structures.

The composition of lithogenic waters develops from rain water and snow through the dissolution of wall-rocks in the presence of atmospheric and biogenic CO_2, and atmo-

Table 2 continued

No. of sample	10	11	12	13
Symbol in diagrams	F	G	O	pp
Locality	Mariánské Lázně Bohemian Massif Spring Ferdinand I	Františkovy Lázně Bohemian Massif Borehole Glauber IV Sandstone, Phyllite	Odrava (Cheb Basin) Bohemian Massif Borehole HV-18c	Hrádek u Pacova Bohemian Massif
Source	Amphibolite	Granite	Phyllite, depth 624 m	Atmospheric wet precipitation in a rural region
References	Škop et al. in Pačes, 1974	Suchý, 1957	Pačes et al., 1981	Moldan 1980
Temperature °C	0.7	13.4	20	
Presence of $CO_{2(g)}$ at surface	2508 mg/L	1191 mg/L	no	
pH	6.3	8.0	7.5	4.27
Na	2870	6479	47040	0.18
K	86.8	136.7	288	0.08
NH_4	0.98		35	1.1
Mg	141.0	129.8	278	0.07
Ca	199.6	482.0	455	0.62
Fe	33.6	8.5	83	
Mn	0.32	0.13	1.5	
Cl	1121	2456	28700	0.41
F	0.07		1.2	
HCO_3	3050	3284	1400	0.0
NO_3	0.0		85	3.0
SO_4	3173	9853	54200	5.7
SiO_2	86.4	62.1	16	
$\log\frac{C+Ca}{Cl+Na}$ (molar concentrations)	−0.45	−0.73	−1.92	−0.10
$\log\frac{S}{C+Na}$ (molar concentrations)	−0.72	−0.51	−0.56	0.88
log TDS (total dissolved solids, g/L)	1.03	1.36	2.12	−1.95

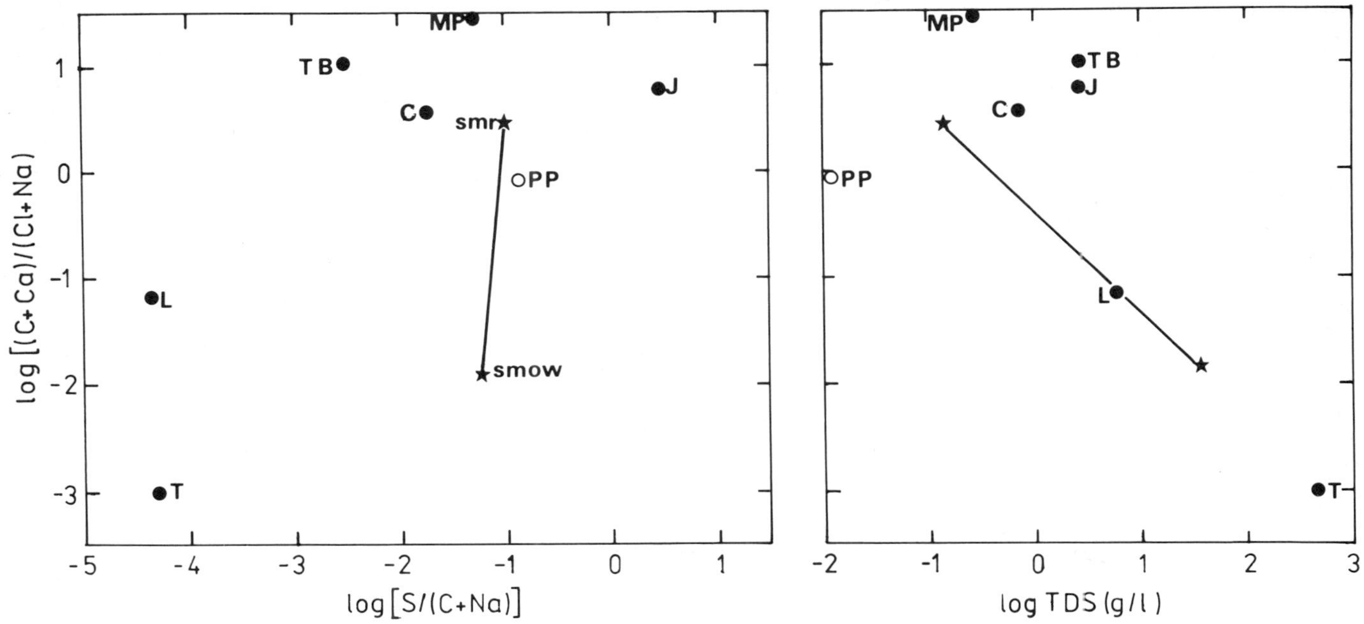

Figure 1. *Hydrochemical-field diagram representing lithogenic waters of meteoric origin; smr – smow is tie line between river water and oceanic water, L is groundwater from granite charged with CO_2 of a deep origin, T is meteoric water from salt deposit, TB is groundwater from limestone charged with CO_2 of a deep origin, MP is groundwater from limestone, C is groundwater from sandstones and marlstones, J is seepage of mine water from a sulphide ore deposit, pp is wet precipitation in the Bohemian Massif.*

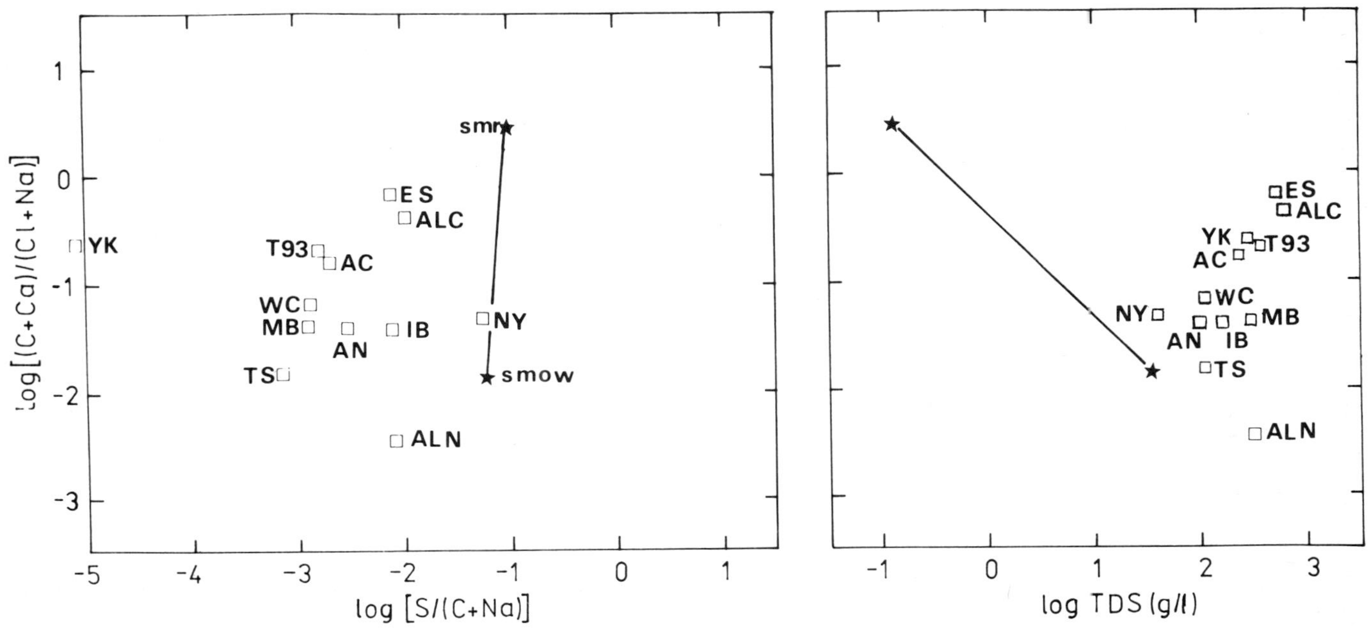

Figure 2. *Hydrochemical-field diagram of oil-field brines. Symbols are explained in Table 1.*

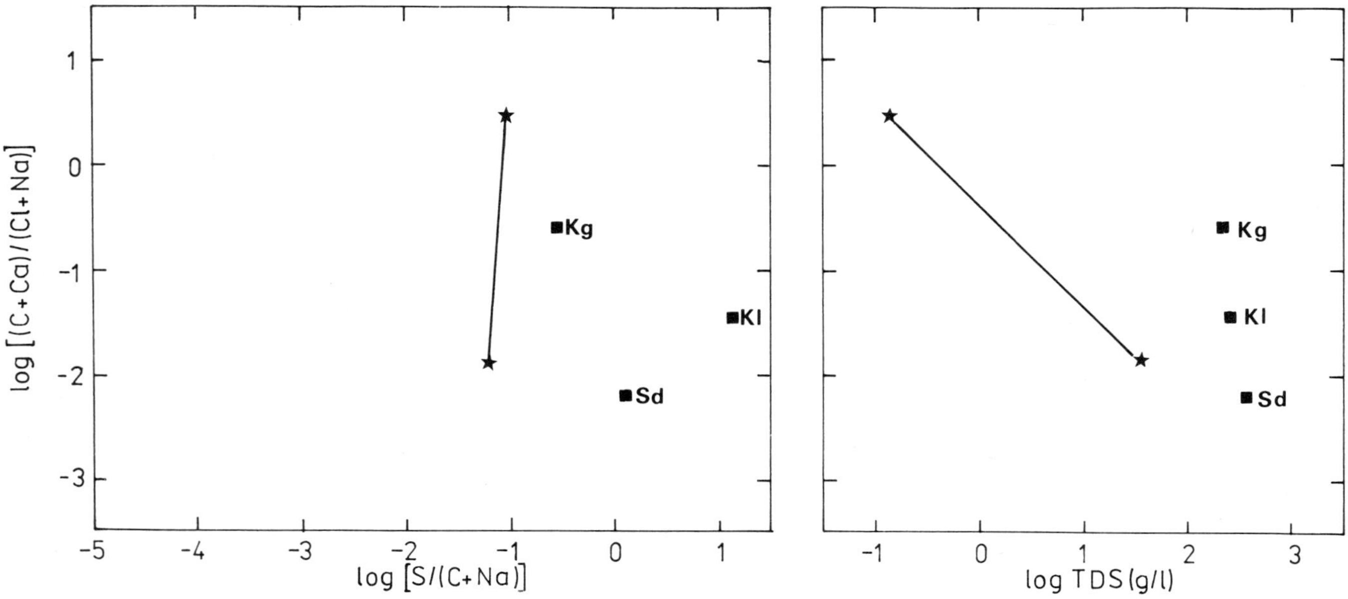

Figure 3. *Hydrochemical-field diagram of waters from volcanic lakes and a brine enriched in sulphate ions. Symbols are explained in Table 1.*

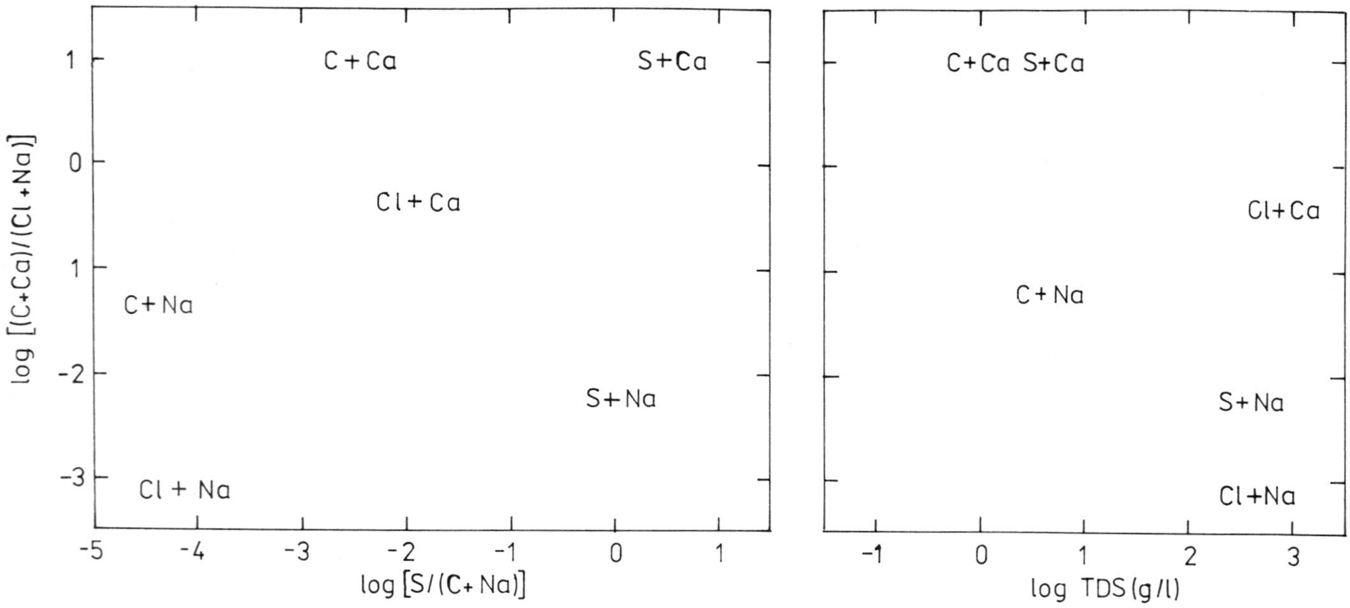

Figure 4. *Position of hydrochemical end members in hydrochemical-field diagram.*

spheric O_2. When these two gaseous components are exhausted, the water-rock interaction slows down and the chemical composition of water reaches a steady state which is a function of velocity of water percolation and reaction rates. When CO_2 of a deep origin enters the steady-state system, hydrolysis of wall-rock can continue at an accelerated rate. This process shifts the chemical composition to the left-hand side of the diagram. The hydrolysis of carbonate rocks shifts it to the upper part of the diagram, and the hydrolysis of aluminosilicate rocks containing albitic component shifts it to the central-left portion of the diagram. The oxidation of pyrite and other sulphides shifts the composition to the upper right-hand side of the diagram. Dissolution of salt deposits by meteoric water results in brines that plot in the lower-left corner. A hypothetical evolution of oil-field brines shown in the diagram assumes that the origin of the water is oceanic. The brines evolve from Cl-Na types to Cl-Ca type with increasing TDS.

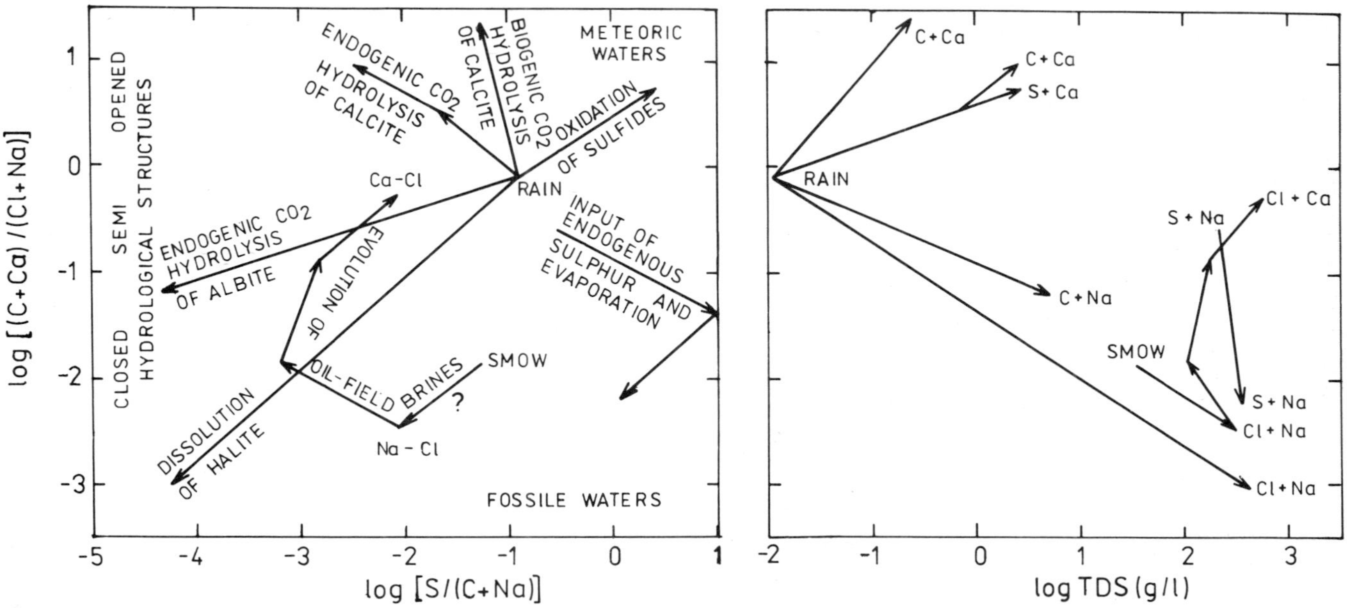

Figure 5. *Genetic interpretation of the hydrogeochemical-field diagram derived from the interpretation of the origin of natural waters listed in Table 1.*

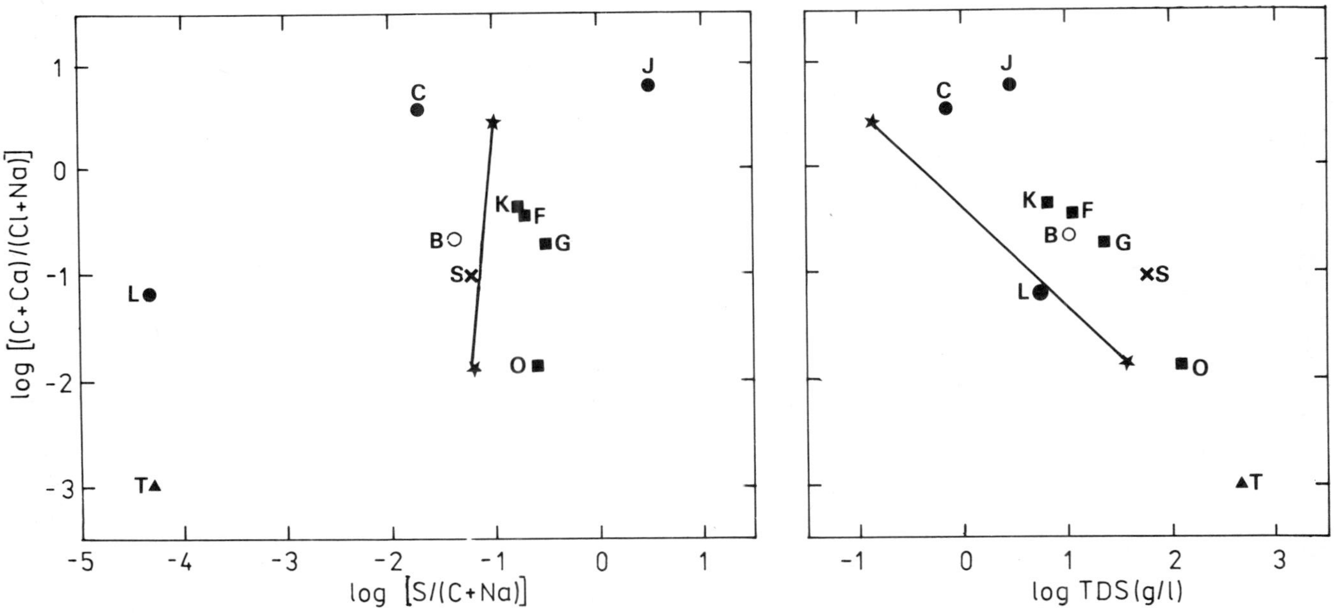

Figure 6. *Saline waters from the Bohemian Massif and related lithogenic waters.* ● *– ground water,* ▲ *– meteoric water from a salt deposit,* ■ *– saline waters enriched in fossil SO_4 + Na component,* × *– saline water containing SO_4+Na and Cl+Na components,* o *– saline water containing Cl+Ca component.*

Finally, saline waters which plot on the right side from the tie-line between "smr" and "smow" are enriched in sulphate. If the sulphate ions are not dissolved from gypsum in the aquifer or they do not result from oxidation of pyrite or other sulphides in aquifers opened to molecular oxygen, then their origin is not obvious. One possible explanation is an input of volcanic sulphur and the concentration of the resulting solution either by evaporation or by some other process such as low-grade metamorphism, membrane filtration or even incorporation of liquid water molecules into hydrous minerals.

The arrows in Figure 5 indicate trends observed in natural waters and not real reaction paths. The diagram must not be used mechanically to explain the origin of other waters by just plotting their chemical composition into certain parts of the diagram. A proper explanation can result only if the observed hydrochemical variations are interpreted together with geological, hydrological and isotopic data.

SALINE WATERS IN THE BOHEMIAN MASSIF

Groundwaters and saline waters in the Bohemian Massif are plotted in Figure 6. The point T represents a sample from the Carpathians. This composition results from simple dissolution of salt. All saline waters in the Bohemian Massif except the water from the Bor Massif are enriched in sulphate ions with respect to the "smr"-"smow" line. The ions are not derived from gypsum deposits because such deposits are not known in this region. When a mineralogical occurrence of gypsum was described it was in places where hydrological conditions were favorable for evaporation of the saline waters so that secondary gypsum was precipitated (Pačes et al., 1981). Sulphate is not derived from the oxidation of pyrite because the isotopic composition of the sulphide sulphur is much heavier ($\delta^{34}S + 16‰$ as compared to $+5.2‰$ for the dissolved sulphate (Šmejkal, 1978). Therefore, an external source of sulphur may be the only explanation for the enrichment. The isotopic composition of sulphur in the mineral waters of the Bohemian spas and in the most saline SO_4-Na water from the basement of the Cheb basin is identical. The occurrence of the waters is determined by the tectonic zones of the Ohře Rift. The latest hypothesis for the origin of the waters (Pačes et al., 1981) is that the Miocene and Oligocene rift lake received volcanic emanations high in sulphur. Dissolved sulphur was oxidized to sulphate ions. Released hydrogen ions reacted with the acid aluminosilicate rocks on the bottom of the lake and enriched the resulting solution in sodium. The solution was concentrated by evaporation in the closed basin and reached a total dissolved solids content of 140 g/L or higher. The solution percolated into open fractures in the bottom of the rift lake and salts diffused into an underlying groundwater body. The saline solution has remained trapped in the joints of the crystalline basement rocks below the Tertiary sediments till today. Where the structure was opened by erosion, and where the water body was saturated with CO_2 of possibly magmatic origin, mineral springs were formed. In deeper parts and in structures closed from the input of CO_2, the most concentrated solution was preserved until it was discovered by the drill hole HV-18c in the Cheb basin. Hydrological, geological, paleontological and hydrochemical data and the occurrence of similar waters in African closed-rift lakes support this hypothesis. However, oxygen and hydrogen in the water are not isotopically heavier as would be expected if evaporation were responsible for the concentration of the original lake water. This would shift the data points on Figure 7 to the right of the global meteoric water line. Independent isotopic data on the brine from the Cheb basin by S. Nuti indicate a shift to the left of the line. A shift in the same direction was observed in saline waters in the Canadian Shield (Frape et al. 1984). The mechanism of isotopic fractionation responsible for the shift in the Bohemian Massif is not well understood yet.

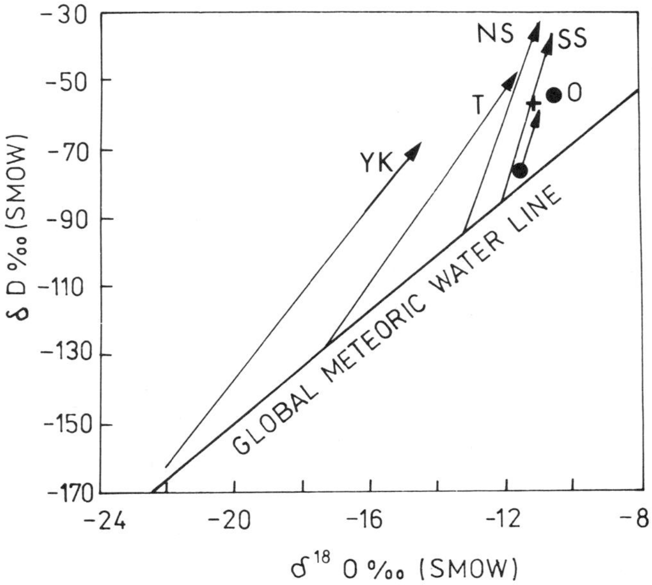

Figure 7. *Isotopic composition of water molecules in the sodium-sulphate brine from the crystalline basement of the Cheb basin (Odrava, borehole HV-18c), and isotopic composition of water molecules in brines of the Canadian Shield (Frape et al., 1984).*

An enrichment in sulphates is also observed in saline waters tapped by drillholes in Permocarboniferous sandstones below the Cretaceous marine sediments. Spacially, the enrichment is related to the Elbe Line. These brines contain more chlorides than the saline waters in the Ohře River Rift. There is no reliable pH determination for these brines, however the high concentrations of bicarbonate suggest that they may be enriched with CO_2 of a deep origin. The increased flux of CO_2 along deep-seated tectonic zones has been documented in other parts of the Cretaceous basin by lower pH values, high alkalinity and high concentrations of calcium; CO_2 is probably consumed during dissolution of calcareous cement in the Cretaceous sandstones. A concentration gradient of Cl^- in Cretaceous aquifers above the Permocarboniferous layer with the most concentrated brine suggests possible diffusion of salts from depth to the surface. If the original brine was Cretaceous, than the diffusion has lasted about 65 million years. The calculated diffusion coefficient for the observed Cl^- gradient at a depth of more than 650 m and for 65 million years is 2×10^{-7} cm^2sec^{-1} (Pačes, 1983). This is a reasonable value considering the tortuosity in the sediments. A Cretaceous origin of the brines is therefore possible. However, the carbon dioxide and sulphate are probably derived from a Tertiary magmatic source.

The third type of brine occurs in the granodioritic body of the Bor Massif. It is a calcium chloride saline water with total dissolved solids not exceeding 11 g/l. The isotopic composition of oxygen is $\delta^{18}O = -10‰$ which is close to the local meteoric waters. The isotopic composition of sulphur is different from the composition in the Ohře River Rift: $\delta^{34}S$ is $+28‰$ (F. Buzek and V. Šmejkal in Pačes, 1983). This water cannot be related to any young marine transgression.

The closest Cretaceous marine sediments are at least 150 km away and there are no geological indications that the Cretaceous transgression reached the Bor. There is not enough chlorine in the granodiorite to explain the high concentrations in the water. At present, there is no probable hypothesis which would explain the origin of the brine. If the saline water were a residual magmatic water, than the isotopic composition of oxygen would be higher than −10‰. The last plutonic activity was older than Variscan and it is highly improbable that any remnants of the old magmatic fluid could be preserved in fractures at a depth of only 400 m.

The above examples of saline waters in the Bohemian Massif indicate that the history of such fluids can be diverse and linked to geological development of the whole region. Water-rock interaction is involved in the evolution of the solutions; however, the ultimate origin of the waters is not known with certainty. It is probable that all the systems were open to inputs of allochtonous components, namely sulphur and carbon dioxide. The origin of chlorides may be volcanic in the Ohře River Rift, marine along the Elbe Line and it is unknown in the calcium-chloride saline water in the Bor Massif.

ACKNOWLEDGEMENTS

The paper contains previously unpublished isotopic data. I thank very much Dr. Sergio Nuti of Instituto Internazionale per le Ricerche Geotermiche, Pisa, Dr. James O'Neil, U.S. Geological Survey, Menlo Park and Prof. Peter Fritz of University of Waterloo, Ontario for the determination of ^{18}O and D in the Odrava brine. The other isotopic analyses were supplied by ing. Václav Šmejkal, Dr. Jana Hladíková and ing. František Buzek, all Geological Survey, Prague. I express my gratitude to all these friends for very useful discussions concerning the interpretation of the isotopic data.

REFERENCES

Arad, A., 1969. Mineral springs and saline lakes of the Western Rift Valley, Uganda: Geochimica et Cosmochimica Acta, v. 33, p. 1169-1181.

Billings, G.K., Hitchon, B. and Shaw, D.R., 1969. Geochemistry and origin of formation waters in the Western Canada sedimentary basin, 2. Alkalimetals: Chemical Geology, 4, pp. 211-223.

Buday, T. (ed.), 1969. Tectonic development of Czechoslovakia. Nakladatelstvi CSAV, Prague, 232 p.

Čermák, V., Halada, S., Jetel, J., Kopecký, L. and Pačes, T., 1981. Geothermal resources in ČSR: Geologický Průzkum, v. 23, p. 12-15 (in Czech).

Collins, A.G., 1975. Geochemistry of oil-field waters: Elsevier, Amsterdam, 496 p.

Franko, O., Gazda, S. and Michalíček, M., 1975. Tvorba a klasifikacia minerálnych vôd Západních Karpat: Geologický Ústav Dionýza Štúra, Bratislava, 230 p.

Frape, S.K. and Fritz, P., 1982. The chemistry and isotopic composition of saline groundwaters from the Sudbury Basin, Ontario: Canadian Journal of Earth Sciences, v. 19, p. 645-661.

Frape, S.K., Fritz, P. and McNutt, R.H., 1984. Water-rock interaction and chemistry of groundwaters from the Canadian Shield: Geochimica et Cosmochimica Acta, v. 48, p. 1617-1627.

Fritz, P. and Frape, S.K., 1980. Comments on the ^{18}O, ^{2}H and chemical composition of saline groundwaters on the Canadian Shield: Proceedings, American Geophysical Union Symposium on Isotope Studies of Hydrological Processes, University of Northern Illinois, DeKalb, IL. p. 57-63.

Gazda, S. and Kullman, E., 1964. Hydrogeochemia podzemných vôd vápencovo-dolomitických komplexov mesozoika Západných Karpat: Geologické Prace, Správy 32, p. 29-46, Geologický Ústav Dionýza Štúra, Bratislava.

Graf, D.L., Meents, W.F., Friedman, I. and Ship, N.F., 1966. The origin of saline formation waters, III: Calcium Chloride waters: Illinois State Geological Survey, Circular 397.

Hitchon, B., Billings, G.K. and Klovan, J.E., 1971. Geochemistry and origin of formation waters in the Western Canada sedimentary basin – III. Factors controlling chemical composition: Geochimica et Cosmochimica Acta, v. 35, p. 567-598.

Hladíkova, J. and Šmejkal, V., 1973. Isotopic analyses of waters: Archive, Ústřední Ústav Geologický, Praha (Ms. in Czech), In Pačes, T. 1974.

Jetel, J., 1970. Hydrogeologie permokarbonu a křídy na opěrném profilu Mělník-Ještěd: Sborník Geologických Věd, HIG, v. 7, p. 7-42.

Krajča, J., 1971. Hydrogeologický průzkum zdroje minerální vody v Lounech: Archive, Ústav Geologického Inženýrství, Brno (Ms in Czech).

Krainov, S.R. and Shvec, V.M., 1980. Osnovy geochimii podzemnych vod: Nedra, Moscow 286 p.

Krotova, V.A., 1957. Conditions of formation of calcium chloride waters in Siberia: Petroleum Geology, v. 2, p. 545-945.

Livingstone, D.A., 1963. Chemical composition of rivers and lakes: U.S. Geological Survey Professional Paper 440-G, 64 p.

Macháček, V. and Šulcek, Z., 1964. Celková chemická analýza minerální vody Vřídlo II v Karlových Varech: Fysiatrický Věstník, v. 43, p. 45-51.

Martin, J.-M. and Meybeck, M., 1978. The content of major elements in the dissolved and particulate load of rivers: In: Biogeochemistry of Estuarine Sediments, Proceedings UNESCO/SCOR Workshop, Melreux, p. 95-110.

Michalíček, M., 1970. K původu chloridonosných solanek v miocénu trebišovské nížiny: Sborník Geologických Věd, HIG, v. 7, p. 107-159.

Moldan, 1980. The analysis of atmospheric precipitation in Czechoslovakia: In: Ecological Impact of Acid Precipitation, p. 124-125 (D. Drablos and A. Tollan, eds.), Jahs. Grefslie Trykkeri A/S, Mysen.

Pačes, T., 1969. Chemical equilibria and zoning of subsurface water from Jachymov ore deposit: Geochimica et Cosmochimica Acta, v. 33, p. 591-609.

Pačes, T., 1972. Flux of CO_2 from the lithosphere in the Bohemian Massif: Nature Physical Science, v. 240, p. 141-142.

Pačes, T., 1974. Springs of carbon-dioxide water in northwestern Bohemia: Field trip guide, International Symposium on Water-Rock Interaction, Ústřední Ústav Geologický, Praha, 82 p.

Pačes, T., 1980. Genetic types of groundwaters and hydrochemical field in the Bohemian Massif: Geologický Průzkum, v. 22, p. 229-231 (in Czech).

Pačes, T., 1983. Principles of Geochemistry of waters: 300 p. Academia, Praha (in Czech).

Pačes, T., Šmejkal, V., Pazdera, A., Kobrová, M. and Barnes, I., 1981. An unique type of a brine in the basement of the Cheb basin: Geologický Průzkum, 23, pp. 196-198, Praha (in Czech).

Pinneker, E.V., 1966. Rassoly Angaro-Lenskogo Artezianskogo bassejna: Nauka, Moscow.

Pištora, Z. (ed.), 1978. Regularities of circulation of Cretaceous groundwaters: Archive Vodní Zdroje, Praha (Ms in Czech).

Škop, Nevoral and Konopač. Archive chemical analysis in Výzkumný ústav balneologický, Mariánské Lásně. In Pačes (1974).

Šmejkal, V., 1978. Isotopic geochemistry of the Cypris formation in the Cheb basin, West Bohemia: Věstník Ústředního Ustavu Geologického, v. 53, p. 3-18.
Suchý, K., 1957. Chemický rozbor a klasifikace minerální vody z pramene Glauber IV ve Františkových Lázních: Fysiatrický Věstník, v. 35, p. 164-166.
Suk, M. (ed.), 1984. Geological history of the territory of the Czech Socialistic Republic: Geological Survey, Prague, published in Academia, Prague, 396 p.
Svoboda, J. (ed.), 1966. Regional Geology of Czechoslovakia, Part I: Czechoslovak Academy of Sciences, Prague, 668 p.
Tkaczenko, R.I. and Zotov, A.V., 1974. Ultrakislye termy vulkanicheskogo proischozhdenia kak rudnye raztvory: In: Gidrotermalnye Mineraloobrazuiuschchie Rastvory Aktivnogo Vulkanizma, Nauka, Sibirskoe Otdelenie, Novosibirsk, p. 86-91.
Turekian, K.K., 1969. The oceans, streams and atmosphere: Handbook of Geochemistry I, pp. 297-323, Springer-Verlag, Berlin.
White, D.E., Hem, J.D. and Waring, C.A., 1963. Chemical composition of subsurface waters: U.S. Geological Survey Professional Paper 440-F.

The Occurrence of Saline Groundwaters and Gases in the Crystalline Rocks of Northern Switzerland

Asaf Pekdeger
Institute of Geology and Paleontology, University of Kiel, Olfhansenstr. 40/60, D-2300 Kiel, F.R. Germany

Werner Balderer
Physics Institute, University of Bern, Fidlerstr. 5, CH/3012, Bern, Switzerland

Pekdeger, A. and Balderer, W., The occurrence of saline groundwaters and gases in the crystalline rocks of Northern Switzerland; in Saline water and gases in crystalline rocks, Editors: Fritz, P., and Frape, S.K.; Geological Association of Canada Special Paper 33, 1987. p. 157–174.

Abstract

Crystalline bedrock in Northern Switzerland has been investigated for its hydrogeological characteristics. Very active flow systems have been encountered and, although saline waters were found at depth, their dissolved loads were much lower than those encountered in other crystalline rock environments at similar depths. Hydrogeological and geochemical evidence indicates that recharge occurs in the Black Forest mountains of Southern Germany and geochemical data show that waters which infiltrate directly into and flow within crystalline rocks do not become saline even at great depths.

Where saline groundwaters are encountered in the crystalline rocks underlying the study area, the waters have passed through sedimentary sequences where they acquired the dissolved solids. The geochemistry of waters from sedimentary sequences are compared with those from crystalline rocks. Evidence exists for crystalline rock-water interactions but this does not result in significant salinity increases such as seen in Canada or the United Kingdom in similar rock types.

Résumé

Le socle cristallin de la Suisse septentrionale a fait l'objet d'une étude visant à établir ses paramètres hydrogéologiques. Des régimes d'écoulement très actifs y ont été observés et, bien que des eaux salées aient été observés en profondeur, leur contenu dissous était beaucoup moins élevé que dans les autres environnements cristallins de profondeur semblable. Des observations hydrogéologiques et géochimiques indiquent que la recharge se produit dans la Forêt-Noire, au sud de l'Allemagne, et des données géochimiques révèlent que les eaux qui s'infiltrent directement et s'écoulent dans les roches cristallines ne deviennent pas salées, même à de grandes profondeurs.

Les eaux salées des roches cristallines de la région étudiée ont circulé dans des unités sédimentaires d'où proviennent leurs solides dissous. La géochimie des eaux des unités sédimentaires a été comparée à celle des eaux des roches cristallines. L'interaction entre l'eau et les roches cristallines a déjà été établie mais n'entraîne pas d'augmentation appréciable de la salinité, comme c'est le cas au Canada ou au Royaume-Uni, dans des roches semblables.

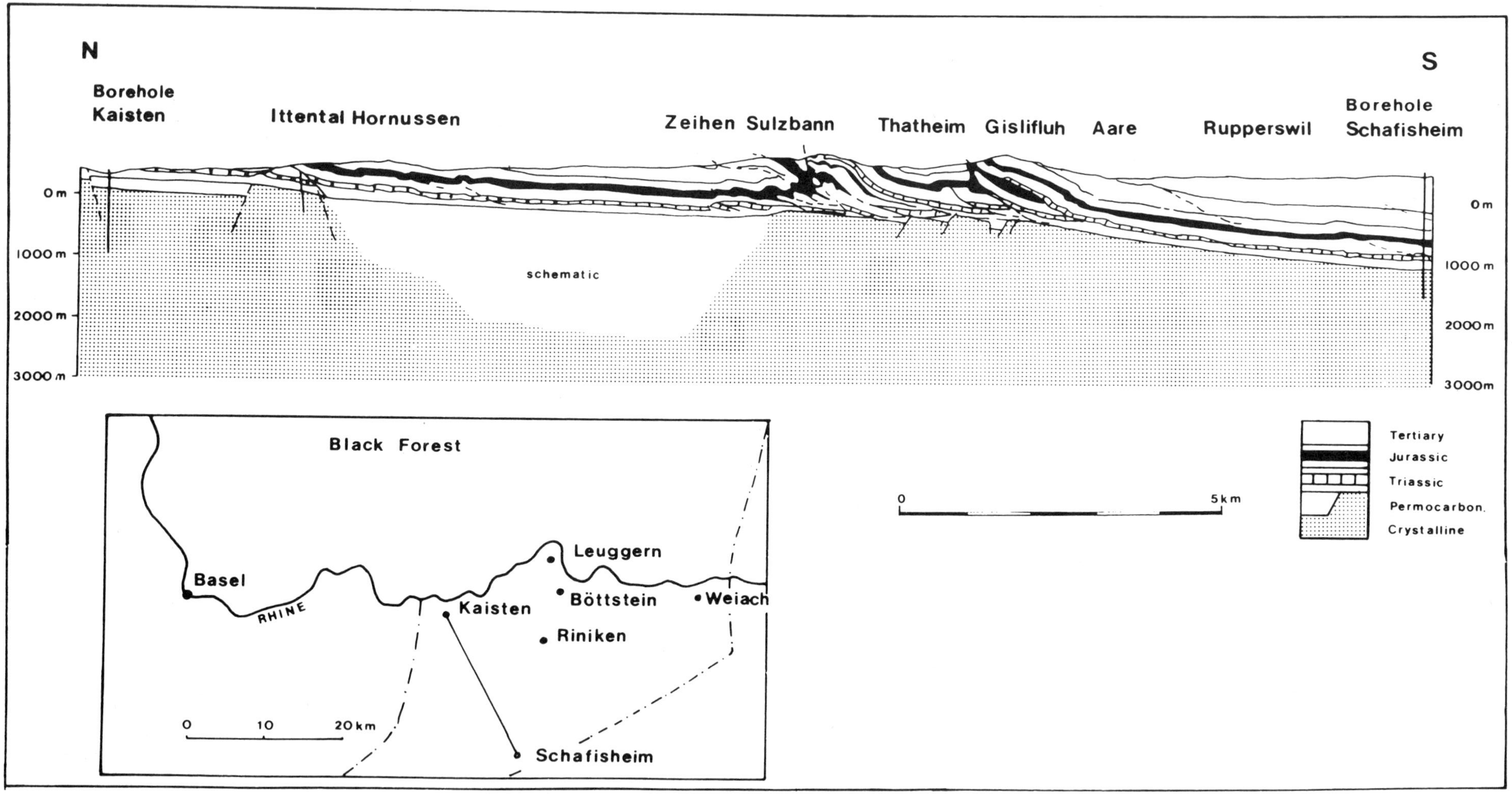

Figure 1. *Schematic geological cross-section of the investigated area (NGB 85-04).*

INTRODUCTION

For the disposal of radioactive waste, Nagra (Nationale Genossenschaft fuer die Lagerung radioaktiver Abfaelle) has recently investigated the hydrogeology and geochemistry of the groundwaters in crystalline rocks and in the overlying sedimentary rocks in Northern-Switzerland (NGB 85/04) on a broad regional scale.

New and very detailed geological and hydrogeological information on Northern Switzerland was obtained from six Nagra boreholes between 1306-2482 m depth, of which five have penetrated crystalline rocks. Sedimentary Permo-Carboniferous troughs with more than 1000 m depth were discovered (Figure 1). These boreholes were evaluated with respect to geology (Peters et al., 1985), permeability and hydraulic potentials (NTB 85-01), chemistry and isotopic composition of the groundwaters (Balderer, 1983, 1985: Pearson, 1985). These data, combined with information from regional hydrogeological programs which involved about 120 springs and wells were used to give an overall picture of the hydrogeological setting (Schmassmann et al., 1984).

GEOLOGY

Figure 2 shows a schematic cross-section from the northern part of the Alps through the Helvetic Nappes and the Swiss Plateau to the Rhine and the granitic massif of the Black Forest.

The crystalline massif from the central part of the Alps is covered by sediments affected by the following tectonic units (from S to N): Autochthonous and paraautochthonous sedimentary complexes (mainly mesozoic sediments forming the so-called Prealps and the Helvetic Nappes) are overthrusted on the tertiary sediments of the subalpine Molasse. The Molasse is in turn overthrusted on the thick stratified complex of the Molasse basin. This basin is made of clastic sediments of tertiary age which form the main part of the morphological feature of the central part of Switzerland, the so-called Molasse Plateau. In the north-eastern part the sedimentary series of jurassic age, folded (Ketten Jura) or tabular (Tafeljura), are outcropping.

As newly explored by some of these recently drilled Nagra boreholes, huge Permo-Carboniferous deposits of up to more than 1000 m thickness have been encountered lying above the ordinary crystalline basement within a trench-zone, limited by faults and where the base of the crystalline bedrock is lowered down to 2500 m below ground-level (see Figure 1).

Further north, near the Rhine valley, the sedimentary cover is eroded, and the crystalline bedrock is appearing in the Black Forest massif.

The investigated crystalline basement rocks are mainly granites and gneisses, penetrated by various types of dikes. The crystalline rocks of the Black Forest mountains to the north of the Permo-Carboniferous trough consist mainly of gneisses and gneiss-anatexites. The boreholes along the Rhine river also encountered mainly gneiss and gneiss-anatexites (Nagra boreholes Kaisten and Weiach) and locally gneiss and granitic rocks (Zurzach and Nagra borehole Leuggern). In the more southern Nagra boreholes Boettstein and Schafisheim only granitic and similar intrusive rocks were found.

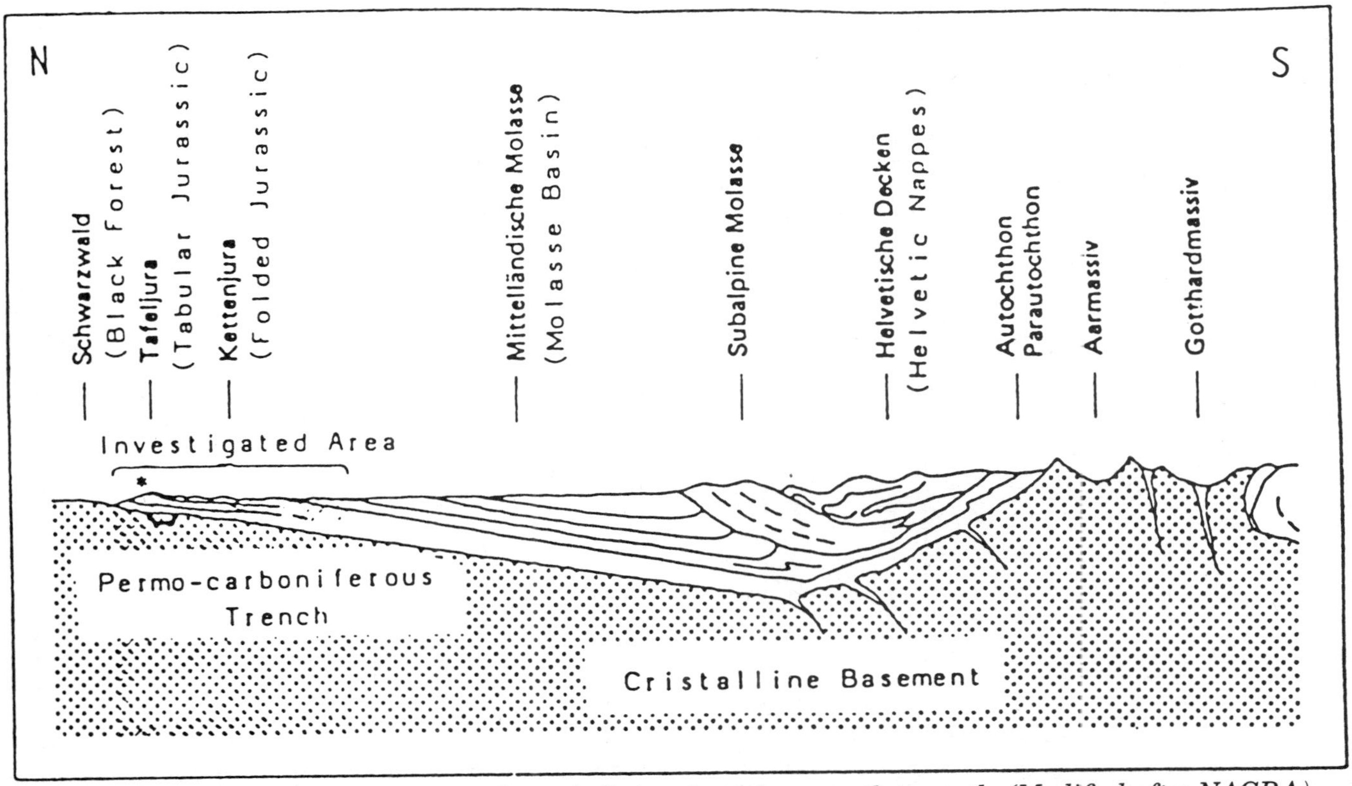

Figure 2. *Geological cross-section through Switzerland from north to south. (Modified after NAGRA).*

HYDROGEOLOGY

In the investigated area in northern Switzerland, depending on the local geological situation, up to 5 aquifers in the sequence of consolidated sediments above the crystalline bedrock may exist as shown in Figure 3.

As indicated by the hydrogeological studies in the Nagra boreholes Boettstein, Kaisten and Schafisheim (NGB 85-04), the groundwater moves in crystalline rocks mainly in the tectonic fracture zones. In the most investigated Nagra borehole, Boettstein (NTB 85-01), two distinct systems of groundwater chemistry and rock permeability can be distinguished due to the weathering intensity of the rock material and the fracture minerals. Below 800 m the permeability and the weathering intensity decreases (NGB 85-04). The permeability of the rocks varies between 10^{-12} and 10^{-5} m/s. The groundwater flow in the crystalline rocks of Boettstein is restricted to permeable fracture zones (so called kakirites) with a hydraulic conductivity greater than 10^{-9} m/s. The conductivity of the undisturbed rock

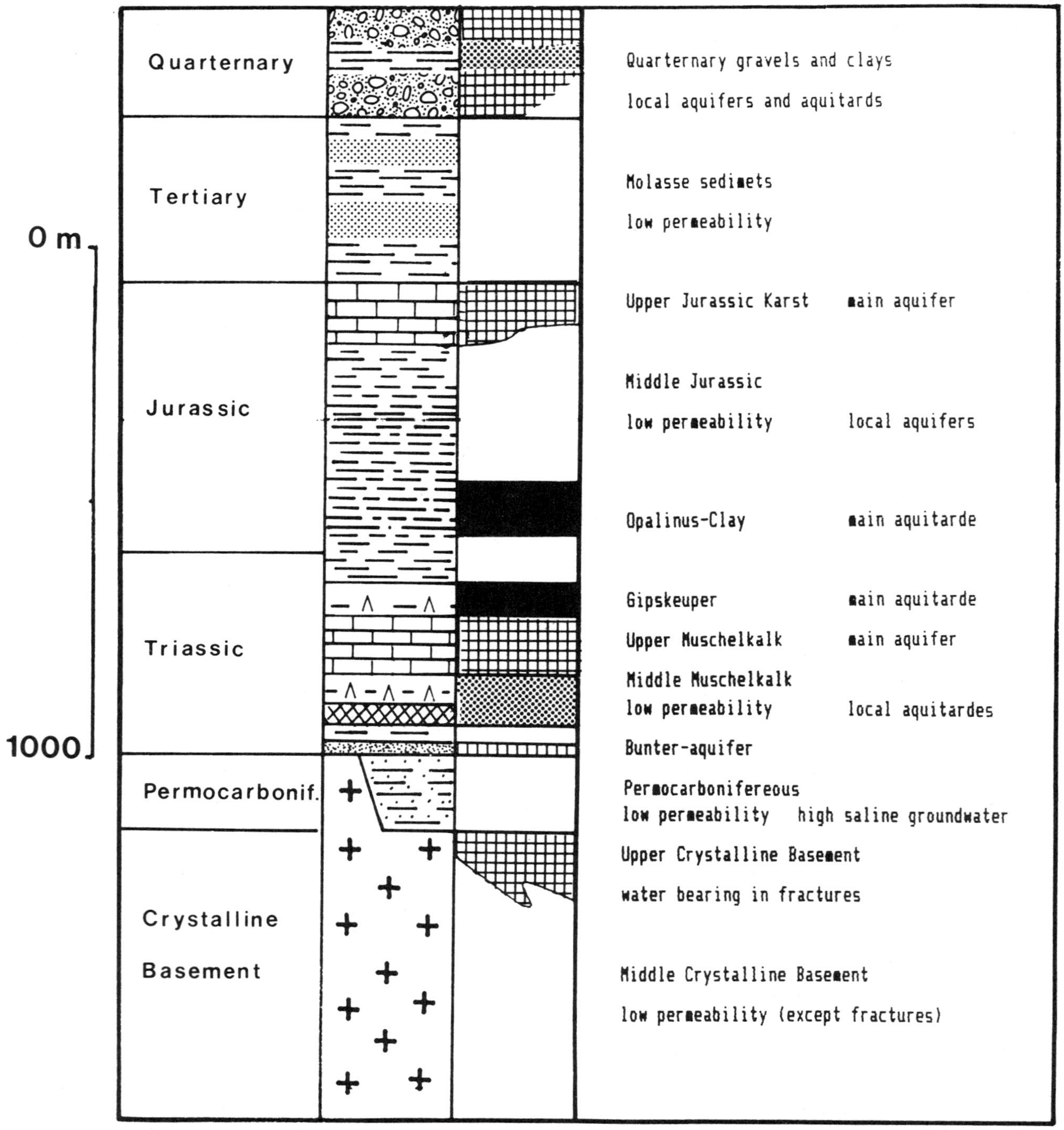

Figure 3. *Generalized lithological profile for northern Switzerland with indication of the main aquifers and aquitards (right column).*

matrix is less than 10^{-12} m/s (NGB 85-04). The low permeability of the plagioclase rich granite below 800 m depth is due to clay mineral formation in the fractures (NGB 85-04).

TYPES OF GROUNDWATERS IN THE SEDIMENTARY AQUIFERS

In the Nagra deep boreholes, the following groundwaters have been encountered in sedimentary aquifers, which are also of interest for the discussion of the origin of the groundwaters found in the crystalline basement rocks (a summary of the different types of groundwaters is given in Table 1).

Lower Freshwater Molasse/Upper Jurassic (Kimmeridge, Malm)

At the borehole Schafisheim in the deposits of the lower fresh-water molasse a groundwater with high salinity (150 meq/l total mineralization) of Na-Cl-type is observed. At the borehole Weiach a groundwater of the same Na-Cl-type with a total mineralization of 118 meq/l was encountered (see Schoeller diagram in Figure 4). From the hydrochemical and the isotope data and also from geological observations on oil exploration wells in Southern Germany, it can be concluded that the groundwater of the Malm aquifers (Kimmeridgian) are hydraulically connected with

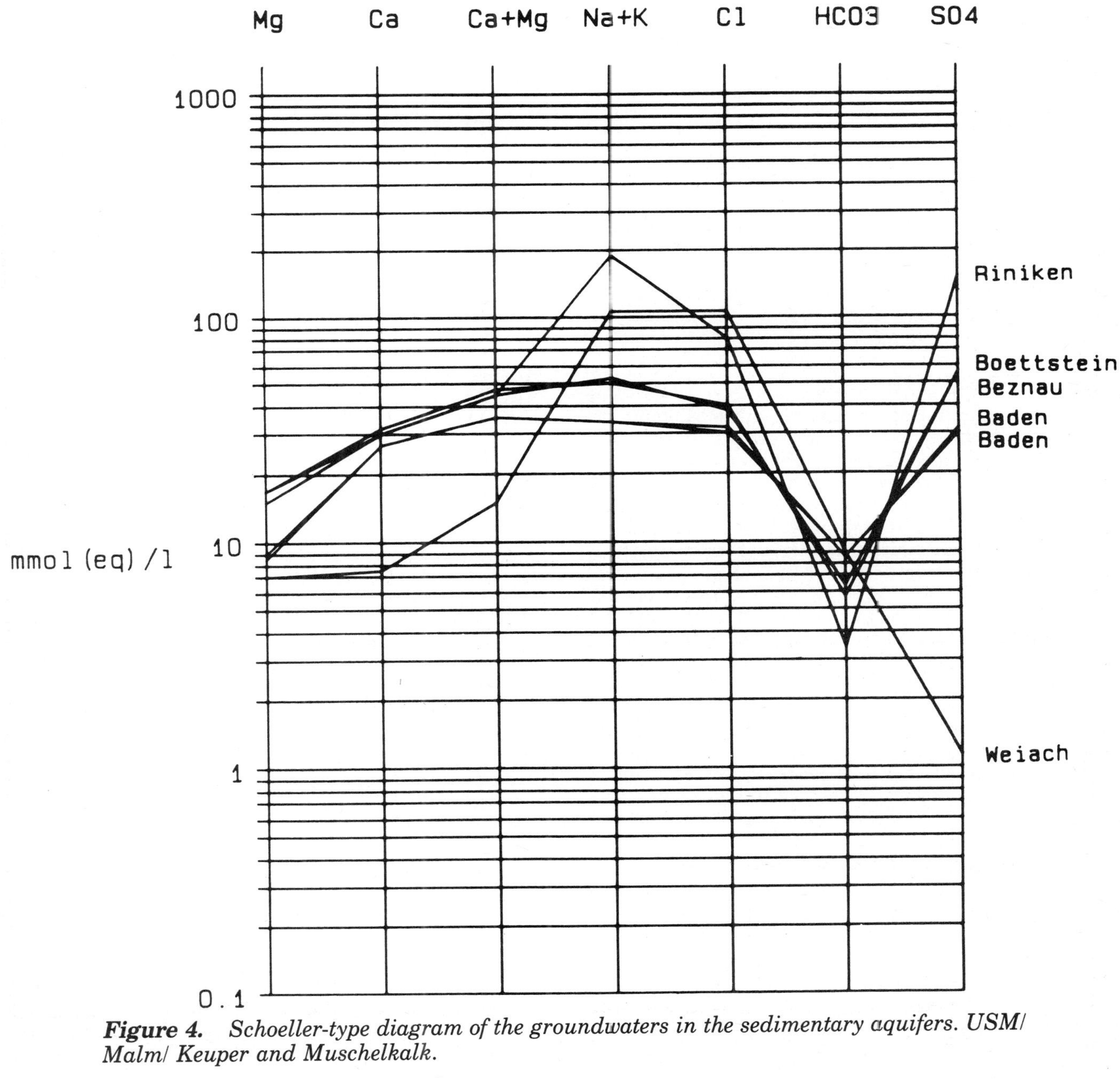

Figure 4. *Schoeller-type diagram of the groundwaters in the sedimentary aquifers. USM/ Malm/ Keuper and Muschelkalk.*

Table 1

Types of groundwaters in the sedimentary aquifers of Northern Switzerland

location	Depth (m)	Aquifer	mmol(eq)/1	g/l	Water Type
Schafisheim	552- 563	lower fresh molasse	150	6.8	Na-Cl
Weiach	243- 267	Kimmeridgian	118	7.2	Na-Cl
Riniken	500- 531	Keuper	234	15.5	$Na-SO_4-Cl$
Weiach	822- 896	Muschelkalk	48	3.4	$Ca-Mg-SO_4$
Boettstein	123- 202	Muschelkalk	98	6.4	$Na-Ca-SO_4-Cl$
Riniken	617- 696	Muschelkalk	237	14.9	$Na-Cl-SO_4$
Schafisheim	1228-1293	Muschelkalk	251	15.9	$Na-Cl-SO_4$
Leuggern	208- 228	Buntsandstein	29	1.8	$Na-Cl-SO_4$
Weiach	981- 990	Buntsandstein	223	14.7	$Na-SO_4-Cl$
Riniken	791- 820	Buntsandstein-Permian	163	10.4	$Na-Cl-SO_4$
Schafishein	1476-1500	Buntsandstein-crystalline	241	16.1	$Na-Cl-SO_4$
Riniken	958- 972	Permian	295	17.9	Na-Cl
Riniken	977-1010	Permian	387	23.4	Na-Cl
Weiach	1109-1124	Permian	619	37.3	Na-Cl
Weiach	1400-1415	Permian	1693	118.6	Na-Cl

the main aquifers of the molasse deposits, especially to the lower freshwater molasse.

A possible explanation of the origin of these waters could be that a brackish-saline water (possible modified formation water), enriched in $\delta^{18}O$ by evaporation effects, is mixed with a groundwater which has infiltrated under colder climatic conditions, as for example the water of the upper marine molasse aquifer.

Upper Triassic (Keuper)

Groundwaters from Keuper deposits, which belong to the undisturbed sedimentary areas above the crystalline bedrock and thus of regional extension, have only been encountered by boreholes at two locations (Beznau Gansinger-Dolomite and Nagra borehole Riniken: Gansinger Dolomite/Schilfsandstein). From the analysis of the groundwater from the two locations the following characterization results: the water from the Riniken borehole has a total mineralization of about 15,700 mg/l of total dissolved solids and is of $Na-Mg-Ca-SO_4-Cl-HCO_3$ type. The water from the Beznau borehole (Gansinger-Dolomite formation) has a total mineralization of 14,600 mg/l and is of $Na-Ca-Mg-SO_4-Cl-HCO_3$ type.

For the Schilfsandstein groundwater from Riniken, the stable isotope values are higher than the modern values which would normally indicate recharge in warm conditions. It is not known at present, how these values should be interpreted with respect to the possible residence time range using the ^{14}C model calculations.

Middle Triassic (Muschelkalk)

Groundwaters from the regional Muschelkalk aquifer are obtained from deep boreholes and from hot springs connected to faults or over-thrust zones. They are of $Ca-Mg-SO_4-HCO_3$ type (Lostorf borehole 3 and 4, and Weiach Nagra deep borehole) or of the $Na-Ca-Cl-SO_4$ type (hot springs at Baden, Beznau borehole and Nagra deep boreholes of Boettstein and Riniken, see representation on Schoeller-type diagram in Figure 4).

Lower Triassic (Buntsandstein)

For the Lower Triassic deposits of the Buntsandstein only groundwaters derived from boreholes, which are intersecting this thin sandstone aquifer (normal thickness in the investigated area of Northern Switzerland between 5 to 25 m) have been studied. Due to its special situation below low permeable limestones and the very low permeable salt and sulphate deposits of the lower Muschelkalk but above and in some cases in direct contact with the crystalline basement of the Permo-Carboniferous deposits of the permian trench, it acts not as an individual aquifer but is in hydraulic connection with the groundwaters of the underlying rock. For instance, where it is lying on the crystalline basement it yields a groundwater which is very similar to the granite waters with lower mineralization and of $Na-SO_4-HCO_3-Cl$ type, but where it is in direct contact to large Permian or even Permo-Carboniferous rocks (Rotliegend and Stephanien) it yields a highly mineralized $Na-(Ca)-SO_4-Cl$ type of water with a total mineralization between 100 and 223 meq l^{-1} (Nagra deep borehole Weiach).

Groundwaters in the Permian deposits

The only groundwaters from Permian deposits within the Permo-Carboniferous trench obtained so far derive from the Nagra deep boreholes Riniken and Weiach.

These waters are highly mineralized (18 g/l-118 g/l) of Na-Cl (Nagra borehole Riniken) to $Na-Ca-SO_4-Cl$ type and contain H_2S and CO_2 (Figure 5). These waters have δ^2H and $\delta^{18}O$ values which are much heavier than all other observed ground- and recharge-waters.

TYPES OF GROUNDWATER IN CRYSTALLINE ROCKS

Shallow groundwaters in crystalline rocks

Groundwaters of near surface type are found in the outcropping crystalline rocks of the Black Forest mountains. The spring Rothaus at about 1.000 m above sea level with a discharge of 10 l/s is a typical example. This shallow groundwater has a tritium content of 76.2 ± 5.0 TU. Remarkable is the relative high ^{222}Rn concentration [155.4 Bq/l] (Schmassmann et al., 1984). The chemical properties of $Ca-Na-(Mg)-HCO_3-Cl-NO_3$ type is typical for the shallow groundwater in crystalline rocks of the area (Figure 7), and similar to the groundwaters found in the upper part of the crystalline bedrock in the Kaisten and Leuggern boreholes (Figures 6 and 7). There is a statistically proved similarity to the deeper crystalline groundwaters of the Black Forest mountains, which despite a higher TDS load (<338 mg/l) are also of $Ca-Na-(Mg)-HCO_3-SO_4$ type (Schmassmann et al., 1984). These groundwaters have temperatures of 17.0-24.2°C. The groundwater from Badenweiler 3 comes from a depth between 293-505.1 m (Sauer, 1974). A contact with sedimentary rocks is probable. These waters are older than Rothaus (Schmassmann et al., 1984): Badenweiler 1 (6.3 ± 7 TU), Badenweiler 3 (<0.9 TU), Buerchau 11.0 ± 1.0 TU).

The dissolved gas concentrations are low in all of these groundwaters (<74 mg/l). The partial CO_2-pressures are low: 0.05-1.0%. The molar ratios of N_2/Ar_2 lie between 27-29 and indicate the atmospheric origin of these gases.

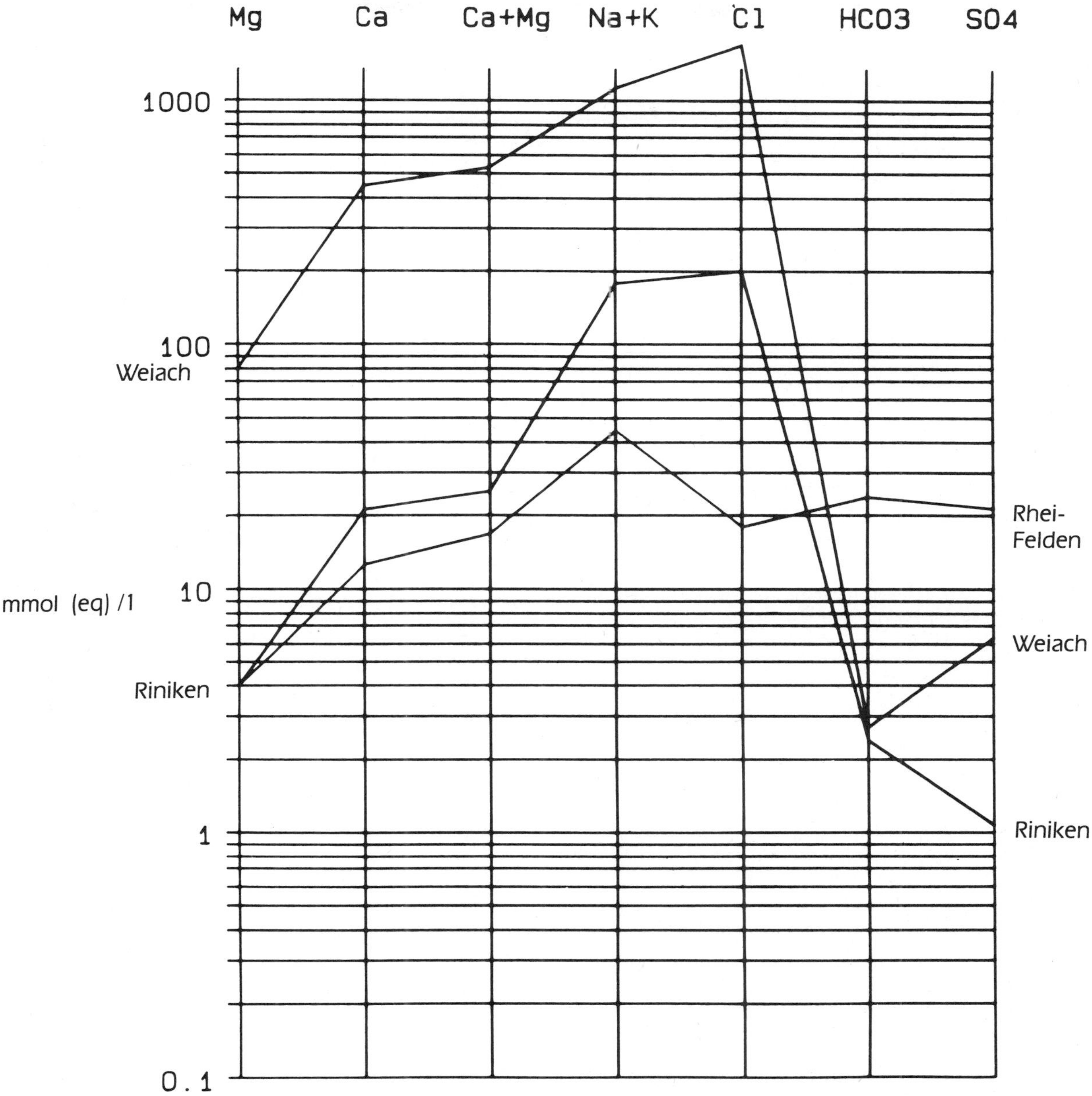

Figure 5. *Schoeller-type diagram of the groundwaters of Buntersandstone, Permian and crystalline rocks.*

Deep groundwaters in crystalline rocks with low salinity

North of the sedimentary troughs, the deep groundwaters in crystalline rocks are low in their dissolved loads (Figure 14). These groundwaters are of Na-SO_4-HCO_3-Cl type (Schmassmann et al., 1984); [Figure 7]. The two boreholes Zurzach 1 and Zurzach 2 are used as thermal springs and are typical for groundwaters of this type. The artesian groundwater flowed during drilling from 402.6 to 469 m at temperatures up to 39.3°C, with a flow rate of 28.7 l/s (Edelmann, 1980).

The groundwaters from the Nagra deep boreholes Boettstein, (393.9-802.7 m), Kaisten (299.6-1305.8 m) and Leuggern (235.0-1209.9) have strong statistical similarity which documents that they are of the same origin and chemical type (Figures 6 and 7). These waters are low in Calcium and Magnesium content with high Ca/Mg molar ratios (11-33) whereas the Ca/Mg molar ratios of the near surface waters in crystalline range between 2 and 7. These waters are high in their fluoride concentrations due to low Ca concentrations (7-14 mg F/l). These waters are in equilibrium with the mineral fluorite.

The argon concentrations [in Zurzach = 0.85 mg/l] (Schmassmann et al., 1984) indicate low infiltration temperatures. These groundwaters contain low dissolved methane concentrations (0.003-0.01 mg/kg).

Isotope measurements of ^{14}C indicate a model age of >15,000 years (Loosli, 1983) for Zurzach groundwater and 4,000->15,000 for different depths in Boettstein borehole (Balderer, 1985).

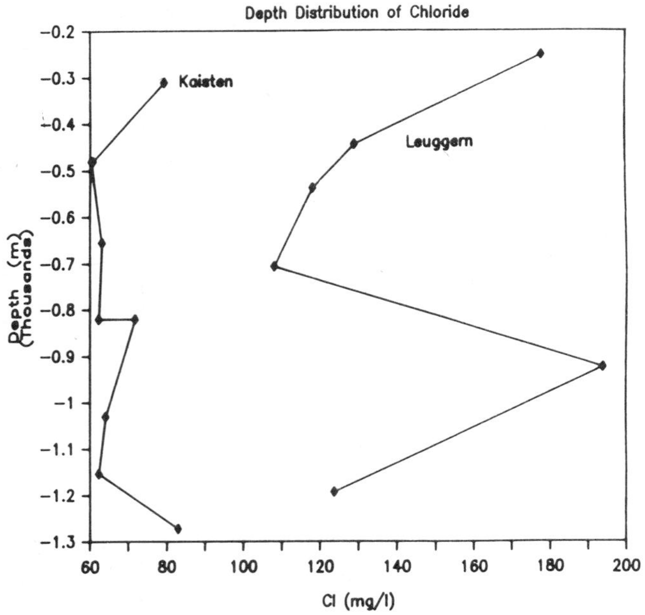

Figure 6. *Depth distribution of chloride in the Nagra boreholes Kaisten and Leuggern.*

Deep groundwaters in crystalline rocks with high salinities

Groundwaters from crystalline rocks in south of the Black Forest mountains in south-western Germany with high TDS loads have been known for a long time [Bad Saeckingen since 1453] (Schmassmann et al., 1984).

The crystalline groundwaters from Saeckingen vary in their salinity between 871-3,498 mg TDS/l depending on dilution from shallow groundwaters. The temperatures range between 17.5-29.4 C. Similar waters of Na-Cl and Na-Ca-Cl type with higher salinities are found in the deepest part of the Boettstein borehole at 1,320.95-1,331.35 m depth. Below the sedimentary troughs saline groundwaters are also encountered in the underlying crystalline rocks (Tab. 2). These waters have similarities in their chemical characteristics to those in the sedimentary troughs. Ca/Mg molar ratios are low (2-7) in these groundwaters with the exception of Boettstein (1,320.95-1,331.35 m). The Ca/Mg molar ratio in the groundwater of Boettstein which is not directly below a sedimentary trough but a few kilometres north of a sedimentary trough, is similar to those of low salinity groundwaters (Ca/Mg molar ratio: 207). The methane concentrations are generally high (in Weiach 27.7-30.6 dissolved mg CH_4/kg). The N_2/Ar ratios are beyond the atmospheric ratio (Figure 10).

CO_2-rich saline groundwaters in crystalline rocks

In a zone of high permeability the groundwater from the borehole Rheinfelden with a high CO_2 concentration flows from 403.8-603.0 m depth. Borehole Rheinfelden lies in an area of high tectonic activity (Rheingraben). This groundwater is of Na-Ca-HCO_3-SO_4-Cl type and is statistically proven to be most similar to those groundwaters from the Permian sediments (see Figures 5 and 7). The high N_2/Ar ratios also show a biogenetic origin of the dissolved gases.

Table 2

Salinity of the groundwater in Nagra Borehole Weiach

Depth (m)	Geology	TDS (mg/l)	Cl (mg/l)
982- 985	Buntersand.	13450.	2542.
1109-1124	Permo-Carb.	37330.	13735.
1400-1415	Permo-Carb.	118560.	59500.
2211-2224	Gneiss	6392.	3382.
2260-2273	Gneiss	7649.	3924.

EVOLUTION OF THE CRYSTALLINE GROUNDWATERS

Groundwater samples taken from the Nagra boreholes Boettstein, Weiach, Schafisheim, Kaisten, Leuggern in crystalline rocks reveal that the distribution of saline waters depends on a complex hydrogeological system. It is assumed that highly saline groundwaters from sedimentary sequences enter the fractures within the crystalline rocks (Figure 6). In Weiach the groundwater salinity drops from higher values in the sedimentary trough to lower values in the underlying crystalline rocks (Tab. 2). In Boettstein the salinity increases in the zone of low permeability (below 800 m). The salinity of this groundwater originates probably from Permian rocks (for the isotopic evidence see Balderer et al., this volume). A clear dependence on depth which has been shown for crystalline rocks in Canada (Fritz & Frape, 1982; Frape et al., 1984) can not be seen (Figure 6).

Unlike the groundwaters of the crystalline formations in shield areas and some plutons, which are highly saline (Fritz & Frape, 1982, Frape et al., 1984) or of moderate salinity (Edmunds et al., 1984; Nordstrom, 1983), the saline groundwaters in crystalline rocks of northern Switzerland seem to be connected with saline groundwaters from low permeable sedimentary rocks. The statistical analyses using factor analyses, discriminant analyses and cluster analyses (Schmassmann et al., 1984) confirm these results. They show that 4 main types of groundwaters in crystalline rocks can be distinguished.

In the crystalline rocks of the Black Forest mountains in the northern part of the investigated area shallow crystalline groundwaters of the type Rothaus, Badenweiler 1, Badenweiler 3 and Buerchau can be observed. The deep groundwaters of Zurzach, Boettstein, Leuggern and Kaisten represent most probably the deeper components of these groundwaters after a residence time of few thousand years (Balderer, 1985; Pearson, 1985; Schmassmann et al., 1984). They are characterized by a low salinity even at greater depths (Figure 4). The chloride concentrations increase from 1.9-10.2 mg/l in the shallow groundwaters and up to 60-123 mg/l in the deeper, low-salinity groundwaters.

Groundwaters with higher salinity are encountered in the vicinity of Permo-Carboniferous sedimentary structures close to main fracture zones (Bad Saeckingen, Weiach, Boettstein, 1300 m) and in the southern part of the investigated area (Schafisheim).

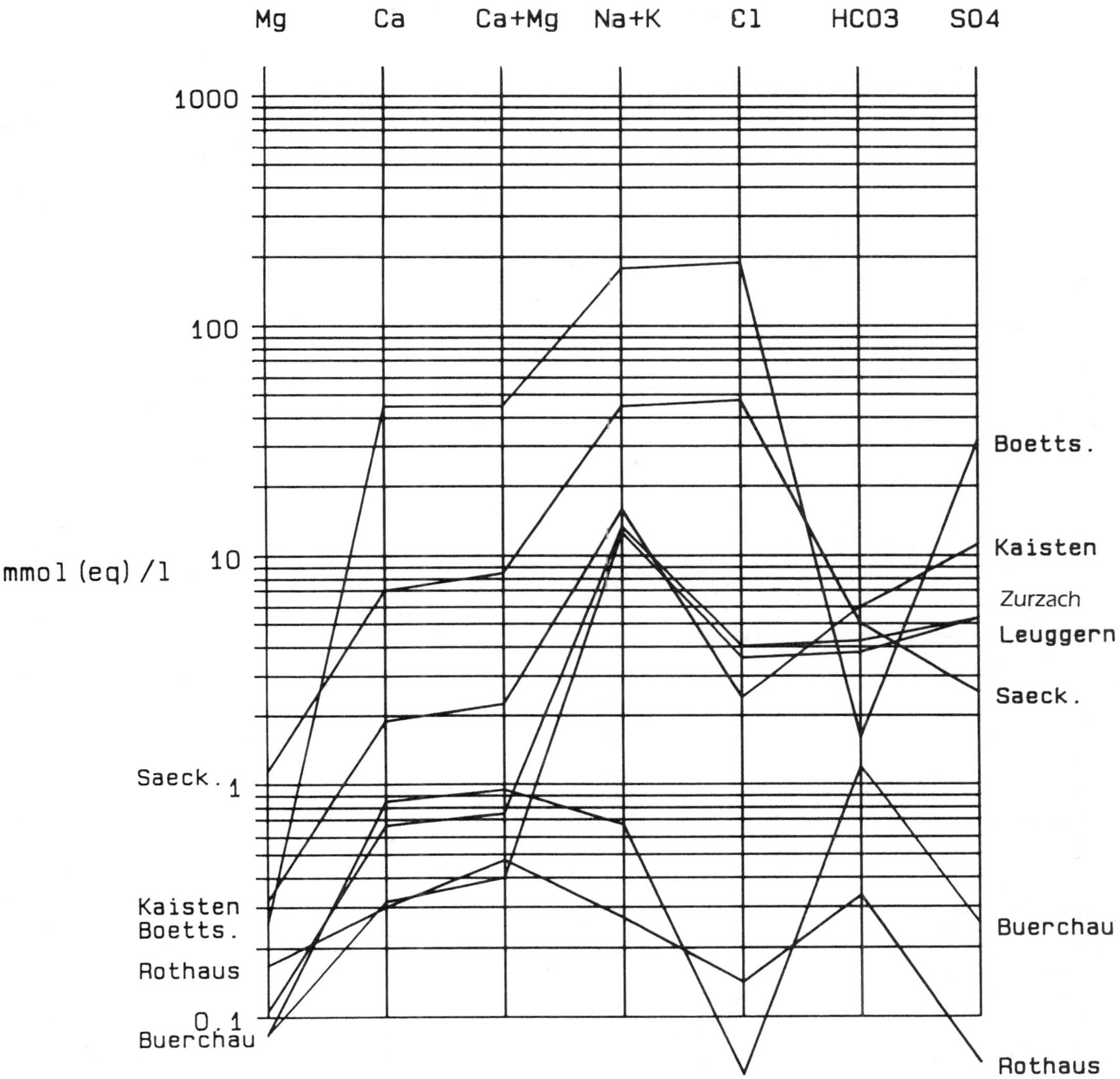

Figure 7. *Schoeller-type diagram of the groundwaters in crystalline rock.*

Along the Rhine valley groundwater with high CO_2 concentrations are found along main fault zones (Rheinfelden, Schmassmann et al., 1984).

Edmunds et al. (1984) used the SMOW (*S*tandard *M*ean *O*cean *W*ater)-dilution line to show that the salinity of the groundwaters in Carnmenellis granite do not originate from sea water but is due to water/rock interaction. In this paper various elements and their concentrations ratios are also compared with the sea water (SMOW) dilution line to indicate the origin of salinity from water/rock interaction in crystalline rocks following Edmunds et al. (1984).

With increasing salinity sodium becomes the dominating cation in all groundwaters in crystalline rocks investigated to date in northern Switzerland, whereas sulphate proves to be the dominating anion in the groundwaters with low to medium salinity (Cl <200 mg/l). In the groundwaters with higher salinity the Na and Cl are the dominating ions (Figure 8). The Na/Cl ratios in saline groundwaters are about 1, similar to the groundwaters in the sedimentary rocks and to SMOW (Figure 9). These results are not consistent with the data of Edmunds et al. (1984), as the Na/Cl ratios in the groundwaters of Carnmenellis granite or in

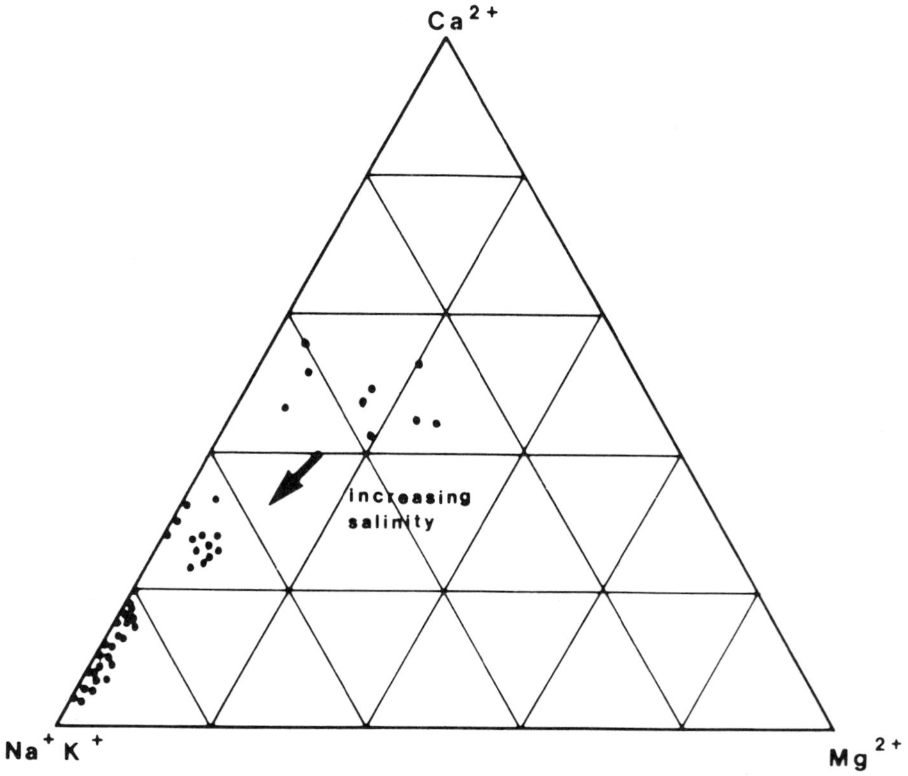

Figure 8. *a) Mole percentage diagram for Ca, Mg, Na and K in groundwaters in the crystalline rocks of the investigated area.*

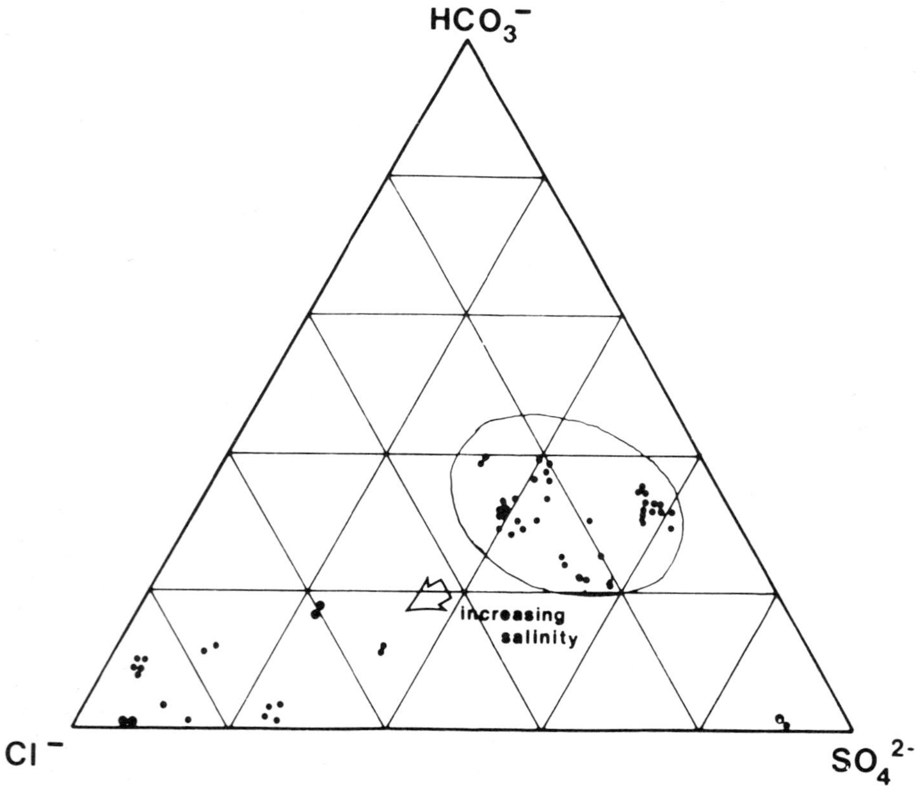

Figure 8. *b) Mole percentage diagram for Cl, SO_4 and HCO_3 in groundwaters in the crystalline rocks of the investigated area.*

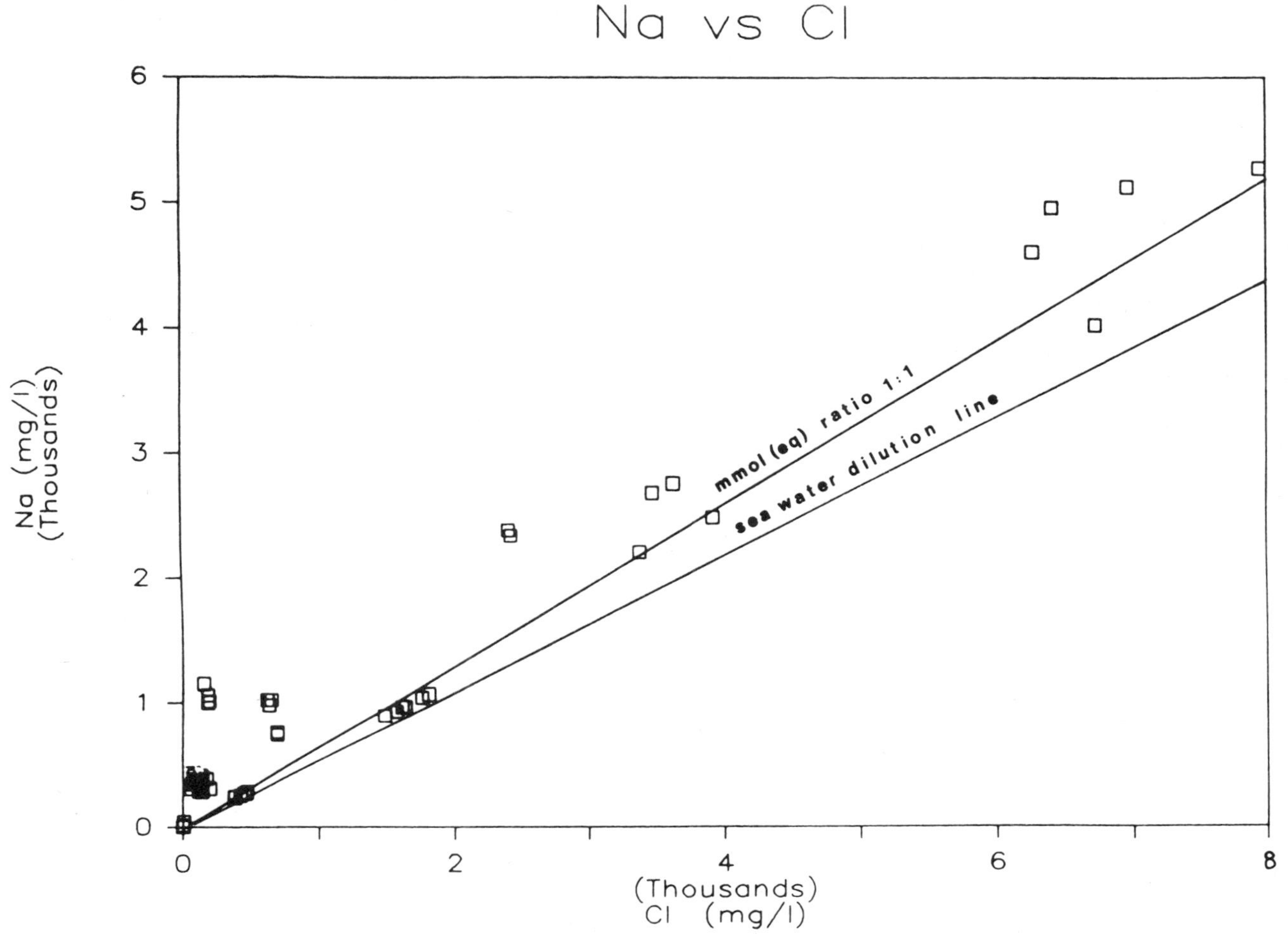

Figure 9. *Na^+/Cl^- concentration ratios in groundwaters in the crystalline rocks of the investigated area.*

the Canadian Shield tend to be lower with increasing salinity compared to SMOW dilution line.

The Ca/Cl, Sr/Cl, Ba/Cl, B/Cl and Rb/Cl ratios are higher than those derived from the SMOW dilution line and similar to those given by Edmunds et al., (1984). The Mg/Cl ratios are lower in comparison to SMOW.

Increasing K/Cl ratios in groundwaters with lower salinity are interpreted by Edmunds et al. (1984) as an effect of recent pollution. A geochemical enrichment seems more probable in the presented case, as some of these groundwaters are of considerable age (>1000 years).

Lithium concentrations increase in the investigated groundwaters with increasing salinity (Figure 10). Cl/Li ratios of about 100 have been interpreted by Edmunds et al. (1985) as evidence of strong congruent reaction of biotite in the crystalline rocks, higher ratios as Cl ⟨= =⟩ OH exchange. The groundwaters with low salinity in crystalline rocks of northern Switzerland have Cl/Li ratios of about 100, the groundwaters with high salinity above 100. Dissolution studies on weathered Boettstein granite shows Cl/Li ratios of about 100.

The low salinity groundwaters in the crystalline rocks have higher Br/Cl ratios than the more saline groundwaters which are in contact with groundwaters from sediments (Br/Cl ratio in SMOW 0.003). These groundwaters have Br/Cl ratios of about 0.003 or less (Figure 11). The Variation of the Br/Cl-ratios in Figure 11 indicate a mixing processes between these two types of groundwaters. However in low concentrated groundwaters the analytical restrictions must be kept in mind, although the concentrations are above detection limits and precision is good enough (0.05 ± 0.05 mg/l, Kussmaul & Antonsen, 1985).

Iodine concentrations in low salinity groundwaters lie below detection limit (0.05 ± 0.05 mg/l, Kussmaul & Antonsen, 1985). In the groundwaters which are in contact with groundwaters from sedimentary rocks, the iodine/chloride ratio lies above the sea water dilution line indicating water/rock interaction.

Silica concentrations in groundwaters of crystalline rocks are controlled by chalcedony (Pearson, 1985), although there are restrictions to the analytical procedure if the samples are not adequately filtrated (<0.1 μm). The stability values calculated with help of PHREEQE (Parkhurst et al., 1980) for crystalline rock forming silicate minerals show that the water chemistry is not much influenced by these minerals. The water reacts more with the fracture minerals, mainly with clay minerals and calcite. Both kinds of lower and higher salinity groundwaters are represented in a well defined stability field of mixed-layer clay

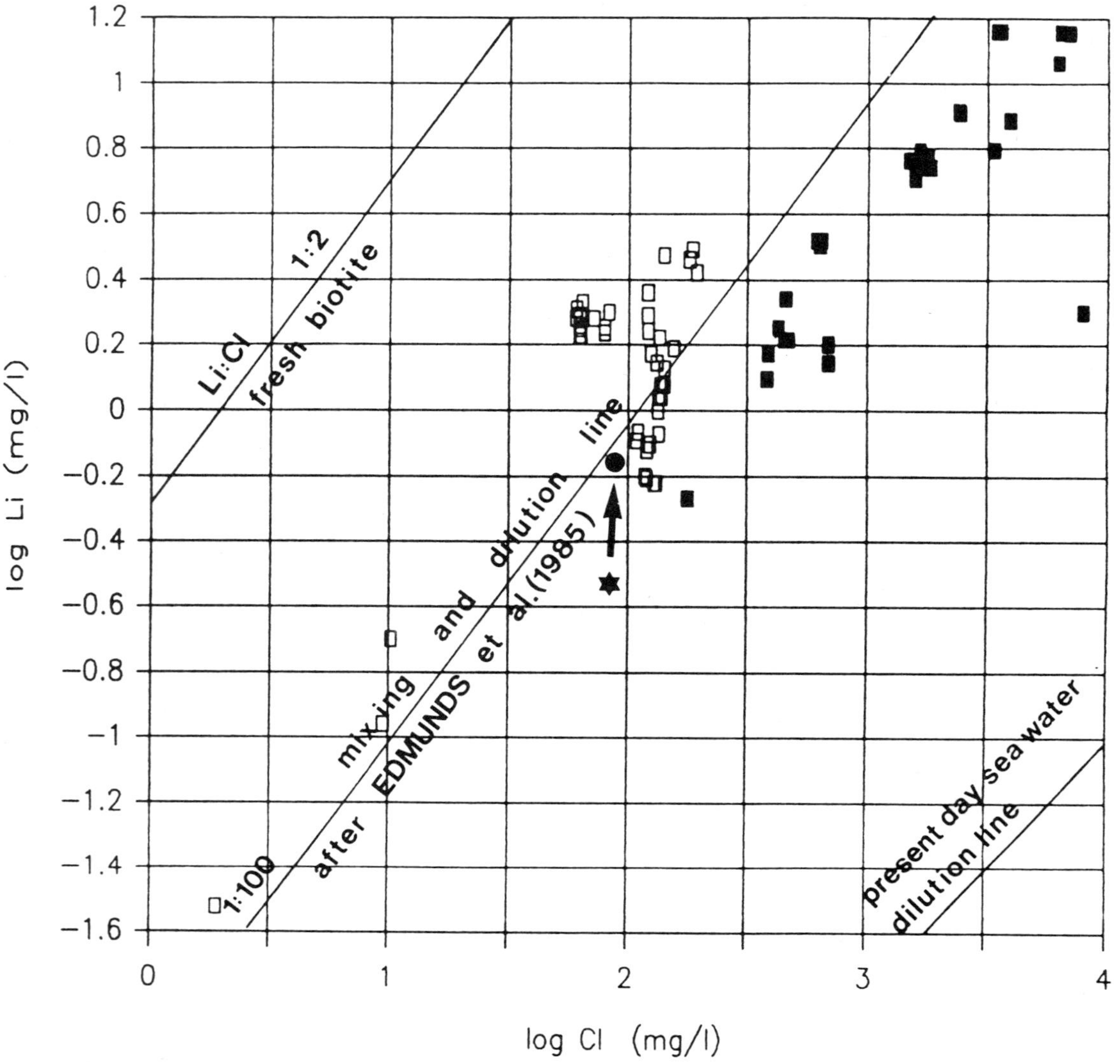

★ experimental dissolution studies (non-weathered Boetstein granite, 1437 m)

● experimental dissolution studies (weathered Boetstein granite, 697 m)

the arrow ⟶ indicates the increase in Li:Cl-ratio probably due to weathering

Figure 10. *Li^+/Cl^- concentration ratios in groundwaters in the crystalline rocks of the investigated area.*

minerals, kaolinite or both (Figure 12) in a stability field diagram published by Aagard and Helgeson (1983).

The content of the dissolved gases in the deep groundwaters is a very suitable tool to distinguish groundwaters of different origin. These data indicate that the most saline groundwaters in the crystalline rocks of northern Switzerland most probably originate from sedimentary rocks. High methane concentrations found in the deep groundwaters of northern Switzerland also seem to be associated with sedimentary rocks. Coal bearing sediments have been found in the Permo-Carboniferous sedimentary troughs (NGB 85/01, NGB 85-04). In the borehole Weiach, in which the crystalline rocks underlay a sedimentary trough, the highest methane concentrations were encountered (27.7-30.6 mg/l dissolved methane). The boreholes Boettstein, Kaisten and Leuggern are lower in their methane concentrations (0.01-0.13 mg/l dissolved methane). Furthermore high salinity and high methane concentrations

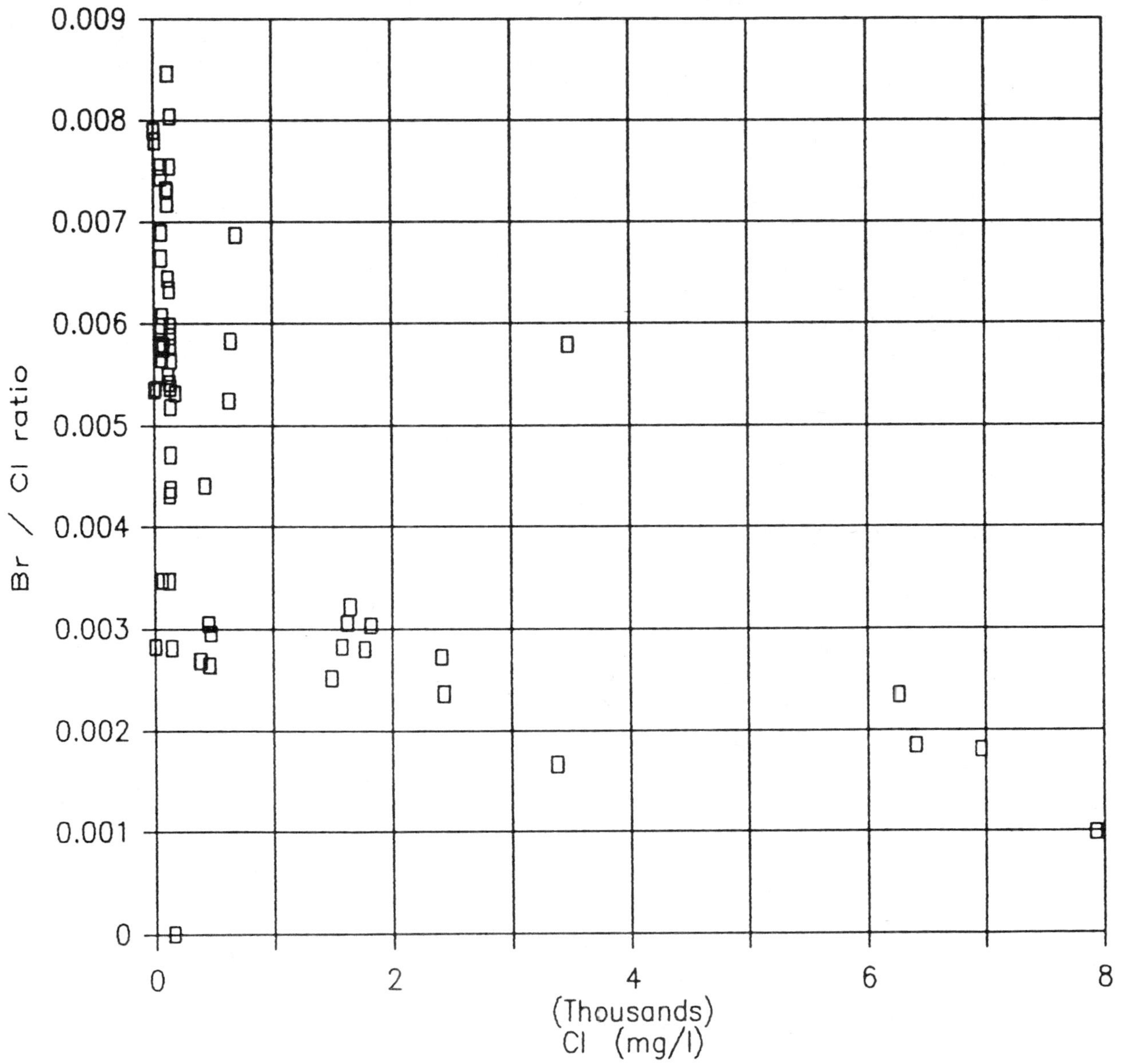

Figure 11. *Decreasing Br^-/Cl^- concentration ratios with increasing salinity in groundwaters in the crystalline rocks of the investigated area.*

(1.8-7.2 mg/l dissolved methane) were observed in the borehole Schafisheim, to which the groundwater probably flows from the sedimentary rocks south of the investigated area (Figure 14).

The dissolved N_2/Ar ratios are mostly higher in the groundwaters than expected from the equilibrium of these gases with the atmosphere. Ratios which are above 28 indicate a biochemical origin of the nitrogen. The deep groundwaters of Northern-Switzerland especially the saline groundwaters from Buntersandstone and Permian rocks have high N_2/Ar ratios (Schmassmann et al., 1984). The groundwaters in crystalline rocks tend to have high N_2/Ar ratios with increasing salinity (Figure 13).

CONCLUSIONS

The origin of chemical composition in saline groundwaters in the crystalline rocks has been discussed by various authors. Fossil sea water, water/rock interaction (Fritz & Frape, 1982; Frape et al. 1984; Edmunds et al., 1984), fluid inclusions (Nordstrom, 1983) were proposed among others. The previously discussed geochemical data as used by Edmunds et al. (1984) and Frape et al. 1984 as evidence of water/rock interaction for the origin of salinity of the groundwaters in crystalline rocks can not be applied to the deep groundwaters in the crystalline rocks in a general sense. The deep groundwaters in crystalline rocks of northern Switzerland are not as highly saline as compared with

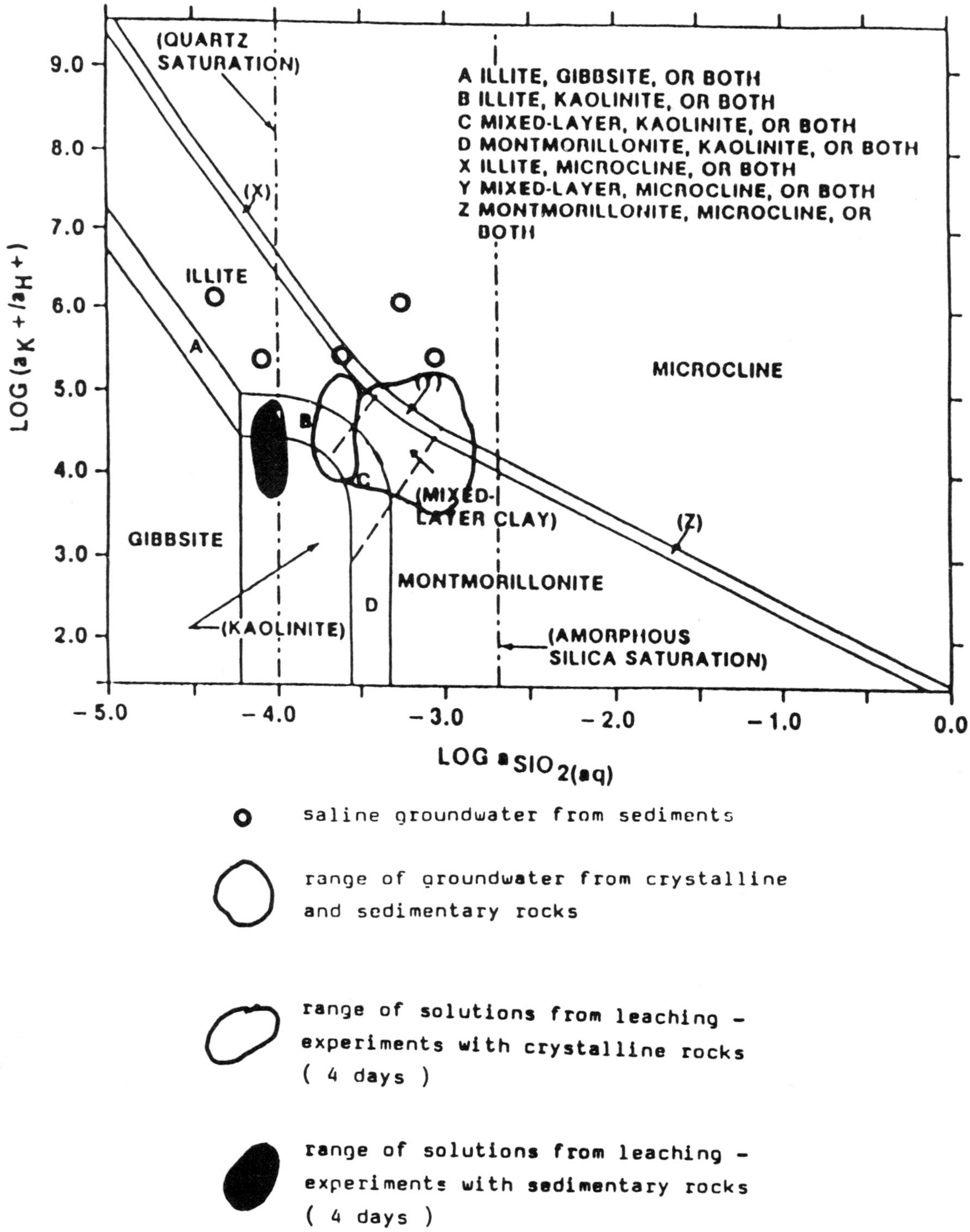

Figure 12. *Clay mineral stability diagram (Aagard and Helgeson, 1983, Fig. 14) with positions of groundwaters in the crystalline rocks of the investigated area.*

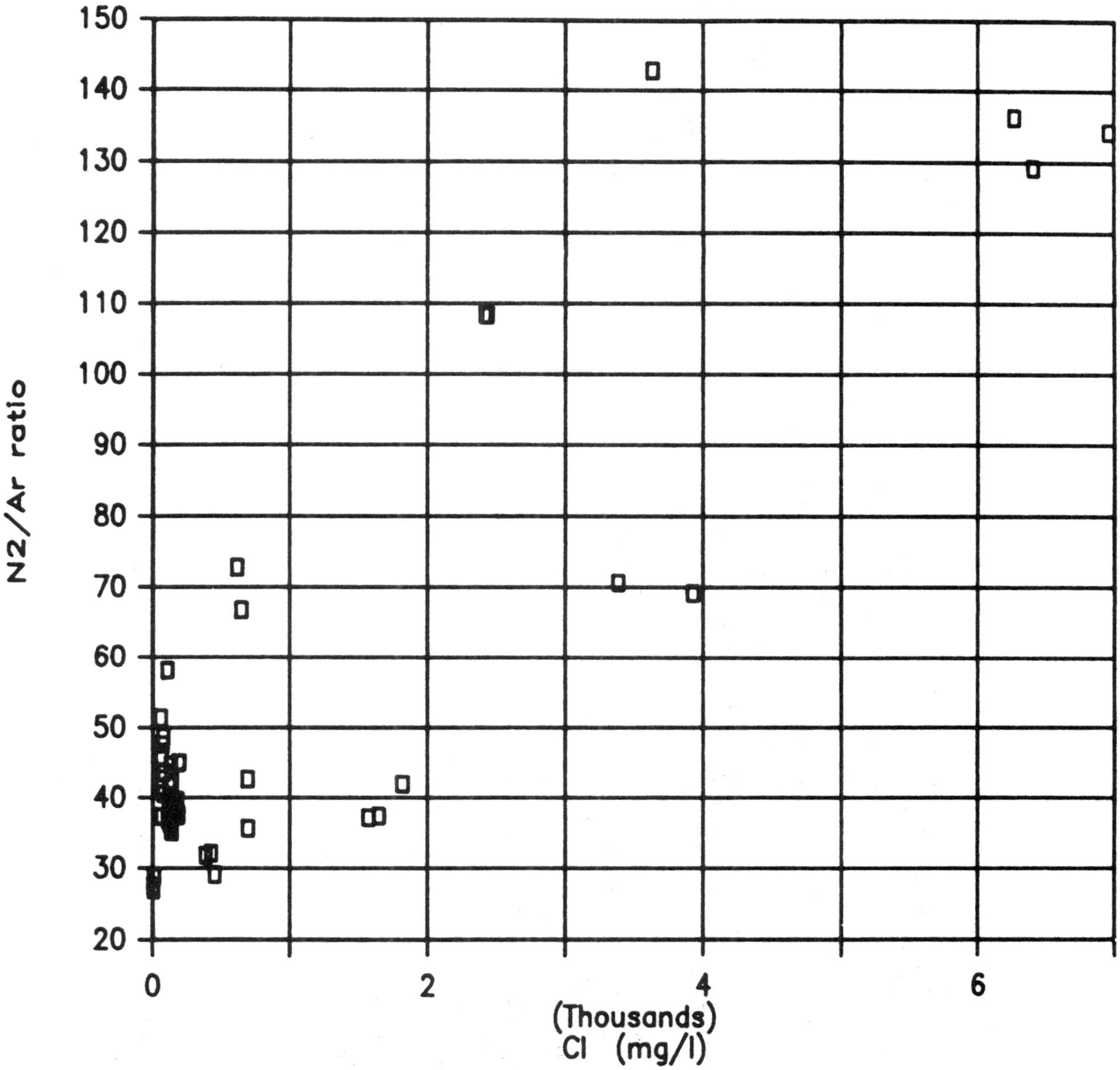

Figure 13. *N_2/Ar concentration ratios versus Cl^- concentrations in groundwaters in the crystalline rocks of the investigated area.*

deep groundwaters in crystalline rocks mentioned elsewhere. There is hydrogeological and geochemical evidence that the groundwaters which infiltrate at the Black Forest mountains and flow in the crystalline rocks are not saline (<200 mg Cl/l) even at greater depths. The saline groundwaters in the crystalline rocks originate most probably from infiltration of groundwater flowing from sedimentary rocks. There is no clear tendency of increasing salinity with increasing depth as described by Edmunds et al. (1984) and Fritz & Frape (1982).

In agreement with the conclusions of the hydrogeochemical evolution of these groundwaters in the crystalline rocks, based on hydrogeochemical and isotope investigations (see Balderer et al., this volume) the results of hydrodynamic model simulations according to Kiraly (1984) and Kimmeier et al. (1984) indicate the following groundwater flow systems (see Figure 14):

a) In crystalline rocks of Northern Switzerland covered by Mesozoic evaporites hydraulic heads are artesian (Figure 15).
b) In the north and north-east (Black Forest mountains) groundwater infiltrates directly into the crystalline rocks and remain primarily in contact with crystalline rocks.
c) From the south deep high-saline groundwater with traces of hydro-carbons which originate from the permo-carboniferous troughs (Figure 3) may enter into the crystalline rocks. An intrusion of these groundwaters from the overlying paleozoic sediments into the crystalline may occur along fault or fracture

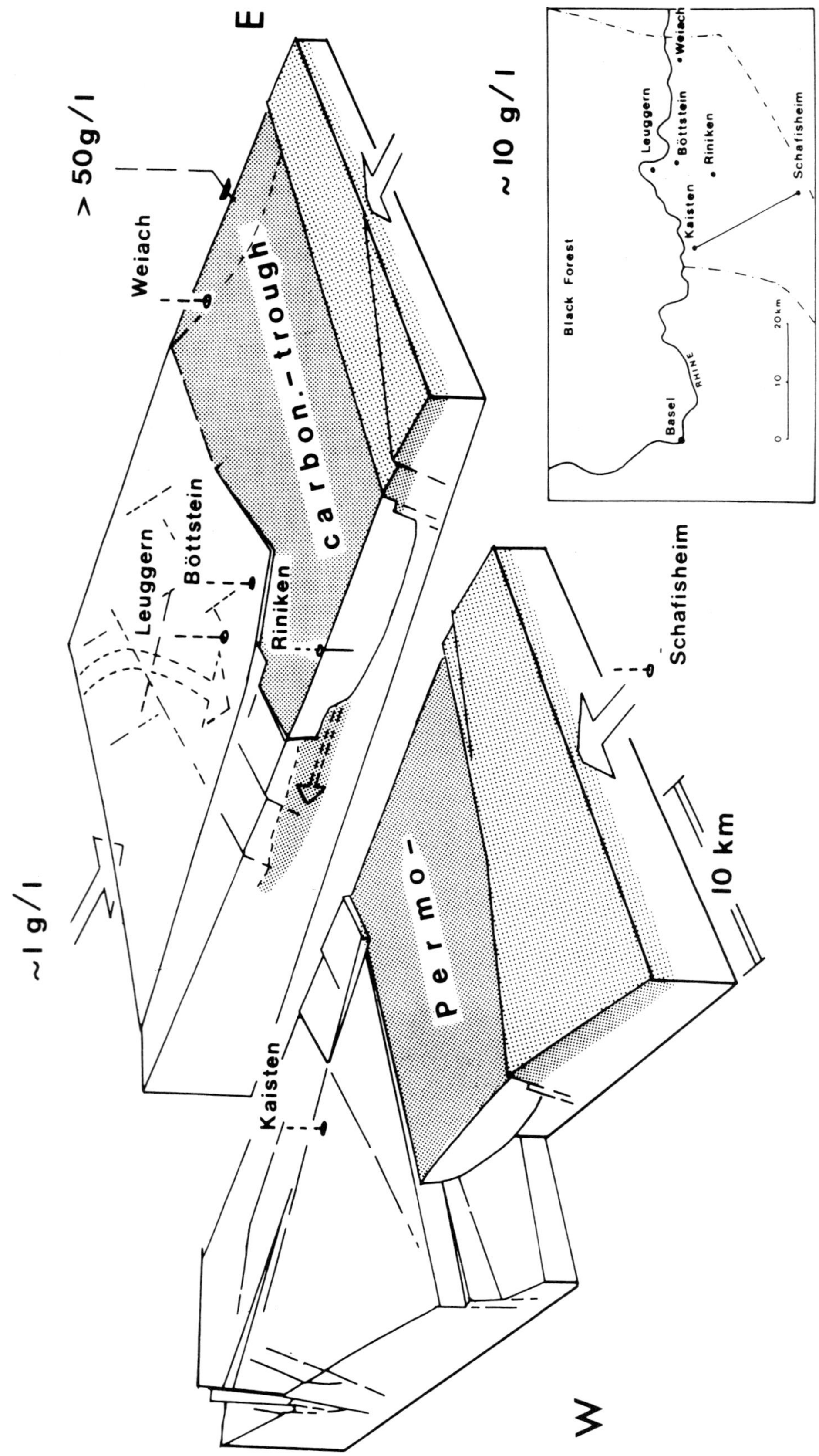

Figure 14. *A schematic view of the groundwater provinces in the upper crystalline rocks and the Permo-Carboniferous trough of Northern-Switzerland (NGB 85-04).*

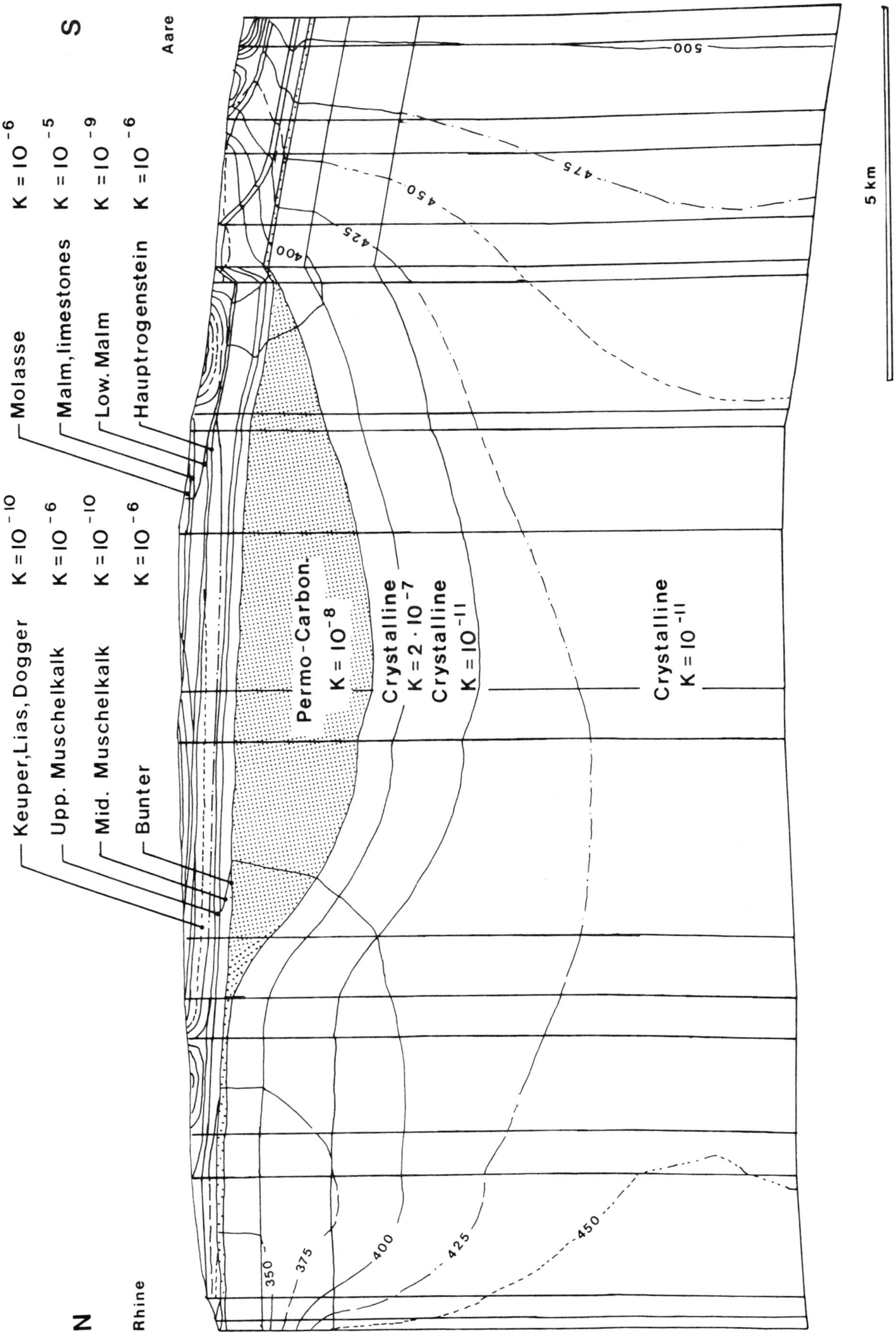

Figure 15. *North-south profile of hydraulic potential distribution from the local hydrodynamic model (hydraulic conductivities in m/s, NGB 85-04).*

systems, like in the case of Bad Saeckingen (Kimmeier et al., 1984). The low permeability of the triassic Mittlerer Muschelkalk aquitard does not allow an appreciable seepage of groundwater from overlying sediments.

d) In Rhine valley between Koblenz and Bad Saeckingen, where these do not exist, groundwater exfiltration from crystalline rocks, Permian and Buntersandstone to upper sedimentary structures occur. In this area groundwater flow from south, east and north is converging (Figure 14).

Geochemical data and the used concentration ratios of some important elements show evidences for water/rock interaction for both groundwater types which are evolved in sedimentary and crystalline rocks. With increasing salinity of the groundwaters however the geochemical characteristics of the groundwater resemble more the sea water dilution line.

In a complicated hydrogeological system as presented, more hydro-geochemical work, supported by isotopes, mineralogy and statistical investigations, is necessary to characterize and distinguish different types of the hydrogeochemical evolution of groundwaters in the crystalline rocks.

ACKNOWLEDGEMENTS

This investigation was supported financially by the Nagra. The authors wish to express their appreciation to Dr. M. Edmunds, Dr. W. Kanz, Prof. Dr. G. Matthess, Dr. F.J. Pearson and Dr. H.J. Schmassmann for many helpful discussions with the authors.

We also express our thanks to Dr. D. Schenk and Professor Dr. Zuber for the critical review of this paper.

The analyses were carried out mainly by the Eidgenoessisches Institut fuer Reaktorforschung, Wuerenlingen, Switzerland; by Institut Fresenius, Taunusstein, FRG; Institut fuer Radiohydrometrie der gsf, Neuherberg, FRG.

REFERENCES

Aagard, P. and Helgeson, H.C., 1983. Activity/composition relations among silicates and aqueous solutions: II. Chemical and thermodynamic consequences of ideal mixing of atoms on homological sites in montmorillonites, illites, and mixed layer clays: Clays and Clay Minerals, v. 31/3, p. 207-217.

Balderer, W., 1983. Bedeutung der Isotopen Methoden bei der hydrogeologischen Charakterisierung potentieller Endlagerorte fuer hochradioaktive Abfaelle: Nagra Technical Report 83-04, 158 p.

Balderer, W., 1985. Sondierbohrung Boettstein: Ergebnisse der Isotopenuntersuchungen zur hydrogeologischen Charakterisierung der Tiefegrundwaesser: Nagra Technical Report 85-06, 260 p.

Edelmann, W., 1980. 25 Jahre Bad Zurzach. Schmid, Zurzach.

Edmunds, W.M., Andrews, J.N., Burgess, W.G., Kay, L.F., Lee, D.J., 1984. The evolution of saline and thermal groundwaters in the Carnmenellis granite: Mineralogical Magazine, v. 48, p. 407-424.

Frape, S.K., Fritz, P. and McNutt, R.H., 1984. Water-rock interaction and chemistry of groundwaters from the Canadian Shield: Geochimica et Cosmochimica Acta, v. 48, p. 1617-1627.

Fritz, P. and Frape, S., 1982. Saline groundwaters on the Canadian Shield – a first overview: Chemical Geology, v. 36, p. 179-190.

Kimmeier, F., Perrochet, P., Andrews, R., Kiraly, L., 1985. Simulation par Modèle Mathématique des Ecoulements Souterrains entre les Alpes et la Foret Noire; Partie A: Modèle Régional; Partie B: Modèdele Local (Nord de la Suisse): Nagra Technical Report 84-50, 166 p.

Kiraly, L., 1985. FEM 301 – A three dimensional model for groundwater flow simulation: Nagra Technical Report 84-49, 96 p.

Kussmaul, H. and Antonsen, O., 1985. Hydrochemische Labormethoden für das Nagra-Untersuchungsprogramm. Nagra Technical Report 85-04, 83 p.

Loosli, H.H., 1983. A dating method with ^{39}Ar: Earth and Planetary Sciences Letters, v. 63, p. 51-62.

Nagra, 1985. Nukleare Entsorgung Schweiz: Konzept und Uebersicht ueber das Projekt Gewaehr 1985: Nagra Technical Report 85-01, 316 p.

Nagra, 1985. Projekt Gewaehr 1985: Endlager fuer hochaktive Abfaelle: Das System der Sicherheitsbarrieren: Nagra Technical Report 85-04.

Nordstrom, K., 1983. Conceptual framework for the chemical processes in the Stripa groundwaters: SKB Stripa Project Report 83-01, p. 106-115.

Parkhurst, D.L., Thorstensson, D.C. and Plummer, L.N., 1980. PHREEQE – a computer program for geochemical calculations: U.S. Geological Survey Water Resources, Investigation 80-96, 210 p.

Pearson, F.J., 1985. Sondierbohrung Boettstein – results of hydrochemical investigations: Analysis and interpretation: Nagra Technical Report 85-05, 116 p.

Peters, T., Matter, A., Blaesi, H.R. and Gautschi, A., 1985. Sondierbohrung Boettstein: Geologie: Nagra Technical Report 85-02, in press.

Sauer, K., 1974. Die Vergroesserung des Thermalwasserdargebots in Badenweiler. Geologische Voraussetzungen und Ergebnisse der Arbeiten 1962-1973: Heilbad und Kurort, v. 21, p. 2-12.

Schmassmann, H.J., Balderer, W., Kanz, W., Pekdeger, A., 1984. Beschaffenheit der Tiefengrundwaesser in der zentralen Nordschweiz und angrenzenden Gebieten: Nagra Technical Report 84-21, 335 p.

Isotopic Investigations of the Water-Rock System in the Deep Crystalline Rock of Northern Switzerland

W. Balderer

Physics Institute, University of Bern, Fidlerstr.5, CH/3012, Bern, Switzerland

J.-Ch. Fontes and J.-L. Michelot

Laboratoire D'Hydrologie et de Créochimie Istopigne,
University of Paris-Sud, 91405 Orsay-Cedex, Paris, France

D. Elmore

Nuclear Structure Research Laboratory, University of Rochester, New York 14627, U.S.A.

Balderer, W., Fontes, J.-Ch., Michelot, J.-L. and Elmore, D.; Isotopic investigations of the water-rock system in the deep crystalline rock of Northern Switzerland; in Saline water and gases in crystalline rocks, Editors: Fritz, P., and Frape, S.K.; Geological Association of Canada Special Paper 33, 1987. p. 175–195

Abstract

This paper presents first results of the isotopic studies on groundwater samples and rock minerals from the deep drilled boreholes in crystalline rocks of the Nagra Investigation Project (for the assessment of repositories for high-level radioactive waste in geological formation). ^{18}O *and* ^{13}C *contents of the secondary calcites in closed and in open fissures, indicate different origins for these calcites, assuming isotope equilibrium with water and* CO_2.

^{34}S *and* ^{18}O *contents of the dissolved sulphates in the groundwater suggest two origins: (i) an oxidation of sulphides from the granitic rock itself and (ii) a sedimentary supply of Permian sulphates. Carbon-14 contents of the Total Dissolved Inorganic Carbon (TDIC) are interpreted according to the possible sources of carbon isotopes (soil* CO_2*, primary and secondary carbonates, deep* CO_2 *production). They cannot be used as independent age indicators, but provide relative ranges of residence times.*

Chlorine-36 data are discussed in terms either of simple decay of cosmic chlorine or of intraformational production. Evaluations of the hydraulic conductivity, calculated from the residence time of the chloride in the rock, range from $2\cdot10^{-13}$ *to* $6\cdot10^{-12}$ $m\ s^{-1}$. *They are in agreement with those obtained from hydraulic tests.*

Résumé

Cet article présente les premiers résultats des mesures isotopiques réalisées sur des échantillons d'eau et de roches prélevés dans les forages profonds en milieu cristallin du programme CEDRA (pour l'évaluation des sites géologiques de stockage des déchèts radioactifs).

Les teneurs en ^{18}O *et* ^{13}C *des calcites secondaires sont discutées en admettant un équilibre isotopique avec les fluides du milieu. Les teneurs en isotopes Pourds des calcites de fractures fermées sont attribuées à des conditions d'hydrothermalisme et sont différentes de celles de fractures ouvertes.*

Les teneurs en ^{18}O *et* ^{34}S *du sulfate aqueux suggèrent une origine mixte: (i) oxydation des sulfures constitutifs du granite (ii) apport de sulfate sédimentaire permien.*

Les teneurs en ^{14}C du carbone inorganique total dissous sont mises en relation avec les différentes sources possibles de carbone (CO_2 du sol, carbonates primaires et secondaries, production profonde de CO_2) et ne semble pas pouvoir fournir d'indication d'âge fiable.

Les teneurs en ^{36}Cl sont interprétées soit en termes de simple décroissance, soit en termes de production profonde in situ. Les temps de résidence du chlorure aqueux dans la roche permettent une evaluation de la conductivité hydraulique (entre $2 \cdot 10^{-13}$ et $6 \cdot 10^{-12} m\ s^{-1}$) comparable à celle obtenue par les méthodes hydrogéologiques.

INTRODUCTION

The Nagra (Swiss National Cooperative for the Storage of Radioactive Wastes) has a mandate to present by 1985 a feasibility study for the underground disposal of high level radioactive wastes. The project is based on the so-called multibarrier concept. This concept includes a storage in highly impermeable (crystalline) rocks and a retardation in case of leakage of radionuclides from the containers. The retardation would involve the sorption properties of the backfill material (clays) and of the natural rock at the storage site, before a long flowpath to the biosphere and a dilution by the groundwaters from sedimentary aquifers.

The region under geophysical, geological, hydrogeological and hydrogeochemical investigations is situated in the northern part of Switzerland.

Main geological features

The studied surface includes about 1200 km^2 on the Swiss Plateau (Fig. 1). The granitic bedrock is outcropping in the Rhine Valley at the northern border of Switzerland and in the Black Forest Massif in FRG. It is covered towards the South by sedimentary deposits of up to 2000 m of thickness from Triassic to Quaternary age. The tectonic complex features (Fig. 2) are described in Pekdeger and Balderer (this volume).

Some of the Nagra boreholes encountered huge Permo-carboniferous deposits of up to more than 1000 m of thickness above the crystalline basement within a trench-zone (see Fig. 2).

Isotopic and hydrochemical investigations were carried out on boreholes tapping different aquifers. As these studies are still in progress and only completed for the deep borehole at Böttstein (Balderer, 1985/2; Pearson, 1985; Pekdeger and Balderer, this volume), most of the discussions will focus on the results from this locality, together with some data available for the other boreholes.

The main objectives of the isotopic study are the origin of the waters and of the dissolved salts. The residence time, and the flow directions within the different aquifers. These informations are crucial for the external calibration of the 3-dimensional hydrodynamic flow models (Kiraly, 1985; Kimmeier et al., 1985).

Hydrogeochemical and isotopic signatures of the main aquifers

In the investigated area, up to 5 aquifers may exist in the consolidated sediments above the crystalline bedrock (W. Balderer, 1985/1; Pekdeger and Balderer, this volume).

The discussions will thus be limited to the possible interactions between the various sedimentary aquifers and the groundwaters from the crystalline basement.

Muschelkalk. Groundwaters from the regional Muschelkalk aquifer are of the Ca-Mg-SO_4-HCO_3 type (Lostorf borehole 3 and 4, and Weiach Nagra deep borehole) or of the Na-Ca-Cl-SO_4 type (hot springs at Baden, Beznau borehole and Nagra deep boreholes at Böttstein and Riniken. Waters are saturated or close to saturation with respect to gypsum, anhydrite, calcite, and dolomite, and have a low redox potential shown by the occurrence of H_2S. The mineralization ranges from 12 to 50 meq l^{-1} for the first type and from 23 meq l^{-1} (Nagra deep borehole Riniken) to 70 meq l^{-1} (Baden hot springs) for the second type.

Buntsandstein. This sandstone aquifer of 5 to 25 m of thickness is located below low permeable limestones and very low permeable salt and sulphate deposits of the Middle Muschelkalk. In some cases it is in direct contact with older deposits and even with the crystalline basement of the Permo-carboniferous deposits of the trench (Fig. 2). When lying on the crystalline basement it yields groundwaters very similar to granite waters (e.g. see Nagra deep borehole, Böttstein) with low mineralization and of Na-SO_4-HCO_3-Cl type. Where it is in direct contact with large Permian or even Permo-carboniferous rocks, waters are of the Na-(Ca)-SO_4-Cl type with a total mineralization between 100 and 223 meq l^{-1} (Nagra deep borehole, Weiach).

Permian. (Nagra deep boreholes Riniken and Weiach.) These waters are highly mineralized

(18 g l^{-1} – 100 g l^{-1}) and contain H_2S and CO_2.

Crystalline. Subsurface groundwaters, representative of the present day recharge in crystalline rocks may be found in the outcropping regions of the Black Forest Massif (Bürchau spring and Rothaus borehole, see Fig. 1) and in the Alps (Grimsel area). Waters are of the Ca-Na-(Mg)-HCO_3 = Cl = NO_3 type (Rothaus) or of the *Ca*-Na-*HCO_3*-(SO_4) type (Bürchau) with total mineralizations of 0.7 meq l^{-1} and 1.7 meq l^{-1}, respectively. In the uncovered crystalline of the Grimsel area (in the Alps), groundwaters have an even lower mineralization of less than 0.6 meq l^{-1} and are of the *Na*-Ca-*HCO_3*-SO_4 type waters.

In other boreholes from northern Switzerland the following types of groundwaters are known from the crystalline basement: an Na-SO_4-HCO_3-Cl type in the boreholes of Zurzach (thermal station) and of Böttstein. These waters, with only 14 to 18 meq l^{-1} seem to be typical of the geological situation, where the crystalline basement is in close contact with the Buntsandstein, i.e. where Permian deposits are either missing or very thin.

Elsewhere (e.g. Kaisten, see Fig. 1), the sulphate content and the total mineralization are high due to a contribution from the Permian deposits. A further chemical type of Na-Cl or Na-Ca-Cl water is found at Säckingen. These waters reflect mixtures of low mineralized recent waters, evidenced by the measurable tritium content, with highly mineralized waters (without tritium). The mixture has a mineralization of up to 60 meq l^{-1}. In the same region, a new borehole (Säckingen Stammelhof) delivered a water of

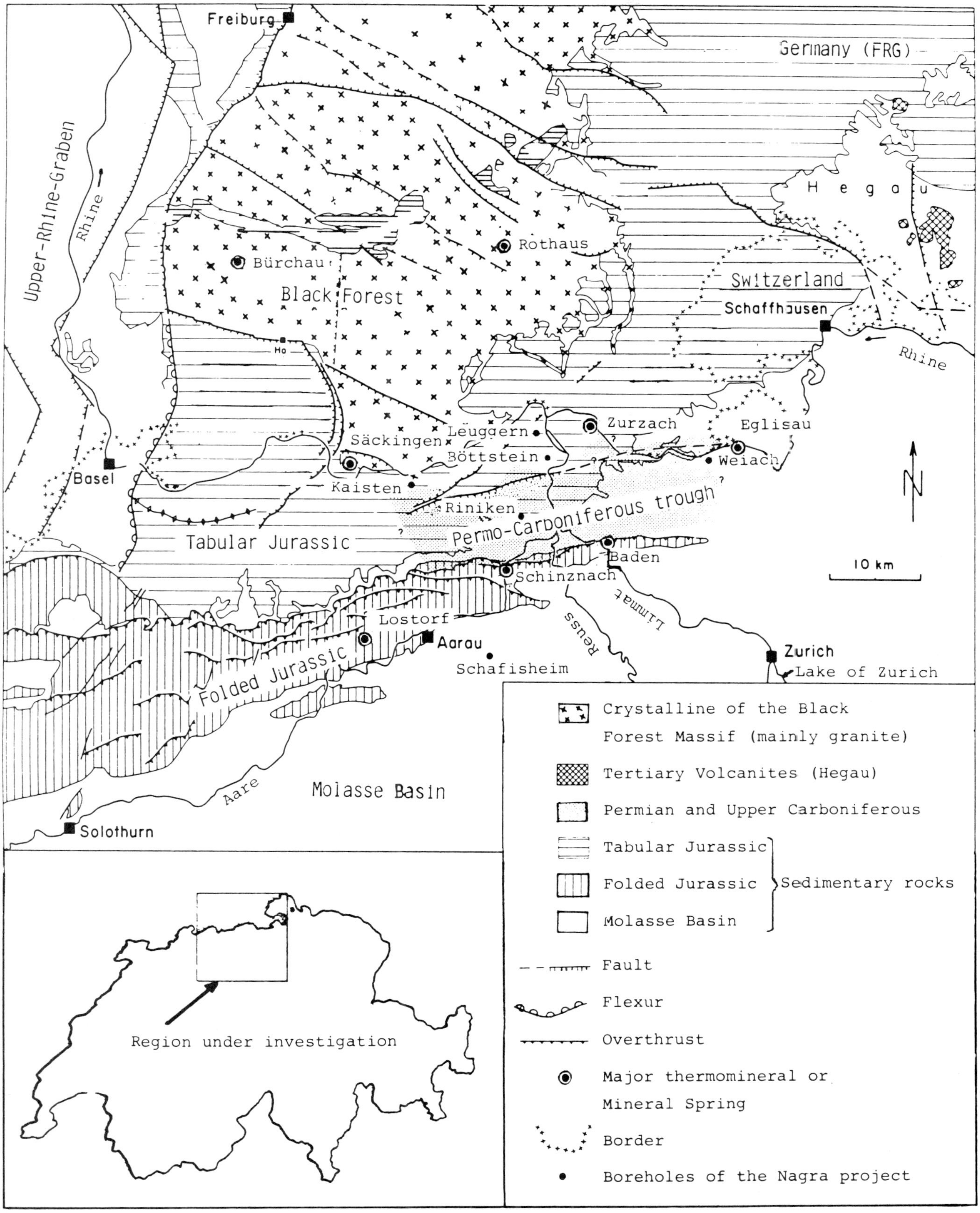

Figure 1. *Tectonic elements of the region under investigation in the Nagra research project for Northern Switzerland (modified after NAGRA NTB 85-01, 1985).*

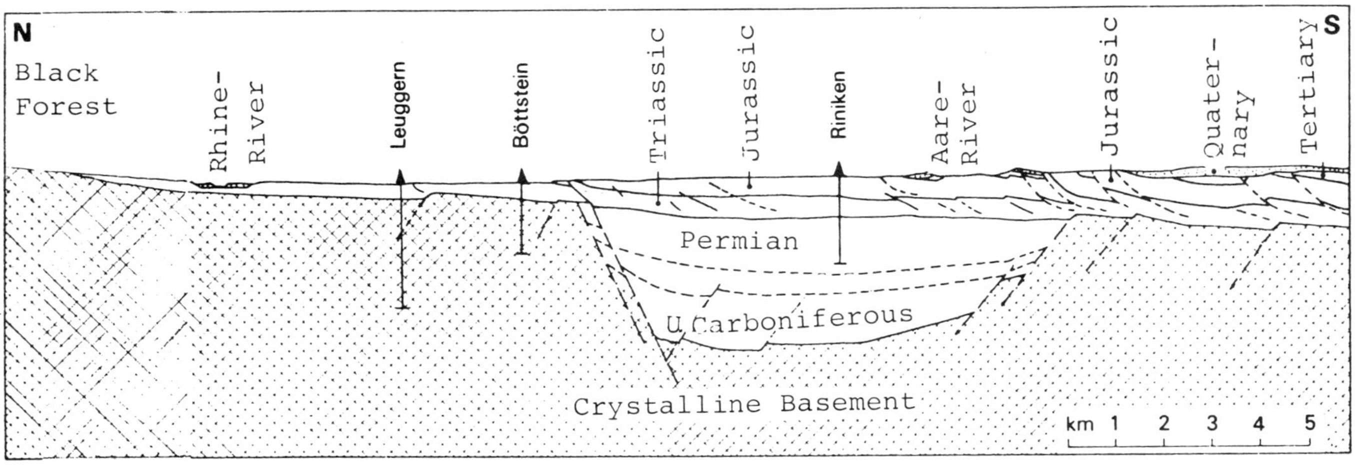

Figure 2. *Detailed cross-section through the region of northern Switzerland with the assumed Permo-Carboniferous trough (maximum depth about 2500 m below ground level) and the location of the NAGRA boreholes Böttstein (Kristal 1), Riniken (Kristal 3), and Leuggern (Kristal 6) (After NAGRA, 1984).*

the same chemical type with a saline load of 119 meq l^{-1}. This groundwater is tritium-free and has a TDIC with low ^{14}C concentration (2.89 ± 0.2% mod. with $\delta^{13}C = -10.9‰$). It could be close to the deep end member of the regional mixing. However, still more saline groundwaters have been found at depth in the crystalline of Böttstein and Schafisheim.

^{18}O and ^{2}H contents of groundwaters

The groundwater from the Muschelkalk shows lower ^{18}O and ^{2}H contents than expected for the supposed recharge areas under present climatic conditions (Figure 3).

In the deep borehole Böttstein, groundwaters from the *Bunter Sandstone* aquifer (lower Triassic) and from the upper part of the crystalline basement (granite) and from the Zurzach thermal station have quite similar heavy isotope contents. For all these groundwaters, a recharge under present climatic conditions in the Black Forest Massif would be possible, despite of the low ^{14}C values. From the $\delta^{2}H$ values of the Zurzach groundwaters, Vuataz (1982) deduced a possible recharge area at the southern part of the Black Forest Massif at an altitude of about 750 to 800 m. From the data of Siegenthaler et al. (1983) under today's climatic conditions, Schmassmann et al. (1984) proposed an altitude of the recharge zone of about 600 m for these groundwaters. This figure is in agreement with ^{18}O values measured in springs from the Black Forest area (unpublished data by Seigenthaler, in: Vuataz, 1982). Calculated altitudes of recharge would lead, under actual climatic conditions, to an average recharge temperature of about 7.5 to 8.0°C (Schmassmann et al., 1984). However, these estimates assume an average infiltration. A correction for evaporation, evapotranspiration, and snow sublimation could probably modify the calculated "isotopic altitudes" of recharge.

In contrast, the ^{18}O and ^{2}H contents calculated for the uncontaminated deep groundwater from the granite at Böttstein (1326 m, Fig. 3) are much higher than in groundwaters formed under todays climatic conditions in the possible recharge areas. The values are only comparable with those found for the groundwaters of the lower Triassic and Permian deposits in the trench area, in the deep boreholes Riniken and Weiach and also of the Keuper sandstone of the Riniken deep borehole.

Several hypothesis must be discussed to explain the origin of the deep brine:

- Residual brine from an acient evaporitic basin.
- Intraformational brine formed through the dissolution of soluble salts.
- Diagenetic brine, i.e. the relictual solution after water-mineral(s) interaction have taken place.
- Result of the mixing of the various processes.

A residual brine should show an evaporitic isotope signature, i.e. far below the meteoric waterline and with a high stable isotope (^{2}H and ^{18}O) content. This is not the case. The dissolution of soluble salts by groundwaters of meteoric origin would have left the point on, or close to, the meteoric waterline. A slight displacement may have been introduced through the contribution of crystallization waters of hydrated salt. However, it does not seem that dissolution of soluble salts may account for the stable isotope content of the deep brine.

Various diagenetic processes may draw the representative point above the meteoric waterline. Some of them were discussed by Frape et al. (1984) for the interpretation of the origin of brines in the Canadian shelf:

- crystallization of clay minerals,
- exchange with H_2S.

Another possibility is the hydration of anhydrite in closed system. Such a process leaves a solution significantly enriched in ^{2}H and depleted in ^{18}O (Fontes and Gonfiantini, 1966) if the hydration process is not limited by saturation.

We favour a diagenetic hypothesis (perhaps multiprocesses) to explain the stable isotope content of the deep brine at Böttstein.

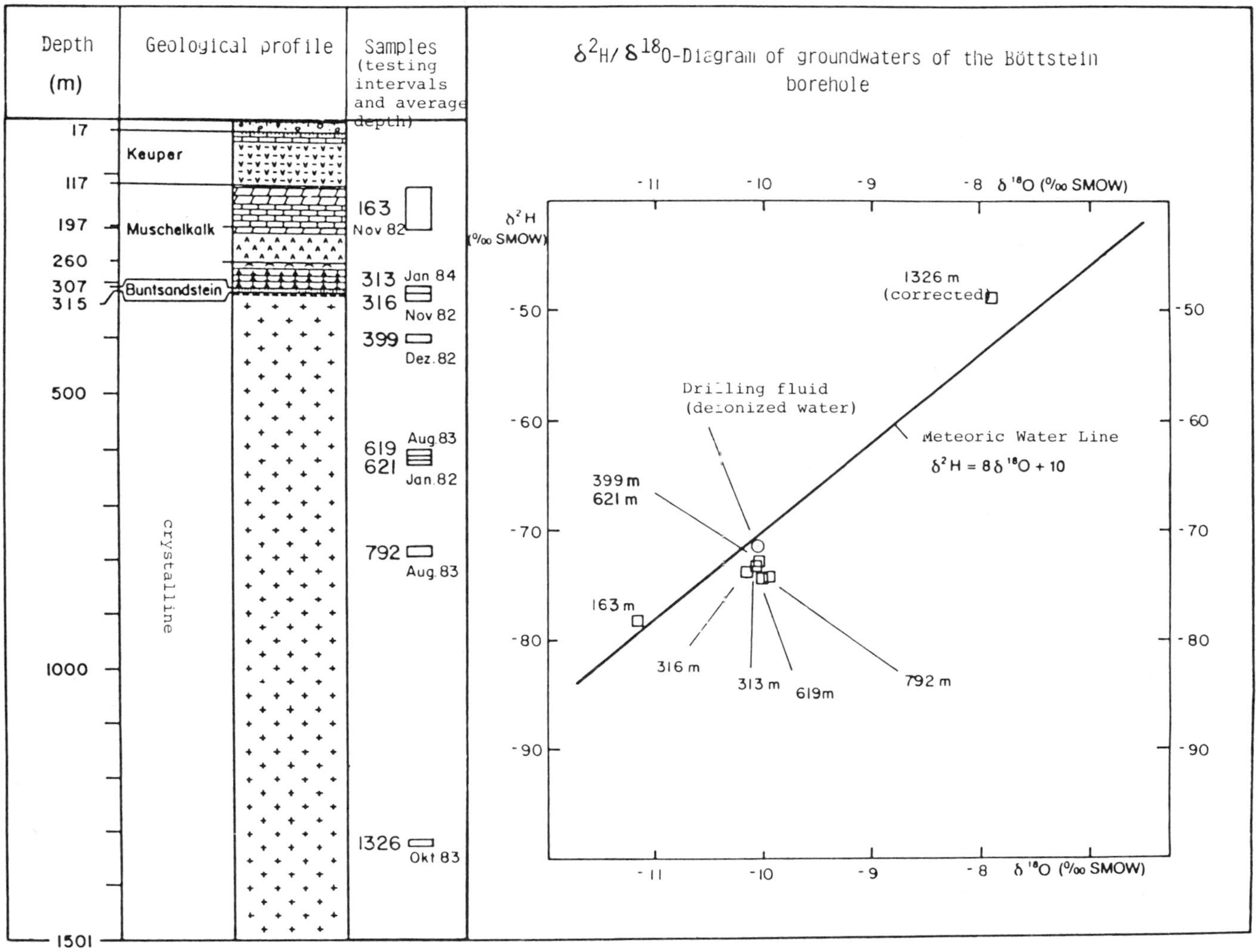

Figure 3. *δ^2H-/$\delta^{18}O$-diagram of groundwater samples from the Böttstein deep borehole (including the isotopic contents of the deionized water used as drilling fluid). The values of the 1326 m sample are corrected for the dilution by the drilling fluid.*

Isotopic evolution of the calco-carbonic system within the crystalline rocks

Conditions of formation of the secondary calcites in Granite. Secondary calcites were sampled on cores from the Böttstein borehole. In the crystalline section, this mineral is frequent in closed fissures but generally rare in open fissures. The ^{13}C and ^{18}O contents from these calcite samples, including 3 samples from open fissures are given in Table 1.

Oxygen-18 contents of fracture calcites. The in-situ temperature and the oxygen isotope composition of the mother solution control the ^{18}O content of the secondary calcite.

The only way to discuss the conditions of formation is to assume equilibrium isotope fractionation between calcite ($CaCO_3$) and water. The thermodependence of this fractionation is expressed, according to O'Neil et al., 1969 (see IAEA, 1983) by:

$$1000 \ln \alpha^{18}O_{calcite\text{-}water} = 2.78\,(10^6\,T^{-2}) - 2.89 \qquad (1)$$

where T is the temperature in Kelvin.

Combining the $\delta^{18}O$ values of the rock carbonates (Table 1) and those of the groundwater samples from the corresponding intervals, equilibrium temperatures of formation of these calcites have been calculated, through the equation (1). For calcite samples from depth-intervals without groundwater samples, the $\delta^{18}O$ values of the water samples of the most neighbouring depth have been used.

These values are then compared to the temperature values of the water (from direct or extrapolated *in-situ* measurements) (Pearson, 1985).

From the results of these calculations (Figs. 5 and 6 and in Table 1), three groups of calcites can be distinguished:

a) Calcites possibly in isotopic equilibrium with groundwater at in-situ temperatures (deviation between calculated and observed temperature is less than 6°C): 638.00 m, 722.32 m, 1000.35 m, 1149.95 m, 1324.27 m, and 1487.98 m (see Table 1 for sample description).

b) Calcites which could be in equilibrium either with

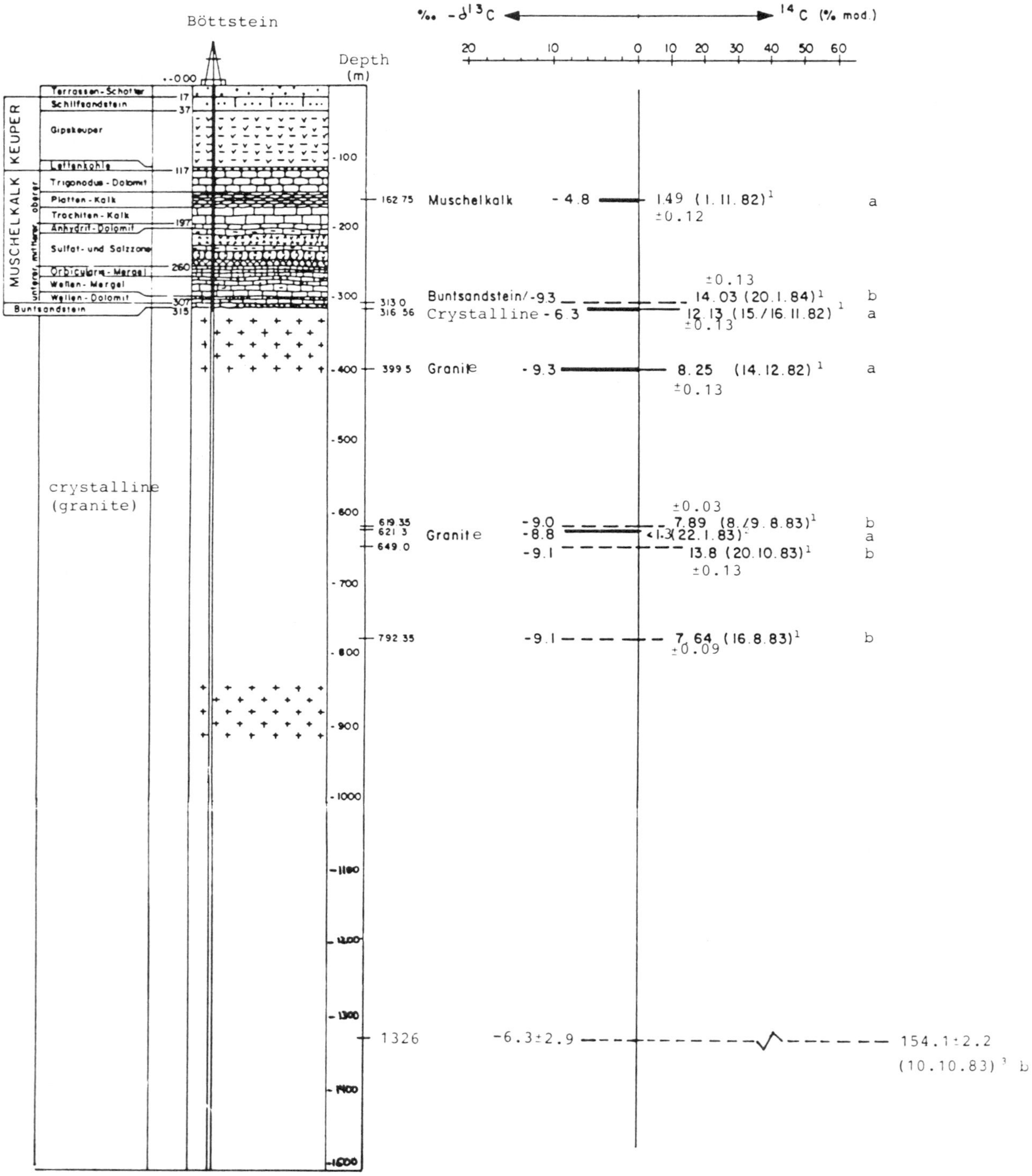

Figure 4. *Lithological profile and ^{13}C and ^{14}C contents of the Total Dissolved Inorganic Carbon in groundwaters of the Böttstein borehole sampled a) during the execution of the borehole and b) after the drilling period (measurements at [1]Bern University, [2]AERE Harwell, [3]AMS measurement Bern University/ETH Zurich).*

Table 1

^{18}O contents of fracture calcites of selected core samples of the Böttstein borehole and results of equilibrium fractionation calculations.

Mineralogical description of the selected samples	Depth (m)	$\delta^{18}O$ of fracture calcites (‰ PDB)	$\delta^{18}O$ of fracture calcites (‰ SMOW)	in-situ water temperature (°C)	$\delta^{18}O$ water (‰ SMOW)	Equilibrium fractionation ϵ water-calcite (‰)	Calculated water temp. for equilibrium fractionation (°C)
Calcite crystals from vugs	122.65	−12.71	17.76	19.7	−11.16	28.92	22.5
Calcite crystals from vugs	131.35	−13.07	17.39	19.7	−11.16	28.55	24.2
Trochitenkalk (Dolomitic-limestone)	192.90	− 9.93	20.62	19.7	−11.16	31.78	10.0
Calcite in fissure in medium altered granite	544.33	−16.48	13.87	33.8	−10.05	23.92	48.9
Wall from open fissure, contains finely dispersed $CaCO_3$	618.72	− 8.87	21.72	36.3	−10.04	31.76	10.1
Strongly altered pegmatite around open fissure (traces of CO_2	619.90	− 8.82	21.77	36.3	−10.04	31.81	9.9
Calcite rich fissure infilling in aplite (+ quartz + clay min.)	638.00	−14.38	16.04	36.9	−10.04	26.08	36.6
Fissure and vein filled with calcite in weakly altered granite	722.32	−15.09	15.30	39.8	− 9.96	25.26	41.1
Kakirite with open fissure (traces of CO_2)	789.90	−11.16	19.36	42.0	− 9.96	29.32	20.6
Medium altered granite in vicinity of kakirite with open fissures (≃ 0.1% CO_2	792.50	−18.52	11.77	42.1	− 9.96	21.73	62.9
Strongly altered aplite with calcites in small veins	816.70	−17.03	13.30	42.9	− 9.96	23.26	52.9
Calcite from filled fissure in medium altered granite	865.25	−20.15	10.09	44.5	− 9.96	20.05	75.0
Calcite-vein in strongly altered granite	1000.35	−16.71	13.63	49.0	−10.0	23.63	50.6
Calcite in fissure in granite with frequent shear-zones	1048.86	−18.11	12.19	50.6	− 9.96	22.15	60.1
Kakiritisized aplitic granite with calcite-veins	1117.73	−19.81	10.44	52.9	−10.0	20.44	72.0
Small calcite vein in granite with shear-zones	1149.95	−17.12	13.21	54.0	−10.0	23.21	53.2
Calcite-rich vein filling in medium altered granite	1324.27	−16.93	13.41	59.8	− 7.9	21.31	65.8
Calcite from small fissure and vein in relatively unaltered granite	1487.98	−19.30	10.96	65.2	−10.0	20.96	68.3

present day groundwater but at temperatures higher by 10 to 30°C than the in-situ values, or with waters with ^{18}O content lower than today: 544.33 m, 792.50 m, 818.70 m, 865.25 m, 1048.86 m, and 1117.73 m.

c) Calcites which could be in isotopic equilibrium with the groundwaters at lower temperatures or with water with an ^{18}O content higher than today: 618.72 m, 619.90 m, and 789.90 m.

For two reasons a temperature effect may be precluded for that group of samples: (i) the drop of more than 20 K is too strong to be explained by a climatic fluctuation, and (ii) a cooler climate would probably have given rise to a lower stable isotope content of the water. Calculated ^{18}O contents of a water in equilibrium with these calcite at in-situ temperatures are high for the region, in the range of $\delta^{18}O$ = −5.7‰ and −4.4‰. These carbonates are originating from open fissures (Table 1). We then proposed that they crystallize in the unsaturated zone from fluids enriched in ^{18}O through evaporation, as already proposed for the circulations from the Mont-Blanc massif (Fontes et al., 1978, Michelot et al., 1983).

Studies of fluid inclusions and of ^{18}O in quartz and of ^2H in clay minerals from closed fissures (illite, biotite, muscovite and chlorite) were performed by Peters et al. (1985).

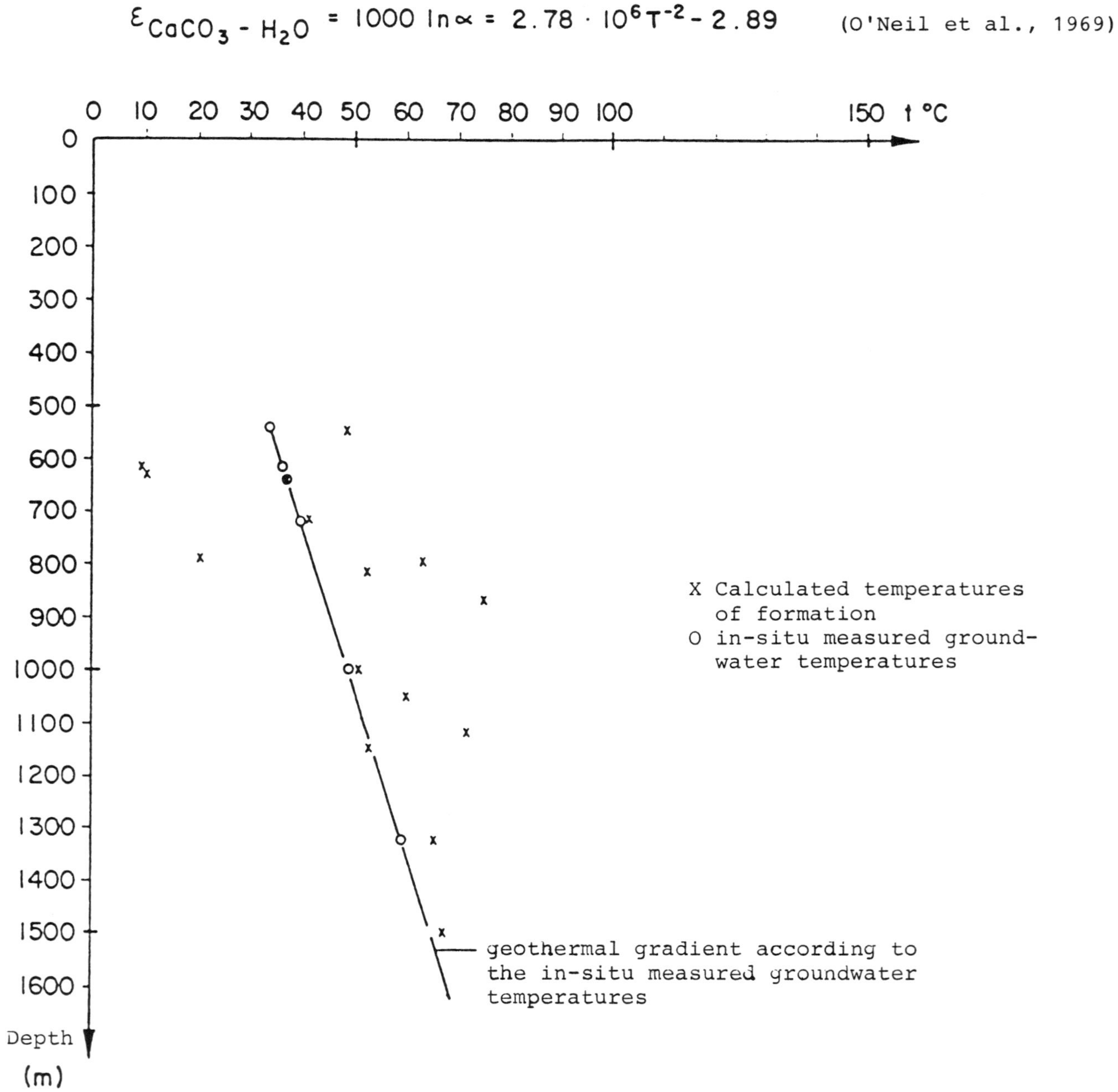

Figure 5. *Representation of the calculated temperatures of formation assuming oxygen isotope equilibrium fractionation between calcites and groundwaters of the deep borehole Böttstein.*

The formation of these minerals would have occurred during a late hydrothermal stage (between Upper Carboniferous and Permian) in the temperature range from 90°C to 230°C. The fluid would have had an isotopic composition in the range of $\delta^2H = -20‰$ to $+10‰$ and $\delta^{18}O = +1‰$ to $+2‰$.

In the parametric representation (adopted from Fritz, 1985), of the data of O'Neil et al., 1969 (Fig. 7), it is possible to explain the formation of the Böttstein calcites from closed fissures by a precipitation from a hydrothermal fluid with $\delta^{18}O$ values between $+2‰$ to $-4‰$ SMOW and a temperature between 90°C to 230°C.

This range contains all carbonate samples from the Böttstein crystalline except, as expected, the 3 samples from the open fissures.

Carbon isotopes of fracture calcites (Figure 8). Three different types of interaction between the dissolved species of the total dissolved inorganic carbon (TDIC) and the carbonate minerals (mainly calcite in the crystalline of Böttstein) must be considered: (i) open system reaction con-

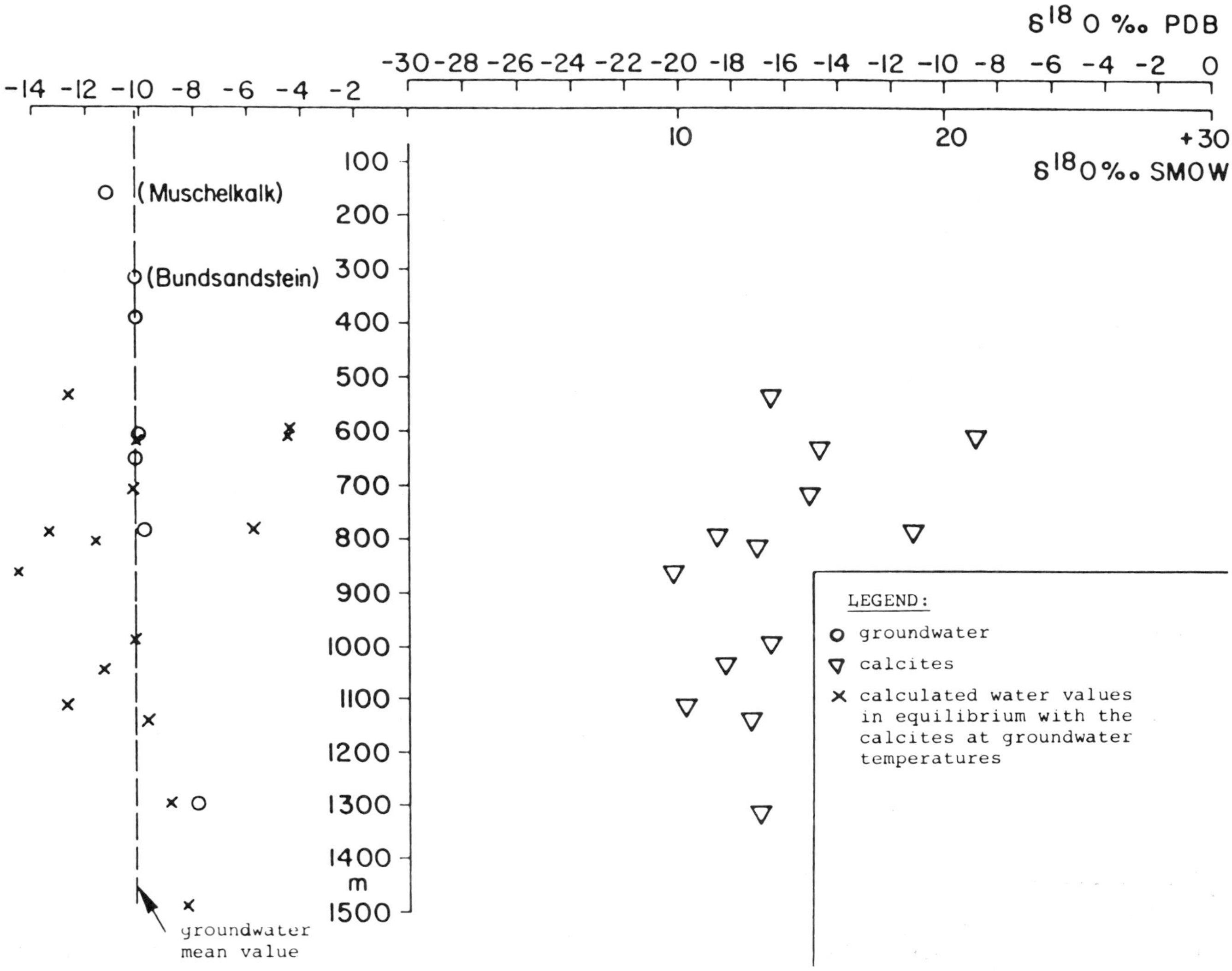

Figure 6. *Deep borehole Böttstein: Oxygen-18 contents of the groundwaters and of the fracture calcites (calculated ^{18}O contents of waters in equilibrium with the calcites at the measured groundwater temperatures).*

trolled by a large (independent) CO_2 reservoir, (ii) the opposite case of a closed system reaction, where the TDIC is formed by mixing of CO_2 and solid carbonate and then enters in partial of total isotopic equilibrium with the gas-phase or with the solid carbonate phase, and (iii) a step-by-step case, where reactions first occur under open system and proceeds under closed system conditions.

The open system allows an isotope fractionation among the following phases: calcite, gaseous carbon dioxide, and various species of the TDIC. This case is described by Bottinga (1969) on base of unpublished measurements of Craig for the carbon (CO_2 – calcite) system.

A similar approach was used for the study of secondary carbonates and TDIC of the deep groundwaters from the Stripa granite by Fritz (1984) and Nordstrom et al. (1985).

No direct measurements of the $\delta^{13}C$ of the gaseous CO_2 are available. The $\delta^{13}C$ (CO_2 gas) value has been calculated at *in-situ* temperatures from the $\delta^{13}C$ values of the TDIC of the corresponding groundwater samples (Table 2), using the species distribution of H_2CO_3, HCO_3^- and CO_3^{2-} calculated from the chemical analyses with the WATEQF model. The following equations were used to describe the open system case (Deines et al., 1974; Wigley, 1975; Fontes, 1983):

$$\delta CO_{2(g)} = \delta_{TDIC} + \frac{\Sigma \epsilon m_i}{\Sigma m_i} \quad (2)$$

where i stands for the species of carbon in solution

$$\Sigma m_i = mH_2CO_3 + mHCO_3^- + mCO_3^{2-}$$

and $\epsilon_i \cong \delta CO_{2(g)} - \delta_i$ (3)

Table 2

^{13}C contents of fracture calcites of selected core samples of the Böttstein borehole and results of equilibrium fractionation calculations (mineralogical sample description in Table 1).

Depth (m)	$\delta^{13}C$ of fracture calcites (‰ PDB)	Case a) equilibrium fractionation between CO_2 and calcite (eq.4) at in-situ water temperature			Case b) equilibrium fractionation between CO_2 and calcite (eq.4) at calculated ^{18}O equilibrium fractionation temperatures		
		in-situ water temperature (°C) (according to Pearson, 1985)	ϵCO_2, $CaCO_3$ (‰) (equation 4)	$\delta^{13}C_{CO_2}$ (‰ PDB calculated (equation 4)	^{18}O water temperature (°C) calculated (equation 1)	δCO_2,$CaCO_3$ (‰) (equation 4)	$\delta^{13}C_{CO_2}$ (‰ PDB) calculated (equation 4)
122.65	–1.24	19.7	–11.2	–12.4	22.5	–10.7	–12.0
131.35	–1.20	19.7	–11.2	–12.3	24.2	–10.5	–11.7
192.90	+0.51	19.7	–11.2	–10.6	10.0	–12.7	–12.1
544.33	–8.80	33.8	–9.20	–18.0	48.9	–7.47	–16.3
618.72	–8.26	36.3	–8.9	–17.2	10.1	–12.6	–20.9
619.90	–7.90	36.3	–8.9	–16.8	9.9	–12.7	–20.6
638.00	–7.97	36.9	–8.82	–16.8	36.6	–8.9	–16.8
722.32	–9.45	39.8	–8.5	–17.9	41.1	–8.3	–17.8
789.90	–9.47	42.0	–8.2	–17.7	20.6	–11.0	–20.5
792.50	–10.73	42.1	–8.2	–18.9	62.9	–6.1	–16.8
816.70	–9.04	42.9	–8.1	–17.2	52.9	–7.1	–16.1
865.25	–6.52	44.5	–7.9	–14.5	75.0	–5.1	–11.6
1000.35	–8.58	49.0	–7.5	–16.0	50.6	–7.3	–15.9
1048.86	–9.40	50.6	–7.3	–16.7	60.1	–6.4	–15.8
1117.73	–8.91	52.9	–7.1	–16.0	72.0	–5.3	–14.2
1149.95	–10.11	54.0	–6.9	–17.1	53.2	–7.0	–17.1
1324.27	–10.91	59.8	–6.4	–17.3	65.8	–5.9	–16.8
1487.98	–9.16	65.2	–5.9	–15.1	68.3	–5.6	–14.8

Temperature dependences for the various ϵ_i are given by Mook (1980).

Using the relationship for the carbon isotope fractionation between CO_2 and calcite (after Bottinga, 1968, as given in IAEA T.R. 228, 1983):

$$\epsilon\,(CO_2, CaCO_3) \cong 10000 \ln\alpha \cong \delta CO_2 - \delta CaCO_3 = -2.9880 \cdot 10^6 \cdot T^{-2} + 7.6663 \cdot 10^3 T^{-1} - 2.4612 \quad (4)$$

where $\alpha = R^{13}(CO_2)/R^{13}(CaCO_3)$, $R^{13} = {}^{13}C/{}^{12}C$

One calculates temperature values which should be equal to the *in-situ* ones in case of an open system equilibrium with environmental CO_2. Another estimation of the temperature can be made through the thermodependence of the carbon-13 enrichment between HCO_3^- and solid calcite (Mook, 1980).

Since the pH is about 8.1 one can consider that all the TDIC is on the form of HCO_3^-:

$$\delta_{TDIC} - \delta_{calcite} \cong \epsilon(HCO_3^- - \text{calcite}) = -4.232 \cdot 10^3\, T^{-1} + 15.10 = \epsilon_s \quad (5)$$

One can then compare the two carbon isotope temperatures, which should be equal since they refer to the same system, the in-situ temperature, and the oxygen isotope temperature.

These calculations lead to a reasonable agreement between calculated and measured temperature for one sample only at 638.00 m (Tables 2 and 3). However, the uncertainty on this type of calculation is illustrated by the difference between the two estimates of the temperature by the ^{13}C equilibria.

For this calcite a crystallization under open system conditions in equilibrium with the formation fluids (water and TDIC) at the *in-situ* temperature, seems possible.

For all the other calcite samples of the Böttstein borehole, where isotopic values for water and TDIC are available, no agreement is observed between the carbon and oxygen isotopic temperatures and the measured values.

This was expected since the calculated values of δCO_2 in equilibrium with the TDIC at formation temperatures appear generally higher (Table 2) than typical soil gases values. This may be due to (i) a variation in the CO_2 global balance, (ii) open system condition on the solid phase, or (iii) closed system (mixing). Since no arguments are available to discuss case (i), we would assume that processes (ii) and/or (iii) may be acting.

If now relation (4) is used with the temperatures calculated from the ^{18}O thermometer (1), there is an agreement between the $\delta^{13}C$ of the CO_2 (gas) in equilibrium with the calcite samples and with the TDIC for the following depths: 544.33 m, 792.50 m, 816.70 m, and 1048.86 m (Table 2, case b and c). This supports the concept of a formation of these calcites in formation fluids at a temperature higher than present by 9.5 to 30.5°C. This is in concordance with the results of the ^{18}O study.

However, the evolution of the inorganic carbon in deep groundwaters must also be considered as a reaction under

Table 2 continued

Case c) equilibrium fractionation calculated from $CO_2/CaCO_3$ relationship (eq.4), according to Bottinga, 1968			Case d) equilibrium fractionation calculated from $TDIC/CaCO_3$ relationship (eq.5)		
$\delta^{13}C_{CO_2}$ (calculated in equilibrium with (TDIC) (‰ PDB) (equation 2)	$\epsilon CO_2,CaCO_3$ (‰) (equation 4)	water temperature (°C) calculated for $CO_2/CaCO_3$ isotopic equilibrium fractionation (equation 4)	$\delta^{13}C_{TDTC}$[1] (groundwater) (‰ PDB)	ϵHCO_3calcite (‰) $^{\delta}TDIC$-$^{\delta}CaCO_3$ (equation 5)	water temperature (°C) calculated for $TDIC/CaCO_3$ isotopic equilibrium fractionation (equation 5)
–11.5	–10.3	25.7	–4.8	–3.56	<0
–11.5	–10.3	25.4	–4.8	–3.60	<0
–11.5	–12.0	13.9	–4.8	–5.31	<0
–16.6	–7.8	45.8	–9.3	–0.50	<0
–15.3	–7.0	53.1	–8.8/–9.0	–0.54/–0.74	<0
–15.3	–7.4	49.5	–8.8/–9.0	–0.90/–1.10	<0
–16.2	–8.2	41.9	–9.1	–1.13	<0
–16.2	–6.8	56.0	–9.1	0.35	13.8
–16.2	–6.7	56.2	–9.1	0.37	14.2
–16.2	–5.5	70.3	–9.1	1.63	41.0
–16.2	–7.2	51.9	–9.1	–0.06	6.0
–16.2	–9.7	30.0	–9.1	–2.58	<0
–10.2	–1.6		–6.3±2.9		
–10.2	–0.8		–6.3±2.9		
–10.2	–1.3	[2]	–6.3±2.9	[2]	[2]
–10.2	–0.1		–6.3±2.9		
–10.2	+0.7		–6.3±2.9		
–10.2	–1.0		–6.3±2.9		

[1] Values of $\delta^{13}C_{TDIC}$ of nearest water yielding zone (represented in Figure 4).
[2] Due to the smaller precision of the AMS result, no further calculation possible.

another case of open system conditions: a complete exchange on the solid phase (Fontes and Garnier, 1979).

In that case, every components of the calco-carbonic system are in isotopic equilibrium with the solid phase as defined by Deines et al. (1974):

$$\delta_{TDIC} = \frac{\Sigma m_i \delta_i}{\Sigma m_i} \qquad (6)$$

$$\delta_{TDIC} = [m_a\, \delta_a + m_b\, \delta_b + m_c\, \delta_c][m_{TDIC}]^{-1} \qquad (7)$$

where $a = H_2CO_3$, $b = HCO_3^-$ and $c = CO_3^{2-}$

Introducing the definition of the fractionation in

$$\alpha_{1-2} = R_1/R_2 = \frac{\delta 1 + 1000}{\delta 2 + 1000}$$

Equation (7) can be transformed by:

$$\frac{\delta H_2CO_3 + 1000}{\delta CaCO_3 + 1000} = \alpha$$

$$\delta H_2CO_3 + 1000 = \alpha\delta CaCO_3 + 1000\,\alpha$$

$$\delta H_2CO_3 = \alpha\delta CaCO_3 + 1000\,(\alpha - 1)$$

introduced into (7) results the following expression:

$$\delta^{13}C_{(TDIC)} = \{m\, H_2CO_3\, [\alpha H_2CO_3, CaCO_3 \cdot \delta CaCO_3 + (\alpha H_2CO_3, CaCO_3 - 1)1000] + m\, HCO_3^-\, [\alpha HCO_3^-, CaCO_3 \cdot \delta CaCO_3 + (\delta HCO_3^-, CaCO_3 - 1)1000] + m\, CO_3^{2-}\, [\alpha CO_3^{2-}, CaCO_3 \cdot \delta CaCO_3 + (\alpha CO_3^{2-}, CaCO_3 - 1)1000]\}/m_{TDIC} \qquad (8)$$

The resulting ^{13}C contents of the Total Dissolved Inorganic Carbon (TDIC) in isotopic equilibrium with the calcites of the fractures are represented in Figure 8 (calculations are made with values of the fractionation factors from Deines et al. (1974) and Mook (1980).

From Figure 8 it appears that calculated values for the TDIC of the crystalline section are higher than the observed values, with the exception of the 865.35 m sample. A possible explanation would be that the precipitation of these calcites would have occurred under temperature higher than present within a fluid similar to the present one.

Conclusions on the stable isotope contents of fracture calcites and groundwater evolution in the crystalline. Most of the calcites are not in equilibrium with present day fluids neither for ^{18}O nor for ^{13}C. The values could be explained by a high temperature crystallization (150°C-200°C) from a hydrothermal fluid (with a ^{18}O content ~–4 to 0‰).

A possibility for calcite dissolution to take place even at calcite saturation could be the occurrence of methane and its oxidation into CO_2. During the drilling operations through the crystalline (granite) at Böttstein, methane was constantly observed, with increasing contents at depth. Methane has been isolated from the granitic core samples (Schenker, personal communication) and was also found in the crystalline rocks from the other deep boreholes (Schafisheim, Weiach).

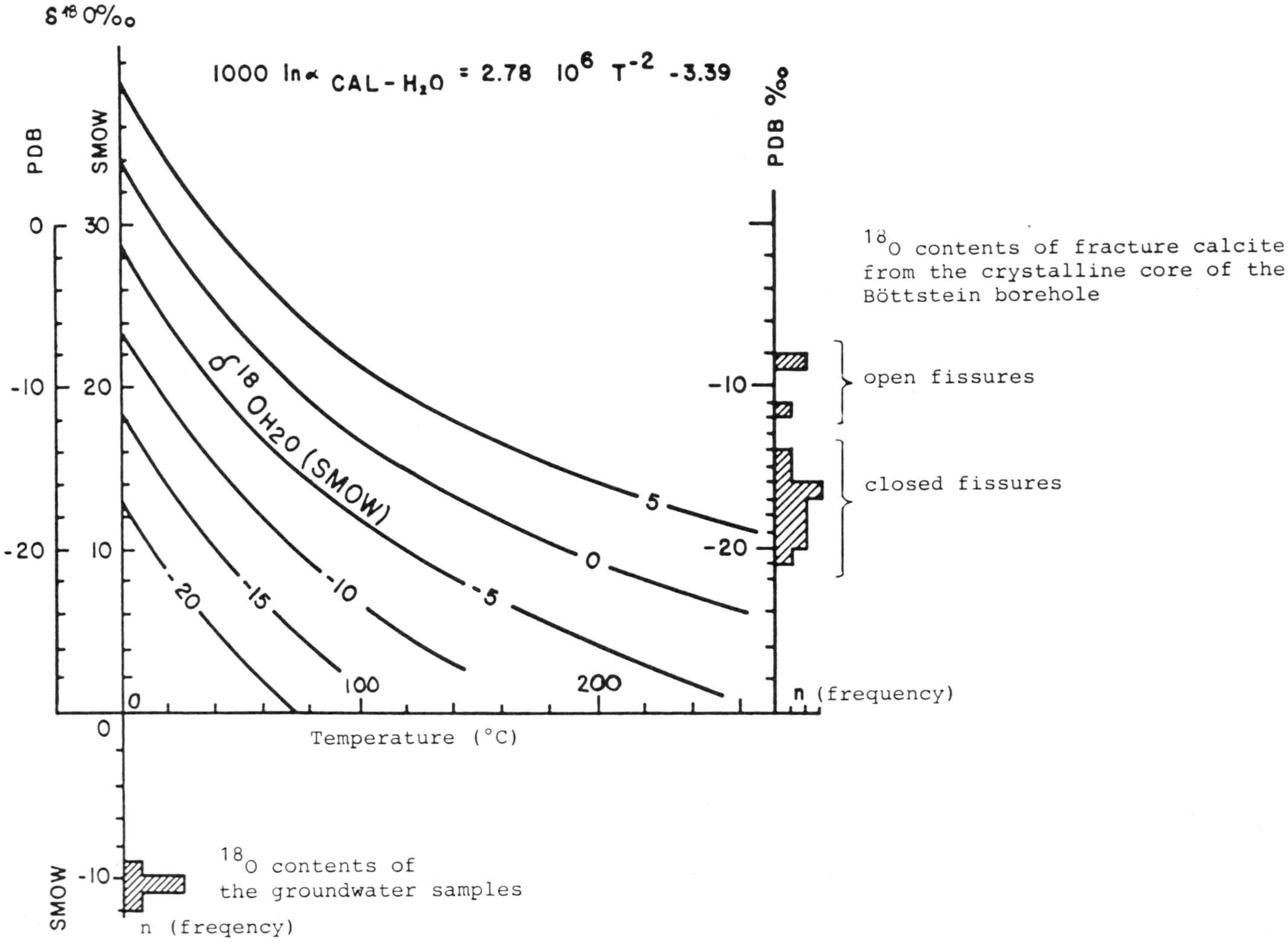

Figure 7. *^{18}O contents of groundwaters and fracture calcites within the crystalline section of the Böttstein borehole (parametric representation of O'Neil et al. 1969 data, adopted from P. Fritz 1984).*

Table 3

Temperatures of ^{18}O and ^{13}C equilibrium fraction of the 638 m sample for the calcite-water and calcite-CO_2 system.

Depth (m)	Sample	Temp. (°C) measured	Temp. (°C) ^{18}O (calcite-water)	Temp. (°C) ^{13}C (CO_2-calcite)
638.00	Calcite infilling in closed fracture in aplite (+ quartz + clay min.)	36.9	36.6	41.9

Table 4

Shallow groundwaters of the Black Forest Massif.

Locality	Chemical type characterization	Total mineralization meq l^{-1}	Temp. °C	pH	TDIC mmol kg^{-1}	^3H TU
Rothaus	Ca-Na-(Mg)-HCO_3-Cl-NO_3	0.7	10	5.88	0.78	76.2±5.0
Bürchau	Ca-Na-HCO_3-(SO_4)	1.7	17.0	8.06	0.98	11.0±1.0

Table 5

Calculated initial conditions for open system evolution.

Locality	HCO_3^- mmol l^{-1}	H_2CO_3 mmol l^{-1}	$\delta^{13}C_{CO_2}$ ‰	$\delta^{13}C_{TDIC}$ (‰) (open system)	$\delta^{13}C_{CO_2}$ ‰	$\delta^{13}C_{TDIC}$ (%) (open system)
Rothaus	0.195	0.585	–23.0	–21.3	–25.0	–23.4
Bürchau	0.96	0.02	–23.0	–14.8	–25.0	–16.8

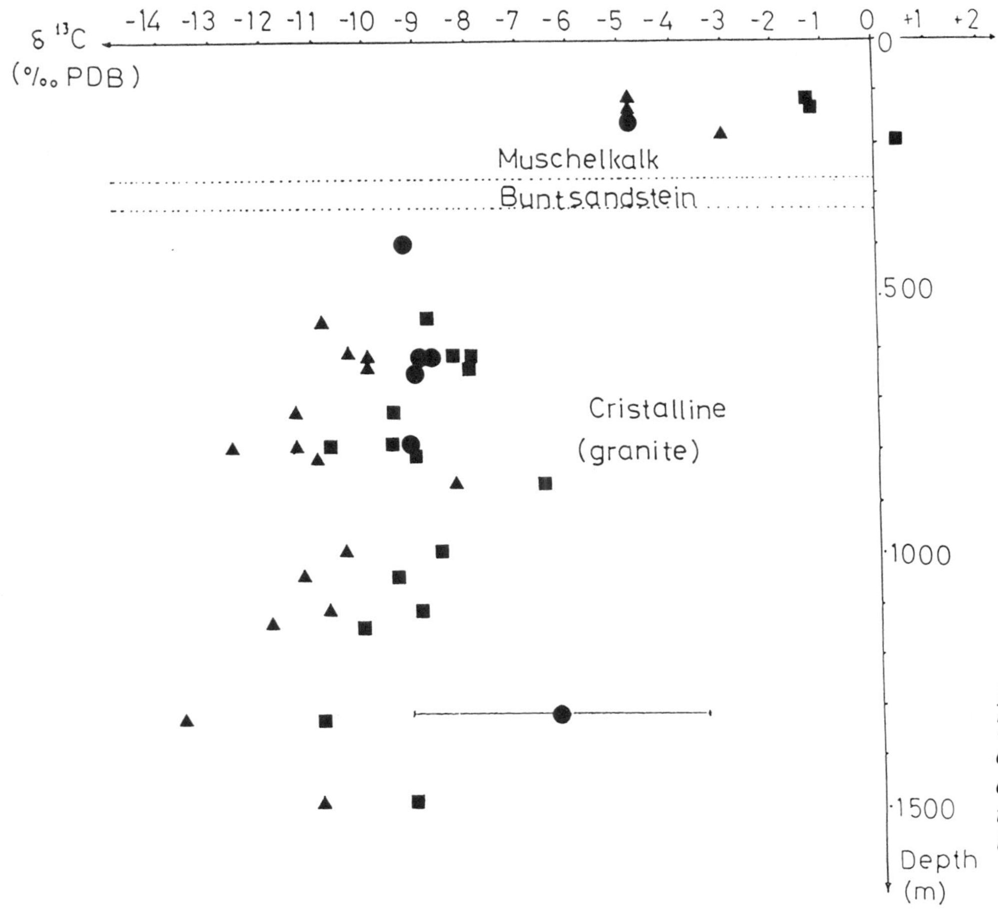

Figure 8. *^{13}C contents of TDIC of groundwaters and analyzed fracture calcites and of resulting TDIC in isotopic equilibrium at the deep borehole Böttstein.*

● $\delta^{13}C_{(TDIC)}$

■ $\delta^{13}C$ (Calcites)

▲ Calculated values of $\delta^{13}C_{TDIC}$ in equilibrium with the Calcites in open system conditions for the case of a complete exchange on the solid phase (Fontes and Garnier, 1979) at in-situ temperatures.

Table 6

Results of combined mass balance and carbon isotope model calculations assuming an evolution from the recharge zone of the Black Forest to the Böttstein groundwaters

Water sample (average depth) (m)	Case a: Evolution from Rothaus groundwater			Case b: Evolution from Bürchau groundwater		
	$\frac{mC_1}{mC_2} = X_2$	$\delta^{13}C^*$ (‰) (of rock carbonate calculated)	^{14}C model age (ka)	$\frac{mC_1}{mC_2} = X_2$	$\delta^{13}C^*$ (‰) (of rock carbonate calculated)	^{14}C model age (ka)
313.0	0.134	–7.44	(modern)	0.170	–8.18	1.7
316.6	0.123	–4.19	0.2	0.155	–4.74	2.2
399.5	0.121	–7.52	3.2	0.152	–8.31	5.2
621.3	0.120	–7.09	>18.4	0.151	–7.73	>20.4
618.4	0.124	–7.25	3.8	0.156	–7.93	5.8
792.4	0.126	–7.35	4.1	0.158	–8.03	6.2
649.0	0.124	–7.37	(modern)	0.156	–8.04	1.2

Table 7

^{14}C model ages according to the different correction models of the Böttstein groundwaters (by admitting the following parameters: atmospheric ^{14}C concentration $A^{14}C_{atm}$ = 100 pmC, ^{14}C activity of carbonate minerals $A^{14}C_m$ = 0 pmC, $\delta^{13}C$ of carbonate minerals (as dissolving phase): 0.0‰ PDB).

Average depth (m)	Sample Nr.	Formation	^{14}C (TDIC) % mod.	$\delta^{13}C$ ‰	$\delta^{13}CO_2$ (calc.) ‰	TAMERS Aa %	TAMERS T (*)	MOOK Aa %	MOOK T (*)
162.75	301/2	Muschelkalk	1.49 ±0.12	–4.8	–11.53	59.11	30.4	–125.6	–
316.56	301/4 301/5	Buntsandstein and granite	12.13 ±0.25	–6.3	–13.80	49.66	11.7	–	–
313.00	301/27	Buntsandstein and granite	14.03 ±0.13	–9.3	–17.35	48.28	10.2	–5.29	–
399.50	301/8	Granite	8.25 ±0.13	–9.3	–16.60	49.57	14.8	10.10	1.7
621.30	301/12	Granite	<1.3	–8.8	–15.40	49.81	30.1	12.92	19.0
2) 619.35	301/16	Granite	7.89 ±0.09	–9.0	–16.09	50.19	15.3	7.62	modern
2) 649.00	301/23	Granite	13.80 ±0.13	–9.1	–16.19	50.6	10.7	8.97	–
2) 792.35	301/18	Granite	7.64 ±0.07	–9.1	–16.20	49.59	15.5	9.89	2.1

1) Sampled during execution of the borehole
2) Sampled after the drilling period
*) T: Model age in 10^3 years (ka)
A_a: Calculated ^{14}C initial concentration (pmC)

Interpretation of the radiocarbon data and groundwater dating in the crystalline

Stable isotope contents (^{18}O and ^{13}C) of the calcites from the open fractures and from some closed fractures, are not in isotopic equilibrium with present day fluids.

It can therefore be advanced that calcite is not currently precipitating. This is in agreement with the almost complete absence of calcite in open fractures.

Some $\delta^{13}C$ values of the analyzed calcites are quite similar to the $\delta^{13}C$ values of the TDIC. In this case, the geochemical and isotopic evolution of these groundwaters may be the result of the interactions with the crystalline rock of the Black Forest Massif, especially through dissolution of fracture calcites.

To check this hypothesis which is of great importance for the calculation of the ^{14}C ages, the following approach was used: groundwaters from the uncovered crystalline area of the Black Forest Massif (Figure 1), at Rothaus (well with pumping station) and Büchau (free flowing subthermal spring), have been considered as representative for groundwater recharge under open system conditions (Table 4). Assuming an evolution under open system conditions, one calculates the initial conditions (Table 5).

For the evaluation of the subsequent evolution, a combined open and closed system evolution was used. In the case of a congruent dissolution process in a closed system, the treatments of Deines (1974) and Wigley et al. (1978) are similar:

$$\delta^{13}C_{TDIC\ final} = [C_{initial} \cdot \delta^{13}C_{initial} + mCa^{2+} \cdot \delta^{13}C_{rock}]/C_{TDIC\ final} \quad (9)$$

initial open system conditions ($C_{initial} \cdot \delta^{13}C_{initial}$); if only calcite is dissolved $mCa^{2+} \approx \Delta mC = C_{initial} - C_{TDIC\ final}$

This expression can be transformed into the expression of the mass-balance approach for congruent dissolution according to Wigley et al. (1978):

$$\frac{C_{initial}}{C_{final}} = \frac{mC_1}{mC_2} = \frac{\delta^{13}C_2 - \delta^{13}C^*}{\delta^{13}C_1 - \delta_{13}C^*} \quad (10)$$

The groundwaters of Rothaus and Bürchau are assumed to have achieved their evolution under open system conditions. The ^{13}C content of the rock carbonate, $\delta^{13}C^*$, can thus be determined according to:

$$\delta^{13}C^* = \frac{\frac{mC_1}{mC_2} \cdot \delta^{13}C_1 - \delta^{13}C_2}{\frac{mC_1}{mC_2} - 1} \quad (11)$$

The calculated $\delta^{13}C^*$ values for the dissolved rock carbonate (assuming an evolution from the recharge zone of the

Table 7 continued

REARDON & FRITZ		INGERSON-PEARSON		GONFIANTINI		FONTES-GARNIER 1979		FONTES-GARNIER 1984	
Aa %	T (*)	Aa %	T (*)	Aa %	T (*)	Aa %	T (*)	Aa %	T (*)
23.26	22.7	20.87	21.8	33.08	25.6	–12.4	–	19.70	21.3
33.30	8.3	27.39	6.7	41.18	10.1	11.0	–	26.36	6.4
47.00	10.0	40.43	8.8	64.23	12.6	33.6	7.2	40.20	8.7
46.94	14.3	40.43	13.1	60.00	16.4	33.7	11.6	39.96	13.0
45.70	29.4	38.26	28.0	53.46	30.7	31.1	26.2	37.40	27.8
46.12	14.6	39.13	13.2	57.29	16.4	31.3	11.4	38.49	13.1
46.12	10.0	39.57	8.7	57.92	11.9	32.5	7.1	38.98	8.6
46.75	15.0	39.57	13.6	57.81	16.7	32.5	12.0	38.98	13.5

Black Forest to the Böttstein groundwaters) are listed in Table 6.

If these values are compared with those found for the calcites from the open fissures in the Böttstein core 618.72 m, $\delta^{13}C$ = –8.26‰; 619.90 m, $\delta^{13}C$ = –7.90‰; 789.90 m, $\delta^{13}C$ = –9.47‰ it can be concluded that an evolution exclusively within the crystalline is possible. But such an evolution cannot be treated by the conceptual ^{14}C correction models based on ^{13}C fractionation and exchange (e.g. Mook, 1980; Fontes and Garnier, 1979) but only by an isotope mass balance along an evolutionary path.

For the proposed evolution from Rothaus and Bürchau under open and closed system conditions the initial carbon-14 concentrations (without decay) have been calculated according to the following equation:

$$\frac{A_{a2} - A_M}{A_{a1} - A_M} = \frac{mC_1}{mC_2} = X_2 \qquad (12)$$

(Wigley, 1978)

with $A_{a1}(\%)$ = initial, open system ^{14}C concentration at the end of the open system evolution versus the gas phase.

$A_{a2}(\%)$ = final, ^{14}C concentration at the end of the closed system process versus the solid phase and after the congruent dissolution evolution = initial value for the ^{14}C age determination.

$A_M(\%)$ = ^{14}C concentration of the rock carbonate, assumed to be zero.

if A_M is equal to 0‰ and if A_{a1} is expressed by the $\delta^{13}C$ evolution (with ^{14}C = 2.3 $\epsilon^{13}C$ according to Saliège and Fontes, personal communication, 1984):

$$A_{a1} = Ag - A_g \cdot 10^{-2} \cdot 2.3 \cdot 10^{-1} (\delta_g - \delta_{TDIC}) \qquad (13)$$

and combined with equation (2) gives a relationship similar to that developed by Wigley (1975):

$$A_{a1} = Ag - \frac{Ag}{100} \cdot \frac{2.3}{10m_{TDIC}} [m_a \cdot \epsilon_{g-a} + m_b \cdot \epsilon_{g-b} + m_c \cdot \epsilon_{g-c}] \qquad (14)$$

where:

A_g = ^{14}C concentration of the gaseous $CO_2(\%)$. The final equation for the calculation of the model-age is:

$$t = \frac{5730}{\ln 2} \cdot \ln \frac{A_{a2}}{A_{TDIC}} = \frac{5730}{\ln 2} \cdot \ln \frac{A_{a1} \cdot X_2}{A_{TDIC}} \qquad (15)$$

The results from this equation are also represented in Table 6.

The adoption of a model of dissolution with a ^{13}C content of the cabonate of about 0‰ (assuming an initial infiltration-evolution in sedimentary rocks) leads to much higher ages than the open system model (see Table 7 for calculations with the correction models of Fontes and Garnier, 1979; Pearson, 1965; and Reardon and Fritz, WATEQF-ISOTOP, 1978).

If no further process is influencing the ^{14}C concentration of the TDIC, the ages listed in Table 6 can be considered as groundwater residence times according to the piston-flow concept. The shorter residence times are shown by the groundwaters from the upper part (upmost 700 m) of the crystalline. These waters would be infiltrated in the outcropping region of the Black Forest massif. The following arguments support this hypothesis:

- The ^{13}C and ^{18}O contents of the fracture calcites suggest that a dissolution of these minerals is probably taking place.

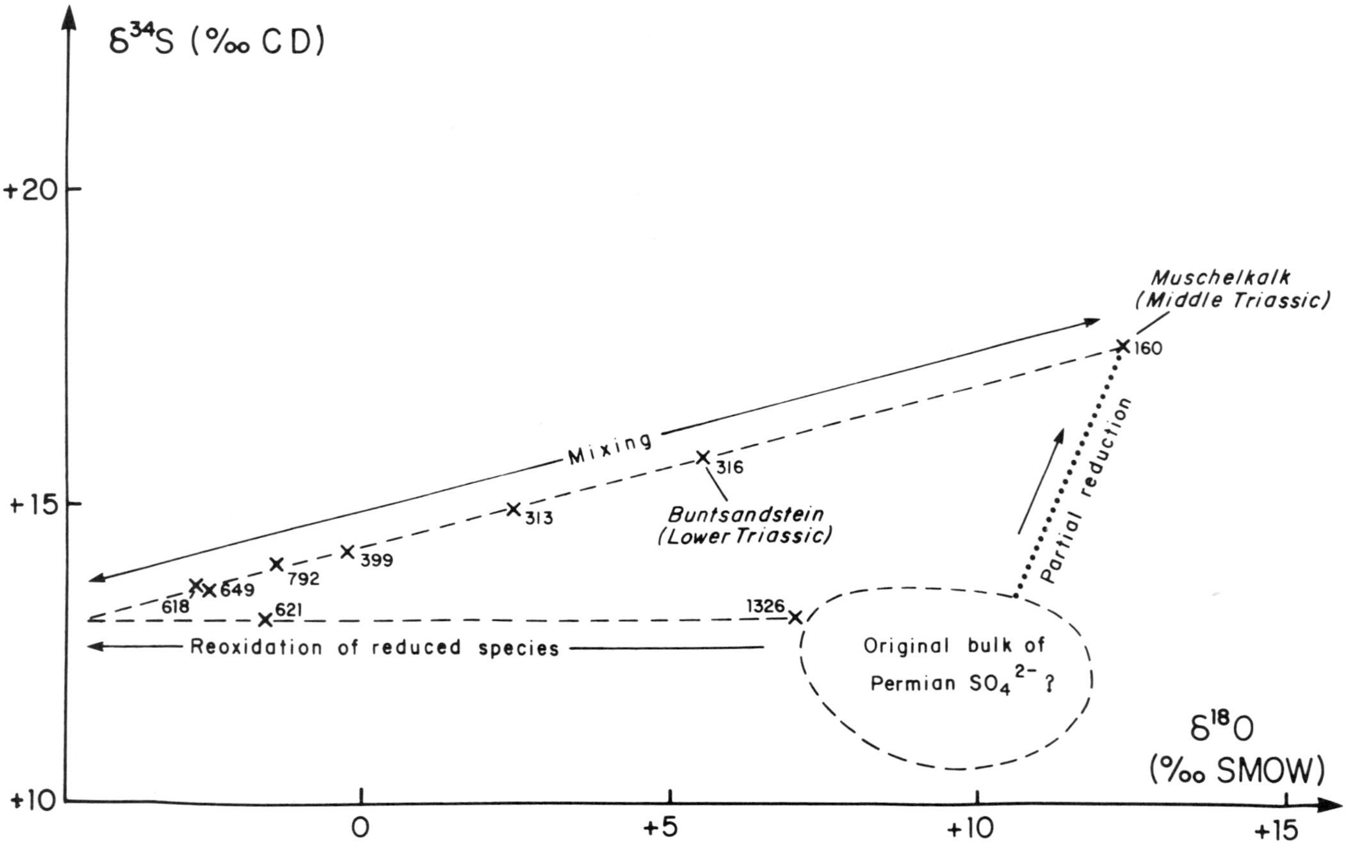

Figure 9. *Interpretation of the stable isotope contents of dissolved sulphate in the Böttstein borehole (Various depths are indicated). A mixing occurs between two sources of SO_4^{2-} : a) Triassic SO_4^{2-} (which could be Permian-derived through dissolution and partial reduction) ; b) SO_4^{2-} resulting from the secondary oxidation of a sulphide derived from total reduction of a Permian source; this reoxidation would have occurred in water with a ^{18}O content similar to that presently found in the granite.*

- The similarity of the ^{13}C contents of the TDIC and of the fracture calcites can be considered as an indication of calcite dissolution.
- The general association of low mineralization and ^{18}O and ^{2}H contents is in agreement with a recharge in the region of the outcropping crystalline of the Black Forest Massif.

For the deeper part of the crystalline at Böttstein, the sample from the depth of 1326 m was contaminated with drilling fluid (deionized water + fluorescein) to about 50-60%. The ^{14}C determination (Fig. 4) of this sample (alkalinity ≈ 1 meq l^{-1}) gave a result which is probably influenced by the contamination (possibly by the organic carbon introduced through alteration of the added tracer). The very high mineralization (13.2 g l^{-1}), the high ^{2}H and ^{18}O contents (as extrapolated for the undiluted end member) and the ^{18}O and ^{34}S contents of the aqueous sulphate ions (as discussed later in more detail) are indications for a distinct origin and a complex evolution of this groundwater. It will be seen from the discussion of the ^{36}Cl determinations that these waters possibly have a much higher residence time than suggested by the ^{14}C of the TDIC.

INTERPRETATION OF ^{34}S AND ^{18}O ANALYSES ON AQUEOUS SULPHATES IN THE CRYSTALLINE

Results are available from the deep borehole Böttstein (Table 8 and Figure 9) and from other groundwaters in Northern Switzerland (Figure 10).

At Böttstein the samples from the upper crystalline (792 m, 649 m, 621 m, 619 m, 339 m) and from the Buntsandstein/weathered crystalline (313 m, 316 m) lie along a linear correlation.

A reduction process cannot account for this relationship since the contents in SO_4^{2-} do not decrease along with the increase in heavy isotopes. The line is also not due to a succession of precipitation stages of gypsum since the solutions are not saturated with respect to this mineral. This correlation may thus be the result of mixing. Several aquifers and sulphate sources may be involved in a mixing process:

- Upper marine molasse: borehole Aqui-Zürich.
- Keuper-Sandstone/Dolomite: deep borehole Riniken.

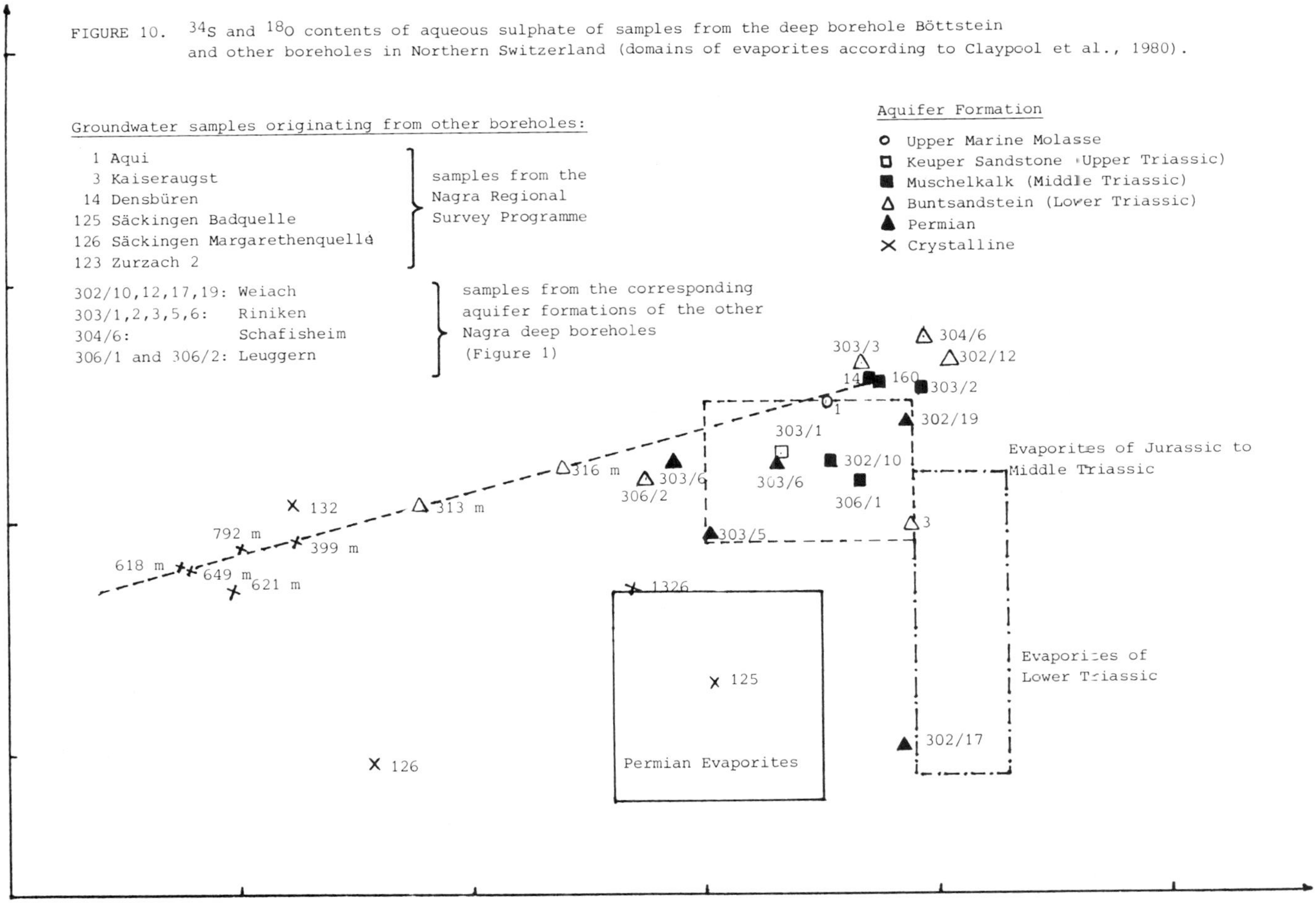

Figure 10. *^{34}S and ^{18}O contents of aqueous sulphate of samples from the deep borehole Böttstein and other boreholes in Northern Switzerland (domains of evaporites according to Claypool et al., 1980).*

Table 8

^{34}S and ^{18}O contents and concentration of aqueous sulphate in groundwaters from the Böttstein borehole (with indication of the saturation index S_I = IAP/KT for gypsum and anhydrite)

Sampling date	Sample No.	Average depth (m)	Aquifer formation	$\delta^{18}O$ (‰ SMOW)	$\delta^{34}S$ (‰ C.D.)	SO_4 (mg l^{-1})	S_I Gypsum	S_I Anhydrite
11/01/82	301/2	163	Muschelkalk	13.70	18.00	2624	0.00	–0.94
01/20/84	301/25b	312.46	Buntsandst./ Crystalline	3.83*	15.35*	287	–1.92	–2.02
11/16/82	301/5	316.56	Buntsandst./ Crystalline	6.92	16.13	359	–1.67	–1.69
12/14/83	301/8	399.35	Crystalline	1.16	14.62	307	–2.13	–2.13
01/22/82	301/12	621.30	Crystalline	–0.15	13.56	336	–2.10	–2.01
08/09/83	301/16	618.40	Crystalline	–1.32	14.09	340	–2.07	–2.05
10/20/83	301/23	649.0	Crystalline	–1.12*	14.01*	339[1]	–2.07[1]	–2.05[1]
08/16/83	301/18	792.4	Crystalline	0.03*	14.45*	339	–2.17	–2.15
08/16/83	301/18	792.4	Crystalline	–0.58	13.80	339	–2.17	–2.15
10/10/83	301/21	1326 (1304)	Crystalline	+8.44*	+13.54*	1560[2]	–0.30	–0.04

[1] Estimated from 301/16 and 301/18 sample
[2] Calculated value for uncontaminated sample (Pearson, 1985)
* SO_4 precipitation and isotope analysis by University of Paris-Sud (Orsay)

For all other samples SO_4 precipitation by Institute Fresenius, isotope analysis by University of Paris-Sud (Orsay).

Table 9

Results of the chlorine-36 analyses of groundwater samples from the Böttstein borehole

Sampling date	Sample average depth (m)	Aquifer formation	$^{36}Cl/Cl_{tot.}$ (10^{-15}at./at.)	Cl^- (mg l^{-1})	^{36}Cl (10^8 at.l^{-1})
30/10-01/11/84	160	Muschelkalk	7.6±3.9	1302	1.7±0.9
20/01/84	312.46	Buntsandst./ crystalline (granite)	13.6±6.9	637.6	1.5±0.8
14/12/82	399.35	Crystalline (granite)	12.9±10.4	120.9	0.3±0.2
09/08/83	618.40	Crystalline (granite)	43.2±9.4	120.5	0.9±0.2
20/10/83	649.0	Crystalline (granite)	48.0±19.2	120.5	1.0±0.4
16/08/83	792.4	Crystalline	29.0±9.7	141.8	0.7±0.2
10/10/83	1326 (1304)	Crystalline (granite)	52.7±10.4	6620	58.4±11.5

- Upper Muschelkalk: deep boreholes Böttstein, Weiach, Riniken, Leuggern, and borehole Densbüren.
- Buntsandstein (in direct contact with the underlying Permian): deep boreholes Weiach, Riniken, Schafisheim, Leuggern, and borehole Kaiseraugst.
- Permian: deep borehole Riniken and Weiach.

Beside these possible end-members for the mixing (Figure 10) the following processes are considered to account for the stable isotope contents of the aqueous sulphates from the upper crystalline in the Böttstein borehole (see Figure 9):

a) A reduction of sulphates of sedimentary origin.
b) An oxidation of the reduced species of sulphur; this oxidation involves the oxygen of the deep groundwater.

Because of the presence of highly impermeable evaporitic series (of anyhdrite and salt deposits) within the normal lithological sequence of the middle Muschelkalk, no direct hydraulic connection is expected between the upper parts (Muschelkalk, Keuper and Molasse) and the lower parts (Buntsandstein and crystalline basement).

A mixing of groundwaters from the Muschelkalk and from the crystalline could only occur in zones, where the hydraulic communication is possible through tectonic fault- or over-thrust-zones. A mixing involving groundwaters from the Permian, the Buntsandstein and the upper part of the Böttstein granite appears probable. The Buntsandstein deposits cover the whole region as a thin layer, about 5 to 15 m layer, sometimes directly overlying the crystalline basement but also lying directly on the Permian deposits where they are present.

At the deepest part of the Böttstein granite (1326 m sample), the aqueous sulphate has an origin which is different from that of the SO_4^{2-} found in the upper parts of the

Table 10

Results of ^{36}Cl age calculations according to various assumptions on the neutron production

Sample average depth (m)	Aquifer formation	Age (ka) according to the following assumptions on the value of the secular equilibrium Req [10^{-15}]					
		53 [10^{-15}]	66 [10^{-15}]	70 [10^{-15}]	140 [10^{-15}]	300 [10^{-15}]	600 [10^{-15}]
312.46	Buntsandstein/ crystalline	129	100	94	44	20	10
399.35	Crystalline (granite)	121	95	89	42	19	9
618.40	Crystalline (granite)	734	462	417	160	68	32
649.0	Crystalline (granite)	1026	565	503	183	76	36
792.4	Crystalline (granite)	344	252	233	101	44	22
1326 (1304)	Crystalline (granite)	in secular equilibrium (~1000)	696	608	205	84	40

crystalline. The ^{34}S and ^{18}O contents of the sulphate of this 1326 m sample are in the range of the primary evaporitic sulphates of Permian and Triassic ages (Figure 10). A direct supply of groundwaters originating from the Permocarboniferous trench into the deep crystalline basement is probable. This explanation is in agreement with the very high mineralization of groundwaters in the deep crystalline and also with the interpretation of their ^{36}Cl contents.

INTERPRETATION OF THE ^{36}Cl MEASUREMENTS

With its half-life of ~300,000 years ^{36}Cl is well suited for the study of old groundwaters. Chlorine determinations have been performed on groundwater samples with very low ^{14}C contents. The results of the ^{36}Cl analyses are reported in Table 9.

For the interpretation of the measured ^{36}Cl concentrations in a groundwater, the respective supplies of different sources have to be evaluated (Fontes et al., 1984):

a) Cosmic ray production: the total fall-out (dry and wet precipitations) is latitude dependent and corresponds to a production rate of 5 to 30 at. m^{-2} s^{-1}.
b) Man-made ^{36}Cl through nuclear tests performed in marine environments from 1953 to 1957.
c) Subsurface spallation by secondary neutrons of the cosmic radiation. This production depends on altitude, latitude, depth in soils, and ^{35}Cl availability for the $^{35}Cl(n,\gamma)$ reaction in pore water from soils.
d) Deep production through nuclear reactions involving neutrons produced in situ.

At the upper part of the crystalline of the Böttstein borehole, all groundwater samples have very low tritium concentrations. This is in agreement with the low amounts of contamination by drilling fluid (deionized water) estimated with artificial tracers (fluorescein, MTFBA). Any significant contribution of ^{36}Cl produced during nuclear tests can thus be neglected.

For the deepest (1326 m) groundwater sample of the Böttstein borehole, the contamination by the drilling fluid was about 50 to 60%. Nevertheless, a man-made ^{36}Cl contribution is precluded because drilling operations were performed with deionized water.

The present day ^{36}Cl cosmic supply (about 26 at. m^{-2} s^{-1} at the latitude of Switzerland) would lead to a concentration in groundwaters of about $5.5 \cdot 10^5$ atoms of ^{36}Cl per liter (with precipitation and evapotranspiration of the order of 2 m and 0.5 m per year, respectively). We assume that the order of magnitude was the same at infiltration time. The ^{36}Cl concentrations in the Böttstein groundwater are about two orders of magnitude higher. These high values can only be explained by a deep production.

The steady state equilibrium value (production equilibrated by decay) of the atomic ratio R_{eq} is reached after a long time (secular equilibrium). It can be calculated according to the following expression:

$$R_{eq} = \frac{10^3\ 0.7577\ \Phi_n \sigma^{35}_{n-\gamma}}{Na \cdot \lambda \Sigma m_i \sigma^{i}_{n-\gamma}} \quad (16)$$

where: N_a = the Avogadro's number
λ = decay constant of ^{36}Cl ($2.3 \cdot 10^{-6}$ a^{-1})
Φ_n = neutron flux production per time unit
0.7577 = isotopic abundance of ^{35}Cl
m_i = concentration of element i (m mol per kg of rock)
$\sigma^{i}_{n-\gamma}$ = neutron cross-section for atom i (b)
$\sigma^{35}_{n-\gamma}$ = 44 b

For the estimation of the neutron production, the detailed chemistry of the rock and especially the rare earth concentrations which have a very high neutron cross-section (Sm, Gd, etc.) should be known. Since the detailed chemical composition of the Böttstein granite is not yet available, the following empirical formula according to Feige et al. (1968) has been applied:

$$P = \rho\,[0.4764\,(U) + 1.57\,(U) + 0.7\,(Th)]\ \text{neutrons cm}^{-3}\ a^{-1} \quad (17)$$

where (U) and (Th) are the concentrations (ppm) in uranium and thorium, and ρ is the rock density (g/cm^{-3}). With the average concentrations of the Böttstein granite (U) = 10 ppm and (Th) = 20 ppm, a production rate of P = $7.56 \cdot 10^4$ $n \cdot kg^{-1}$ a^{-1} is calculated.

With this figure for the neutron production rate, a secular equilibrium value of $R_{eq} = 66 \cdot 10^{-15}$ is estimated. This is close to the observed value in the deep granite of Böttstein (1326 m sample). In the deep crystalline the chloride from the groundwaters is in or close to secular equilibrium with the rock. The residence time of the chloride in the hard rock would be of the order of 10^6 years at least.

For the other groundwater samples from the granite, the chloride is not in secular equilibrium with the granite. The ratio between the measured value and the calculated secular equilibrium value depends on the residence time of the chloride within the Böttstein granite.

The time dependence is given by

$$R = R_o e^{-\lambda t} + R_{eq}(1-e^{-\lambda t}) \quad (18)$$

where, as seen before, R_o, the initial cosmic supply, can be neglected.

Age calculations according to various assumptions on the neutron production are given in Table 10. Groundwater migration from the Permo-carboniferous trench into the crystalline basement is in agreement with the local geological and hydrogeological situations (Figures 1 and 2). Such a flow is also consistent with the hydraulic potential distribution resulting from hydrodynamic simulation models (Kimmeier et al., 1985; Nagra, 1985). The distance is about 2 km along a flow line between the Böttstein well and the Permo-carboniferous trench. An average permeability of the deep granite along this distance can thus be estimated:

- With a residence time of about 10^6 years (achieved secular equilibrium) an average effective flow porosity (storativity) of $4 \cdot 10^{-6}$ given by the hydraulic tests ($8 \cdot 10^{-6}$ to $5 \cdot 10^{-7}$) and a hydraulic gradient of 10^{-3}, a hydraulic conductivity of about $2.5 \cdot 10^{-13}$ m s^{-1} is calculated.
- For a minimum value of $40 \cdot 10^3$ years for the residence time of the chloride within the granite (Table 10) the same calculation gives a maximum value of about $6 \cdot 10^{-12}$ m s^{-1} for the hydraulic conductivity.

These values of the hydraulic conductivity, obtained from ^{36}Cl measurements, compare rather satisfactorily with the data from hydraulic tests within the deeper part of the granite of the Böttstein borehole. Furthermore, this interpretation implies that the chloride is of sedimentary origin through a migration from the Permo-carboniferous trench. This would be in agreement with the interpretation of the ^{34}S and ^{18}O contents of the aqueous sulphate.

CONCLUSIONS: INTERPRETATION OF THE FLOW SYSTEMS WITHIN THE CRYSTALLINE BASEMENT OF NORTHERN SWITZERLAND

The isotopic and hydrogeochemical results on the groundwaters from the crystalline of the investigated deep boreholes and the regional hydrogeological survey lead to the following conclusions on the hydrodynamic flow systems:

1) Waters with low TDS at the upper part of the crystalline:
   - Groundwaters from the upper 700 m of the crystalline basement are recharged in the granitic outcrops of the Black Forest massif.
   - An evolutionary path of the total dissolved inorganic carbon starting from an open system in the recharge zone is suggested by ^{13}C and ^{18}O contents of fracture calcites.
   - Corrected ^{14}C activities of the TDIC indicate residence time values within the range of 2000 years to greater than 20,000 years.
   - The $\delta^{34}S$ and $\delta^{18}O$ contents of the dissolved sulphates indicate a mixing between (i) a sulphate produced by the reoxidation of a previously reduced sulphate, and (ii) a sulphate of primary evaporitic origin. The latter is of general occurrence in regional groundwaters: Upper marine Molasse, Keuper sandstone, Upper Muschelkalk, Buntsandstein, and Permian at the deep borehole Weiach and Riniken. Permian deposits are the original source of this sedimentary sulphate which migrated in solution according to local groundwater flow paths. The regional aquifer Buntsandstein plays a major role in the achievement of the mixing which may hve occurred in regions where the Permian deposits are, and probably were, absent.
2) Waters with high TDS at the lower part of the crystalline. These groundwaters are of the Na-Ca-Cl-SO_4 type, with ^{18}O and ^{2}H contents far out of the range of the present regional meteoric waters. They could represent migrated groundwaters of sedimentary origin, e.g. brines from the Permo-carboniferous trench with long residence times according to the ^{36}Cl measurements.

These conclusions are in agreement with the very low permeability values determined from the in-situ hydraulic test.

This general hydrodynamic flow pattern is also confirmed by the simulation models and the hydrochemistry of the various groundwaters of Northern Switzerland (see Pekdeger and Balderer, this volume). A relatively fast flow of low mineralized waters occurs from the Black Forest Massif at NW to and beneath the Rhine river valley in the upper part of the crystalline of the deep borehole Böttstein at SE. A very low flow of highly mineralized groundwaters occurs from the Permo-carboniferous trench at SE to the deeper crystalline basement at NW. Another flow component coming from South, beyond the Permo-carboniferous is predicted by the hydrodynamic model but was not identified from the presented isotopic survey.

REFERENCES

Balderer, W., 1985. The NAGRA investigation project for the assessment of repositories for high-level radioactive waste in geological formations: Mineralogical Magazine, April 1985, v. 49, p. 281-288.

Balderer, W., 1985. Sondierbohrung Böttstein, Ergebnisse der Isotopenuntersuchungen zur hydrogeologischen Charakterisierung der Tiefergrundwässer: Nagra Technical Report 85-06, January 1985.

Bottinga, Y.J., 1968. Calculation of fractionation factors for carbon and oxygen exchange in the system calcite-carbon dioxide-water: Journal of Physical Chemistry, v. 72, p. 800-808.

Bottinga, Y.J., 1969. Calculated fractionation factors for carbon and hydrogen isotope exchange in the system calcite-carbon dioxide-graphite-methane-hydrogen-water vapor: Geochimica et Cosmochimica Acta, v. 33, p. 49-64.

Claypool, G.E., Holser, W.T., Kaplan, I.R., Sakai, Hitoshi and Zak, Israel, 1980. The age of sulphur and oxygen isotopes in marine sulphate and their mutual interpretation: Chemical Geology, v. 28, p. 199-260.

Cortecchi, G., 1973. Oxygen-isotope variation in sulphate ions in the water of some Italian lakes: Geochimica et Cosmochimica Acta, v. 37, p. 1541-1542.

Craig, H., 1953a. The geochemistry of the stable carbon isotopes: Geochimica et Cosmochimica Acta, v. 3, p. 53-92.

Craig, H., 1953b. Isotopic geochemistry of hot springs: Geological Society of America Bulletin, v. 64, p. 1410.

Deines, P., Langmuir, D. and Harmon, R.S., 1974. Stable carbon isotope ratios and the existence of a gas phase in the evolution of carbonate groundwaters: Geochimica et Cosmochimica Acta, v. 38, p. 1147-1164.

Deines, P., 1980. The isotopic composition of reduced organic carbon: In: Handbook of Environmental Isotope Geochemistry (Eds. P. Fritz and J.-Ch. Fontes), Vol. I, Elsevier, Amsterdam.

Fontes, J.-Ch., 1983. Dating of groundwater: In: Guidebook on Nuclear Techniques in Hydrology 1983. IAEA, Vienna, 1983.

Fontes, J.-Ch. and Gonfiantini, R., 1967. Fractionnement isotopique de l'hydrogène dans l'eau de cristallisation du gypse: Comptes Rendus, Séance de l'Academie des Sciences, Série D, v. 265, p. 4-6.

Fontes, J.-Ch., Bortolami, G.C. and Zuppi, G.M., 1978. Hydrologie isotopique du massif du Mont-Blanc: Isotope Hydrology 1978 (Proc. Symp. Neuherberg, 1978), Vol. 2, Vienna.

Fontes, J.-Ch. and Garnier, J.M., 1979. Determination of the initial ^{14}C activity of the total dissolved carbon: A review of the existing models and a new approach: Water Resources Research, v. 15, p. 399-413.

Fontes, J.-Ch. and Michelot, J.L., 1983. Stable isotopes geochemistry in groundwater systems from Stripa: In: Geochemical and Isotope Characterization of the Stripa Groundwaters-Progress Report. SKB Stripa Project Report 83-01, p. 74-98.

Fontes, J.-Ch. and Garnier, J.M., 1984. Unpublished manuscript.

Fontes, J.-Ch., Brissand, I. and Michelot, J.L., 1984. Hydrological implications of deep production of chlorine-36. Nuclear Instruments and Methods in Physics Research B5, North-Holland, Amsterdam, p. 303-307.

Frape, S.K., Fritz, P. and McNutt, R.H., 1984. Water-rock interaction and chemistry of groundwaters from the Canadian Shield. Geochimica et Cosmochimica Acta, v. 48, p. 1617-1627.

Friedman, I. and O'Neill, J.R., 1977. Compilation of stable isotope fractionation factors of geochemical interest: In: Data of Geochemistry (Ed. M. Fleischer). U.S. Geological Survey Professional Paper, 440-KK: 1-12 (6th ed.).

Fritz, P., 1984. The carbon and oxygen isotopic composition of aqueous carbonate and calcites in Stripa groundwater and rocks: SKB Stripa Project Report 85-06, Stockholm, Sweden, p. 8:1-8:20.

IAEA, 1983. Isotope techniques in the hydrogeological assessment of potential sites for the disposal of high-level radioactive wastes: Technical Report Series No 228, Vienna.

IAEA, 1983. Palaeoclimates and palaeowaters: A collection of environmental isotope studies: Proceedings of an Advisory Group Meeting, Vienna, 25-28 Nov. 1980.

Ingerson, E. and Pearson, F.J., 1964. Estimation of age and rate of motion of groundwater by the ^{14}C method: In: Recent Researches in the Fields of Hydrosphere, Atmosphere and Nuclear Geochemistry, p. 263-283, Maruzen, Tokyo.

Kimmeier, F., Perrochet, P., Andrews, R. and Kiraly, L., 1985. Simulation par modèle mathématique des écoulements souterrains entre les Alpes et la Forêt Noire; Partie A: Modèle régional, Partie B: Modèle local (nord de la Suisse): Nagra Technical Report 84-50, January 1985.

Kiraly, L., 1985. A three dimensional model for groundwater flow simulation: Nagra Technical Report 84-49, January 1985.

Lloyd, R.M., 1968. Oxygen isotope behaviour in the sulphate-water systems: Journal of Geophysical Research, v. 73, p. 6099-6110.

Michelot, J.L., 1982. Evolution temporelle des teneurs en isotopes stables et en tritium des eaux du tunnel du Mont-Blanc. Estimation des paramètres hydrauliques à l'échelle d'un massif cristallin: Thèse de doctorat, Université de Paris-Sud, Orsay.

Michelot, J.L., Bentley, H.W., Brissaud, I., Elmore, D. and Fontes, J.-Ch., 1983. Proceedings International Symposium on Isotope Hydrology in Water Resource Development. IAEA, Vienna, 1983.

Mook, W.G., 1980. Carbon-14 in hydrological studies: In: Handbook of Environmental Isotope Geochemistry (Eds. P. Fritz and J.-Ch. Fontes), Vol. 1, Elsevier, Amsterdam.

NAGRA, 1985. Nukleare Entsorgung Schweiz: Konzept und Uebersicht über das Projekt Gewähr 1985: Nagra Technical Report 85-01, Baden, January 1985.

NAGRA, 1985. Endlager für hochaktive Abfälle: Das System der Sicherheitsbarrieren: Nagra Technical Report 85-04, Baden, January 1985.

NAGRA, 1985. Sondierbohrung Böttstein – Untersuchungsbericht: Nagra Technical Report NTB 85-01.

Nielsen, H., 1979. Sulphur isotopes. In: Lectures in Isotope Geology (Eds. E. Jäger and J.C. Hunziker). Springer-Verlag.

Nordstrom, D.K., Andrews, J.N., Carlsson, L., Fontes, J.-Ch., Fritz, P., Moser, H. and Olsson, T., 1985. Hydrogeological and hydrogeochemical investigations in boreholes – final report of the phase I geochemical investigations of the Stripa groundwaters: SKB Stripa Project Report 85-06, July 1985, 250 p.

O'Neil, J.R., Clayton, R.N. and Mayeda, T.K., 1969. Oxygen isotope fractionation in divalent metal carbonates: Journal of Chemical Physics, v. 51, p. 5547-5558.

Pearson, F.J., 1965. Use of C-13/C-12 ratios to correct radiocarbon ages of material initially diluted by limestone: In: Radiocarbon and Tritium Dating. Proceedings of Sixth International Conference on Radiocarbon, Pullman, Washington, p. 357-366.

Pearson, F.J. and Rightmire, C.T., 1980. Sulphur and oxygen isotopes in aqueous compounds: In: Handbook of Environmental Isotope Geochemistry (Eds. P. Fritz and J.-Ch. Fontes), Vol. 1, Elsevier, Amsterdam, p. 227-258.

Pearson, F.J., 1984. Geochemistry of waters from the Böttstein borehole. Prepared for Nagra, 5 October 1984 (unpublished).

Pearson, F.J., 1985. Sondierbohrung Böttstein – Results of hydrochemical investigations: Analysis and interpretation: Nagra Technical Report 85-05, January 1985.

Pekdeger, A. and Balderer, W., 1986. The occurrence of saline groundwaters and gases in the crystalline rocks of Northern Switzerland: (this volume).

Peters, T.J., Matter, A., Bläsi, H.R., Gautschi, A., Thury, M., Schlanke, S. and Müller, W.H., 1985. Sondierbohrung Böttstein: Geologie: Nagra Technical Report 85-02.

Plummer, L.N., Jones, B.F. and Truesdell, A.H., 1976. WATEQF, a fortran IV version of WATEQ, a computer program for calculating chemical equilibrium of natural waters: U.S. Geological Survey Water Resources Investigation, 76-13.

Reardon, E.J. and Fritz, P., 1978. Computer modelling of groundwater ^{13}C and ^{14}C isotope compositions: Journal of Hydrology, v. 36, p. 201-224.

Saliège, J.F., 1979. Détermination expérimentale du fractionnement isotopique $^{14}C/^{13}C$ au cours des processus naturels: Mémoire CNAM, Université de Paris, 80 p.

Savin, S.M., 1980. Oxygen and hydrogen isotope effects in low temperature mineral-water interactions: In: Handbook of Environmental Isotope Geochemistry (Eds. P. Fritz and J.-Ch. Fontes), Vol. 1, Elsevier, Amsterdam, p. 283-328.

Schmassmann, H., Balderer, W., Kanz, W. and Pekdeger, A., 1984. Regionale Untersuchungen über die Beschaffenheit der Tiefengrundwässer: Nagra Tecnical Report 84-21.

Siegenthaler, U., Schotterer, U. and Oeschger, H., 1983. Sauerstoff-18 und Tritium als natürliche Tracer für Grundwasser: Gas-Wasser-Abwasser, v. 63, p. 477-483.

Tamers, M.A., 1967. Surface-water infiltration and groundwater movement in arid zones of Venezuelas: In: Isotopes in Hydrology, IAEA, Vienna, p. 339-395.

Tamers, M.A. and Scharpenseel, H.W., 1970. Sequential sampling of radiocarbon in groundwater. Isotope Hydrology 1970, IAEA, Vienna, p. 241-256.

Tamers, M.A., 1975. Validity of radiocarbon dates on groundwater: Geophysical Surveys, v. 2, p. 217-239.

Vuataz, F.-D., 1982. Hydrogéologie, géochimie et géothermie des eaux thermales de Suisse et des régions alpines limitrophes: Matériaux pour la Géologie de la Suisse, Hydrologie, 29.

Wigley, T.M.L., 1975. Carbon-14 dating of groundwater from closed and open systems: Water Resources Research, v. 11, p. 324-328.

Wigley, T.M.L., 1976. Effect of mineral precipitation on isotopic composition and ^{14}C dating of groundwater: Nature, v. 263, p. 219-221.

Wigley, T.M.L., Plummer, L.N. and Pearson, F.J., 1978. Mass transfer and carbon isotope evolution in natural water systems: Geochimica et Cosmochimica Acta, v. 42, p. 1117-1139.

Radiolytic Salt Enrichment and Brines in the Crystalline Basement of the East European Platform

I.F. Vovk

Division of Research and Laboratories, International Atomic Energy Agency, A-1400 Vienna, Austria

Vovk, I.F., Radiolytic salt enrichment and brines in the crystalline basement of the East European Platform; in Saline water and gases in crystalline rocks, Editors: Fritz, P., and Frape, S.K.; Geological Association of Canada Special Paper 33, 1987. p. 197–210.

Abstract

Saline waters and brines have been encountered repeatedly in the crystalline basement of the East European Platform, in particular within the Ukrainian and Baltic shields and the Russian Plate. Samples of brines and associated gases were obtained from several dozen mines and boreholes, two of which, in Tatar Dome and Kola Peninsula, pass through shear zones and fractures filled with brine to the depth of 5099 m and 12,000 m, respectively. Highly concentrated brines and gases are also often found as fluid inclusions in minerals and igneous and metamorphic rocks. The concentration of total dissolved solids (TDS) in water sampled from mines and boreholes increases with depth from 0.1 to 100 g l^{-1} at 0 to 1000 m to more than 300 g l^{-1} at 2000 to 3000 m and deeper. The brines are of two major chemical types: Na-Cl and Na-Ca-Cl. The Na-Cl brine is found mainly along the Krivoy Rog regional fault crossing the Ukrainian Shield, while the Na-Ca-Cl brine is more widely distributed within the platform. The major component of gas dissolved in Na-Cl waters from iron mines in Krivoy Rog is CO_2, accompanied by H_2 and He, while Na-Ca-Cl brine from the borehole in Tatar Dome contains mainly N_2 and CH_4, accompanied by He and H_2. Free gases associated with brines and filling the porous media of crystalline rocks are predominantly of H_2-CH_4 or CH_4-H_2 composition with smaller amounts of N_2 and He. The chemical and isotopic data, as well as geological position suggest that saline waters of Na-Cl type from the Krivoy Rog region may be diluted Paleozoic sea brine. The source of the Na-Ca-Cl brines and major gases, particularly H_2, is not well understood but could be residual metamorphic fluids originated from Precambrian sea water highly modified radiolytically and/or in some other way. Radiolytic salt enrichment through the decomposition of water molecules owing to the natural irradiation of water-rock systems during spans of geological time is discussed on the basis of experimental data and geochemical evidence.

Résumé

Des eaux salées et sursalées ont été observées à de nombreuses reprises dans le socle cristallin de la plateforme de l'Europe de l'est, en particulier à l'intérieur des boucliers ukrainien et baltique et de la plaque de la Russie. Des échantillons d'eaux salées et de gaz ont été prélevés dans plusieurs dizaines de mines et de trous de forage. Deux d'entre eux, provenant du dôme des Tatars et de la péninsule de Kola, traversent des zones de cisaillement et des fractures contenant de l'eau sursalée situées à des profondeurs respectives de 5099 m et 12 000 m. Assez fréquemment, des gaz et des eaux sursalées très concentrés se rencontrent aussi dans les inclusions fluides des minéraux des roches ignées et métamorphiques. La concentration de l'ensemble des solides dissous (TDS) dans l'eau des mines et des trous de forage augmente avec la profondeur; de 0,1 à 100 g l^{-1} entre 0 et 1000 m, elle atteint plus de 300 g l^{-1} entre 2000 et 3000 m et à des profondeurs supérieures. Les eaux sursalées se divisent en deux groupes principaux d'après leur composition chimique,

Na-Cl et Na-Ca-Cl. Les eaux sursalées riches en Na et Cl se trouvent principalement en bordure de la faille régionale Krivoy Rog, qui traverse le bouclier ukrainien, tandis que les eaux riches en Na, Ca et Cl sont réparties plus uniformément sur la plateforme. Le gaz dissous dans les eaux du type Na-Cl des mines de fer de Krivoy Rog est principalement constitué de CO_2. On y trouve aussi du H_2 et du He. Par ailleurs, les eaux sursalées du type Na-Ca-Cl des forages effectués sur le dôme des Tatars sont surtout constituées de N_2, de CH_4 et de quantités moindres de He et H_2. Les gaz libres associés aux eaux sursalées et comblant les pores des roches cristallines sont surtout composés de H_2-CH_4 ou de CH_4-H_2, ainsi que de quantités moindres de N_2 et He. Les données chimiques et isotopiques, ainsi que le contexte géologique, indiquent que les eaux salées du type Na-Cl de la région de Krivoy Rog pourraient être des eaux marines diluées d'âge paléozoïque. On ne connaît pas encore avec certitude la source des eaux sursalées du type Na-Ca-Cl ni des principaux gaz, surtout H_2, mais elle pourrait être associée à des fluides métamorphiques résiduels dérivés d'une eau marine précambrienne ayant subi de profondes modifications, d'origine radiolytique ou non. L'enrichissement salin d'origine radiolytique, résultant de la décomposition des molécules d'eau à la suite de l'irradiation naturelle des régimes eau-roche durant certains intervalles géologiques, sera traité en fonction des données expérimentales et des observations géochimiques.

INTRODUCTION

Saline waters with concentration of total dissolved solids (TDS) 10-100 g l^{-1} in crystalline rocks of the East European Platform, particularly in the Ukrainian Shield (Krivoy Rog iron mines), were known long ago (Makov, 1948). Lately it was established, by drilling of a number of deep boreholes, that saline waters and brines (with TDS up to 333 g l^{-1}) are widespread geological phenomena in the basement of the Platform. Besides the Ukrainian Shield, they have been encountered within the Baltic Shield and the part of the Platform covered by sedimentary rocks, known as the Russian Plate. Samples of very saline waters and brines and associated gases were obtained from several dozens of mines and boreholes, two of which, one on the Tatar Dome (Russian Plate) and the other on the Kola Peninsula (the Baltic Shield) passed through shear zones and fractures filled with brines right up to the depths of 5099 m and 12,000 m, respectively.

The borehole on the Kola Peninsula continues its successful penetration into the earth's crust (project depth 15 km) after reaching a depth of 12 km at the end of 1984. It has surprised geologists with a number of unexpected discoveries (Kozlovsky, 1984). For example, on the basis of seismic studies, it was implied that at the depth of 7 km the rocks would change from granitic to basaltic. But approximately at this depth metamorphic rocks of clearly sedimentary origin change to granites which continue to the present depth of 12 km. The geothermal gradient in the borehole also appeared to be significantly higher than assumed for such a stable shield area (the measured temperature at the depth of 11 km is about 200°C). But perhaps the most unexpected discovery is that the borehole met shear zones and fractures in crystalline rocks filled with brines at very large depths. Actually the drilling revealed the occurrence of gas-saturated brines at all intersected depths. At present the geochemical study of these brines has not been accomplished, and results are to be published elsewhere.

Even the available qualitative information on the occurrence of brines at the great depths in the Kola borehole has a very important scientific value, because it provides evidence that the brines in the crystalline basement may be deep crustal fluids which should occur at other localities within the earth's crust in comparatively large amounts. In the light of this discovery, the often encountered fluid inclusions in minerals of igneous and metamorphic rocks also rise in importance as a source of geochemical information on residual crust fluids and as a tool to distinguish them from saline waters and brines, which could penetrate in the upper part of the basement from the overlying sedimentary cover.

The problem of formation (origin and chemical evolution) of brines in the crystalline basement of the earth's crust is fundamental not only for geochemistry but also for other related geological sciences. It is also becoming more and more important for the solution of some of the urgent practical tasks, particularly the study of radioactive waste disposal. More data are required for its solution and this is one of the reasons for drilling very deep boreholes in crystalline rocks. Drilling of a new deep well (15,000 m) has been started in the USSR in the Krivoy Rog region. It is apparent that all the available data and existing approaches to their interpretation should be reviewed and developed into an integrated theory before or parallel with embarking on such an expensive undertaking as drilling of super deep boreholes.

The author's objective is to outline in this article the major results of the study of the saline waters and brines within the crystalline basement of the East European Platform and to continue the discussion of their genesis and geochemical evolution (Vovk, 1979a, 1981, 1982; Frape et al., 1984) with special emphasis on radiolytic salt enrichment caused by the natural radioactivity as a factor of water decomposition and water-rock interaction. The description of the geological and hydrogeological conditions of saline water and brine occurrences and of the analytical methods used for their studies can be found in Greiser et al. (1970), Voytov (1974), Zorkin et al. (1977), Vovk (1982). A brief summary of the description of sampling sites is presented below.

SAMPLING SITES

Deep boreholes and mines have penetrated the crystalline basement in a variety of locations (Fig. 1, Table 1) within three major structural units of the Platform: the Baltic Shield (A, B, 2-4), the Ukrainian Shield (C, D, 12, 13), and the Russian Plate (1, 5-11).

On the Baltic Shield, sampling sites are related to mines of the Sopcha copper-nickel deposit and three boreholes on the south-eastern margin of the Shield. The Sopcha massive is a part of a large gabbro pluton intruded into Archean and Proterozoic metamorphic complex (schists, gneisses). Groundwaters in the mine workings are fresh to the depths of 300-400 m, at depths of 400-700 m they are brackish, and deeper they change to saline. The boreholes on the south-

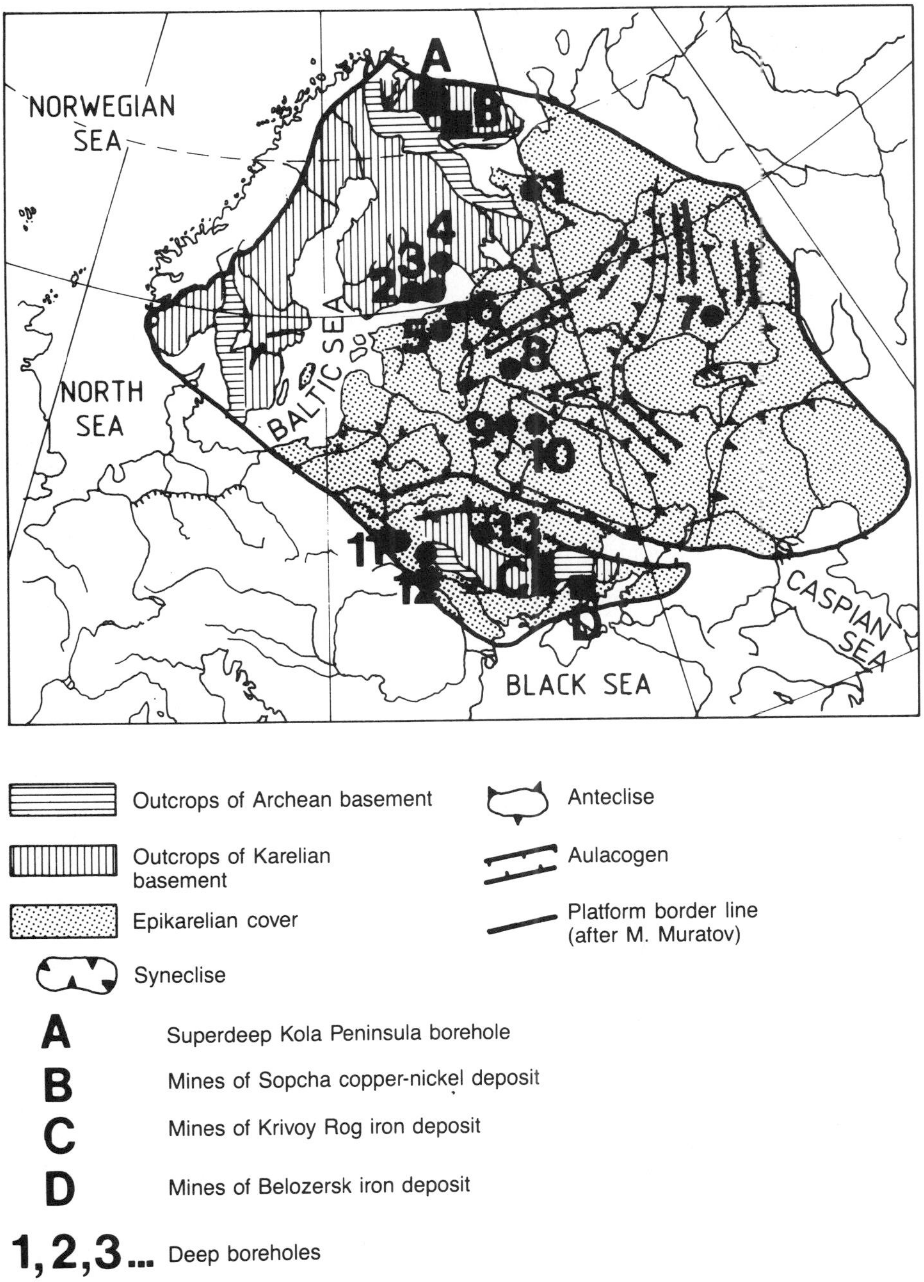

Figure 1. *Location of boreholes and mines sampled for saline waters and brines in the crystalline basement of the East European Platform.*

eastern margin of the Shield encountered the basement at depths of 185-220 m and the saline waters and brines at depths of 583-735 m (Table 1, locations 2-4).

On the Ukrainian Shield, the main location of saline waters and brines is the Krivoy Rog iron basin, which geologically represents an Archeozoic/Proterozoic geosyncline broken by a regional fault, crossing the Ukrainian Shield and the basement of the adjacent Dnieper-Donetsky aulacogen (location C).

Though mine workings have influenced the natural hydrogeological situation in the basin, three vertical hydrogeochemical zones can be distinguished: 0-200 m – bicarbonate-sulfate fresh water zone; 200-400 m – sulfate-chloride brackish water zone; 400 m and deeper – sodium chloride saline water and brine zone (Table 2). Three other locations are 12 – a borehole on the western slope of the Shield, 13 – the same on the eastern one, D – mines of Belozersk iron deposit on the southern slope. Location 13 is a deep borehole (1927 m), which encountered the basement at a depth of290 m. Here water sampling showed an increase of TDS concentration with a depth from 17 g l^{-1} (360 m) to 200 g l^{-1} (1416 m).

Within the Russian Plate, the boreholes encountered the basement at depths of 600 m (location 1) – 2500 m (location 11) and penetrated it down to a depth of 5099 m (location 7). At location 7, hydrogeological and geochemical investigation of brine and dissolved gases was carried out within the interval 4700-5099 m (Zorkin et al., 1977). In this interval,

Table 1

Major chemical features of saline waters and brines in the crystalline basement of the East-European Platform.

Location (see Fig. 1)	Depth (m)	TDS* g l^{-1}	Chemical composition eq %
B (Sopcha massive)	800	57	$\frac{Cl\ 100}{(Na + K)\ 85\ Mg\ 12\ Ca\ 3}$
1 (Archangelsky region)	670	69	$\frac{Cl}{(Na + K)\ Ca}$
2 (Sertolovo, Leningradsky region)	583	41	$\frac{Cl\ 100}{(Na + K)\ 53\ Ca\ 32\ Mg\ 15}$
3 (Mednoye Ozero, Leningradsky region)	700	94	$\frac{Cl\ 100}{(Ca\ 65\ (Na + K)\ 30\ Mg\ 5}$
4 (Pesochnoye, Leningradsky region)	735	138	$\frac{Cl\ 100}{(Na + K)\ 93\ Ca\ 6\ Mg\ 1}$
5 (Nevel, Pskovsky region)	936	97	$\frac{Cl\ 98\ SO_4 2}{(Na + K)\ 78\ Ca\ 16\ Mg\ 6}$
6 (Pestovo, Novgorodsky region)	1612	228	$\frac{Cl\ 100}{(Na + K)\ 74\ Ca\ 20\ Mg\ 6}$
7 (Tatar ASSR)	5099	333	$\frac{Cl\ 100}{Ca\ 78\ (Na + K)\ 22}$
8 (Radkino, Kalininsky region)	1758	246	$\frac{Cl\ 96\ SO_4 4}{(Na + K)\ 85\ Mg\ 11\ Ca\ 4}$
9 (Smolensky region)	1128	136	$\frac{Cl\ 98\ SO_4 2}{(Na + K)\ 82\ Cl\ 10\ Mg\ 8}$
10 (Kaluzhsky region)	1000	66	$\frac{Cl\ 95\ SO_4 5}{(Na + K)\ 79\ Ca\ 18\ Mg\ 4}$
11 (Gorokhov, Volynsky region)	2535	164	$\frac{Cl\ 100}{Ca\ 60\ (Na + K)\ 37\ Mg\ 3)}$
12 (Satanov, Kmelnitsky region)	643	38	$\frac{Cl\ 100}{(Na + K)\ 62\ Ca\ 33\ Mg\ 5}$
13 (Kievsky region)	800	62	$\frac{Cl\ 98\ SO_4 2}{(Na + K)\ 64\ Ca\ 33\ Mg\ 2}$
C (Mine Rodina, Krivoy Rog)	1090	133	$\frac{Cl\ 98\ SO_4 2}{(Na + K)\ 86\ Mg\ 11\ Ca\ 3}$
D (Belozersk iron mine)	480	63	$\frac{Cl\ 97\ SO_4 3}{(Na + K)\ 58\ Ca\ 27\ Mg\ 15}$

* Total Dissolved Solids (After Greiser et al., 1970; Panova 1976; Zorkin et al., 1977; Vovk, 1982).

the brine-bearing rocks are represented by garnet-pyroxene cataclastic gneisses, intensively altered by hydrothermal processes. The piezometric level of brine reached the steady state at the depth of 300 m after a 7-day relaxation period. The temperature of the water, measured with an electrothermometer, was 90°C at a depth of 4700 m. Samples of brine and dissolved gas were obtained from depth 4680 m after stabilization of specific gravity of the brine in the borehole (its laboratory-measured value at 20°C = 1.22 kg dm^{-3}).

The discharges of saline waters and brines from fractures in mines and boreholes vary within very broad limits – from less than 1 litre per hour to more than 300 litres per minute. Brief test pumping of the brine from the deep borehole in the Tatar Dome with the drawdown of 300 m (600 m dynamic level below land surface) resulted in the flow of 80 litres per minute. At most sites, the discharges decreased drastically in time thus showing that discovered accumulations of brine are isolated from the upper fracture zone of fresh water or saline waters and brines of the sedi-

Table 2

Geochemistry of representative groundwaters from mines in the Krivoy Rog Iron Basin (the Ukrainian Shield).

Sample no.	Location*	Depth (m)	pH	Na mg l^{-1}	K mg l^{-1}	Ca mg l^{-1}	Mg mg l^{-1}	Cl mg l^{-1}	SO_4 mg l^{-1}	HCO_3 mg l^{-1}	TDS mg l^{-1}	$\delta^{18}O$‰ (SMOW)
1	L	900	7.1		6706	481	669	12467	809	134	21267	− 6.7
2	L	975	5.5		11318	832	1173	21264	1457	18	36062	− 7.3
3	L	975	7.2		10592	732	1137	19732	1531	140	33863	− 6.8
4	L	900	7.4		1766	140	170	2801	633	342	5852	− 8.9
5	L	750	7.2		1356	140	164	2198	608	293	4760	−11.8
6	R	1090	6.8		41034	1323	2790	72598	1542	43	119331	− 1.9
7	R	1090	6.1	33529	530	1343	2180	62050	1689	35	101372	− 4.9
8	R	1090	5.6	45263	771	1443	2943	80842	1825	5	133119	− 3.3
9	R	1090	7.5	22876	347	952	1392	40421	1493	111	67607	− 7.1
10	R	1015	6.6		28720	1273	1933	50980	1582	61	84549	− 5.1
11	D	427	6.9		7953	501	578	13963	925	323	24244	−11.7
12	C	540	7.9	6007	100	526	398	11488	604	245	19384	− 9.8
13	C	540	8.0	6157	103	446	426	11559	614	130	19467	−11.3
14	CI	1123	7.0	31885	365	1373	1332	56675	1237	403	91326	− 5.5
15	CI	1123	7.1	33811	415	1563	1556	58646	1313	354	97730	− 4.0
16	C	100	8.0	45	5	53	20	50	142	107	423	− 9.3

*Mines: L – Lenin, R – Rodina, D – Drenazhnaya, C – Communar, CI – Commintern.

mentary cover (in the Tatar Dome borehole it changed to the Archean crystalline rocks at the depth of 1848 m).

CHEMICAL COMPOSITION

The number of chemical analyses of saline waters and brines in the Krivoy Rog iron basin now exceeds 1000, but only several dozens of such water samples have been collected at other localities and analyzed for their chemical composition. Only about 60 samples have been analyzed for isotope (mainly ^{18}O) content, almost all of them from the Krivoy Rog basin.

All available chemical data, presented partly in Tables 1-3 and Fig. 2-3, show that concentration of TDS in groundwaters in crystalline basement increases irregularly with depths from 0.1-1.0 g l^{-1} (on the Baltic Shield even from 0.05-0.1 g l^{-1}) in shallow wells first slowly to 3-10 g l^{-1} at depths 400-800 m and then sharply to 100-200 g l^{-1} and more (Table 1, locations 6-8).

The highly saline waters and brines are of two main chemical types: Na-Cl and Na-Ca-Cl (or Ca-Na-Cl). All samples of saline waters and brines from the Krivoy Rog basin without exception belong to the Na-Cl type with no major scattering in the relative content of cations (Fig. 2). Samples from the boreholes, brines encountered in other localities, show the wide range of variation of both absolute and relative content of Na and Ca – from purely Na-Cl to amost purely Ca-Cl type.

The chemical data presented in Table 1 and Fig. 3 suggest that the most saline brines at great depths will be Ca-Na-Cl waters. This is an important characteristic feature of brines in the crystalline basement which has long been known for the Canadian Shield under the special term of the North American type of brines (Gurevich, 1964; Frape et al., 1984), and reported for several locations on other platforms of the world (Gavrilenko and Derpholz, 1971), so that now it seems to be global.

As to the other constituents, particularly sulfates, potassium and minor elements, their concentrations in brines are strongly dependent on the local rock and fluid geochemistry and vary from traces to very high amounts. As a rule, due to high concentrations of calcium, causing precipitation of calcium sulfates, the sulfate ions content in groundwater of the crystalline basement is very low. But in a favourable geochemical environment it can reach dozens of grams per litre. For example, very high concentrations of sulfate ions (13 800-75 600 ppm), accompanied by equally high contents of potassium (18 000-42 300 ppm), have been reported for fluid inclusions in granites of Bornholm Island within the Baltic Shield (Roedder, 1972). Among the daughter minerals in fluid inclusions in quartz from pegmatites in the Ukrainian Shield, cryolite, sphalerite, chlorargyrite,

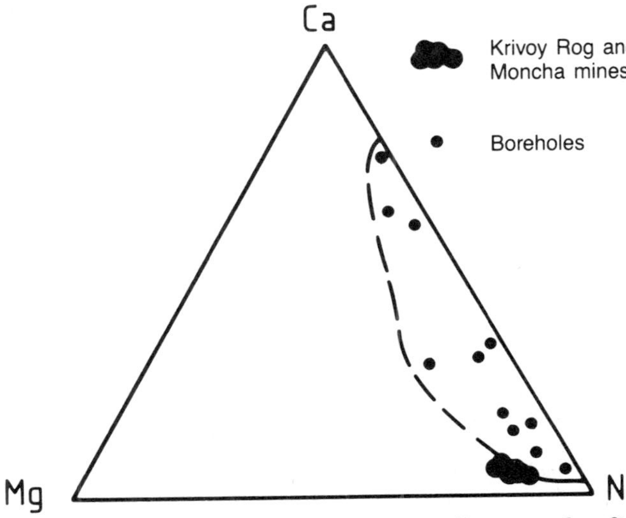

Figure 2. *Equivalent percentage diagram for Ca, Na and Mg concentrations in saline groundwaters and brines in the crystalline basement of the East European Platform.*

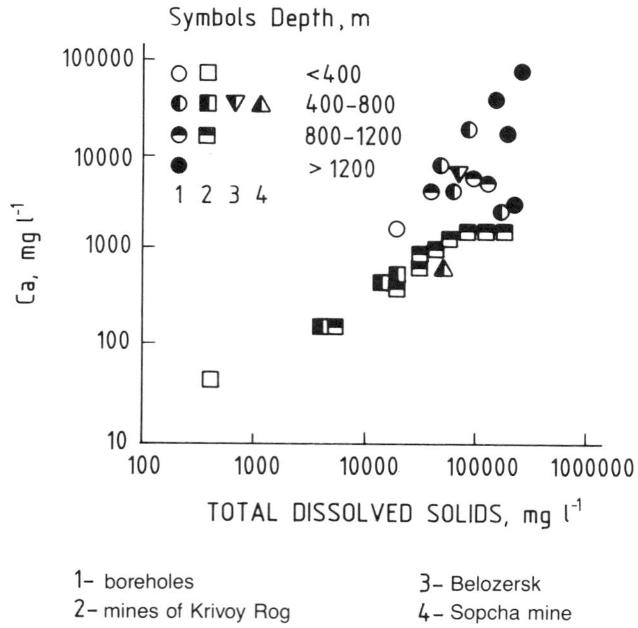

Figure 3. *Calcium concentrations versus TDS in groundwaters of the crystalline basement of the East European Platform.*

halite, sylvite, calcium and iron chlorides as well as other minerals containing ore elements have been identified (Boyarskaya et al., 1977).

ASSOCIATED GASES

The content of the major constituents of gases associated with saline waters and brines in the crystalline basement of the East European Platform varies with locations and geochemical media (Table 4). Nevertheless, one regularity, that is the increase of groundwater gas saturation factor with depth, followed by changes in chemical composition of the dissolved gases, can be easily noticed. The major component of gas, dissolved in low gas-saturated Na-Cl waters from iron mines in the Krivoy Rog basin, is CO_2, accompanied by H_2 and He with only traces of CH_4 (Table 4, locations CLa, CLb). Much higher (two orders of magnitude) gas-saturated brine from Tatar Dome (Table 4, location 7a) contains mainly N_2 and CH_4, also accompanied by H_2 and He with small relative amounts of CO_2 and O_2. Free gases from the porous parts of crystalline rocks and flowing out in gas jets are predominantly of H_2-CH_4 or CH_4-H_2 composition with relatively high amounts of He and low N_2.

The combustible gases in free gas jets from crystalline basement both in the Ukrainian and Baltic Shields are substantially methanic. The CH_4 content in the gas jet in the Lenin mine (Table 4, location CLc) is 71.40% out of 73.2% of the methane series content. Free gas jet from the

Table 3

Geochemistry of representative saline waters and brines from deep boreholes in the crystalline basement of the East-European Planform. (After Greiser et al., 1970; Panova, 1976; Zorkin et al., 1977, Vovk, 1982)

Location (see Fig. 1 + Table 1)	Depth (m)	pH	Na + K mg l^{-1}	Ca mg l^{-1}	Mg mg l^{-1}	Cl mg l^{-1}	SO_4 mg l^{-1}	HCO_3 mg l^{-1}	Br mg l^{-1}	J mg l^{-1}	TDS mg l^{-1}	$\frac{r\,Na}{r\,Cl}$	$\frac{Br}{Cl}$
2	583	6.7	9070	4730	1340	26240	6	35	56	1	41478	0.53	0.0021
3	700	6.7	11640	21600	1080	59260	44	29	31	2	93686	0.30	0.0004
4	735	—	50577	2790	250	83320	4	20	781	11	137753	0.93	0.0094
7	5099	—	29941	92000	0	209450	0	110	1870	8	333384	0.22	0.0090
13	800	7.1	15802	7129	298	37849	798	40	—	—	61992	0.64	—

Table 4

Geochemistry of representative gases from crystalline basement of the East-European Platform. (After Voytov, 1974; Zorkin et al., 1977; Vovk, 1982)

Location	Depth (m)	Sampled medium	Gas cotent $cm^3\ kg^{-1}$	Chemical composition of gas, % (vol.)					
				CO_2	O_2	H_2	$C_n\ H_{2n+2}$	N_2+Ar	He
7a	5099	Groundwater	370.0	0.1	2.2	8.9	16.1	67.2	5.66
7b	2937	Granite gneiss	1.3	—	—	95.6	4.4	—	—
7c	2750	Dike diabase	57.8	—	—	96.1	3.9	—	—
7d	1918	Drilling mud		23.0	—	1.3	24.4	51.2	0.05
7e	2740	Drilling mud		0.0	—	18.7	1.8	79.5	0.04
7f	2760	Drilling mud		0.0	—	96.9	3.0	0.0	0.15
9a	1079	Groundwater	28.0	3.7	—	13.5	—	80.7	2.08
9b	1128	Groundwater	31.3	2.0	2.7	—	23.6	69.6	2.12
CLa	750	Groundwater	1.8	75.2	—	23.2	0.4	—	1.26
CLb	825	Groundwater	2.2	79.6	—	17.7	0.2	—	2.49
CLc	825	Free gas jet		—	—	13.1	73.2	12.4	1.12

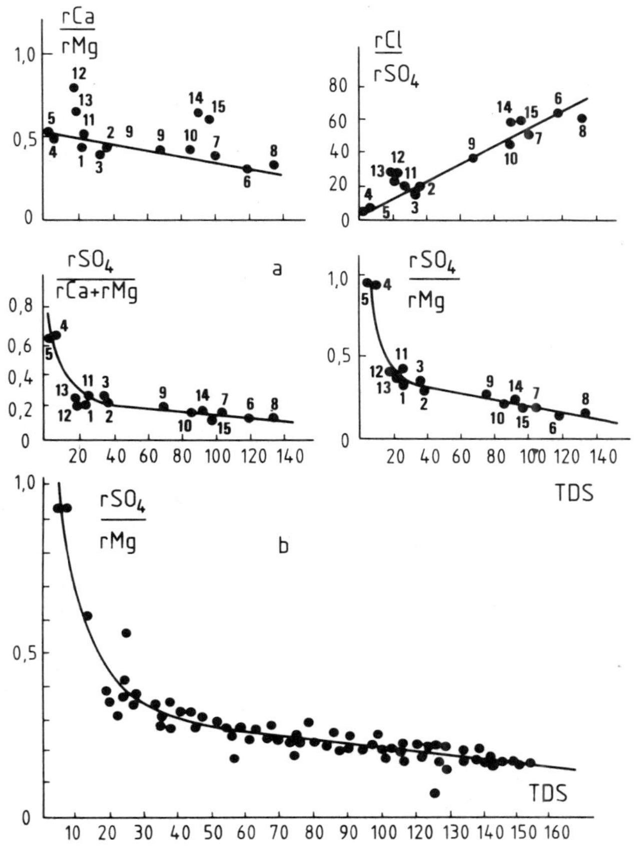

Figure 4. *Major component equivalent ratios versus TDS for Krivoy Rog groundwaters.*

horizontal borehole, drilled at the depth of 300 m in an iron mine in Northern Sweden contained (%): C_2H_{2n+2} – 71.2 (CH_4 – 70.0); N_2 – 19.7; He – 6.0; H_2 – 0.4; CO – 0.8; CO_2 – 1.2 (Kudryavtsev, 1959).

The gas content in the closed porous space of rocks and its chemical composition is dependent to some extent on the rock type. It has been established that gas from the schists in the Krivoy Rog basin contains, besides H_2 and CH_4, elevated amounts of CO_2 and CO. The gas content of the iron-bearing rocks there varies from 0.06 to 21.24 $cm^3\ kg^{-1}$, its major constituents being H_2, CO_2, CO and CH_4 (Vovk, 1982). In the metamorphic rocks of granulite faces in the Kola Peninsula it varies within 2-4 $cm^3\ kg^{-1}$ with H_2 as the major gas, while for igneous rocks the limits of variation are 0.5-15 $cm^3\ kg^{-1}$ and H_2 remains the dominant gas (68-88%). In igneous rocks of some agpaitic massives, particularly of the Khibine massive, the dominant gas is methane and its content reaches up to 200 cm^3 per kilogramme of rock (Petersilye and Pripachkin, 1979; Petersilye et al., 1979).

The gas composition of the crystalline basement encountered in the deep Tatar Dome borehole shows approximately the same features as those of the gases in the Kola Peninsula rocks (Table 4, location 7b). Of special interest here is the comparison of this background of gases from the diabase dike, passed through within the depths 2740-2760 m, and from samples of drilling mud in the vicinity of the dike and in the remote interval (Table 4, locations 7c-7f). One can see that the dike itself and the rock interval washed by the drilling mud immediately under the dike are strongly enriched in H_2 content (more than one order of magnitude). The most probable reason for the enrichment is that the dike is a monolithic barrier for gas flow causing accumulation of H_2 comparatively easily migrating from the greater depths of the basement.

The isotopic composition of the methane carbon from gas jets in Krivoy Rog basin, Kola Peninsula and from gas dissolved in brine of Tatar Dome appeared to be rather light in comparison with marine CO_3, with $\delta^{13}C$ = –4.7 to –5.7% (Voytov, 1974, Zorkin et al., 1977), thus the organic source of the carbon may be assumed.

ORIGIN AND CHEMICAL EVOLUTION

The ^{18}O content and chemical composition of groundwater in the Krivoy Rog basin (Table 2, Fig. 4, where 4a is based on data from Table 2 and 4b – on 72 additional chemical analyses) are assumed to reflect mixing of brines and local fresh waters, caused by natural hydrodynamic processes as well as by mining activities. This assumption is supported by the results of correlation analysis presented in Table 5, which, in turn, form a basis for the reconstruction of the chemical composition of the original brine diluted to the concentrations actually encountered.

Proceeding from the values of Na/Cl and Br/Cl ratios, which in Krivoy Rog brines are close to 0.87 and 0.003, respectively (Vovk, 1982), it can be further assumed that the saturation of brine with respect to halite was not surpassed and thus TDS of the original brine was not higher than 330 $g\ l^{-1}$. Another set of evidences for the reconstruction of the genetic type and chemical composition of the original brine in the Krivoy Rog basin is represented by the isotopic composition of groundwater, which shows a correlation between δD and $\delta^{18}O$ values as well as between isotopic and chemical data (Vovk and Vysotsky, 1982; Sobolev, 1984). On the regression line (the first equation in Table 5),

Table 5

Correlation between ^{18}O and chemical components of groundwater in Krivoy Rog basin: $\delta^{18}O$ – ‰ SMOW, M (TDS) – gl^{-1}, ions – (mg-eq) 1^{-1}

Components	Correlation Coefficient	Regression equation
$\delta^{18}O$ – M	0.885	$\delta^{18}O$ = 0.0623M – 10.65
$\delta^{18}O$ – [Cl^-]	0.884	$\delta^{18}O$ = 0.0036 [Cl^-] – 10.57
$\delta^{18}O$ – [SO_4^{2-}]	0.803	$\delta^{18}O$ = 0.2675 [SO_4^{2-}] – 13.72
$\delta^{18}O$ – [Na^+ + K^+]	0.882	$\delta^{18}O$ = 0.0041 [Na^+ + K^+] – 10.52
$\delta^{18}O$ – [CA^{2+} + Mg^{2+}]	0.906	$\delta^{18}O$ = 0.0292 [Ca^{2+} + Mg^{2+}] – 11.36

TDS of 330 g l^{-1} corresponds to $\delta^{18}O$ = + 9.9‰, which is in agreement with the calculated model values for a large evaporative basin in arid climate (Vlasova and Bresgunov, 1978). A similar relationship between isotopic composition and TDS of groundwaters was also reported for a real sedimentary basin by Hitchon and Friedman (1969).

The composition of brine, calculated by using the equations of Table 5, values of $\delta^{18}O$ (+ 9.9‰) and equivalent ratios Na/Cl (0.87), Ca/Mg (O, received by the extrapolation of the line on the graph in Fig. 4), is presented in Table 6. On the basis of comparison with experimental brine obtained by seawater evaporation and natural brine from the Sivash Lake (contemporary Black Sea lagoon), the original brine in the crystalline rocks of the Krivoy Rog basin can be related to brines of surface evaporative basins. The geological structure and the history of the region (Precambrian regional fault, existence of large masses of salt and salt facies on the vast territory of the adjacent Dnieper-Donetsky aulacogen and on the slope of the Ukrainian Shield) lead to the conclusion that it is Paleozoic sea brine, penetrated into the crystalline basement of the Shield.

The source of the Na-Ca-Cl brines and major gases, particularly H_2, in the crystalline basement of the earth's crust is not well understood. Proceeding from the ubiquity of brines and the fact that the basement of the continental crust is constituted to a large extent of metamorphic sedimentary rocks, it can be assumed that they are residual metamorphic fluids originated from Archeozoic and Proterozoic sea water highly modified in the process of water-rock interaction. This process could cause the evolution of the initial chemical composition of the fluids toward a Ca-Cl end member (Frape et al., 1984). Yet, the multi-fold increase of salinities, which would require the removal of the major portion of water under closed system condition, is not easy to explain. Though several approaches to the solution of this problem have been suggested (Gurevich, 1964; Gavrilenko and Derpholz, 1971; Shwartsev, 1979; Fritz and Frape, 1982; Vovk, 1982; Frape et al., 1984), we see our task in discussing here only radiolytic effects which can be noticed in old brines and rocks of the crystalline basement on the background of modern advances in radiation chemistry.

RADIOLYSIS

The water-rock system under consideration is not at all or not completely closed with respect to H_2 (Roedder, 1972; Braester and Thunvik, 1982), therefore, a logical step to explain the removal of water from the system could be the search for a mechanism of water decomposition with the formation of free hydrogen gas. The author believes that, under the thermodynamic conditions of the brine occurrences in the crust, the most probable mechanism of water decomposition is radiolysis. During the span of geologic time, radiation from mainly uranium, thorium and daughters provided this mechanism to decompose water and thus caused several major geochemical consequences, among them an increase in salinities of the residual fluids (Vovk, 1979a, 1981, 1982).

The process of groundwater radiolysis has recently been intensively studied within the programmes of geological radioactive waste disposal (Glass, 1981; Christensen and Bjergbakke, 1982; Erikson and Jacobsson, 1983; Neretnieks, 1983; Panno and Soo, 1984). These studies support the radiolytic approach in the geochemical evaluation of the residual fluids in the crystalline basement. They show that radiolysis of brines and saline waters may be especially efficient, resulting in considerable hydrogen storage although its accumulation in deionized water is insignificant. One of the experiments is particularly important. It demonstrates the importance of high concentrations of OH scavengers contained in the solutions (Br^-, Cl^-) and synergistic effects between the various ions on the radiolytic process, due to which, in a complex salt brine, hydrogen and oxygen have been generated at an equilibrium pressure of about 100-140 bar (Gray, 1983; Gray and Simonson, 1984).

Based on experimental data and mathematical models, J. Dubessy has come to the conclusion that fluids abnormally rich in H_2 and O_2, with pressure of O_2, estimated at room temperature, reached 60 bar, which he pin-pointed in association with uranium mineralization in the Canadian Shield and the Oklo deposit, come from alpha radiolysis of an aqueous phase before entrapment in quartz crystals. In one case (the Oklo uranium deposit), it was concluded that the molecular hydrogen under pressure up to 20 bar resulted from joint radiolysis of organic matter and water (Dubessy, 1983, 1985). This and other examples of the application of the radiolytical model for the explanation of geochemical and mineralogical peculiarities of the uranium deposits (Vovk, 1979b; Myers, 1981; Cramer, 1985) show that besides experimental physico-chemical basis there is also geological basis for its application to some natural water-rock systems. Brines in the depths of the earth's crust may be one of them.

Radiolysis of water by alpha, beta or gamma rays consists initially of excitation and ionization of water molecules ($H_2O \rightsquigarrow H_2O^*$, $H_2O \rightsquigarrow H_2O^+ + e^-$) followed by ion-molecule reactions (e.g., in spur regions $H_2O^+ + H_2O \rightarrow H_3O^+ + OH$), dissociation of molecules ($H_2O^* \rightarrow H + OH$, $H_2 + O$), solvation of ions ($e^- + nH_2O \rightarrow e^-_{aq}$), reactions of species (e.g., $H + H \rightarrow H_2$, $H + OH \rightarrow H_2O$).

There is a concensus that ~10^{-9}s after the absorption of energy by neutral water, the predominant species are

$$H_2O \rightsquigarrow H_2,\ H_2O_2,\ H,\ OH,\ e^-_{aq},\ H^+_{aq},\ OH^-_{aq}.$$

The values for the molecular and radical yields (G values = number of molecules per 100 eV of absorbed energy) vary somewhat with the kind of absorbed radiation (alpha, beta or gamma). However, the sum of the oxidizing radicals created is equal to the sum of the reducing radicals and the total number of molecules of water decomposed per 100 eV of absorbed energy, known in radiation chemistry under

Table 6

Comparison of brines

	1	2	3
Cl^-	201818	190000	180704
SO_4^{2-}	4224	25600	8886
Na^+	111826	103000	94835
K^+	5044	5000	1543
Mg^{2+}	8906	15700	13413
Ca^{2+}	0	240	331
TDS	331818	339549	299702

1) calculated original brine for the Krivoy Rog basin,
2) obtained in experiments of sea water evaporation,
3) natural brine from Sivash Lake, mg l^{-1}.

the notation of $G(-H_2O)$, for homogeneous radiolysis of pure (deionized) water is about 4 (3.6 + 4.6).

In water solutions, as the products of radiolysis drift towards each other, they encounter and react with solute molecules before encountering a counterpart and recombining back into water molecules, thus the efficiency of radiolysis increases. Besides, in such heterogeneous systems as water-saturated rocks, the efficiency of radiolytic water decomposition can be increased by the process of energy transfer from solid to liquid phase and due to the presence of oxide catalysts, many of which show their action only under irradiation. $G(-H_2O)$ values for homogeneous radiolysis of water solutions with high concentrations of scavengers of radicals (OH or H) as well as of water solutions containing clutrate compounds ("gas-hydrates") can reach 12-15 (Shubin and Kabakchy, 1969). For heterogeneous radiolysis of water absorbed on the oxides $G(H_2)$ values as high as 10 to 30 molecules per 100 eV have been reported (Garibov, 1983).

The authors' data (Vovk, 1979a) obtained in experiments of gamma irradiation of complex geological objects are partly presented in Table 7 and Table 8, where samples 1-9 and 12 are bottom sediments of the Black Sea and samples 10 and 11 are from the Mediterranean. They demonstrate

Table 7

Chemical composition of interstitial water of the sea bottom sediments

Sample No.	(m–eq) l^{-1} Na^+	K^+	Ca^{2-}	Mg^{2+}	Cl^-	SO_4^{2-}	HCO_3^-	CO_3^{2-}	TDS g l^{-1}
				non-irradiated					
1	297.5	8.1	15.0	72.5	350	35.6	7.5	0	23.1
2	290.0	7.8	15.0	67.5	340	32.6	4.3	3.4	22.3
3	296.7	8.3	15.0	67.5	350	28.7	6.7	2.1	22.7
4	274.7	7.2	15.0	67.5	330	28.1	8.3	0	21.3
5	292.3	8.3	15.0	67.5	350	24.4	3.9	4.8	22.3
6	288.9	8.1	15.0	60.0	340	23.6	3.8	4.5	21.8
7	284.8	8.3	15.0	72.5	340	33.7	6.9	0	22.3
8	300.3	7.4	25.0	70.0	350	46.3	3.0	3.4	23.7
9	301.1	5.6	10.0	66.5	340	26.6	13.1	3.6	22.3
10	551.0	14.3	15.0	120.0	640	40.3	6.4	13.6	40.7
11	552.7	15.2	5.0	122.5	650	6.4	13.3	25.7	39.7
				irradiated, Dose 1.10^6 Gy					
1	323.1	8.3	30.0	80.0	370	42.4	29.0	0	26.6
2	318.3	8.6	25.0	85.0	380	38.6	14.3	4.0	26.0
3	315.1	8.2	20.0	85.0	370	37.3	17.0	4.0	25.4
4	289.0	12.7	32.5	82.5	360	37.7	19.0	0	24.6
5	327.4	10.1	22.5	77.5	380	36.0	14.5	7.0	25.7
6	284.5	7.2	22.5	72.5	240	26.2	16.2	4.3	22.7
7	308.7	8.1	30.0	80.0	360	40.3	26.5	0	25.7
8	315.3	9.3	45.0	80.0	370	64.3	15.3	0	27.3
9	309.3	7.9	17.5	82.5	346	37.3	33.8	0	24.5
10	581.6	15.0	15.3	125.0	660	44.6	19.8	12.5	42.8
11	578.4	18.2	7.5	122.5	670	26.6	19.2	10.8	42.7

Table 8

Composition of gas in ampules containing sea bottom sediments

Sample No.	Gas pressure 10^5 Pa	initial gas volume cm^3	Produced gas volume cm^3	vol. % O_2	N_2	CO	CO_2	H_3 + Ar	$CH_4.10^{-3}$	$C_2H_6 10^{-3}$
				non-irradiated						
1	~1.0	~100	—	12.5	84.5	0.5	1.6	1.0	0	0.5
3	~1.0	~100	—	14.5	82.8	—	1.7	1.0	—	—
6	~1.0	~100	—	20.0	78.1	—	0.9	1.0	—	—
12	~1.0	~100	—	15.0	82.7	—	1.3	1.0	—	—
				irradiated, Dose 1.10^6 Gy						
1	2.1	100	110	0.15	20.3	2.3	8.6	68.7	7.6	
3	1.8	102	82	3.10	31.5	2.1	8.7	55.6	1.4	1.7
6	1.5	100	50	1.10	28.6	1.2	8.4	60.7	4.0	
12	2.0	114	114	0.30	20.4	3.6	15.0	60.7	2.8	2.1

clearly that at least three major geochemical processes are caused by radiolysis of the water-rock systems: salt enrichment of interstitial water (increase in TDS), production and accumulation of hydrogen and consumption of oxygen in oxidation of organic carbon, sulfur and perhaps other variable valence elements. They also show that salt enrichment is a complex function both of the radiolytic decomposition of water molecules, proved by the increase of Cl^- concentrations in interstitial water, and of accumulation of such TDS constituents as bicarbonates, sulfates and cations, particularly calcium, as a result of irradiation and interaction of water radiolysis products with rock matrix.

The dose rate in these experiments was 4.2 Gy s^{-1}. In laboratory conditions dose rates vary from 10^{-4} to 10^6 Gy s^{-1} and in natural water-rock systems from about 10^{-10} to 10^{-3} Gy s^{-1} (Vovk, 1982). It has been established that within equal integral doses, lower dose rates of irradiation are chemically more efficient, as with the decreasing dose rate overlapping of spurs and of their diffusion zones decreases which in turn "saves" initial radiolytic products from recombination into water molecules. For example, radicals escaping the widely spaced spurs are nearly completely scavenged in the solutions above about 10^{-6}M, whereas the radicals in the closely spaced spurs are only beginning to react with the radical scavenger above 10^{-5}M of solute (Trumbore and Youngblade, 1983). Thus, the chemical action of the natural irradiation is at least as efficient as the chemical action in laboratory experiments and the final results can be determined by multiplication of experimentally established yields and integral absorbed doses.

The absorption of energy of the radioactive decay in liquid and solid phases of rocks has been evaluated (Vovk, 1979a) to be approximately equal per unit of mass. Thus, to calculate the absorbed doses in groundwater, it is sufficient to calculate them for water-rock systems as a whole. This was done for the major types of crust rocks on the basis of abundance of the radioactive elements (Table 9) and energy contribution from each element (Vovk, 1982). Rock porosity and density were also taken into account. The results of calculations are presented in Table 10. Actual absorbed doses might be considerably higher at the early stage of the formation of the earth's crust when numerable radioactive isotopes with the periods of half-decay 10^6-10^8 years (^{26}Al, ^{10}Be, ^{107}Pd, ^{129}I, ^{146}Sm, ^{205}Pb, ^{236}U, ^{244}Pu, ^{247}Cm) were present in the rocks.

Table 9

Abundance of the radioactive elements in the major rock types of the earth's crust (U, Th n.10^{-4} %, K, %) after Ronov and Yaroshevsky (1976):

	1	2	3	4	5
U	2.9	1.9	1.1	2.7	0.7
Th	7.8	9.2	1.8	11.6	2.4
K	2.33	1.61	0.50	2.40	1.10

1 – clay,
2 – sandstone,
3 – limestone,
4 – granite,
5 – basalt.

Net radiolytic salt enrichment of groundwater (without contribution from rock matrix material) can be easily found, taking into account that the increase in concentration of a conservative TDS constituent (Cl^-) corresponds to the amount of decomposed water:

$$\frac{C_1}{C_2} = 1 - \frac{\Delta m}{m}$$

where C_1/C_2 = ratio of initial and final concentration of Cl^-, $\Delta m/m$ = ratio of decomposed and initial amount of water. The latter can be expressed as a function of radiolytic yield $G(-H_2O)$ and absorbed dose (D):

$$\frac{\Delta m}{m} = \frac{G(-H_2O)\ MD}{100N}$$

where M and N are water molecular weight and Avogadro's Number, respectively.

The calculated values of integral absorbed doses of irradiation in natural water-rock systems are more or less accurate, at least the actual ones are undoubtedly not smaller than those presented in Table 10, but the $G(-H_2O)$ values for heterogeneous radiolysis of water solutions in rocks are

Table 10

Absorbed doses of irradiation (Gy) in rock matrix and groundwater, calculated for Table 9. Rock porosity n (%), rock density δ (kg dm^{-3}), and time periods t from 1 to 4.5.10^9 years (Vovk, 1982).

		1	2	3	4	5
n		20	10	5	1.0	1.0
δ		2.30	2.35	2.50	2.71	2.71
			Absorbed doses			
	/4.5-3.5/.10^9	5.8.10^7	4.4.10^7	2.0×10^7	—	
	/3.5-1.6/.10^9	5.9.10^7	4.9.10^7	2.0.10^7	—	2.1.10^7
	/1.6-0/.10^9	3.2.10^7	2.7.10^7	1.1.10^7	3.9.10^7	1.0.10^7
Time	/3.5-0/.10^9	—	—	—	—	3.1.10^7
	/1.0-0/.10^9	1.8.10^7	1.5.10^7	0.7.10^7	2.3.10^7	0.6.10^7
	1.0-0/.10^8	1.7.10^6	1.4.10^6	0.6.10^6	2.1.10^6	0.5.10^6
	/1.0-0/.10^6	1.7.10^4	1.4.10^4	0.6.10^4	2.1.10^4	0.5.10^4
	1.0	1.7.10^{-2}	1.4.10^{-2}	0.6.10^{-2}	2.1.10^{-2}	0.5.10^{-2}

still very obscure. On the basis of the above considerations, $G(-H_2O)=5$ was accepted as a representative value for calculation of radiolytic salt enrichment in crust fluids originated from sea water of normal salinity in clay sediments, metamorphosed later into crystalline rocks of basaltic and granitic layers. The results of calculation are given in Table 11.

One can see from Table 11 that in argillaceous sedimentary rocks during the period of formation of basaltic layer (4.5-3.5).10^9 years ago, initial water salinity 35 g l^{-1} (20 g l^{-1} of Cl^-) could rise to 73 g l^{-1} (in NaCl equivalent) and up to 200 g l^{-1} (119.8 g l^{-1} of Cl^-) in the present, if it remained in that layer. In the granitic layer (3.5-0).10^9 years), initial sea water could reach salinity 400 g l^{-1} (241 g l^{-1} of Cl^{-1}) and water penetrated there from basaltic layer with Cl^- content 43.8 g l^{-1} could reach salinity up to 800 g l^{-1} (527.7 g l^{-1} of Cl^-).

Thus, there is a reasonable possibility to consider actual concentrations of salts in crust fluids depending on the radiolytic accumulation and precipitation processes. Numerous geological evidences in the form of salt crystals dispersed in Precambrian rocks (Bell and Jackson, 1974), in fluid inclusions (Roedder, 1972), and even in the form of veins of halite, filling the fractures in the diabase and other crystalline rocks of the shields (Searls, 1956) are consistent with this approach.

The radiolytic salt enrichment in brines accompanied by halite precipitation may lead to the formation of Ca-Cl type of brines with increased Br/Cl ratio which has actually been observed both in the East European Platform (Table 3, Location 7) and in the Canadian Shield (Frape et al., 1984).

It is obvious that as a global phenomenon, radiolysis of groundwater must play a multifold geochemical role. This subject has been discussed in previous publications by the author (Vovk, 1979a,b, 1981, 1982). Some of its major effects are worthy of mentioning here without going into detail as geological evidence of the process of groundwater radiolysis.

From the above discussion of gases associated with brines it follows that a considerable portion of radiolytic hydrogen could have been accumulated in fluid inclusions and porous media of rocks but as an easily migrating gas, free hydrogen diffuses through the rocks accompanied by radiogenic helium and thus the groundwater radiolysis maintains its constant flow to the atmosphere long known as hydrogen-helium "breath" of the earth's crust.

The other water constituent – oxygen – is practically all consumed in the oxidation of organic matter, iron, sulfur and other elements, evidence for which can be found in occurrences of oxidized iron, sulfates and other oxidized secondary minerals in crystalline basement under the geochemically dominating reducing conditions. Oxidized state is a characteristic feature of fluid inclusions in minerals of old crystalline rocks. Besides carbon dioxide and carbonates, they often contain other oxidized daughter minerals which are not dissolved even under temperatures much exceeding the homogenization temperature (Roedder, 1972). In cases when radiolytic oxygen is produced in the oxidized geochemical environment, for example in the red-beds, or if there were no variable valence elements available (in some of the fluid inclusions) it can be accumulated in quantities much exceeding its solubility limit in water (Vovk, 1979a). Findings of geological objects like fluid inclusions, described by J. Dubessy (1983, 1985), where oxygen and hydrogen are detected simultaneously, are rare but are very important to prove that in nature back reaction does not eliminate the elemental gas molecules produced in water radiolysis similarly to what was established in laboratory experiments with irradiation of complex brine (Gray, 1983; Gray and Simonson, 1984).

Kinetic and other effects of isotope fractionation of such elements as hydrogen, carbon, nitrogen, oxygen, sulfur and uranium can be predicted on the basis of radiolytic decomposition of water molecules and subsequent oxidation and reduction reactions. Thus, the radiolytic approach may render a good service for better understanding of the isotope compositioin of the above mentioned elements in brines, associated gases and secondary minerals in the crystalline rocks of the Precambrian shields as well as of all kinds of uranium deposits. At present this problem requires further experimental and field studies.

Water-rock interaction is affected by irradiation not only through water radiolysis but also through the radiation damages in the rock matrix. The experimental results of irradiation of the most widespread natural alumosilicate minerals (quartz, feldspars, caolinite, chlorite and others) show that the coherence of minerals and particles in the structure formation depends both on the radiolysis of water and the changes which take place in the sub-surface layers of the mineral grains at appropriate irradiation doses. These changes are capable of creating tension in the structure, modifying the density and mechanical stability of minerals, thus giving rise to their disintegration. A dose of 10^5-10^6 Gy has been determined as the "threshold dose", marking the beginning of these effects (Spitsyn et al., 1981); hence it seems reasonable to look for their manifestation in nature.

Indeed, natural radiation damage in minerals, particularly in zircon, in the form of radial and circular fractures formed around radioactive grains is common for many uranium deposits and occurrences in crystalline rocks (Vovk, 1982). Initial alteration of minerals usually begins at fractures and proceeds toward the centre of a grain. For example, in the Baie Johan Beetz area, Quebec, scanning electron micrographs revealed the fractured and

Table 11

Calculated concentration of $Cl^-(C_2)$ in relict radiolytically decomposed waters originated from sea sediments (clays) in Epikarelian cover (A) granitic (B) and basaltic (C) layers for two values of initial (C_1) Cl^- concentrations: 20 gl^{-1} and 43.8 gl^{-1}.

Layer	t, 10^9 years	Absorbed dose 10^7 Gy	$\frac{\Delta m}{m}$	C_2 gl^{-1} $C_1=20$	C_2 gl^{-1} $C_1=43.8$
A	1.6 – 0	3.2	0.300	28.5	—
B	3.5 – 1.6	5.9	0.552	44.6	97.8
	3.5 – 0	9.8	0.917	241.0	527.7
C	4.5 – 3.5	5.8	0.543	43.8	—
	4.5 – 1.6	7.9	0.739	76.6	—
	4.5 – 0	8.9	0.833	119.8	—

heterogeneous nature of ore and accessory minerals, implying that between 10 and 80% of the uranium and other original radionuclides and non-radioactive ions have been leached from the minerals during alteration and replacement. On the basis of mineralogical and geochemical studies, J. Rimsaite came to the conclusion that fracturing and shattering of minerals in the Baie deposit results in part from radiation damage. The heterogeneity of the accessory minerals is due to subsequent alteration along fractures, where primary minerals either alter to secondary oxides, hydroxides, carbonates, silicates or they are replaced by phyllosilicates, oxides and other constituents (Rimsaite, 1982).

Similar alterations are also common in old granitic rocks of the Precambrian shields, where natural radiation created microfractures and other damages in crystalline lattices of minerals and thus prepared them for penetration of water molecules into their volume with hydration and weathering as further consequences (Vovk, 1982). An example to demonstrate it is given in Fig. 5 where, in borehole 50, the increase in thickness of residual kaolin (from 20-30 to 60 m) corresponds to the increase of the radioactivity of non-weathered granite (from 20-30 to 50-60 10^{-13} A kg^{-1}). Thus, natural radioactivity of rocks must not be overlooked in the discussions of the formation of brines in the crystalline basement of the shields in the water-rock interaction approach, particularly through hydration of silicates (Schwartsev, 1979; Fritz and Frape, 1982), as an alternative possibility of increasing TDS concentrations in groundwater.

SUMMARY AND CONCLUSIONS

Saline waters and brines of Na-Cl and Na-Ca-Cl chemical types are ubiquitous in the crystalline basement of the East European Platform as elsewhere in the earth's crust. They are accompanied by free and dissolved gases greatly enriched in hydrogen and helium. Origin and chemical evolutions of the crust brines is a scientific problem important for a number of practical applications in geochemistry, economic geology, environment protection, etc. Further study of this problem is required to come to definite conclusions, but some of them can be suggested at present as working hypotheses and ideas for examination.

Brines of the Na-Cl chemical type, occurring in regional fault areas similar to Krivoy Rog basin, originate from the surface sea water evaporative basins, penetrated into the crystalline basement through shear zones from the Epikarelian (Epicambrian) sedimentary cover. Na-Ca-Cl brines found across the East European Platform, particularly in the Tatar Dome as well as across the Canadian Shield, may represent residual fluids preserved in metamorphic and magmatic rocks since Archeozoic sedimentation.

Both radiolysis of water molecules and radiation damages of the rock matrix owing to the decay of natural radioactive elements during the geological time spans are likely to significantly contribute to water decomposition and water-rock interaction, resulting jointly in the increase of TDS concentrations and changes in cation ratios of the residual fluids. Molecular hydrogen and helium produced in this process are subjected to gravitational segregation and form a constant flow of free hydrogen-helium gas into the atmosphere, while oxygen, the other constituent released from water molecules, may be completely consumed in oxidation of rock matter or accumulated in limited quantities in natural gases and confined containers like fluid inclusions in minerals or some of the redbed aquifers.

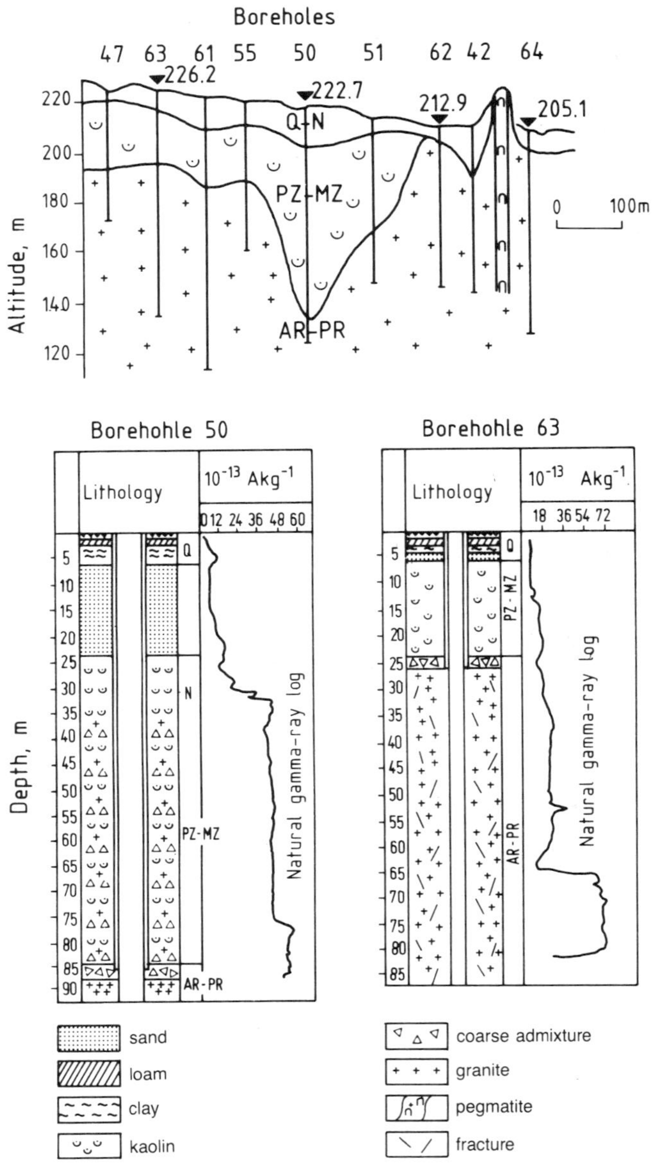

Figure 5. *Cross-section of a granite massive in the Ukrainian Shield with results of radioactivity measurements in boreholes.*

ACKNOWLEDGEMENTS

The author would like to express his appreciation to P. Fritz and S.K. Frape for identifying the need for this paper, to E. Roedder for the discussions of some of the geochemical aspects of fluid inclusion related to the radiolysis of water, and to Mrs. Helen Fadlaoui for her secretarial assistance.

This paper was prepared as a contribution to the Joint Annual Meeting of the Geological Association of Canada and the Mineralogical Association of Canada, Fredericton, 15-17 May 1985.

REFERENCES

Bell, R.T. and Jackson, G.D., 1974. Aphebian halite and sulphate indications in the Belcher group, Northwest Territories: Canadian Journal of Earth Sciences, v. 11, p. 722-728.

Boyarskaya, R.V., Dolomanova, E.I., Nosik, L.P. and Fadyukov, E.M., 1977. Morphology and chemical composition of fluid inclusions in quartz from pegmatite in Volyn based on data of scanning electron microscopy and mass-spectrometry: In: "Problems of regional and genetic mineralogy", Ukrainian Academy of Science, Kiev, p. 98-108.

Braester, C. and Thunvik, R., 1982. An analysis of the conditions of gas migration from a low level radioactive waste repository: SKB Stripa Project Report 83-21, 44 p.

Christensen, H. and Bjergbakke, E., 1982. Radiolysis of groundwater from spent fuel. SKB Stripa Project Report 82-18, 28 p.

Cramer, J.J., 1985. Natural analogs to nuclear fuel waste disposal in crystalline rocks: an overview. To be published as SKB Stripa Project Report.

Dubessy, J., 1983. A contribution towards the knowledge of fluids associated in the formation of uranium deposits. Revue Générales Nucleaires v. 6, p. 472-474.

Dubessy, J., 1985. Contribution a l'étude des interactions entre paleo-fluides et mineraux a partir de l'étude des inclusions fluides par micro-spectrometrie Raman. Consequences metallogeniques: Thèse docteur es Science. L'institut national polytechnique de Lorraine, VI – Les deséquilibres chimiques d'origine radiogénique, p. 119-153.

Eriksen, T. and Lind, J., 1983. Formation of free hydrogen during radiolysis in a bed of clay: Brookhaven National Laboratory Technical Report 1000, Upton, NY, USA 20 p.

Eriksen, T. and Jacobsson, 1983. Radiation effects on the chemical environment in a radioactive waste repository: SKB Stripa Project Report 83-27, 24 p.

Frape, S.K., Fritz, P. and McNutt, R.H., 1984. Water-rock interaction and chemistry of groundwaters from the Canadian Shield: Geochimica et Cosmochimica Acta, v. 48, p. 1617-1627.

Fritz, P. and Frape, S.K., 1982. Saline groundwaters on the Canadian Shield – a first overview: Chemical Geology, v. 36, p. 179-190.

Garibov, A.A., 1983. Water radiolysis in the presence of oxides: In: "Proc. 5th Symposium on Radiation Chemistry", Budapest, 1982, Akademiai Kiado, p. 377-384.

Gavrilenko, E.S. and Derpholz, V.F., 1971. Hypogene hydrosphere of the Earth: Ukrainian Academy of Science, Kiev, 272 p.

Glass, R.S., 1981. Effects of radiation on the chemical environment surrounding waste canisters in proposed repository sites and possible effects on the corrosion process: Sandia National Laboratory Report 81-1677, Albuquerque, NM, USA, 89 p.

Gray, W.J., 1983. Gamma radiolysis of groundwaters found near potential radioactive waste repositories: Pacific Northwest Laboratory Report PNL-SA-10928, Richland, WA, USA, 10 p.

Gray, W.J. and Simonson, S.A., 1984. Gamma and alpha radiolysis of salt brines: Pacific Northwest Laboratory Report PNL-SA-12746, Richland, WA, USA, 8 p.

Greiser, E.L., Koslov, V.B. and Pavlov, A.N., 1970. Brines in crystalline rocks of the baltic Shield: Izvestiya Akademii Nauk SSSR, Seriya Geologicheskaya, v. 3, p. 141-143.

Gurevich, M.S., 1964. Types and palaeochemical conditions of formation of groundwaters of the chloride and calcium composition: In: "Earth's crust chemistry", Volume 2, Nauka, Moscow, p. 489-507.

Hitchon, B. and Friedman, I., 1969. Geochemistry and origin of formation waters in the western Canada sedimentary basin — I. Stable isotopes of hydrogen and oxygen: Geochimica et Cosmochimica Acta, v. 33, p. 1321-1349.

Kozlovsky, Ye. A., 1984. The world's deepest well. Scientific American, v. 251, p. 98-104.

Kudryavtsev, N.A., 1959. Petroleum, gas and solid bitumens in igneous and metamorphic rocks: Gostoptechizdat, Leningrad, 278 p.

Makov, K.I., 1948. Hydrogeological conditions in the Ukrainian Shield: Geologicheskij Zhurnal, v. 9, no. 3, p. 65-69.

Myers, W.B., 1981. Genesis of uranium-gold pyritic conglomerates: In: "Proceedings of a Workshop on Genesis of Uranium and Gold Bearing Precambrian Quartz-Pebble Conglomerates", Golden, Colorado, 1975, U.S. Geological Survey Professional Paper N.1161-A-BB, p. AA1-AA26.

Nash, K.L., Fried, S., Friedman, A.M., Susak, N., Rickert, P. and Sullivan, J.C., 1983. Radiation effects in solution and on the solid/liquid interface: Nuclear Technology, v. 60, p. 257-266.

Neretnieks, 1983. Movement of a redox front downstream from a repository for nuclear waste: Nuclear Technology, v. 62, p. 110-115.

Panno, S.V. and Soo, P., 1984. Potential effects of gamma irradiation on the chemistry and alkalinity of brine in high-level nuclear waste repositories in rock salt: Nuclear Technology, v. 67, p. 268-281.

Panova, S.P., 1976. Saline and brine groundwaters of the Baltic Shield (within the USSR): In: Ground brines of the USSR. VSEGEI, Leningrad, p. 98-101.

Petersilye, I.A. and Pripachkin, V.A., 1979. Hydrogen, carbon, nitrogen and helium in gases of igneous rocks: Geokhimiya, v. 7, p. 1028-1034.

Petersilye, I.A., Fedkova, G.A. and Pavlova, M.A., 1979. Gases and organic matter in Archean granulitic rocks of the Kola Peninsula: Geokhimiya, v. 12, p. 1883-1888.

Rimsaite, J., 1982. The leaching of radionuclides and other ions during alteration and replacement of accessory minerals in radioactive rocks: In: Current Research, Part B, Geological Survey of Canada, Paper 82-1B, p. 253-266.

Roedder, E., 1972. Composition of fluid inclusions: Data of Geochemistry (6th ed.). – U.S. Geological Survey Professional Paper, N440-JJ, 164 p.

Ronov, A.B. and Yaroshevsky, A.A., 1976. A new model of chemical structure of the earth's crust: Geokhimiya, v. 12, p. 1763-1795.

Searls, F. jr., 1956. Halite in the Undersill gold mine, Ontario: Economic Geology, v. 51, p. 192-196.

Shubin, V.N., Kabakchy, S.A., 1969. Theory and medthods of radiation chemistry of water: Nauka, Moscow 215 p.

Shwartsev, S.L., 1979. The role of aluminosilicate rocks in formation of chemical composition of groundwaters: In: Problems of theoretical and regional hydrogeochemistry: Moscow University, p. 115-119.

Sobolev, V.I., 1984. New data on chemical composition of chloride ground water in the Krivoy Rog iron-ore basin: Geokhimiya, v. 9, p. 1364-1377.

Spitsyn, V.I., Balukova, V.D., Kosareva I.M. and Kabakchi, S.A., 1981. Experimental evaluation of changes in properties of natural minerals under irradiation: In: Scientific Basis for Nuclear Waste Mangement, v. 3, Plenum Press, New York and London, p. 429-433.

Trumbore, C.N. and Youngblade, W. 1983. Recent calculation modelling data from the pulse radiolysis and gamma radiolysis of water and aqueous solutions: In: Proceedings of the Workshop on the Interface between Radiation Chemistry and Radiation Physics, Argonne, Illinois, 1982, Argonne National Laboratory Report ANL-82-88, p. 82-90.

Vlasova, L.S. and Bresgunov, V.S., 1978. Formation of isotope composition of hydrogen and oxygen in natural brines based on model calculations: In: Isotopy of natural waters. Nauka, Moscow, p. 140-164.

Vovk, I.F., 1979a. Radiolysis of ground water and its geochemical role: Nedra, Moscow, 231 p.

Vovk, I.F., 1979b. Some mineralogical and geochemical peculiarities of uranium deposits in sedimentary rocks as a result of the radiolysis of underground waters: Zapiski Vserossiiskogo Mineralogicheskogo Obshchestva No. 4, p. 397-407.

Vovk, I.F., 1981. Radiolytic model of formation of brine composition in the crystalline basement of shields: Geokhimiya, v. 4, p. 467-480. Translated in Geochemistry International 1981: 80-93.

Vovk, I.F., 1982. "Brines in the crystalline basement of the shields": Ukrainian Akademy of Science, Kiev, 163 p.
Vovk, I.F. and Vysotsky, V.I., 1982. Oxygen isotope composition and geochemical characteristics of groundwaters in Krivbass as indicators of their inter-connection and origin: Vodniye Resursy, v. 1, p. 39-47.
Voytov, G.I., 1974. Estimation of gas exchange in the shields (on the example of the Ukrainian Shield): Geologicheskij Zhurnal, v. 34, no. 2, p. 68-82.
Zorkin, L.M., Voitov, G.I. and Stadnik, E.V., 1977. Gases and waters in the crystalline basement of the Tatar Dome: Izvestiya Vysshikh uchebnykh zavedenij – Geologiya i Razvedka, v. 6, p. 82-95.

Methane in the Crystalline Rocks of the Canadian Shield

P. Fritz, S.K. Frape and M. Miles
Department of Earth Sciences, University of Waterloo, Waterloo, Ontario, Canada N2L 3G1

Fritz, P., Frape, S.K. and Miles, M. Methane in the crystalline rocks of the Canadian Shield; in Saline water and gases in crystalline rocks, Editors: Fritz, P., and Frape, S.K.; Geological Association of Canada Special Paper 33, 1987. p. 211–223.

Abstract

Methane in association with minor amounts of higher hydrocarbons, nitrogen, helium, argon and hydrogen discharge from numerous boreholes in mines across the Canadian Shield. Isotopic analyses on methane, ethane and aqueous carbonate dissolved in associated salt waters and brines show that these hydrocarbons have most likely an abiotic origin. It is remarkable that each mining district releases methane with uniform ^{13}C and ^{2}H contents which are not readily related to a specific carbon source or generating mechanism.

The isotope data provide evidence which suggests that the methane and associated hydrocarbons did not form in a high temperature environment. Instead, especially hydrogen isotope data point towards kinetic and low or intermediate temperature processes. Further support for this interpretation is found in the free hydrogen associated with some occurrences. It is very strongly depleted in deuterium as is characteristic for hydrogen gas generated in other areas through the reduction or radiolytic decomposition of water at low, near surface temperatures.

Geochemical and isotopic data show that these methanes are not active participants in geochemical reactions which affect water chemistries and the generation of brines. However, their formation in crystalline rock environments does occur and, therefore, contributions to commercial hydrocarbon reservoirs cannot be excluded.

Résumé

Le méthane de même que des quantités mineures d'hydrocarbures plus légers, d'azote, d'hélium, d'argon et d'hydrogène s'échappent de nombreux trous de forage pour s'infiltrer dans les mines du bouclier canadien. Des analyses isotopiques effectuées sur le méthane, l'éthane et le carbonate aqueux dissous dans les eaux salées et sursalées révèlent que ces hydrocarbures sont fort probablement d'origine inorganique. Il importe de constater que dans chaque secteur minier, le méthane possède des teneurs en ^{13}C et ^{2}H uniformes qui ne sont pas nécessairement associées à une source de carbone ou un mode de formation particuliers.

L'interprétation des données isotopiques indique que le méthane et les hydrocarbures ne se sont pas formés à haute température. Les donnés isotopiques, surtout celles de l'hydrogène, indiqueraient plutôt des processus cinétiques se produisant à température peu élevée ou intermédiaire. La présence d'hydrogène libre dans certains échantillons appuie cette hypothèse. Ce dernier est très fortement appauvri en deutérium, tout comme l'hydrogène de régions différentes produit par réduction ou radiolyse de l'eau à des températures faibles, se rencontrant près de la surface.

Les données géochimiques et isotopiques indiquent que ces types de méthane ne participent pas activement aux réactions géochimiques modifiant la composition chimique de l'eau et la genèse des eaux sursalées. Toutefois, il faut tenir compte de leur formation dans les roches cristallines et, par conséquent, de leur présence dans les réservoirs commerciaux d'hydrocarbures.

INTRODUCTION

Methane is an ubiquitous gas in crystalline rocks. It is found in fluid inclusions of hydrothermal minerals (Guha *et al.*, 1979; Ramboz *et al.*, 1985), it has been identified as an important component in metamorphic systems (Brown, 1982; Kreulen and Schuiling, 1982) and discharges in considerable quantities at active oceanic rifts (Welhan and Craig, 1983). In deep crystalline rock boreholes in the USSR methane is encountered in association with calcium chloride brines (Bogomolov *et al.*, 1980), with saline waters it occurs in boreholes in the crystalline rocks of Finland (Nurmi and Kukkonen, in press), and it is found discharging freely in many mines on the Canadian Shield.

The oldest record known to us is found in the Annual Report (1887/88) of the Geological and Natural History Survey of Canada which a.o. describes the Silver Islet Mine in Lake Superior; on pg. 29H it reads: ".... A curious feature of the vein is the combustible gas which has been met with in large quantities in the workings. This gas, according to Mr. Lowe, is accompanied by water containing calcium chloride solution. He says: 'Two gallons of this water furnished roughly measured, nearly a pint of very acrid and deliquescent salt, chloride of calcium'. He further states 'the gas and water are principally confined in large vugs or cavities in the vein under great pressure....it' (the gas) 'is met with in the deepest workings ... above the 8th level all water infiltrating into the mine is pure lake water, whilst below that level is a zone of highly mineralized water....the gas is unequally distributed through the lower workings.......it was previously met with in the 3rd and 6th levels."

A much more recent, internal report (1974) from an operating mine in Yellowknife (NWT) records that: "... I noticed water (in a deep drill hole) welling intermittently out of the hole...... After one of the wells of water I held my lighter near the T. A ball of flame about a foot across and perhaps two feet high resulted ..." Thus methane discharges are known but no specific documentation of these occurrences has been prepared.

This paper is a first summary of a survey of methane discharges from exploration boreholes in mines of the Canadian Shield. The survey was carried out during studies of the nature and occurrence of salt waters in these rocks and were based on analyses of ^{13}C and ^{2}H contents and chemistries (Fritz and Frape, 1982; Frape and Fritz, 1982; Frape *et al.* 1984). Environmental isotope analyses have been used to classify genetically methanes of different origins (Schoell, 1983) and it was felt that a combination of chemical and isotope analyses could contribute to a better understanding on the origin and geochemistry of these gases and their significance in terms of interactive geochemical systems which involve brines, gases and the host rock. In addition, the results should contribute to discussions on the significance of mantle discharges and hydrocarbon accumulations (Gold and Soter, 1982).

Samples were collected exclusively from exploration boreholes drilled from underground mine levels. Figure 1 depicts sampling locations where methane in association with other gases is frequently encountered, often in boreholes which also discharge brines or saline waters.

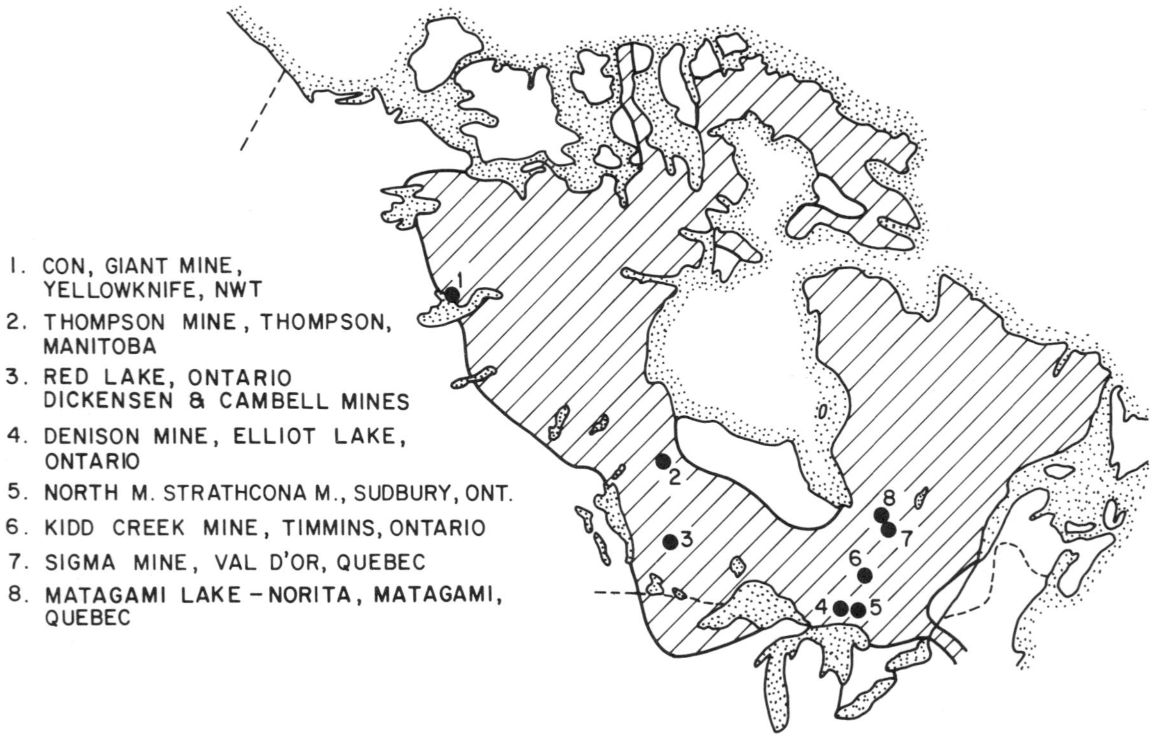

Figure 1. *Location of sampling sites on the Canadian Shield.*

METHODOLOGY

Sampling

Gases were collected only from freely discharging boreholes by two methods: If no water but gas flow occurred, the gas were trapped in a simple inverted bottle arrangement. Only glass bottles were used. In flowing boreholes, however, a "gas stripper" was employed. Figure 2 is a diagrammatic representation of the device and indicates that evacuated sample flasks, flow through flasks or soft copper tubing were used to retain the gas. Although the device did not permit sampling at formation pressures i.e. at the point of gas/fluid entry into the borehole, it provided good samples of the free gas phase at mine level atmospheric pressures and also permitted some pressurization for the collection of rare gas in the copper tubes. Rare gas data will be discussed elsewhere.

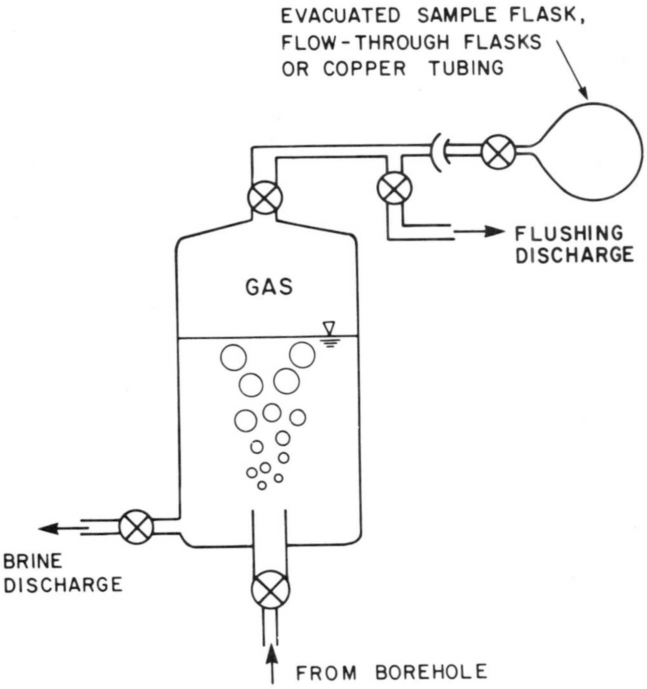

Figure 2. *Sampling device for the collection of gas samples from flowing boreholes.*

Isotope analyses

A detailed description of analytical procedures will be presented elsewhere (Fritz *et al.*, in prep.) but they are summarized here. Isotope analyses were done after separation of methane, carbon dioxide and other gases on a gas chromatograph which is connected to a copper oxide oxidation system. This permitted the collection of carbon dioxide, methane converted to carbon dioxide and water, as well as light hydrocarbon derived water and carbon dioxide. The CO_2 fractions were purified and directly analyzed for ^{13}C contents whereas the water which was generated during oxidation was passed over hot uranium for conversion to hydrogen gas and mass-spectrometric analyses.

Carbon isotope results of repeat analyses of different sample portions (some of which were stored for over one year for repeat analyses) usually were within 1‰; final results are expressed as permil difference from the PDB standard. The overall reproducibility for hydrogen analyses was somewhat lower and repeat analyses typically are within 10‰; deuterium contents are expressed as permil differences from *S*tandard *M*ean *O*cean *W*ater (SMOW). Results are summarized in Table 1.

Chemical analyses

Samples were submitted to Atomic Energy Canada Ltd. in Pinawa (Manitoba) for analyses of chemical compositions. This contribution is gratefully acknowledged as is any other related assistance provided by AECL through the salt water programme. Duplicate analyses were done at the University of Waterloo.

Some gas samples contained small amounts of O_2. It was assumed that these were due to air contamination and necessitated minor corrections of the data set. Unfortunately, neither pressure data nor accurate discharge measurements are available. The lack of this information as well as the high salinities of the associated fluids (up to in excess of 300 gr l^{-1} *T*otal *D*issolved *S*olids, TDS) which affect activity coefficients (see below) makes it at present impossible to back-calculate the solubility of the gases at formation pressures. Results are summarized in Table 2 as volume % concentrations. However, estimates based on solubility data provided by Duffy *et al.* (1961) suggest that in most, if not all cases the free gases collected at the borehole collar are present as dissolved constituents at formation pressures.

RESULTS AND DISCUSSION

Methane may be produced in inorganic systems and at elevated temperatures by reactions such as

$$CO_2 + 4\,H_2 \rightleftharpoons CH_4 + 2\,H_2O$$

This has been documented both in "terrestrial" geothermal systems as well as hydrothermal vents in midoceanic regions (Welhan, 1981; Welhan and Craig, 1983).

It could, potentially, also be produced during the hydrolysis of mafic rocks, i.e. for example during serpentization. Apps (1985) proposes that in such systems the following overall reaction may occur.

$$15\,FeO_{(silicates)} + 3H_2O + CO_2 \rightleftharpoons 5\,Fe_3O_2 + H_2 + CH_4$$

Geochemical modelling showed that, indeed, such reactions are possible. Furthermore, thermocatalytic methane generated in the cracking and decomposition of organic matter is produced at elevated temperatures in many sedimentary basins.

In the geothermal systems of the East Pacific Rise, carbon isotopic equilibria in the CH_4-CO_2 system have been inferred for temperatures above about 500°C (Welhan, 1981). At lower temperatures, the chemical reaction and similar reactions still proceed (Gerlach, 1980). However, carbon isotope equilibria are not necessarily established (Welhan, 1981, Welhan and Craig, 1982). Sackett and Chung (1979) report that no isotope exchange was recognizable in laboratory experiments with temperatures up to 500°C. Whether such exchange would have been visible if the experiment were done over months and years rather than days is open

Table 1.

The isotopic composition of methane, ethane and dissolved inorganic carbonate in salt waters and brines of the Canadian Shield.

LOCALITY and SAMPLE	CHLORIDE	ALKAL	CH_4-13C	CH_4-2H	C_2H_6-13C	C_2H_6-2H	DIC-13C	H_2O-2H
THOMPSON, Manitoba								
65124			−52.0	−280				
65178	123000	5					6.9	−77
44135	12000	26					−.9	
T3-27394-84	89200	20	−53.9	−299				−102
T3-27377-84	74400	24	−54.5	−285				−107
T3-STA 330-84	36100	13	−57.5	−279				−118
SUDBURY								
N3640A, JUNE 82	166200	43	−26.5	−162				−44
N3640B, APRIL 83	153000	0	−26.6	−155				−42
N3644B, APRIL 83	168000	0	−27.7	−203				−45
N3645, APRIL 83	55600	25	−25.0	−147				−73
N3646A, APRIL 83	162700	58	−28.4	−223				−39
FR33192	11000	9					−24.7	−78
L40140	15500	42					−14.2	−79
L40141	4160	95					−14.1	−80
CS508421	38	152					−11.2	−80
CS508481	11	129					−13.7	−80
CS205002	1470	107					−23.8	−82
CS205003	1280	19					−24.6	−98
CCS4000, '85			−36.4	−438	−24.8	−326		
STR31178, APRIL 83	67700	0	−29.0	−133				
YELLOWKNIFE – GIANT								
2000(NEAR C-SHAFT floor, '85)			−51.3	−330	−40.5			−144
YELLOWKNIFE – CON								
4301-1,N end drift,	28900	95	−51.2	−328	−42.5		−13.8	−152
4301-2,N drift, '85)	34130	87	−51.4	−360	−42.5		−14.1	−150
4500(1A), 1980	66400	27					−17.7	−125
4500(2), 1980	31900	69					−15.4	−145
4500-2,N drift '85)	38590	108	−49.6	−360	−44.8			−135
4500(3,1985)	18000	34	−48.4	−322			−19.4	−160
4500(6A), 1980	65900	65					−14.5	−123
4500(LAT-14650, '85)	—	—	−47.4		−41.7			—
ELLIOT LAKE								
JV-906	8035	30.6	−36.1	−308				−100
Denison 30876, 1985	1425	36	−37.4	−297				
MATAGAMI QUEBEC								
2002(90, R46)			−54.5	−396				
2004(90, R36, '84)	140680	54.6	−50.9	−437			−18.0	−63
2004(50, R28, '85)	1500	39.3					−16.3	−111
2000(90, R36, '85)			−53.7		−43.5		−20.2	−55
NORITA – QUEBEC								
N256 – 1985			−48.6	−470				
N18-4W-85¢-1985			−47.7	−423				−69
UN-243 (28)-1985	3417		−55.5	−398				
UN-257(28)-1985								−96
VAL D'OR, QUEBEC								
Sigma 10680, 1985	4950	376	−46.3	−301			−15.5	−97
Sigma 12042, 1984	10400	37.1	−46.3	−332			−16.3	−153
TIMMINS ONTARIO								
Kidd Creek 2185	37560	34.9	−39.8	−349	−37.5	−345	−24.0	−73.5
Kidd Creek 3172	38570	35.1	−41.8		−42.1			−68.4
RED LAKE ONTARIO								
Dickenson 25-320			−32.9	−276				
Dickenson 25-324			−38.8	−309				
Dickenson 26-102			−37.0	−289				
Dickenson 27-119			−38.7	−310				
Campell 2733, 1985	108700	<2	−40.5		−40.1			−60

Chemical concentrations are in mgle; ^{13}C-contents are δ‰ PDB values, ^{2}H-contents are δ‰ SMOW values.

Table 2.

The composition of gases discharging in mines on the Canadian Shield

	GAS COMPOSITION, volume %									
LOCALITY and SAMPLE	O_2	N_2	CO_2	CH_4	C_2H_6	C_3H_8	C_4+	He	Ar	H_2
THOMPSON, Manitoba										
65124	.06	30.7	<.06	64.8	1.5	.1	<.01	1.8	.35	.76
T3-27394-84	<.04	45	.3	54.3	2.23	.17	<.01	1.59	.54	N/A
T3-27377-84	.3	35.6	.46	64.3	3.05	.39	<.01	.79	.7	N/A
T3-STA 330-84	<.04	48.8	.26	49.1	1.85	.14	<.01	1.55	.53	N/A
SUDBURY										
N3640A, JUNE 82	N/A	79.9	<.1	9.75	N/A	N/A	N/A	5.3	N/A	N/A
N3640B, APRIL 83	<.3	79.2	<.3	8.7	.075	<.03	.025	6.8	3.6	N/A
N3644B, APRIL 83	N/A	77	<.1	14.4	.09	<.02	<.01	4.7	N/A	N/A
N3645, APRIL 83	<.5	77.2	<.3	12.5	.087	.04	.04	6.3	3.9	N/A
N3646A, APRIL 83	N/A	74.6	<.1	15.6	.09	<.01	<.01	6.7	N/A	N/A
CCS4000, '85	.17	3.28	<.02	55.1	9.43	.66	.22	5.17		30.3
STR31178, APRIL 83	<.1	83.3	<.1	.2	N/A	N/A	N/A	10.2	2	N/A
YELLOWKNIFE – GIANT										
2000 (NEAR C-SHAFT f	<.04	19.5	.46	64.4	4.99	.51	.08	7.5	.53	<.01
YELLOWKNIFE – CON										
4301-1,N end drift,	.52	24.5	.29	66.7	.93	N/A	N/A	6.4	.57	N/A
4301-2,N drift, '85)	.76	24.3	.11	61.1	2.7	.17	<.01	6.78	.05	<.01
4500-2,N drift, '85)	<.04	50.6	.43	40	.63	.03	<.01	6.64	.87	<.01
4500(3, 1985)	2.02	73.2	.07	15.6	.25	.02	<.01	7.8	1.01	<.01
4500(LAT-14650, '85	<.04	8.15	.58	87.6	1.68	.03	<.01	1.31	.35	<.01
ELLIOT LAKE										
JV-906	<.07	26.7	.11	54.3	.11	.02	.03	19.3	.75	N/A
Denison 30876, 1985	2.77	38.2	.08	44.2	.04	N/A	N/A	14.2	.55	N/A
MATAGAMI QUEBEC										
2002(90, R46)	.72	21.4	.22	75.9	1.55	.12	<.03	1.1	1.51	N/A
2004(90, R36, '84)	.24	11.1	.28	85.1	2.08	.24	<.03	1.6	.23	N/A
same sample, rep. '85	.48	11.3	<.02	82.1	1.81	.14	<.07	1.88	.01	N/A
NORITA – QUEBEC										
N256 – 1985	1.59	28.6	.46	62.3	2.1	.26	<.03	N/A	3.17	N/A
N18-4W-85-1985	<.07	6.73	.26	89.5	2.71	.18	<.03	1.03	.52	N/A
UN-243 (28)-1985	<.07	16.7	.1	78.8	1.91	.09	<.03	1.11	1.5	N/A
VAL D'OR, QUEBEC										
Sigma 10680, 1985	.07	55.1	.27	37.8	.66	N/A	N/A	N/A	5.61	.51
TIMMINS ONTARIO										
Kidd Creek 3185	.43	12.9	.1	76.1	3.72	.48	.12	1.41	.16	N/A
Kidd Creek 3172	4.12	31.1	.08	73.3	5.2	.87	.31	1.62	N/A	N/A
RED LAKE ONTARIO										
Dickenson 25-320	.28	54.38	.33	41.9	1.05	.22	.05	3.97	.24	N/A
Dickenson 25-324	.12	44	.39	52.7	1.37	.25	.06	3.62	.22	N/A
Dickenson 26-102	6.27	70.2	1.02	17.9	.29	.02	<.01	1.55	.45	N/A
Dickenson 27-119	<.01	61.9	2.55	30.4	1.55	.83	.22	3.34	.38	N/A
Campell 2733, 1985	<.04	34	.02	52.5	.63	.12	<.01	10.9	.53	<.01

to question. This contrasts with earlier observations by Lancet and Anders (1970) who note an approach towards equilibrium in the Fischer-Tropsch catalytic synthesis of methane at about 125°C or Barker (1979) who appeared to establish CO_2-CH_4 isotope equilibria to temperatures as low as about 300°C in the presence of a Ni-catalyst. Thus, it appears that if appropriate catalysts (and time?) are available, isotope exchange might be possible at rather low temperatures.

Hydrogen isotope equilibria are much easier established and equilibration between CH_4-H_2-H_2O in geothermal systems does occur to temperatures as low as 100°C (Welhan, 1981). No data are available for lower temperatures. However, equilibration does not exist in normal groundwaters in which methane is generated by bacteria or added through diffusion from other reservoirs, etc.

In most low temperature environments and normal groundwaters, the occurrence of methane is related to bacterial activities. However, as Stumm and Morgan (1981) point out (pg 456):

"with the exception of CH_4 no organic solutes encountered in natural waters are thermodynamically stable. For example, the disproportionation of acetic acid

$$CH_3\,COOH \rightleftharpoons 2H_2O + 2C\text{solid} \quad \log K = 18$$

$$CH_3\,COOH \rightleftharpoons CH_4\text{gas} + CO_2\text{gas} \; \log K = 9$$

is thermodynamically favoured but prevented by slow kinetics".

Thus, although thermodynamics would permit the inorganic production of methane, bacterial activities are usually required to promote methane generation in low temperature environments.

Methane-producing bacteria are often active in environments in which sulfate reduction (bacterially mediated) has gone to completion. Thus, a sequence of reduction reactions is established by bacterial populations. This sequence tends to coincide with a "theoretical" sequence expected in systems which become increasingly reducing.

It is thus interesting to note that several researchers have commented that in low-temperature systems where biogenic methane is produced isotopic equilibration does appear to occur. (Nakai, 1960; Galimov, 1969). However, the breakdown (fermentation) of organic matter produces through kinetic processes CO_2 and CH_4 which are isotopically far apart ($\delta^{13}C\text{-}CO_2 > 0$ ‰ $\delta^{13}C\text{-}CH_4 < -50$‰.) Although these differences are close to equilibrium values they probably do not reflect an isotope-equilibration process.

The observation, that biologically-mediated systems also appear to obey thermodynamic controls was the basis for a study by Thorstenson (1970) in which he investigated the "equilibrium distribution of small organic molecules in natural waters". He concluded that in marine environments (sediments) the predominant dissolved species associated with the decomposition of organic matter (including C, N and S species) tend to approach equilibrium concentrations with the exception of methane whose abundance was found to be often considerably higher than the equilibrium values.

Barker (1979), Barker and Fritz (1981a) made similar observations during a survey of methanes in groundwaters from Ontario, Manitoba and North Dakota. They note, that calculated PCH_4 values are many orders of magnitude smaller than measured ones.

The Canadian Shield methanes can almost certainly not be directly related to the marine system analysed by Thorstenson, nor the groundwater data of S.W. Ontario. However, in previous studies (Fritz and Frape 1982, Frape *et al.* 1984) we have suggested that the ^{18}O contents of the saline waters and brines are possibly controlled by low temperature exchange between rock-minerals and fluids. Since this must be an exceedingly slow process, low-temperature equilibria could potentially be important for these methanes and a short discussion of the present geochemical environments is warranted, with special emphasis on redox reactions.

Geochemical considerations

Thorstenson (1970) noted that in different marine sediments methane tended to be out of equilibrium with other species. However, other researchers suggested that such equilibria might exist and, for example, Edmunds et al. (1984) write about groundwater in the U.K. that

"the low Eh in these waters possibly reflects the decrease in Fe-total and the dominance of CH_4 and/or H_2 as the controlling redox species, effectively setting up a lower redox boundary".

Thus, these authors assume that methane actively participates in redox reactions which could be described in terms of the following equation:

$$CH_4 + 3H_2O \rightleftharpoons HCO_3^- + 9H^+ + 8e^-, \log K = -30 \quad (2)$$

$$pe = 0.125\,[30 - 9pH - 3\log aH_2O + \log \frac{aHCO_3^-}{aCH_4}] \quad (3)$$

In systems for which activity coefficients are known, it is then possible to determine whether observed pe, aCH_4 and $aHCO_3$ values correspond to those expected for equilibrium distributions. The Ionic Strength of Canadian Shield Brines varies between about 6 and 8, values which are similar to those found in the Dead Sea. There, Krumgalz (1982) determined simple ion activity coefficients for major ions which permit to estimate that the activity coefficient for HCO_3^- would be close to 2.0, although one must realize that the Dead Sea brines are Na dominated, whereas Canadian Shield brines have a Ca-Na-Cl chemistry. This may increase carbonate complexing and thus lower the activity coefficients (E.J. Reardon, pers. comm.).

For the activity coefficient γCH_4 Thorstenson (1970) assumed that $\gamma CH_4 = \gamma N_2 = 1.1$ is valid for "marine" systems, yet in the Shield brines, the value may be higher than 5 (Garrels and Christ, 1965). A value of $\gamma CH_4 = 5$ will be used in the present calculations.

For many purposes, the distribution of a gas between vapour and solution phases is described by Henry's law which states that

$$f_{gas} = K \times N_{gas\text{-}aq}$$

where f_{gas} = fugacity of the gas
K = Henry's law constant
N = Mol fraction of the gas

Henry's law constants for methane dissolving in saline solutions can be estimated from data presented by Duffy *et al.* (1961). Extrapolation yields for 10^5 Pa (1 bar) total pressure at the sampling point and salinities encountered in the Shield brines a value in the order of $K = 10^5$. For lower salinities this value will decrease. Assuming further that $fCH_4 \simeq pCH_4$ and that $aH_2O = 0.85$ the following relationship is obtained:

$$pe \rightleftharpoons 4.07 - 1.125\,pH - 0.125\log \frac{pCH_4}{mHCO_3}$$

This equation shows that these systems are very insensitive to changing methane concentrations or alkalinities but depend strongly on pH. Measurements of pH in lower salinity waters in the Canadian Shield are close to or above pH = 7 and tend to decrease to values close to 6 as brine chemistries are approached (Frape and Fritz, 1985). However, no estimates of the reliability of these measurements have been attempted.

Assuming a pH = 6 for the brines, with measured pCH_4 values in the free gas phase and titrated alkalinities (assumed| to be HCO_3^-) in the range of 3×10^{-4} moles l^{-1} then a pe value of −3.1 is calculated. This value is close to the pe value expected for the $CH_4 - HCO_3$ system. However, this "agreement" is not necessarily evidence for equilibrium conditions because equations used are rather insensitive to methane and bicarbonate concentrations.

Similar calculations are possible for the $CO_2 - CH_4$ system but, again, methane and carbon dioxide concentrations have much less effect than pH on calculated pe values.

A remarkable feature of the Canadian Shield brine and saline groundwaters is the coexistence of methane and sulfate. Although, from a thermodynamic point of view this does not have to be surprising it nevertheless does define a relatively narrow area of stability with respect to pe-pH conditions, if equilibrium were to exist.

In most groundwater systems the reduction of sulfate to H_2S or HS^- is thought to be biologically mediated. Whereby the "kinetics" of the reaction depend largely on the environmental conditions within which the bacteria have to exist. (Edmunds *et al.* 1984). Furthermore, bacterial sulfate reduction usually precedes the activities of methane producing bacteria. In lake sediments, for example, one typically observes the production of methane once sulfate reduction has gone close to completion. Thermodynamic data support the same sequence (Champ *et al.*, 1979), yet the similarity of pe^0 values for sulfate reducing and methane producing reactions emphasize that a fair degree of overlap does exist between the different systems.

We have undertaken a search for bacteria in the brines on the Canadian Shield and must conclude that it is safe to assume that bacterial activities in the deep groundwaters on the Canadian Shield are virtually non-existing. It is also important to note that H_2S was never detected in brines with salinities >100 g l^{-1} and can only rarely be found in boreholes with waters that have a component of fresher water, where bacterial activities may play a role. However, one has to consider that the formations of sulfide minerals can exert a dominant control on sulfide concentration. The absence of H_2S in the brines does thus not necessarily signify the non-existence of bacterial reduction.

Nevertheless, active biological reduction of sulfate is not likely to be important because otherwise it would be necessary to assume that the aqueous sulfate can be continuously replenished through dissolution of gypsum – which is the only ubiquitous sulfate mineral in these rocks. Instead, the reason for its presence is either a kinetic situation whereby low reaction rates maintain disequilibria during geological time spans or, more likely, sulfate is indeed a stable ionic species and redox conditions permit the coexistence of methane and SO_4.

Isotope data

General considerations

The carbon and hydrogen isotope analyses on methane summarized in Table 1 are graphically shown in Figure 3. A close grouping of samples from each locality and a positive correlation between $\delta^{13}C$ and δ^2H do appear to exist.

The distinct differences between methanes from different mining sites is a strong indication that regional rather than local (i.e. the immediate borehole environment) phenomena control the isotopic composition of the methane. If, for example, bacterial processes within individual boreholes were responsible for the methane production, then it would be very unlikely that the observed isotope distribution could be obtained, because bio-kinetic reaction rates and associated isotope effects tend to be very variable. This does not totally rule out bacterial processes in the groundwater around the mine workings, which could be fairly similar for a particular system.

However, the distribution also clearly indicates that no single source (not affected by secondary processes) can account for the grouping of the isotope data; i.e. a simple non-fractionating migration of methane from one source to the various regional scale of two isotopically different methanes, with Matagami and Sudbury type methanes being end members. However, at present, an interpretation based on mixing appears not to be warranted since data available show that the methane concentrations in the free gas phase of North-Sudbury boreholes are rather low (Table 2) so that extreme mixing ratios would have to be invoked, unless the gas moves as free gas rather than in association with its host fluids. (Note: North-Sudbury refers to mines in the North Range of the Sudbury basin, i.e. Strathcona, Fraser and North Mines).

The lower methane concentration in North-Sudbury waters as compared to areas with high methane concentrations such as Matagami-Norita could also reflect an evolutionary trend whereby, for example, methane "consumption" rather than differential migration would be important. Bacterial consumption of methane is a common process and might occur in these systems. However, in order to result in the regional distribution, it cannot be a process which is linked to individual boreholes but must be a regional phenomenon.

We have thus tentatively concluded that the isotopic composition of these methanes is most likely determined by their genesis *and* secondary processes. Time is thereby of special importance because reactions and isotope exchange which appear ineffective in laboratory experiments or during short geological timespans, might become important if many millions of years are available. To explore this further the following section discusses the biogenic and abiogenic formation and secondary modification of methanes in Shield environments.

At this point the reader may ask: what role plays geology and can specific relationships between host rocks and gas compositions be considered. We have attempted to explore this aspect but so far have not been able to define clear relationships. However, attention is paid to this aspect and further sampling and analyses may provide insights not recognized today.

Bacterial origin

Schoell (1983) proposed a classification of methanes based on their isotopic compositions and the amounts of associated higher hydrocarbons (C_{2+}). He notes that bacterial gases are devoid of most of these hydrocarbons (typically < 1%) and are depleted in ^{13}C. Methanes with $\delta^{13}C$ < −60‰ PDB were considered to be biogenic in nature whereas thermogenic methanes typically fall into the range of $\delta^{13}C$ = –15 to –55‰. Thus a "grey", undiagnostic area exists between about –50 and –60‰. Barker and Fritz (1981b), Woltemate *et al.* (1984) and others have shown that biogenic methanes can have $\delta^{13}C$ values above –60‰, if bacterial oxidation or diffusive loss has occurred (Coleman *et al.*, 1981; Edwards and Klassen, 1984).

Chemical data summarized in Table 2 show that many of the methanes are associated with higher hydrocarbons which exist in concentrations in excess of those usually

Figure 3. Comparison of S¹³C and S²H values of methane from different mining districts on the Canadian Shield.

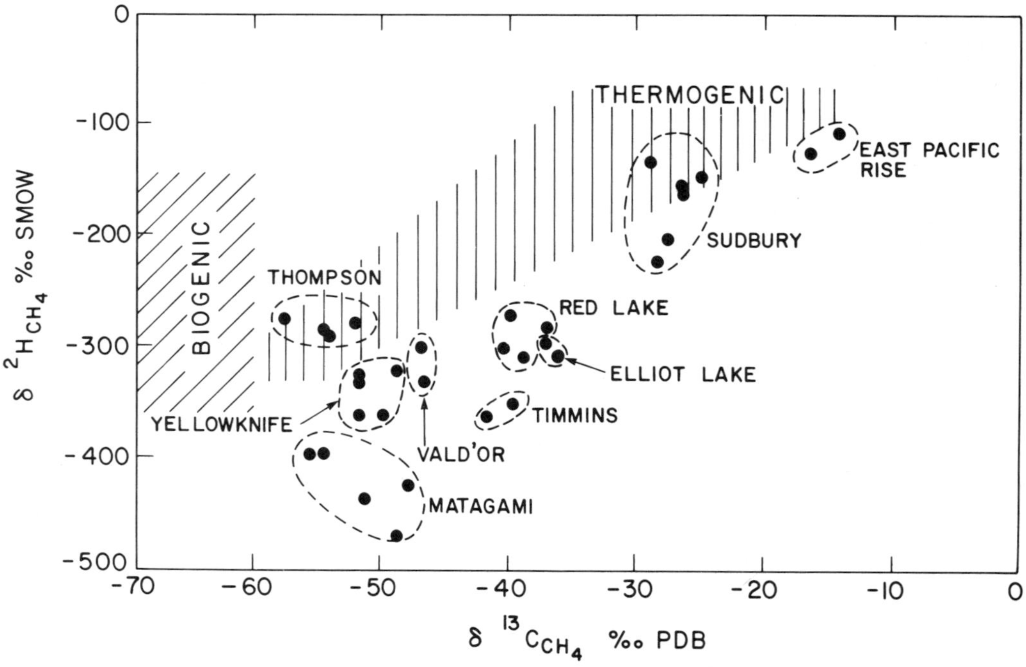

found in bacterial gases. This is a first argument against a biologic genesis of these gases.

It has also been shown that the deuterium contents in biogenic methanes are related to their genesis: Woltemate *et al.* (1984) showed that in a shallow fresh-water lake, approximately 70% of the biogenic methane is produced by fermentation of acetate whereas the remainder is generated by the reduction of CO_2. Accordingly, methanes produced by acetate fermentation receive the bulk of their hydrogen from the methyl-group of acetate, whereas the rest originates in the formation water. Thus, if acetate fermentation is the basis for methanogenesis then about 25% formation water hydrogen should be found in the methane, and Woltemate *et al.*, (1984) propose on the basis of field and experimental dates that:

$$\delta^2H_{CH_4} = 0.4\ \delta H_2O - 323‰$$

The δ^2H values of water associated with the Canadian Shield methanes range from –50 to –130‰. The lowest δ^2H values in the shield methanes should thus be close to –350‰. Most of the Shield gases fall into this range. However, those from Matagami are clearly below this tentative limit. It appears unlikely that the methanes at Matagami formed under conditions much different from those found elsewhere in the Canadian Shield and, therefore, one must tentatively conclude that, based on present knowledge, it is unlikely that these methanes have a biogenic origin.

The relatively narrow range of isotopic composition at the different mining sites is also of significance and provides additional evidence for an abiotic origin of this gas – provided the methanes preserved their primary isotopic compositions and have not been affected by secondary processes. The possible impact of secondary modifications is discussed below.

Thermogenic origin – Carbon-13

Abiotic formation of methane occurs primarily at elevated temperatures and involves either the alteration of organic matter or inorganic reactions in which, among others, carbon dioxide, hydrogen gas and water participate. This requires that methane must have formed either at depth (i.e. the lower crust or mantle) or during metamorphic/hydrothermal events. Both appear to be possible because methane discharges actively at oceanic rift zones and is an important and ubiquitous constituent in fluid inclusions of ores and minerals in the rocks of the Canadian Shield. In case of a metamorphic genesis the gas would have been trapped since these events – which would have consequences for discussions on the origin of the associated brines.

The deep gas-hypothesis has in the recent past been strongly advocated by T. Gold (e.g. Gold, 1979; Gold and Soter, 1982) who argue that it is possible to explain major accumulations of methane and associated hydrocarbons through mantle degassing and diffusive migration of methane to the surface.

"Degassing" is known to occur at ocean ridges (Welhan, 1981; Welhan and Craig, 1983). At the East Pacific Rise methanes are characterized by $\delta^{13}C$ values between –15 and –20‰ and are thought to be in isotopic equilibrium with associated carbon dioxide for temperatures between 500 and 700°C. Geochemical consideration clearly suggest an abiotic origin for this methane, although it is not yet known "whether H_2 and/or CH_4 are produced within the basalt system by chemical reactions or extracted from the basalts as an intrinsic mantle component" (Welhan and Craig, 1979).

The carbon isotopic composition of thermocatalytic

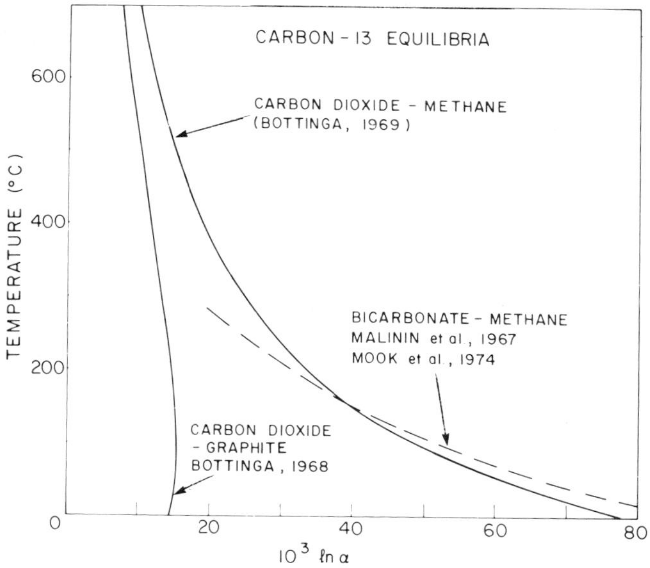

Figure 4. *Carbon isotope fractionation effects in the methane – carbon dioxide-bicarbonate system.*

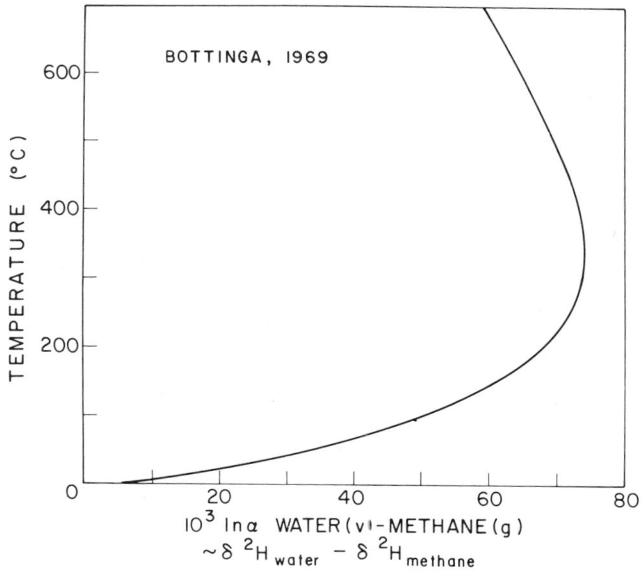

Figure 5. *Hydrogen isotope fractionation in the system methane – water.*

methane will depend on the carbon sources as well as thermal and pressure conditions. Carbon sources could include carbonate, carbon dioxide, graphite or even organic matter. The reaction would be similar to a Fischer-Tropsch synthesis, which produces abiotic methanes in reactions between CO, H_2 and NH_3 in the presence of a catalyst (Lancet and Andres, 1970). Similar to what is known for biological systems, kinetic effects dominate the abiotic formation of methane and it is not possible to predict the $\delta^{13}C$ values which could be generated in different environments.

However, it is instructive to investigate whether at present isotopic equilibria do exist between methane and carbon compounds, especially dissolved in organic carbon. Existence of such equilibrium would imply that these gases had a very long residence time in their present or similar host rocks.

The fractionation effects in the CH_4-CO_2-HCO_3^- system are very large (Bottinga, 1969; Richet *et al.*, 1977), and Figure 4 shows the isotopic differences to be expected under equilibrium conditions. A value of about 70‰ for the difference between the $\delta^{13}C$-CH_4 and $\delta^{13}C$-CO_2 or $\delta^{13}C$-HCO_3^- can be estimated for present environmental conditions i.e. temperature < 50°C.

No isotopic compositions for carbon dioxide in the Shield gases are available but the $\delta^{13}C$-DIC data listed in Table 1 and comparison with associated methanes clearly show that equilibration did not occur at ambient temperatures. In all cases the isotopic differences are much smaller and would reflect much higher equilibration temperatures than those existing today. The N. Sudbury gases would then be those formed at highest temperatures – provided equilibration occurred in all systems with the same gas phase. If average magmatic carbon dioxide were present then the upper temperature limit would be close to 300°C, i.e. substantially lower than temperatures at which relative fast equilibration does occur.

The lack of isotopic equilibrium does not preclude a quasi-chemical equilibrium, yet based on these isotope data it is obvious that these methanes are unlikely to be active participants in low temperature geochemical reactions.

Thermogenic origin – Deuterium

The deuterium contents of biogenic and abiogenic methanes overlap over a wide range. No clear separation between the two groups is possible, although biogenic methanes tend to have lower δ^2H contents than high temperature ones, Schoell (1980). Figure 5 depicts isotopic differences between water and methane over the 0-700°C temperature range and documents that maximum fractionation does not occur at low temperatures but rather at 300-400°C. However, even at these temperatures the isotopic difference is less than 80‰.

Therefore, if equilibration had occurred (and in geothermal systems hydrogen isotope equilibria are established faster and to lower temperatures than carbon isotope equilibria; Welhan, 1981), much higher δ^2H values than those measured should be expected. The Canadian Shield methanes should have δ^2H values above –200‰ and, for equilibria with the brines at present day temperatures values above –100‰ would be seen. Such is not the case, and isotopic equilibria between methane and water clearly have not been established for most if not all Shield methanes.

The exception could be the gases found at Sudbury, yet their rather variable composition between –130 and –220‰ also appears to preclude such equilibration.

One other process could explain the isotopic composition of these methanes, and especially their low δ^2H contents: Such low values are typical for hydrogen gases from the East Pacific rise and even lower values are found in geothermal areas such as in the Salton Sea Trough, at Yellowstone (Welhan, 1981) or on Iceland (Arnason, 1977). Hydrogen gas discharging in alkaline springs in Ophiolitic sequences in Oman has values close to –700‰ (Neal and Stanger, 1983), as has hydrogen found in wells penetrating the Kansas section of the American rift systems (Goebel et al., 1984).

Free hydrogen in geothermal systems is usually thought of as being associated with CO_2-H_2O-CH_4-H_2 reactions. However, the free hydrogen in Oman is probably being generated through reactions in the $Fe(OH)_2 - Fe_2O_3/Fe_3O_4 - H_2O$ system. For those discharging in Kansas, no specific mode of generation has been proposed and although a deep seated origin is favoured, a genesis similar to those found in Oman appears likely.

Free hydrogen has been encountered in several localities (see Table 2) but the high concentrations at Sudbury (~30 vol %) are surprising and could be seen as evidence that either geothermal hydrogen or hydrogen produced by the oxidation of reduced iron plays a role in the Canadian Shield methanes. As mentioned, the generation of free hydrogen in ultramafic rocks leads to very low δ^2H values for such gas. The deuterium contents of the Sudbury hydrogen has not yet been measured, but based on a combined H_2 + CH_4 isotope analysis of one sample, similarly low δ^2H values are to be expected. Thus, the presence of such deuterium depleted gas and its participation in the methane genesis could explain the very low δ^2H values of the Canadian Shield methanes.

A very different mechanism to generate hydrogen has been proposed by Vovk (1981) who suggests that over long timespans (> 10^8 years) the radiolytic decomposition of water in crystalline rocks could be important. Hydrogen would be produced (almost certainly strongly depleted in deuterium) and could lead to the generation of methane through a reaction with reduced carbon or carbon dioxide. At low temperatures, kinetic effects could be expected for both hydrogen and carbon isotopes. Again, the reaction could be compared to the Fischer-Tropsch synthesis for which very strong depletions in ^{13}C in the methane (e.g. Thompson) are known, although no hydrogen isotope effects have been measured. (Lancet and Andres, 1970).

This could explain the observed isotope distributions based on a combination of isotope effects and variable carbon sources, although this discussion is rather speculative and this system must be investigated further before more concrete statements are possible. Furthermore, the question about the preservation of the primary isotope contents is of crucial concern to the above discussions. A brief discussion of possible effects of bacterial oxidation and diffusion are, therefore, warranted.

Secondary processes – Bacterial oxidation

Bacterial oxidation is one secondary process which could profoundly modify the composition of the residual methane (Coleman *et al.*, 1981; Edwards and Klassen, 1984). Methane oxidizers use preferentially the light isotopes and thus enrich the remaining gas in both ^{13}C and ^2H.

Figures 6 and 7 compare methane concentrations with $\delta^{13}C$ and δ^2H values, respectively. Also indicated are evolutionary paths for residual methane following bacterial oxidation assuming that Rayleigh conditions exist where

a) changes in methane concentrations reflect only loss of methane and not addition of other gases. (This assumption is approximately correct, although helium concentrations are quite variable. For example, helium might have been added at Elliot Lake during a long geochemical evolution in a uranium rich environment. In these gases helium comprises up to 20% of the free gas phase).
b) the bacterial isotope effects are constant. Barker and Fritz (1981b) determined experimentally the kinetic effects on $\delta^{13}C$ during methane oxidation. A depletion factor of close to –25‰ was found. However, depending on environmental conditions, this value could be somewhat lower or higher. Thus the assumption of constancy may not be correct. This is also true for deuterium effects which Coleman *et al.*, (1981) determined to be 8 to 14 times higher than those for ^{13}C.

In both cases the data available do not fit a simple Rayleigh distillation model even if only the highest methane concentration (least affected by dilution with other gases) are compared. Considering the regional origin of these samples and the assumptions made, this is not surprising and it may be more instructive to compare ^2H and ^{13}C values directly. In Figure 8 evolutionary "paths" define an area in which the ^2H enrichment is between 8 and 14 times greater than ^{13}C enrichment. Using a $\delta^{13}C$ = –60‰

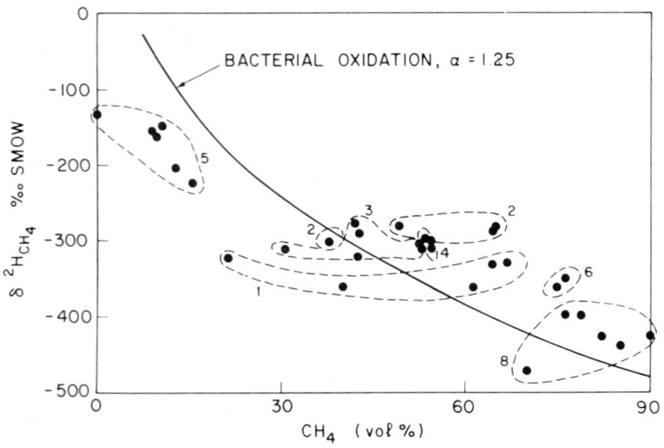

Figure 6. *Methane concentrations (vol %) versus S²H SMOW values. Numbers refer to Fig. 1.*

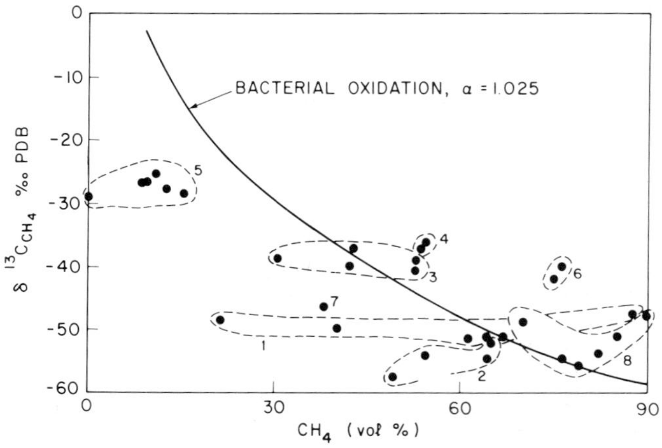

Figure 7. *Methane concentrations (vol %) versus S¹³C PDB values. Numbers refer to Fig. 1.*

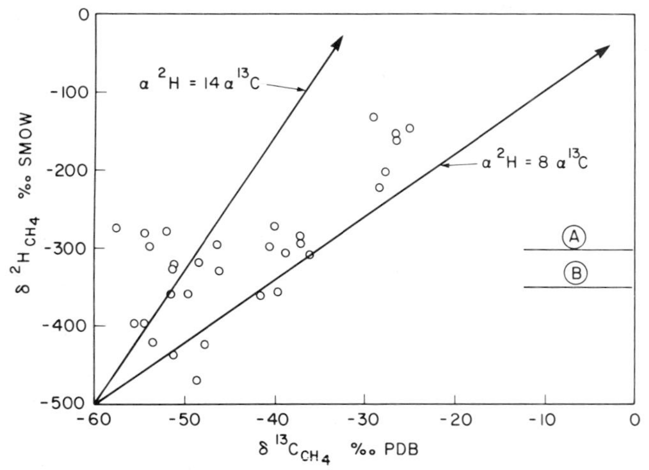

Figure 8. *Isotope effects expected in the residual methane following partial bacterial consumption. Consumption of deuterium bearing methane is 8 to 14 times faster than the use of carbon-13 methane. This is indicated by the two trend lines.*

and δ^2H = –500‰ as "primary" source value one must conclude that the majority of the samples *could* have been affected by bacterial oxidation.

However, this interpretation must also explain the regional distribution pattern of the isotope contents and methane concentrations. Biological systems are dominated by kinetic isotope effects, and it would be very surprising if the observed regional pattern were maintained if bacterial consumption would occur in boreholes following drilling activities rather than in undisturbed systems. As pointed out above, we have no evidence that biological consumption occurs before or after drilling but if it occurs in undisturbed systems a quasi steady state between the ratio of consumption and supply must have been reached. Although not impossible, this appears to be an unlikely concept.

Secondary processes – Diffusion

Diffusion and biological oxidation remove preferentially the isotopically light species from a given methane reservoir. It is usually assumed that the isotope effects are inverse proportional to the square root of the masses. However, for methane this may not be true (Edwards and Fritz, in prep.) although the general direction of the effects does not change.

Thus, if a mantle origin were invoked to explain the origin of the Shield gases, and if mantle gases are represented by methanes discharging at the East Pacific Rise, then most gases collected in this study would represent accumulations of isotopically depleted methane which had migrated from a reservoir, whose "heavier" portion would have to be found at greater depths.

The data available cannot be used to discuss this aspect in detail because nothing is known about migration path lengths. However, the shorter in time and length the migration path, the less effective would be the isotope separation process. For example, the gases at Sudbury could be closer to a mantle source or be less fractionated because of higher rock permeabilities than found at Thompson or Matagami. (Consider, that the Sudbury basin is postulated to be an impact crater.) In this case a positive correlation between isotopic enrichments and methane concentrations would have to be expected, yet for the data available a clear negative correlation is observed between δ^2H or $\delta^{13}C$ values of methane and its concentration (Figure 6 and 7). This makes it difficult to use diffusion from a mantle source as transport mechanism especially since continued accumulations and high methane contents (i.e. low concentration gradients *not* absolute amounts per rock mass) would have to reduce the isotope effects. One must conclude that such diffusion alone is unlikely to account for the observed concentration/isotope relationships.

This negative correlation could, however, suggest a different diffusion process: rather than considering the samples as accumulations of gas coming from a deep source they could reflect residuals of a larger pool from which light species were preferentially lost. This could explain not only the ^2H and ^{13}C values but also the isotope-concentration relationship. Note, that bacterial consumption of methane would also produce this effect. It would imply that the Thompson-Matagami methanes have better preserved their primary compositions than the gases at Sudbury..., an observation which is not substantiated by isotope data from other crystalline rock or high temperature environments.

CONCLUSIONS

Methane occurs in mines across the Canadian Shield. Isotopic and geochemical analyses show that this gas has an abiotic genesis, whereby each mining district releases gases with an unique isotopic composition which is not readily related to a specific source or generating mechanism. It is unlikely that one specific environment (e.g. the mantle) is the source of these methanes.

If an unique source with isotopically unique methane were to exist, then secondary processes such as bacterial oxidation and/or diffusion would have to modify concentrations and isotopic compositions. An apparent negative correlation does exist between ^{13}C contents and methane concentrations. Should this be verified in future studies, then it is very unlikely that bacterial oxidation with variable isotope effects plays a major role. Not excluded, however, are diffusion effects.

If diffusion does modify isotopic compositions and methane concentrations, then the most simple model would assume that the source gas is similar in composition to the Thompson-Matagami methanes. However, these gases are strongly depleted in ^{13}C and do *not* resemble "known" mantle or deep crustal gases. Unless the basic assumption or isotopic composition to deep crustal or mantle gases are wrong, their genesis is not as easily explained as would be the case for the Sudbury gases which have a typical thermocatalytic signature.

The assumption which would most easily reconcile most present data would state that in each mining district an isotopically unique gas was generated during metamorphic/hydrothermal reactions. This, however, does not explain the apparent positive correlation between δ^2H and $\delta^{13}C$ because the hydrogen and carbon cycle in these systems are not implicitly linked, especially if one argues that radiolysis played a major role in the genesis of the low ^2H gases.

Geochemical considerations show that the CH_4-HCO_3-CO_2 system could be at least close to chemical equilibrium and that redox reactions would thus play an active role in the distribution of these species. However, isotope analyses document that methane composition is far from equilibrium with respect to CO_2-HCO_3^- and H_2O.

The significance of these observations is not yet fully explored and ongoing studies will utilize rare gas chemical and isotopic analyses, ^{15}N determinations and a closer comparison with the sulfur and carbon isotope data from associated compounds in an attempt to answer the question of origin and/or geochemical history of these methanes.

ACKNOWLEDGEMENTS

The authors are greatly indebted to Dr. M. Schoell and Dr. J. Welhan for a critical review of this manuscript. All isotope analyses are done at the University of Waterloo with assistance of R.J. Drimmie, T.W. Edwards, R. Heemsherk and G. Nower. Chemical analyses were done at Waterloo (R. Kellerman) and by AECL in Pinawa, a contribution for which we are most grateful. This project was carried out with a NSERC Strategic grant to P. Fritz and S.K. Frape and received additional support from NSERC operating grants to P. Fritz and S.K. Frape.

REFERENCES

Apps, J.A., 1985. Methane formation during hydrolysis by mafic rocks: Lawrence Berkeley Laboratory, University of California, Annual Report 1984, p. 13-17.

Arnason, B., 1977. The hydrogen-water Isotope Thermometer applied to geothermal areas in Island: Geothermics, v. 5, p. 75-80.

Barker, J.F., 1979. Methane in groundwaters – a carbon isotope geochemical study: Ph.D thesis, University of Waterloo, 1979, manuscript.

Barker, J.F. and Fritz, P., 1981a. The occurrence of methane in some groundwater flow systems: Canadian Journal of Earth Sciences, v. 18, p. 1802-1816.

Barker, F.J. and Fritz, P., 1981b. Carbon isotope fractionation during bacterial methane oxidation: Nature, v. 293, p. 289-291.

Bogomolov, G.V., Atroschenko, P.O., Bogomolov, Y.G. and Tsalko, P.B. 1980. Nature of brines and gas and vol shows in blocks of crystalline rock: Doklady Akademiya Nauka SSSR, v. 238, p. 1192-1195 (translated).

Bottinga, Y., 1969. Calculated fractionation factors for carbon and hydrogen isotope exchange in the system calcite-CO_2-graphite-methane-hydrogen: Geochimica et Cosmochimica Acta, v. 33, p. 49-64.

Brown, M., 1982. Fuids in metamorphism – a conference report: Episodes, v. 1982, p. 37-38.

Champ, D.R., Gulens, J. and Jackson, R.E., 1979. Oxidation reduction sequences in groundwaters: Canadian Journal of Earth Sciences, v. 16, p. 12-23.

Coleman, D.O., Risatti, J.B. and Schoell, M., 1981. Fractionation of carbon and hydrogen isotopes by methane-oxidizing bacteria: Geochimica et Cosmochimica Acta, v. 45, p. 1033-1037.

Duffy, J.R., Smith, N.O. and Nagy, B., 1961. Solubility of natural gases in aqueous salt solutions – I: Geochimica et Cosmochimica Acta, v. 24, p. 23-31.

Edmunds, W.M., Miles, D.C. and Cook, J.M., 1984. A comparative study of sequential redox processes in three British aquifers: Proceedings of Symposium Hydrochemical Balances of Freshwater Systems (E. Eriksson, Editor). IAHS – Symposium, Uppsala, 1984. IAHS Publication No. 150, p. 55-70.

Edwards, T.W.D. and Fritz, P., in prep. The isotopic composition of methane in organic sediments of small lakes in Ontario, Canada.

Edwards, T.W.D. and Klassen, R.A., 1984. Preliminary isotopic and chemical characterization on natural gas in sediments of Lake Harrington (Quebec) and Golden Lake (Ontario): Geological Survey of Canada, Paper 84-1B, p. 237-243.

Frape, S.K. and Fritz, P., 1982. The chemistry and isotopic composition of saline groundwaters from the Sudbury Basin, Ontario: Canadian Journal of Earth Sciences, v. 19, p. 645-661.

Frape, S.K. and Fritz, P., 1985. Redox measurements in groundwaters of the Canadian Shield: Final Report to AECL, May 1985, unpublished manuscript.

Frape, S.K., Fritz, P. and McNutt, R.H., 1984. The role of water-rock interaction in the chemical evolution of groundwaters from the Canadian Shield: Geochimica et Cosmochimica Acta, v. 48, p. 1617-1627.

Fritz, P. and Frape, S.K., 1982. Saline groundwaters in the Canadian Shield – a first overview: Chemical Geology, v. 36, p. 179-190.

Fritz, P., Frape, S.K., Andrews, J.N. and Macko, S. (in prep.). The geochemistry of deep gases in the Canadian Shield: For submission to Applied Geochemistry.

Galimov, E.M., 1969. Isotopic composition of carbon in gases of the crust: International Geology Review, v. 11, p. 1029-1104.

Garrels, R.M. and Christ, C.L., 1965. Solutions, Minerals and Equilibria: Harper and Row, New York, 1965.

Gerlach, T.M., 1980. Chemical characteristics of the volcanic gases from Nyiragongo lava lake and the generation of CH_4-rich fluid inclusions in alkaline rocks: Journal of Volcanology and Geothermal Research, v. 8, p. 177-189.

Goebel, E., Conveney Jr., R.M., Angino, E., Zeller, E.J. and Dreschhoff, G.A.M., 1984. Geology, composition, isotopes of naturally occurring H_2/N_2 rich gas from wells near Junction City, Kansas: Oil and Gas Journal, March 7, 1984, p. 215-221.

Gold, T., 1979. Terrestrial sources of carbon and earthquake outgassing: Journal of Petroleum Geology, v. 1, p. 3-19.

Gold, T. and Soter, S. 1982. Abiogenic methane and the origin of petroleum: Energy, Exploration and Exploitation, v. 1, p. 89-104.

Guha, J., Leroy, J. and Guha, D. 1979. Significance of fluid phases associated with shear zone Cu-Au-mineralization in Dore Lake complex. Chibougamau, Quebec. Bulletin Mineralogie, v. 102, p. 569-576.

Kreulen, R. and Schuiling, R.D. 1982. N_2-CH_4-CO_2 fluids during formation of the Dome de l'Agout, France: Geochimica et Cosmochimica Acta, v. 46, p. 193-203.

Krumgalz, B., 1982. Physico-chemical study of Dead Sea waters: Marine Chemistry, v. 11, p. 209-222.

Lancet, H.S. and Anders, E., 1970. Carbon isotope fractionation in the Fischer-Tropsch synthesis of methane: Science, v. 170, p. 980-982.

Malinin, S.D.; Kroptova, D.I. and Grinenico, V.A., 1967. Experimental determination of carbon isotope exchange constants in the CO_2-HCO_3^- system under hydrothermal conditions: Geokhimiya, v. 8, p. 927-935.

Mook, W.G., Bommerson, J.C. and Staverman, W.H., 1974. Carbon isotope fractionation between dissolved bicarbonate and gaseous carbon dioxide: Earth and Planetary Science Letters, v. 22, p. 169-176.

Neal, C. and Stanger, G., 1983. Hydrogen generation from mantle source rocks in Oman: Earth and Planetary Science Letters, v. 60, p. 315-321.

Nakai, N., 1960. Carbon isotope fractionation in natural gas in Japan: Journal of Earth Sciences, v. 8, p. 174-180.

Nurmi, P.A. and Kukkonen, I.T., 1986. Geochemistry of water and gas from deep drill holes: a new sampling technique: Canadian Journal of Earth Sciences, v. 23, in press.

Ramboz, C., Schnapper, D. and Dubessy, J., 1985. The P-V-T-X-fO_2 evolution of H_2O-CO_2-CH_4-bearing fluid in a wolframite vein: Reconstruction from fluid inclusion studies: Geochimica et Cosmochimica Acta, v. 49, p. 205-220.

Ritchet, P., Bottinga, Y. and Javoy, M., 1977. A review of hydrogen, carbon, nitrogen oxygen sulfur, chlorine stable isotope fractionation among gaseous molecules: Annual Review of Earth and Planetary Sciences, v. 5, p. 65-110.

Sackett, W.M. and Chung, H.M., 1979. Experimental confirmation of the lack of carbon isotope exchange between methane and carbon oxides at high temperatures: Geochimica et Cosmochimica Acta, v. 43, p. 273-276.

Schoell, M., 1980. The hydrogen and carbon isotopic composition of methane from natural gas of various origins: Geochimica et Cosmochimica Acta, v. 44, p. 649-662.

Schoell, M. 1983. Genetic characterization of natural gases: American Association of Petroleum Geologists Bulletin, v. 67, p. 2225-2238.

Stumm, W. and Morgan, J.J., 1981. Aquatic Chemistry: 2nd ed: New York, John Wiley and Sons, p. 780.

Thorstenson, D.C., 1970. Equilibrium distribution of small organic molecules in natural waters: Geochimica et Cosmochimica Acta, v. 34, p. 745-770.

Vovk, I.F., 1981. Radiolytic model of formation of brine compositions in the crystalline basement of shields: Geochimica, v. 4, p. 467-480, translated in Geochemistry International, 1981, p. 80-93.

Welhan, J.A. and Craig, H., 1979. Methane and hydrogen in East Pacific Rise hydrothermal fluids: Geophysical Research Letters, v. 6, p. 829-831.

Welhan, J.A., 1981. Carbon and hydrogen gases in hydrothermal systems: the search for a mantle source: Ph.D thesis, University of California, manuscript.

Welhan, J.A. and Craig, H., 1982. Abiogenic methane in mid-ocean ridge hydrothermal fluids: In: Proceedings of the Deep Source Gas Workshop, (W.J.G. Williams, Editor). Morgantown, Virginia, 1982. DOE/METC 82-50, p. 122-129.

Welhan, J.A. and Craig, H., 1983. Methane, hydrogen and helium in hydrothermal fluids at 21°N on the East Pacific Rise: In: Hydrothermal Processes at Seafloor Spreading Centers (Eds. P.A. Rona, K. Bostrom, L. Laubier and K.L. Smith, Jr.) Plenum Press, New York, 1983, p. 391-409.

Woltemate, I, Whiticar, M.J. and Schoell, M., 1984. Carbon and hydrogen isotopic composition of bacterial methane in a shallow fresh water lake: Limnology and Oceanography, v. 29, p. 985-992.

Characteristics of Abiotic Methane in Rocks

John A. Welhan[1]

Geological Research Division, Scripps Institution of Oceanography, La Jolla, California 92093, U.S.A.

[1]Present address: Department of Earth Sciences, Memorial University of Newfoundland, St. John's, Newfoundland, Canada A1B 3X5

Welhan, John A.; Characteristics of abiotic methane in rocks; in Saline water and gases in crystalline rocks, Editors: Fritz, P., and Frape, S.K.; Geological Association of Canada Special Paper 33, 1987. p 225–233

Abstract

Methane in mid-ocean ridge (MOR) hydrothermal fluids can be shown to be derived abiotically from MOR basalts, together with carbon dioxide and helium of mantle origin, by the leaching action of fluids circulating in the crust. Carbon isotope ratios in this methane are distinctive (–15 to –18 ‰ vs PDB), and, in light of experimental evidence for high-temperature carbon isotope exchange between carbon dioxide and methane, very likely reflect isotopic equilibrium with mantle CO_2 (δ^{13} = –5 to –7 ‰ vs PDB) at temperatures of 500-800°C within the basalt. Lower temperature equilibration appears not to affect the methane isotope ratios significantly due to the sluggishness of carbon isotope exchange at T <500°C. Available data on igneous rocks which are rich in ^{13}C-enriched methane, suggest that high-temperature, abiotically-synthesized methane preserves its unusual isotopic signature at low temperatures for very long times. It is postulated that the ^{13}C-enriched character of such high-temperature methane, and the characteristic carbon isotope compositions of co-existing light hydrocarbon gases, could be a distinctive tracer for abiotically-synthesized methane, and that the high $^{13}C/^{12}C$ ratios of methane found in deep brines of the Canadian Shield may reflect an abiotic origin within these rocks.

Resume

Le methane dans les fluides hydrothermales des raies de la mer originait dans les basaltes, ainsi que l'acide carbonique et helium qui sont derive du manteau de la terre. Les rapports isotopiques du carbon dans cet methane sont distinct (–15 ‰ a – 18 ‰ PDB), et les concentrations d'ethane et des autres hydrocarbones sont tres faibles. L'evidence experimental pour l'echange isotopique du carbon indique ces valeurs representent l'equilibre isotopique avec CO_2 du manteau a 500-800°C, dans les basaltes. Les donnees sur les roches igne qui contenait le methane enriche en ^{13}C, indique le methane inorganique, produit par les temperatures elevees, retenait longtemps son rapport isotopique du carbon. Il est conjecture que le methane enriche en ^{13}C dans les saumures en profondeur du Bouclier Canadien, avec rapports des hydrocarbones et leur compositions isotopiques characteristiques, peut indique une origine inorganique du methane dans ces roches granitiques.

INTRODUCTION

The occurence of methane in deep brines of the Canadian Shield, with widely variable carbon isotope and hydrogen isotope ratios (Fritz and Frape, this volume), has raised the question of its origin. Low $\delta^{13}C$ values (<–50 ‰ vs PDB) are characteristic of biologically-synthesized (biogenic) gas, whereas thermocatalytic (thermogenic) natural gases produced from pre-existing organic matter have –50 < $\delta^{13}C$ < –20 ‰ vs PDB (Stahl, 1977). The existence of deep brine methane with $\delta^{13}C$ approaching –20 ‰ also raises the possibility of an abiogenic origin, via inorganic chemical reactions in the mantle or deep crust, or possibly within the rocks which enclose these brines. Abiotic synthesis is poorly defined in terms of the chemical and isotopic characteristics of the synthesized gas. Thus, in what follows, data on abiotic methane in hydrothermal fluids and rocks are reviewed, and some general characteristics of abiotically-synthesized gas are summarized. This review, it is hoped, will help place in perspective some of the results which have been obtained on light hydrocarbon gases in deep brines, and suggests some diagnostic tests for determining the origin(s) of these gases.

CARBON GASES IN ROCKS

Data on carbon in mantle-derived rocks has produced conflicting evidence on the nature of primordial carbon in the mantle. Early studies pointed out the presence of a so-called "reduced" carbon component in igneous rocks with $\delta^{13}C \sim -20$ to -25 ‰ vs PDB (Craig, 1953; Fuex and Baker, 1973), which suggested either an organic carbon contamination, or a dispersed hydrocarbon-like component presumably of abiotic nature. However, recent studies utilizing crushing and step-heating experiments have shown that oceanic tholeiites contain predominantly oxidized carbon (ie. CO_2) with an isotopic composition of ca. -5 to -10 ‰ vs PDB (within the range of mantle carbon; Pineau et al, 1976; Moore et al, 1977; DesMarais and Moore, 1984), and that the isotopically light carbon component is most likely a contaminant (DesMarais and Moore, 1984; Sakai et al, 1984). In addition, methane has been shown to be present in small quantities, together with traces of carbon monoxide (Pineau et al, 1976; Zolotarev et al, 1981).

Therefore, it appears that MORB carbon is characterized by: 1) primarily oxidized carbon species, with minor methane; and 2) carbon isotope ratios near that of the postulated primordial mantle carbon (although recent hypotheses concerning recycling of carbon, water, etc. through the upper mantle suggest that the primordial nature of this carbon may long ago have been lost). The CO_2 found in these MORB rocks has been analyzed for its carbon isotope ratio, and found to be in the range of primordial carbon, but the isotopic composition of the associated methane has not been directly measured. Thus, what is undoubtedly *the* best example of a terrestrial abiotic hydrocarbon has not been isotopically characterized directly, due to the experimental difficulties associated with such small quantities of gas in these rocks.

Fortunately, indirect isotopic measurements of this methane are available, from studies of high-temperature East Pacific Rise hydrothermal fluids which contain relatively large quantities of methane not obviously related to sedimentary/crustal sources, in close association with CO_2 and He of mantle derivation, where numerous lines of evidence support an abiotic origin within the mid-ocean ridge basalt itself. In the following section, pertinent data and evidence for the abiotic origin of this methane are reviewed, as a background to subsequent discussion of methane in continental igneous rocks, particularly in hydrocarbon-rich igneous rocks.

ABIOTIC METHANE AT MID-OCEAN RIDGES

The mid-ocean ridge hydrothermal system at 21°N on the East Pacific Rise (EPR) was the first high-temperature hydrothermal system that was discovered and sampled by manned submersibles (Speiss et al, 1980). It is also the most intensely studied of the high-temperature oceanic spreading ridge hydrothermal systems, of which there are now about a dozen known. The 21°N system is the type example for abiotic methane, and in what follows, data from this system will be reviewed (see Welhan and Craig, 1982, 1983, for details).

Several dissolved tracers have been used to locate high-temperature venting in the oceans, among them manganese, helium (eg: Lupton et al, 1980) and methane (Kim et al, 1983). What is most conspicuous about these hydrothermally-injected compounds is their high degree of covariation in the water column, reflecting their derivation in the high-temperature fluids debouching into the ocean (eg: Lupton et al, 1980). Helium has two natural isotopes – ^3He and ^4He – of which only the latter is produced in appreciable quantity by radioactive decay and nuclear interactions in the earth's crust and mantle (Craig and Lupton, 1981). Helium which is injected into the oceans at hydrothermal sites has a $^3He/^4He$ isotopic ratio 8-10 times higher than atmospheric helium, and represents a true primordial helium source, enriched in the rare isotope, which is still degassing from the mantle some 4.5 Gyrs after accretion. The tremendous enrichment of the mantle helium isotope ratio compared to atmospheric or crustal (radiogenic) helium makes it an extremely sensitive tracer of mantle outgassing even in continental crustal environments (eg: Craig et al, 1978).

Helium in hydrothermal fluids at MOR spreading centres is derived from leaching of fresh MOR basalt by high-temperature fluids, and possibly from direct magma outgassing, as well (Craig and Lupton, 1981; Lupton, 1983). This helium is accompanied by high concentrations of CO_2, H_2 and CH_4 (Craig et al, 1980; Welhan and Craig, 1979; Figure 1). Like helium, the isotopic composition of MOR hydrothermal CO_2 mimics that found in MOR basalt vesicles, with $\delta^{13}C \sim -5$ to -7 ‰ vs PDB (Pineau et al, 1976;

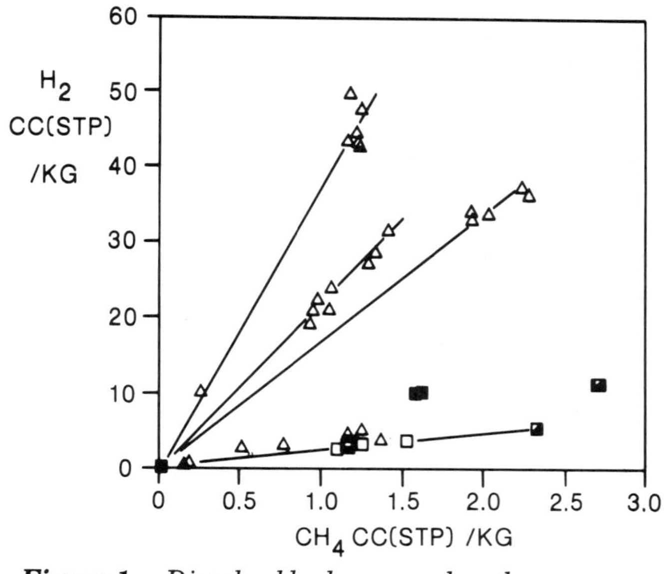

Figure 1. *Dissolved hydrogen and methane concentrations in 21°N (triangles), 13°N (open squares) and 11°N (diagonally marked squares) vent fluids, showing range of gas concentrations in high-temperature mid-ocean ridge hydrothermal vents. Lines show differences in H_2/CH_4 ratios among different vent fields. Data from Welhan et al (1984) and unpublished data.*

Moore et al, 1977). Thus, leaching of MOR basalts by the hydrothermal fluid extracts not only helium but CO_2 as well. Unlike the proportion of H_2 in the fluid, which varies by an order of magnitude relative to helium (Welhan and Criag, 1983), the $CH_4/^3He$ ratio in four major hydrothermal systems on the EPR varies by only a factor of two, suggesting a common source (Welhan et al, 1984).

Source of methane

Hydrogen isotope ratios in these hydrothermal gases indicate that isotopic equilibrium has proceeded down to temperatures near the exit temperatures of these hydrothermal vents, ca. 350°C (Table 1). Thus, hydrogen isotope data are of little value in determining gas origins (Welhan and Craig, 1983). In contrast, C-isotope ratios of 21°N CH_4 are very high compared to methane of continental hydrothermal systems and natural gases (Figure 2; DesMarais et al, 1981; Lyon and Hulston, 1984). These isotopic ratios indicate either that high-temperature carbon isotope exchange with CO_2 has occured, an hypothesis which has lost credibility in light of recent continental geothermal gas studies (eg: DesMarais et al, 1981; see next section), or that methane is derived from more conventional thermocatalytic processes acting on buried organic matter, which must have undergone extreme thermal maturation in this case. Derivation of these high carbon isotope ratios by microbial methane oxidation (eg: Barker and Fritz, 1981) must be discounted because of the high temperatures involved as well as the observed D/H ratios in methane (δD = –126 to –102 ‰ vs SMOW; Welhan and Craig, 1983), which are too high to be explained by such a process.

In order to gain an appreciation of the relative contribution of thermocatalytic or "thermogenic" gas in the 21°N system, it is instructive to compare the hydrocarbon gases in a seafloor spreading centre of similar temperature, in which thermogenic gas production is known to be the predominant source of methane and higher hydrocarbons. The Guaymas Basin spreading centre in the Gulf of California (Figure 3) is such an environment. There, basaltic volcanic material is intruded into the thick, organic-rich sediments of the Gulf, rather than extruded onto virgin oceanic crust, as at 21°N. A complete spectrum of hydrocarbons, up to and beyond the gasoline range, has been identified in the sediments and hydrothermal fluids (Simoneit, 1983), and is directly attributable to thermal decomposition of sedimentary organic matter due to recent intrusion of sills into the organic-rich sediments (Einsele et al, 1980).

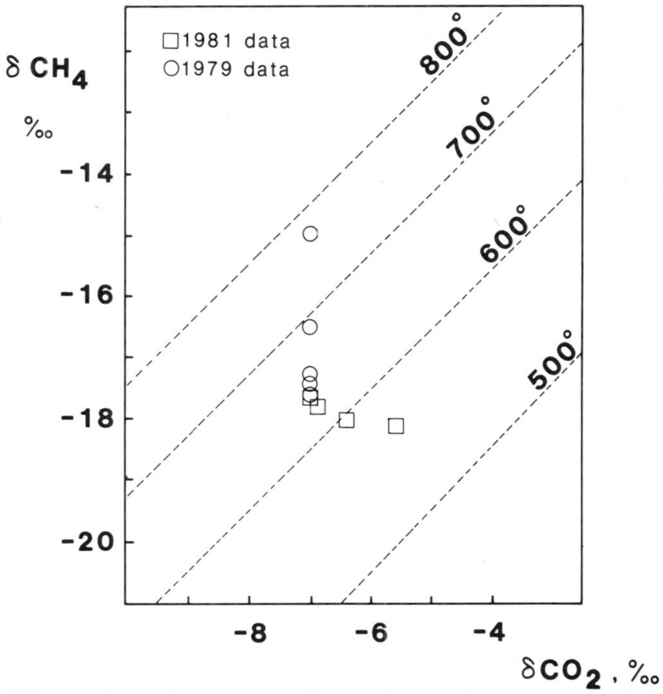

Figure 2. *Summary of isotopic compositions of 21°N East Pacific Rise methane and carbon dioxide. Data from Welhan and Craig (1983), Craig et al (1980) and unpublished work. Isotherms calculated from partition functions of Richet et al (1977).*

A comparison of light hydrocarbon gases and methane carbon isotopes in Guaymas Basin and 21°N fluids is shown in Table 2. The isotopic composition of CH_4 and the hydrocarbon concentrations in Guaymas Basin fluids reflect an overwhelming input of thermocatalytically-derived hydrocarbons (Welhan and Craig, 1982). Hydrocarbon concentrations, including CH_4, are up to 100 times higher than in 21°N fluids, and C_2^+/CH_4 ratios are ten times higher. Relatively high contents of C_2^+ hydrocarbons are characteristic of thermogenic hydrocarbons, and the low carbon isotope ratios at Guaymas (–43 to –51 ‰ vs PDB) are typical of thermogenic natural gas (Stahl, 1977).

These characteristics of Guaymas Basin hydrothermal hydrocarbons are in sharp contrast to the low C_2^+/CH_4 and

Table 1

Summary of isotopic data and isotope geothermometer results for 350°C vent fields at 21°N, East Pacific Rise.

Vent[1]	$\delta D(H_2)$, ‰ SMOW	$\delta^{13}(CH_4)$, ‰ PDB	$T(H_2\text{-}H_2O)$[4], °C	$T(CH_4\text{-}CO_2)$[4], °C
A[2]	–396	–15 to –17.7[3]	310	640-770
B	–383	–17.8	330	630
C	–343	–18.0	370	600
D	n.m.[5]	–18.1	n.m.	570

[1] Data from Welhan and Craig (1982; 1983); unpublished data.
[2] Vents arranged geographically from northeast to southwest along ridge axis. Spacing between vents typically 1-7 km.
[3] This vent displayed significant changes in exit temperature from 1979 (350°C) to 1981 (273°C), as well as in carbon isotope ratios.
[4] Temperatures calculated from tabulated partition functions ratios of Richet et al (1977), using observed $\delta D(H_2O)$ and $\delta^{13}C(CO_2)$ values (Craig et al, 1980; unpublished data).
[5] Not measured.

Table 2

Comparison of light hydrocarbon gases in 300+ °C hydrothermal fluids from 21°N and Guaymas Basin (after Welhan and Craig, 1982).

	T, °C	$C_2^+/CH_4 \times 10^4$ C_2H_6	C_2H_4	C_3H_8	$\delta^{13}(CH_4)$
EPR 21°N					
A	273	10.4	n.d.	8.6[1]	−17.7
B	350	6.2	7.4	6.2	−17.8
C	350	9.2	4.0	n.d.	−18.0
D	350	4.0	3.0	n.d.	−18.1
Guaymas Basin	315	134	0.16	9.1[1]	−44 to −50

[1] Also contains C_4 and C_5 gases.

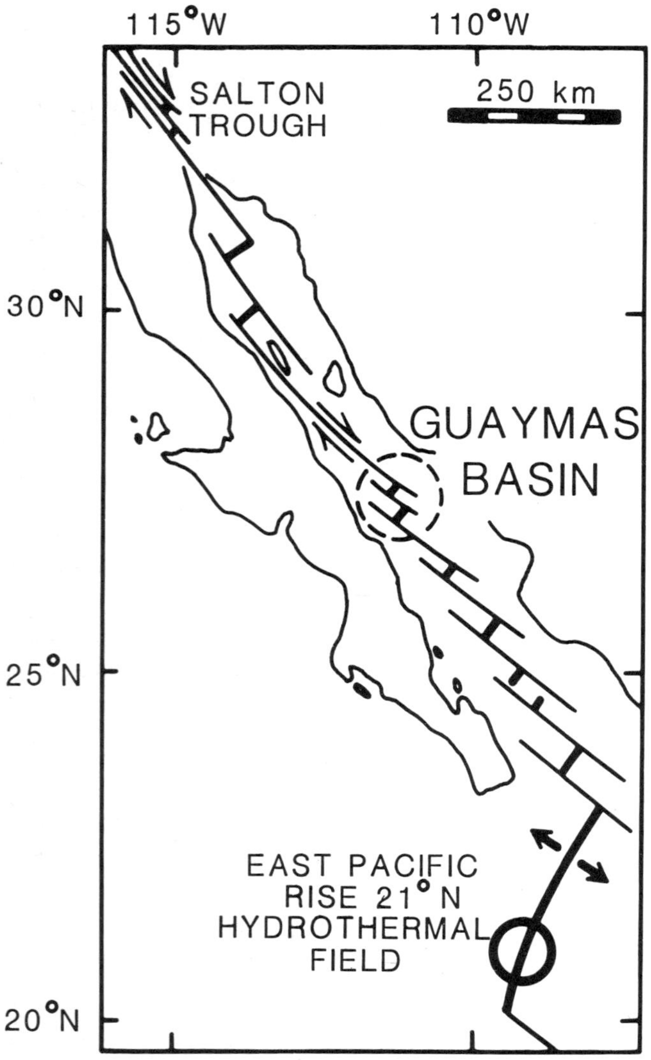

Figure 3. *Location of Guaymas Basin hydrothermal field relative to 21°N area, on the East Pacific Rise-Gulf of California rift/transform fault system.*

$\delta^{13}C$ values seen at 21°N, and are mutually consistent with a predominantly thermogenic origin of Guaymas methane (Welhan and Criag, 1982). The very low C_2^+/CH_4 ratios in 21°N gases are similar to biogenic gases, such as the low-temperature Orca Basin brine, Gulf of Mexico (Sackett et al, 1979), but the extreme enrichment of ^{13}C in 21°N methane compared to such biogenic gases (eg: $\delta^{13}CH_4 = -73.5$ ‰ vs PDB in Orca brines) rules out any possibility of a biogenic origin for the 21°N gases. Although the carbon isotope data and light hydrocarbon composition data are not conclusive proof of gas origin, taken together they reinforce the difference in CH_4 origins between the Guaymas Basin and 21°N hydrothermal systems. The similarities in hydrothermal expression, temperature, and the tectonic and volcanic setting of these two high-temperature systems, and the extreme differences in the role of sedimentary organic carbon, as expressed in methane carbon isotopes and light hydrocarbon composition, points to a predominantly non-thermogenic, abiogenic origin for hydrothermal methane at 21°N.

Perhaps the strongest evidence for an abiotic methane source at 21°N is found in the proportion of methane to other gases (H_2, CO_2 and He) in the fluids and in the MOR basalts. As shown in Table 3, CH_4/CO_2 and CH_4/H_2 ratios in MOR basalts are remarkably similar to those in 21°N fluids, and preliminary data on $CH_4/^3He$ ratios in MOR basalts indicate ratios very similar to those in the fluids, as

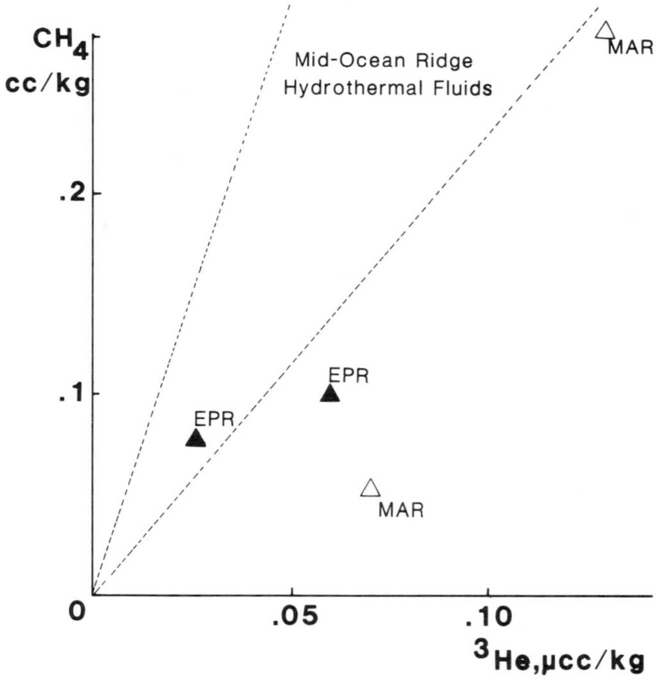

Figure 4. *Data on methane and helium in mid-ocean ridge basalt glasses, analyzed by crushing and gas chromatography (updated from Welhan and Craig, 1983). Helium normalized to mantle helium-3. Range of $CH_4/^3He$ ratios in 11°N, 13°N and 21°N hydrothermal fluids shown as dashed lines. MAR = Mid-Atlantic Ridge; EPR = East Pacific Rise.*

well (Welhan and Craig, 1983; Figure 4). Furthermore, the very narrow range in $CH_4/^3He$ ratios found in several different high-temperature hydrothermal systems along the EPR (Welhan et al, 1984) suggests a common basaltic source. Short of a direct comparison between methane carbon isotope ratios in rocks and fluids, the above data provide a very strong argument for an abiotic methane source, originating within MOR basalt, which is leached from the rocks by fluids circulating within the crust.

Interpretation of carbon isotope ratios

The evidence for an abiogenic origin for CH_4 (as well as CO_2) begs the question of the meaning of the "temperature" derived from the distribution of carbon isotopes between CH_4 and CO_2. As shown in Figure 2, these temperatures, as calculated from theoretical partition functions (Richet et al, 1977), range from ca. 500-800°C. In comparison, this geothermometer commonly yields temperatures of ~ 300-400°C in continental high-temperature geothermal systems, (eg: Lyon and Hulston, 1984). It has been suggested (Craig, 1963) that $\delta^{13}C$ values of hydrothermal methane may reflect an organic source, so that the calculated temperature based on the CH_4-CO_2 isotopic geothermometer is only fortuitously close to actual reservoir temperatures. More recently, DesMarais et al (1981) have shown that observed $CH_4/C_2{}^+/CO_2$ proportions do not indicate chemical equilibrium, and that the isotopic compositions of $C_1 - C_6$ hydrocarbon gases in high-temperature geothermal systems are wholly consistent with a thermogenic origin, thereby casting strong doubt on the validity of CH_4-CO_2 isotopic geothermometry. In such systems, then, the dual origin of the carbon gases (organic CH_4 vs volcanic/magmatic CO_2) precludes a rational interpretation of calculated isotopic equilibrium temperatures, except possibly in a qualitative sense (DesMarais et al, 1981).

Furthermore, experimental evidence supporting the very existence of carbon isotope exchange between CH_4 and CO_2 has been lacking (Sackett and Chung, 1979). More recently, Harting and Maass (1980) published data which support the concept of carbon isotope exchange, but only at high temperatures (Figure 5). Using ^{13}C-enriched methane, they heated CO_2 and CH_4 together in closed glass vials for 16-hour periods, and measured the change in isotopic composition of the CO_2. Their experimental data show that carbon isotope exchange is very slow (ie. immeasurable over the duration of their runs) up to temperatures of about 500°C, but with increasingly favourable kinetics at T >600°C (Figure 5). Also shown in Figure 5 are exchange times for 50% attainment of equilibrium, calculated by Giggenbach (1982). Giggenbach's calculations indicate that for 99% attainment of isotopic equilibration, times of the order of 10^7 years would be required at 300°C. This suggests that most continental geothermal methanes cannot represent isotopic equilibrium with associated CO_2, and that high-temperature exchange events would tend *not* to be masked by low-temperature re-equilibration.

On the premise that methane, CO_2 and H_2O are present in the basalt, and that gaseous equilibrium occurs in discrete vesicles and inclusions at sub-solidus temperatures, then the CH_4/CO_2 ratio represents a temperature of equilibration at a given oxygen fugacity. Gerlach (1980) has shown that, in C-H-O gas mixtures in which oxygen fugacity is internally-controlled by gas equilibria, CH_4 does not become appreciable even at very low temperatures. On the other hand, in a system in which oxygen fugacity is controlled externally by a silicate oxygen buffer, CH_4/CO_2 ratios increase rapidly with decreasing temperature. Such calculations show that the CH_4/CO_2 ratios observed in 21°N hydrothermal fluids and in MOR basalts (~ 0.01) are consistent with C-H-O gas-phase chemical equilibration in vesicles and inclusions within the basalt at temperatures of ~ 500-700°C in a quartz-fayalite-magnetite fO_2-buffered system. These temperatures are remarkably similar to those inferred from carbon isotope distributions between hydrothermal methane and carbon dioxide, and further support a high-temperature origin for these gases within the basalt itself.

Thus, if MOR hydrothermal methane is synthesized within the basalts, it would have sufficient time, on kinetic grounds, to equilibrate isotopically with CO_2 at high temperatures; the correspondence between carbon isotopic temperatures and those inferred from CH_4/CO_2 ratios suggests that the carbon isotope geothermometer is valid and representative of a high-temperature methane origin in MOR hydrothermal systems. Furthermore, the ^{13}C-enriched character of high-temperature abiotic methane would tend to be preserved at lower temperatures, due to the unfavourable kinetics of carbon-isotope exchange at low temperature.

Table 3

Comparison of relative CH_4, CO_2 and H_2 concentrations in mid-ocean ridge basalts and hydrothermal fluids.

Source (no. of samples)	CH_4/CO_2	CH_4/H_2
Fluids		
21°N EPR (n=33)[1]	0.0091-0.0114	0.027-0.18
13°N EPR (n=7)[2]	n.m.	0.25-0.38
11°N EPR (n=5)[2]	n.m.	0.15-0.25
Basalts		
MAR (n=1)[3]	0.0127	0.0525
MAR (n=7)[4]	0.0040-0.0167	0.051-0.137
EPR (n=2)[1]	n.m.	0.11-0.29

[1] Welhan and Craig (1983); unpublished data.
[2] Welhan et al (1984).
[3] Pineau et al (1976).
[4] Zolotarev et al (1981).

ABIOTIC METHANE IN IGNEOUS ROCKS

In the preceeding pages, the evidence for methane of inorganic origin in some hydrothermal systems has been summarized. An important conclusion arising from this work has been that abiotic methane is present in MOR basalt, and that the carbon isotopic composition of methane coexisting with CO_2 at high temperature in the rock is controlled by isotopic equilibrium. An important consequence of this conclusion is that, in systems where the carbon gases are dominated by CO_2 with $\delta^{13}C$ greater than about −10 ‰ vs PDB, methane in high-temperature (>400°C) isotopic equilibrium with such CO_2 would be characterized by $\delta^{13}C$ values > − 20 ‰ vs PDB, and would be isotopically distinguishable from thermogenic or biogenic gases of low-temperature origin, which generally have $\delta^{13}C < -20$ ‰ vs PDB.

Factors which would tend to obliterate this distinction include: a) alteration of methane and/or CO_2 isotopic compositions by secondary processes such as bacterial methane oxidation; and b) dilution of the abiotic methane isotopic signal by biologic and thermocatalytic methane sources. Consideration of these factors, together with associated evidence for a deep magmatic gas flux beneath one of the resurgent domes of the Yellowstone caldera, has provided convincing evidence for a high-temperature, ^{13}C-enriched magmatic methane source ($\delta^{13}C = -18.9$ ‰ vs PDB) in a continental hydrothermal system quite different from the MOR environment, and has demonstrated that detection of abiotic methane is possible in a geochemically complex continental setting through its carbon isotope signature (Welhan et al, 1983; Welhan et al, in prep.).

The obvious question, from the standpoint of deep continental brine gases, is how much abiotic methane can be expected in crystalline rocks of the continental crust, and can it reasonably be expected to retain its isotopic signature as a means of source identification? Occurences of heavy hydrocarbons showing some evidence for abiogenic origins have been classified by Sylvester-Bradley and King (1963) as: 1) vein and hydrothermal hydrocarbons (many of which show strong evidence for biogenic origins), such as those associated with Au, U, and Hg ores; and 2) igneous hydrocarbons, such as those of the Kola Peninsula, Russia, which appear to be the best candidates for an abiogenic derivation. In what follows, published data from the Khibiny and Lovozero massifs on the Kola Peninsula and the Ilimaussaq massif in southern Greenland are examined in order to assess the possibility of detecting abiotic methane in crystalline igneous rocks, as well as generalizing about its chemical and isotopic characteristics.

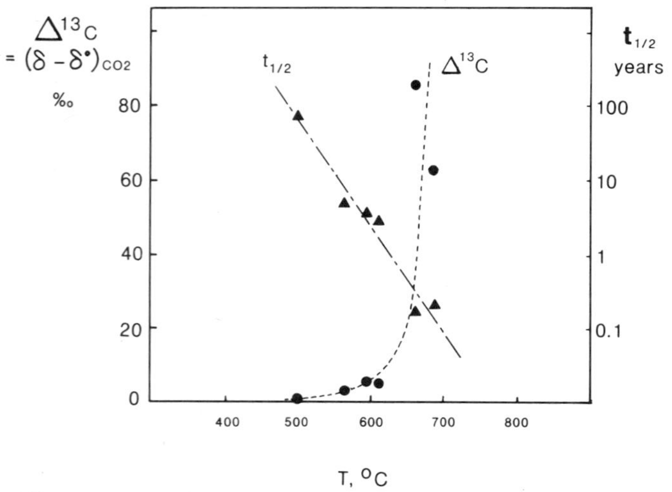

Figure 5. *Plot of experimental data of Harting and Maass (1980), supporting high-temperature carbon isotope exchange between 66% carbon-13 enriched methane and natural carbon dioxide. Also shown are calculated half-times for attainment of exchange equilibrium, based on these data, as calculated by Giggenbach (1982).*

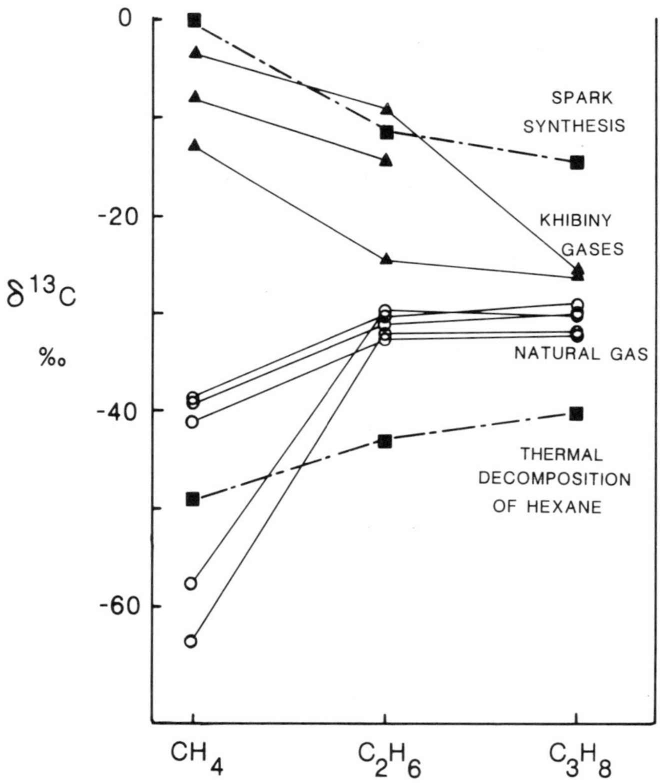

Figure 6. *Comparison of relative carbon isotope ratios between $C_1 - C_3$ hydrocarbons in igneous rocks (triangles) and natural gases (circles), after Khitarov et al (1979) and Galimov (1969). Also shown are experimental data (squares) from DesMarais et al (1981) for synthesis of hydrocarbons by a) sparking in a methane atmosphere; and b) thermal decomposition of hexane. For clarity, the scales for the experimental data have been arbitrarily shifted to avoid overlap.*

The Khibiny massif on the Kola Peninsula is classified as an agpaitic alkaline intrusive complex (Sorensen, 1974), characterized by modal nepheline and sodalite and very high hydrocarbon contents in inclusions of primary mineral phases. Light hydrocarbon gases, up to 200 cc (STP)/kg of rock, have been reported in these rocks (Petersil'ye and Pripachkin, 1979), as well as heavier hydrocarbons up to bituminous compounds. Geologic relationships such as concentric zonation of hydrocarbon concentrations within the pluton and a correlation of $CO_2/CH_4/C_2H_6$ ratios with pegmatite mineralogy and ore type, suggest that methane and other hydrocarbons are produced within the cooling alkaline magma, rather than introduced from without (Petersil'ye and Pripachkin, 1979). A case has been made for a deep, possibly mantle source for free gas emanations associated with this complex (Khitarov et al, 1979), since the present gas flux from the massif would deplete the intrusive of its occluded gas in about 1 million years, whereas the age of the complex is 280-300 million years.

Published hydrocarbon isotopic data for the Khibiny rocks are summarized in Table 4, together with published data on other agpaitic alkaline rocks. The most unusual feature of these data is emphasized in Figure 6, where the carbon isotope ratio of $C_1 - C_3$ gases is plotted versus number of molecular carbon atoms. Unlike thermogenic hydrothermal gases and natural gases, which show low $\delta^{13}CH_4$ values and an increase in $^{13}C/^{12}C$ with increasing alkane molecular weight (due to kinetic fractionations accompanying thermal cracking), the Khibiny gases demonstrate the reverse effect. This trend is quite contrary to that expected for thermal cracking, but is consistent with a polymerization process whereby the heavier alkanes are synthesized from methane; experimental verification of this effect has been given by DesMarais et al (1981; Figure 6). Thus, the carbon isotope data give strong support to an abiotic origin for these hydrocarbons, although the actual processes and mode of origin remain unclear. Methane and trace amounts of CO_2 in these inclusions may be related isotopically, since CO_2 is up to 19 ‰ enriched in ^{13}C relative to CH_4 (Galimov, 1969), as would be expected if it were in isotopic equilibrium with methane at temperatures of ~ 300°C. Such a temperature is comparable to late-stage fluid inclusion filling temperatures of 200-500°C (Sobolev et al, 1974), and provides further support for a late stage magmatic or hydrothermal origin for methane in these rocks.

The similarity of methane carbon isotope ratios in several different agpaitic intrusive complexes (Table 4) suggests a common origin for these hydrocarbons. Based on fluid inclusion studies of the Kola and Ilimaussaq intrusives (eg: Sobolev et al, 1974; Konnerup-Madsen et al, 1981), common genetic factors in the development of these plutons appear to be: 1) relatively low-temperature, late-stage fluid participation; 2) low oxygen fugacities (below QFM during most or all of the sub-solidus cooling history; and 3) predominance of magmatic CO_2 in early, high-temperature inclusions, as opposed to methane and higher hydrocarbons which dominate lower-temperature, late-stage inclusions, in which CO_2 is a minor constituent.

It is important to note that in a closed system, such as a gaseous inclusion, where low-temperature chemical equilibration increases the proportion of methane relative to CO_2, the isotopic composition of methane approaches that of the total carbon in the system as temperature decreases. Thus, at high temperature, CO_2 is virtually the sole carbon species present, with, for example, $\delta^{13}C$ = –7 ‰; as temperature decreases, methane becomes the dominant carbon species, but by mass balance,

$$\delta(CH_4) = \delta(\Sigma C) - \Delta(1 - CH_4/\Sigma C)$$

where $\Delta = \delta(CO_2) - \delta(CH_4)$ and $\Sigma C = CH_4 + CO_2$. Thus, as $CH_4/\Sigma C \rightarrow 1$, $\delta(CH_4) \rightarrow -7$‰. In this sense, the high $^{13}C/^{12}C$ ratios in Kola Peninsula and Ilimaussaq rocks can be explained by chemical and isotopic equilibration of C-H-O gases at low temperatures, leading to high CH_4/CO_2 ratios and high methane carbon isotope ratios. Such an interpretation is consistent with the inferred low temperatures of late-stage inclusions in these rocks, their low apparent oxygen fugacities, and the predominance of CO_2 in early, high-temperature inclusions.

CONCLUSIONS

The study of hydrocarbon-rich agpaitic igneous rocks indicates that methane of abiotic origin, characterized by high $^{13}C/^{12}C$ and CH_4/CO_2 ratios, is a constituent of continental igneous rocks, although its importance in more siliceous granitoid rocks is unknown. However, compared with the MOR basalts, which are characterized by very low methane contents (reflecting higher oxygen fugacities and/or temperatures than the agpaitic rocks) and apparently similar high carbon isotope ratios (reflecting high-T isotopic exchange with CO_2), these data indicate that abiotically-synthesized methane may occur in a wide spectrum of petrologic systems. Certainly, available data on granitic and pegmatitic rocks show no paucity of detectable methane: published analyses of gases in normal granitic rocks and pegmatites show average methane contents ranging from 0.3 cc(STP)/kg to 4.2 cc(STP)/kg in granites and pegmatites, respectively (Roedder, 1972).

If abiotic methane occurs in granitic rocks of the Precambrian shield, and if its isotopic composition reflects high-temperature synthesis from igneous CO_2, the kinetic data of Harting and Maass (1980; Figure 5) suggest that its isotopic signature should be retained at low temperatures

Table 4

Comparison of carbon isotope ratios in hydrocarbons in Khibiny, Lovozero and Ilimaussaq agpaitic alkaline intrusions.

Source	CO_2/CH_4	$\delta^{13}(CH_4)$	$\delta^{13}(C_2H_6)$	$\delta^{13}(C_3H_8)$
Khibiny				
Urtite[1]	~0	−12.8	−24.5	−26.0
Khibinite[1]	n.m.	−3.2	−9.1	−25.7
Khibinite[2]	0	−13.2		
Eudialyte[1]	n.m.	−7.9	−14.2	
Eudialyte[2]	0.005	−8.4		
Pegmatite feldspar[2]	0.086	−14.6		
Ijolite[2]	0	−8.4		
Gneiss[2]	n.m.	−6.0		
Gas seeps from ijolite/urtite[1]	0	−11.8 to −19.3		
Lovozero[2]	0.0015	−5.3		
Ilimaussaq[2]	~0	−7.7		

[1] Khitarov et al (1979).
[2] Galimov (1969).

for a considerable time. In fact, the carbon isotope data on Ilimaussaq and Kola rocks suggest that the high-temperature, ^{13}C-enriched character of abiotic methane *does* survive long periods of time (> 10^7 years) at low-temperature.

By analogy with the MOR environment, wherein fluids acquire high concentrations of rock-derived gases, including methane, through interaction with the rock, such a source of abiotically-synthesized methane within the granitoid rocks of the Precambrian shield could be leached by deep groundwaters (and brines) over much longer periods of time at low temperature. In the absence of bacterial fractionations, such a hydrocarbon source should remain identifiable through its carbon isotope signature (CO_2 liberated in such a manner need not be represented in the fluid, due to calcite deposition in Ca-rich brines).

However, given the isotopic compositions of hydrocarbons which have been observed in deep brines (Fritz et al, this volume) and the ubiquity of thermocatalytically-produced hydrocarbons in the earth's crust, verification of an abiotic component will be difficult. For example, the data presented by Fritz et al. (this volume) on the Sudbury deep brine gases (which contain the lowest hydrocarbon concentrations), could be interpreted to represent minimal augmentation by thermogenic gas. Hence, of the Canadian Shield brine methanes, they show the highest $\delta^{13}C$ values (to –25 ‰ PDB) and the lowest C_2^+ concentrations (<0.3 volume %), consistent with a predominantly abiotic derivation. But as pointed out by Fritz et al, these data can also be interpreted in more conventional terms. Furthermore, it should be stressed that in light of the isotopic complexities introduced by bacterial processes, a much better understanding of microbiologic factors in the biogenesis and biodegradation of hydrocarbons will be required, if we are to understand the origins of these deep brine gases, and ultimately of the brines themselves.

ACKNOWLEDGEMENTS

The author wishes to thank those at the Scripps Institution of Oceanography who participated in the mid-ocean ridge diving programs during the collection of hydrothermal samples, noteably H. Craig, K. Kim, and J. Lupton, and who contributed so much to the understanding of mantle outgassing beneath the oceans. Gratitude is also due the U.S. National Science Foundation which funded the author's work while at Scripps, and to the National Science and Engineering Research Council of Canada for support to continue these studies of mantle gases in hydrothermal systems and in rocks.

REFERENCES

Barker, J.F. and Fritz, P., 1981. Carbon isotopic fractionation during microbial methane oxidation: Nature,v. 293, p. 289-291.

Craig, H., 1953. The geochemistry of the stable carbon isotopes: Geochimica et Cosmochimica Acta, v. 3, p. 53-92.

Craig, H., 1963. The isotopic geochemistry of water and carbon in geothermal areas: In: Nuclear Geology on Geothermal Areas, Spoleto, Italy, p. 17-53.

Craig, H. and Lupton, J.E., 1981. Helium-3 and mantle volatiles in the ocean and the oceanic crust: In: The Sea, C. Emiliani ed., Volume 7, p. 391-428.

Craig, H., Lupton, J.E., Welhan, J.A. and Poreda, R., 1978. Helium isotope ratios in Yellowstone and Lassen Park volcanic gases: Geophysical Research Letters, v. 5, p. 897-900.

Craig, H., Welhan, J.A., Kim, K.R., Poreda, R. and Lupton, J.E., 1980. Geochemical studies of the 21°N East Pacific Rise hydrothermal fluids, Transactions of the American Geophysical Union, v. 61, p. 992.

DesMarais, D.J. and Moore, J.G., 1984. Carbon and its isotopes in mid-ocean basaltic glasses: Earth and Planetary Science Letters,v. 69, p. 43-57.

DesMarais, D.J., Donchin, J.H., Nehring, N.L. and Truesdell, A.H., 1981. Molecular carbon isotopic evidence for the origin of geothermal hydrocarbons: Nature, v. 292, p. 826-828.

Einsele, G. and eighteen others, 1980. Intrusion of basaltic sills into highly porous sediments, and resulting hydrothermal activity: Nature, v. 283, p. 441-445.

Fuex, A.N. and Baker, D.R., 1973. Stable carbon isotopes in selected granitic, mafic and ultramafic rocks: Geochimica et Cosmochimica Acta, v. 37, p. 2509-2521.

Galimov, E.M., 1969. Isotopic composition of carbon in gases of the crust: International Geology Review, v. 11, p. 1092-1104.

Gerlach, T.M., 1980. Chemical characteristics of the volcanic gases from Nyiragongo lava lake and the generation of CH_4-rich fluid inclusions in alkaline rocks: Journal of Volcanology and Geothermal Research, v. 8, p. 177-189.

Giggenbach, W.F., 1982. Carbon-13 exchange between CO and CH_4 under geothermal conditions: Geochimica et Cosmochimica Acta, v. 46, p. 159-165.

Harting, P. and Maass, I., 1980. New results on carbon isotope exchange in the methane-carbon dioxide system (in German): In: Mitteilungen zur 2. Arbeitstagung 'Isotope in der Natur', November 1979, Vol 2b, Leipzig, p. 13-24.

Khitarov, N.I., Kravtsov, A.I., Voitov, G.I., Fridman, A.I., Ortenberg, N.A. and Pavlov, A.S., 1979. Free emanation gases of the Khibiny massif (in Russian): Sovetskaya Geologiya 1979(2), 62-73.

Kim, K.R., Craig, H. and Horibe, Y., 1983. Methane: a real-time tracer for submarine hydrothermal systems: Transactions of the American Geophysical Union, v. 64, p. 724.

Konnerup-Madsen, J., Rose-Hansen, J. and Larsen, E., 1981. Hydrocarbon gases associated with alkaline igneous activity: evidence from compositions of fluid inclusions: Rapport Grønlands Geologiske Undersøgelse, v. 103, 99-108.

Lupton, J.E., 1983. Terrestrial inert gases: isotope tracer studies and clues to primordial compositions in the mantle: Annual Review of Earth Planetary Sciences, v. 11, p. 371-414.

Lupton, J.E., Klinkhammer, G.P., Normark, W.R., Haymon, R., MacDonald, K.C., Weiss, R.F. and Craig, H., 1980. Helium-3 and manganese at the 21 N East Pacific Rise hydrothermal site: Earth and Planetary Science Letters, v. 50, p. 115-127.

Lyon, G.L. and Hulston, J.R., 1984. Carbon and hydrogen isotopic compositions of New Zealand geothermal gases: Geochimica et Cosmochimica Acta, v. 48, p. 1161-1171.

Moore, J.G., Batchelder, J.N. and Cunningham, C.G., 1977. CO_2-filled vesicles in mid-ocean ridge basalt: Journal of Volcanology and Geothermal Research, v. 2, p. 309-327.

Petersil'ye, I.A. and Pripachkin, V.A., 1979. Hydrogen, carbon, nitrogen and helium in gases from igneous rocks: Geochemistry International, v. 16(4), p. 50-55.

Pineau, F., Javoy, M. and Bottinga, Y., 1976. $^{13}C/^{12}C$ ratios of rocks and inclusions in popping rocks of the Mid-Atlantic Ridge: their bearing on the problem of deep-seated carbon: Earth and Planetary Science Letters, v. 29, p. 413-421.

Richet, P., Bottinga, Y. and Javoy, M., 1977. A review of hydrogen, carbon, nitrogen, oxygen, sulphur and chlorine stable isotope fractionation among gaseous molecules: Annual Review of Earth Planetary Sciences, v. 5, p. 65-110.

Roedder, E., 1972. Composition of fluid inclusions: In: Data of Geochemistry Chapter JJ, U.S.G.S. Professional Paper 440-JJ, 164 p.

Sackett, W.M. and Chung, M., 1979. Experimental confirmation of the lack of carbon isotope exchange between methane and carbon oxides at high temperatures: Geochimica et Cosmochimica Acta, v. 43, p. 273-276.

Sackett, W.M., Brooks, J.M., Bernard, B.B., Schwab, C.R., Chung, H. and Parker, R.A., 1979. A carbon inventory for Orca Basin brines and sediments: Earth and Planetary Science Letters, v. 44, p. 73-81.

Sakai, H., DesMarais, D.J., Ueda, A. and Moore, J.G., 1984. Concentrations and isotope ratios of carbon, nitrogen and sulfur in ocean-floor basalts: Geochimica et Cosmochimica Acta, v. 48, p. 2433-2441.

Simoneit, B.R.T., 1983. Effects of hydrothermal activity on sedimentary organic matter: Guaymas Basin, Gulf of California-Petroleum genesis and protokerogen degradation: In: Hydrothermal Processes at Seafloor Spreading Centers, P.A. Rona et al. ed.; Plenum Press, p. 451-471.

Sobolev, V.S., Bazarova, T.Yu. and Kostyak, V.P., 1974. Inclusions in the minerals of some types of alkaline rocks: In: The Alkaline Rocks, H. Sorensen ed.; Wiley, New York, p. 389-401.

Sorensen, H. ed., 1974. The Alkaline Rocks: J. Wiley, New York, 622 p.

Speiss, F.N. et al (RISE Project Group) 1980. East Pacific Rise: Hot springs and geophysical experiments: Science, v. 207, p. 1421-1433.

Stahl, W.J., 1977. Carbon and nitrogen isotopes in hydrocarbon research: Chemical Geology, v. 20, p. 121-144.

Sylvester-Bradley, P.C. and King, R.J., 1963. Evidence for abiogenic hydrocarbons: Nature, v. 198, p. 728-731.

Welhan, J.A. and Craig, H., 1979. Methane and hydrogen in East Pacific Rise hydrothermal fluids: Geophysical Research Letters, v. 6, p. 829-832.

Welhan, J.A. and Craig, H., 1982. Abiogenic methane in mid-ocean ridge hydrothermal fluids: In: Proceedings of the Deep Source Gas Workshop, W.J. Gwilliam ed., Morgantown, W. Virginia; DOE-METC 82-50, p. 122-129.

Welhan, J.A. and Craig, H., 1983. Methane, hydrogen and helium in hydrothermal fluids at 21°N on the East Pacific Rise: In: Hydrothermal Processes at Seafloor Spreading Centers, P.A. Rona et al. eds.; Plenum Press, New York, p. 391-409.

Welhan, J.A., Kim, K.R. and Craig, H., 1984. Hydrothermal gases at 11°N and 13°N on the East Pacific Rise: Transactions of the American Geophysical Union, v. 65, p. 973.

Welhan, J.A., Poreda, R., Rison, W., Kim, K.R. and Craig, H. (in prep.) Mantle and crustal gases of the Mud Volcano geothermal system, Yellowstone National Park, Wyoming.

Welhan, J.A., Rison, W., Poreda, R. and Craig, H., 1983. Geothermal gases of the Mud Volcano Area, Yellowstone National Park: Transactions of the American Geophysical Union, v. 64, p. 882.

Zolotarev, B.P., Voytov, G.I., Sarkisyon, I.S. and Cherevichnaya, L.F., 1981. Gases in basaltoid rocks of the Mid-Atlantic Ridge, as shown by data from the 45th voyage of the GLOMAR CHALLENGER: Doklady Akademiya Nauka SSSR, Earth Sciences Section, v. 243, p. 207-210.

Noble Gases in Groundwaters from Crystalline Rocks

J.N. Andrews
School of Chemistry, University of Bath, Bath BA2 7AY, U.K.

Andrews, J.N.; Noble gases in groundwaters from crystalline rocks; in Saline water and gases in crystalline rocks, Editors: Fritz, P., and Frape, S.K.; Geological Association of Canada Special Paper 33, 1987. p. 234–244.

Abstract

Groundwaters in crystalline rocks acquire dissolved gases due to air-equilibration and excess air entrainment at recharge and from additional sources during their migration. These additional sources include the diffusion of radiogenic He from depth along fluid-filled fractures, the release of radiogenic ^{40}Ar by mineral alteration and the production of biogenic or thermogenic CH_4 and N_2 from organic matter.

$^3He/^4He$ isotopic measurements may be used to identify the source rock of radiogenic He which dilutes the atmospheric He in groundwaters during their migration.

The noble gas contents of groundwaters are related to their recharge temperatures which can enable groundwater age to be placed in its palaeoclimatic context. Flow dispersion and excess air entrainment in fracture flow systems complicate the estimation of recharge temperatures for groundwaters in crystalline rocks.

Some examples of the use of dissolved gas measurements for the interpretation of groundwater flow in fractured systems are discussed.

Résumé

Les eaux souterraines dans les roches crystallines acquièrent des gaz en solution par moyen d'équilibrage avec l'atmosphère et de l'entraînement d'un excès d'air dans la zone de recharge, et les gagnent également d'autres origines pendant leur migration. Entre ces origines supplémentaires on peut citer la diffusion de He radiogène monté de profondeur à travers des fractures remplies de liquide, l'émission de ^{40}Ar radiogène qui résulte de la dégradation de minéraux, et la création biogène ou thermogène de CH_4 et de N_2 à partir de matières organiques.

La mesure des isotopes $^3He/^4He$ permet d'identifier la roche d'où provient le He radiogène qui dilue le He atmosphérique dans l'eau souterraine pendant sa migration.

La teneur en gaz nobles des eaux souterraines montre une correspondance à la température de recharge, ce qui permet de placer l'âge de ces eaux dans son propre contexte paléoclimatique. La dispersion du débit d'eau et l'entraînement d'un excès d'air dans les systèmes d'écoulement dans les roches fracturées compliquent le calcul des températures de recharge des eaux dans les roches crystallines.

Nous considérons ici des exemples de l'utilisation de la mesure du teneur en gaz nobles pour l'interprétation de systèmes d'écoulement d'eaux souterraines dans les roches fracturées.

INTRODUCTION

The major sources of dissolved gas in groundwaters include dissolution by equilibration with the atmosphere at recharge, the incorporation of additional or "excess" air by entrainment of air during migration of the groundwater through the unsaturated zone, plus gases produced by radiogenic and biogenic mechanisms within the aquifer. The dissolution of atmospheric gases is dependent upon temperature and the partial pressure of the gases in the atmosphere and for dilute solutions obeys Henry's law. The Henry's law constants for atmospheric gases have been determined experimentally (Morrison and Johnson, 1954; Benson and Krause, 1976) and may be used to estimate temperatures which prevailed during groundwater recharge after correction for the "excess" air present in solution. The "excess" air component of dissolved gas may vary from 10 to almost 100% of the air-equilibration component and must be determined experimentally (Mazor, 1972; Andrews and Lee, 1979; Heaton and Vogel, 1981). Its magnitude is probably dependent upon the morphology of flow channels in the unsaturated zone and upon the frequency of recharge which combine to permit air and air-saturated water to be alternately admitted to the flow paths. Such excess air dissolves under increased hydrostatic pressures as it migrates with the groundwater in an aquifer. Relatively large amounts occur in flow systems within fractured rocks.

During groundwater migration, the radiogenic gases, principally ^{4}He and ^{40}Ar, may be dissolved from the rock matrix, thus supplementing the amounts of these gases dissolved at recharge. Radiogenic ^{4}He has been observed to increase in the direction of groundwater flow (Andrews and Lee, 1979; Bath, et al., 1979; Heaton and Vogel, 1979; Andrews et al., 1985; Torgersen and Clarke, 1985) in predominantly porous media aquifers. Biogenic CO_2 produced within the soil horizon is dissolved by groundwaters in the unsaturated zone and biogenic or thermogenic CH_4 and N_2 may also be generated in the aquifer system. The amounts of these gases that are generated during groundwater migration, together with any temperature increase due to geothermal heating of the groundwater at depth, determine the extent to which the water is supersaturated with respect to solubilities at their atmospheric partial pressures. Many groundwaters exsolve gases when depressurised to atmospheric pressure because of such supersaturation. For this reason, it is generally necessary to collect groundwater samples for dissolved gas analyses under pressures up to several bars (more if large amounts of methane are present). The pressures necessary to retain all gases in solution and the problem of reconstructing in-situ groundwater gas contents from analysis of exsolved gases and degassed waters are discussed by Andrews and Wilson (1986).

The gas contents of groundwaters may also be influenced by structural relationships. Fault-aided diffusion may enhance the amounts of gas in solution, especially for the relatively diffusive noble gases ^{4}He and ^{222}Rn (Scholz et al, 1973; Bulashevich and Bashorin, 1973; Banwell and Parizek, 1985). Crustal stresses prior to earthquakes are known to cause enhancement of He/Ar and N_2/Ar ratios in groundwaters associated with fault zones (Sugisaki, 1978, 1981), suggesting that N_2 as well as He is migrating from the deep crust in such situations.

GROUNDWATER RECHARGE TEMPERATURES FROM NOBLE GAS SOLUBILITIES

The isotopic composition of atmospheric noble gases and their partial pressures in the atmosphere are listed in Table 1. The amounts of these gases dissolved on equilibration of water with the atmosphere are given by Henry's law as:

$$n_i = p_i/k_i \tag{1}$$

where n_i is the number of moles of gas dissolved in 1 mole of water at a gas partial pressure, p_i, and k_i is the Henry's Law constant (atm^{-1}). This can be more conveniently expressed as:

$$S_i = \beta_i p_i/1000 \tag{2}$$

where S_i is the amount of gas 'i' dissolved (cm^3STP/cm^3H_2O) at its atmospheric partial pressure p_i (atm.) and β_i is the Bunsen coefficient for the gas ($cm^3.STP.litre^{-1}\ atm.^{-1}$). The Bunsen coefficient and the Henry's Law constant are related by the equation:

$$\beta_i = 1.244 \times 10^6 \times \rho/k_i \tag{3}$$

where ρ is the solution density (g/cm^3). The solubilities of the gases vary with temperature (Fig. 1) and Bunsen coefficients are therefore dependent upon temperature. The temperature at which air-equilibration of a groundwater occurs in the unsaturated zone is close to the mean annual air temperature for the geographical region of recharge. It is therefore possible to derive the "recharge temperature" of a groundwater from quantitative analysis of the noble gases in solution. A difficulty arises because of the almost universal presence of "excess air" in solution, which occurs because of entrainment of air with the groundwater as it enters the saturated zone. The relative compositions of noble gases dissolved by air-equilibration and of those in air itself (Table 2) are sufficiently different to permit the correction of analytical data for the presence of excess air. Small aliquots of noble gases in their air-abundance ratios are sequentially subtracted from the measured gas contents until the recharge temperatures estimated for the Ne, Ar, Kr and Xe remainders are in best agreement. The average of these temperatures is the estimated recharge temperature of the groundwater. Helium is omitted from this procedure as it is produced by radioactive decay of the U and Th series elements in the rock matrix of the aquifer. It is also necessary to measure the $^{40}Ar/^{36}Ar$ ratio to determine if the Ar present is solely of atmospheric origin. Leakage of radiogenic ^{40}Ar due to feldspar alteration, for example, results in increased $^{40}Ar/^{36}Ar$ ratios. In such cases, corrections should be made to the Ar analysis for the radiogenic ^{40}Ar present before the excess air corrections are applied to determine the recharge temperature.

The recharge temperature derived from noble gas measurements allows the time of groundwater recharge to be placed in its palaeoclimatic context (Andrews and Lee, 1979; Bath et al., 1979; Bath, 1984; Andrews et al., 1985). The isotopic composition (δ^2H and $\delta^{18}O$) of a groundwater also reflects the climatic conditions at recharge (Bath, 1984). The relationship between the noble-gas estimated recharge temperature and the isotopic composition of groundwaters has been determined for several aquifers (Andrews et al., 1984, 1985; Rudolph et al., 1984). Figure 2 shows plots of recharge temperature against $\delta^{18}O$ for some European aquifers. All of these aquifers have zones which were recharged under cooler climatic conditions than presently prevail and these have been attributed to recharge in the interstadial prior to the last glaciation. A strong continental rain-out effect on the isotopic composition is evident in figure 2, groundwaters becoming much lighter over continental Europe than for maritime Europe. Because of such effects, the isotopic composition of a groundwater cannot uniquely determine its recharge temperature, and noble gas measurements are needed for an absolute estimate of this temperature. For a given location, however,

Table 1

Composition of atmospheric and fissiogenic noble gases

Isotope	volume % in atmosphere	moles % of element	^{235}U thermal neutron fission yield %	moles % in fissiogenic element
^{3}He	6.8×10^{-10}	10^{-4}		
^{4}He	5.239×10^{-4}	100		
Total He volume percent in the atmosphere	$= 5.239 \times 10^{-4}$			
Total He partial pressure in the atmosphere	$= 5.239 \times 10^{-6}$atm.($\pm$ 0.004)			
^{20}Ne	1.65×10^{-3}	91.05		
^{21}Ne	4.7×10^{-7}	.01		
^{22}Ne	1.62×10^{-4}	8.94		
Total Ne volume percent in the atmosphere	$= 1.818 \times 10^{-3}$			
Total Ne partial pressure in the atmosphere	$= 1.818 \times 10^{-5}$atm. ($\pm$ 0.004).			
^{36}Ar	3.15×10^{-3}	0.33		
^{38}Ar	5.9×10^{-4}	0.06		
^{40}Ar	0.934	99.6		
Total Ar volume percent in the atmosphere	= 9.34			
Total Ar partial pressure in the atmosphere	$= 9.34 \times 10^{-3}$atm.($\pm$ 0.001)			
^{78}Kr	4.0×10^{-7}	0.354		
^{80}Kr	2.59×10^{-6}	2.27		
^{82}Kr	1.317×10^{-5}	11.56		
^{83}Kr	1.316×10^{-5}	11.55	0.544	15.26
^{84}Kr	6.48×10^{-5}	56.90	1.00	28.06
^{86}Kr	1.98×10^{-5}	17.37	2.02	56.68
Total Kr volume percent in the atmosphere	$= 11.39 \times 10^{-5}$			
Total Kr partial pressure in the atmosphere	$= 1.139 \times 10^{-6}$atm.($\pm$ 0.01)			
			^{235}U thermal neutron fission yield %	moles % in fissiongenic element
^{124}Xe	8.3×10^{-9}	0.096		
^{126}Xe	7.7×10^{-9}	0.090		
^{128}Xe	1.7×10^{-7}	1.919		
^{129}Xe	2.3×10^{-6}	26.44	0.8	3.53
^{130}Xe	3.5×10^{-7}	4.08		
^{131}Xe	1.8×10^{-6}	21.18	2.93	12.95
^{132}Xe	2.3×10^{-6}	26.89	4.38	19.39
^{134}Xe	9.0×10^{-7}	10.44	8.06	35.62
^{136}Xe	7.6×10^{-7}	8.87	6.46	28.55
Total Xe volume percent in the atmosphere	$= 8.60 \times 10^{-6}$			
Total Xe partial pressure in the atmosphere	$= 8.60 \times 10^{-8}$atm.($\pm$ 0.01)			

Noble gas abundance data from Glueckauf (1946) and Gleuckauf and Kitt (1956).

Table 2

Noble gas compositions in air; air-saturated water (ASW) and in air-saturated water containing excess air (ASW/EA)

Gas	Gas composition (Volumes/10^6 volumes of Ar)		Concentrations of dissolved gas, (10^{-8}cm^{3}STP/cm^{3}H$_2$O)		
	air	ASW, 10°C	ASW, 10°C	ASW/EA, 10°C CI = 1.2	ASW/EA, 10°C CI = 2.0
He	560	122	4.75	5.94	10.70
Ne	1,950	531	20.65	24.78	41.30
Ar	1,000,000	1,000,000	38910.0	41030.0	49520.0
Kr	122	238	9.24	9.50	10.54
Xe	9	34	1.32	1.34	1.43

Contamination index (CI) = $[Ne]_W/[Ne]_{ASW}$

Where $[Ne]_W$ = Ne concentration in the water sample

$[Ne]_{ASW}$ = Ne concentration in air saturated water at 10°C.

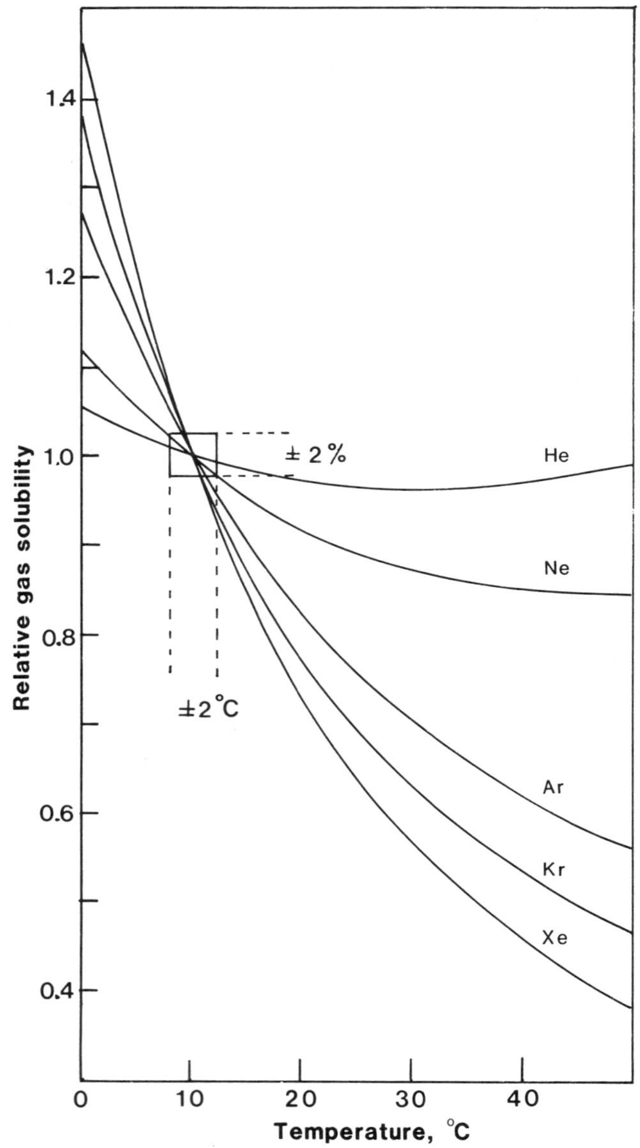

Figure 1. *The variation of noble gas solubilities with temperature relative to solubility at 10°C.*

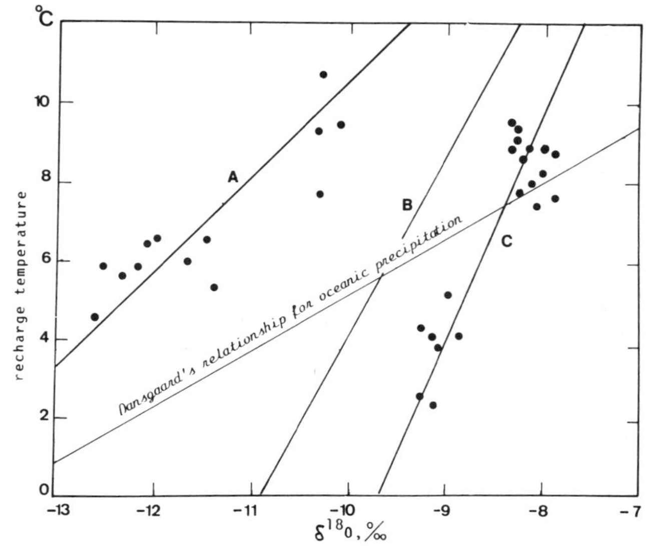

Figure 2. *The relationship between the noble-gas derived recharge temperature and the oxygen isotope composition of groundwaters from the Miocene (A) of Upper Austria (Andrews et al., 1985), the Triassic Sandstone (B) near Nuremberg, F.R.G. (Rudolph et al., 1984), and the Triassic Sandstone (C) of East Midlands, U.K. (Andrews et al., 1984).*

isotopic lightening is accompanied by a decrease in the recharge temperature. The continental effect was qualitatively similar in its effect on recharge temperature and isotopic composition during the last interstadial in Europe. Since the paleoclimatic record has been established in detail by isotopic studies on foraminifera (Hays et al., 1976), the determination of groundwater recharge temperatures can place groundwater residence times within possible climatic eras.

HELIUM IN THE ATMOSPHERE, ROCKS AND GROUNDWATERS

The isotopes ^3He and ^4He are both produced by radiogenic processes in terrestrial rocks. ^3He is generated in the Earth's crust by the action of neutrons on Li, the crustal neutron flux being the result of (α,n) reactions on light nuclei. ^4He is generated by decay of U and Th and their daughter isotopes. These reactions may be summarised as:

$$^6Li(n,\alpha)^3H \xrightarrow[\beta\text{-decay}]{} {}^3He \tag{4}$$

$$^{238}U \longrightarrow {}^{206}Pb + 8\,{}^4He \tag{5}$$
$$^{235}U \longrightarrow {}^{207}Pb + 7\,{}^4He \tag{6}$$
$$^{232}Th \xrightarrow[\alpha,\ \beta\text{-decay}]{} {}^{206}Pb + 6\,{}^4He \tag{7}$$

The $^3He/^4He$ ratio of atmospheric He is 1.3×10^{-6} (Coon, 1949) and this relatively high abundance of ^3He was first explained by Libby (1946) as a consequence of the reaction of cosmic-ray produced neutrons with nitrogen in the atmosphere:

$$^{14}N(n,{}^3H)^{12}C;\ {}^3H \xrightarrow[\beta\text{-decay}]{} {}^3He + \beta^- \tag{8}$$

Teegarden (1967) estimated the ^3He production rate due to neutron interaction with atmospheric N_2 to be 0.2 atoms $cm^{-2}s^{-1}$. Johnson and Axford (1969) showed that the possible solar-wind contribution to the atmosphere is 4.0 atoms $cm^{-2}s^{-1}$, and also concluded that the other possible sources of ^3He in the atmosphere are relatively unimportant and that the rate of thermal and polar-wind escape of ^3He from the atmosphere is 7.15 ± 3.6 atoms $cm^{-2}s^{-1}$.

Previous work on the $^3He/^4He$ isotopic ratio in terrestrial samples has been reviewed by Rankama (1954, 1963) and more recently, by Tolstikhin (1978), Mamyrin and Tolstikhin (1984) and by Ozima and Podosek (1983). Craig and Lupton (1976) found that the $^3He/^4He$ ratio for He in the-oleiitic glasses from the mid-oceanic ridges is $\simeq 10^{-5}$, that is, greater than that of atmospheric He. They regard this He as being remnant He of primordial origin, which has been stored in the mantle. Excess ^3He has also been found in deep ocean waters adjacent to oceanic ridges (Clarke, et al., 1969; Craig et al., 1975; Jenkins et al., 1980) and in volcanic and hot-spring gases (Nagao et al., 1981). Craig et al. (1975) have estimated the flux of ^3He from the deep oceans into the troposphere to be 4 atoms $cm^{-2}s^{-1}$ so that ^3He in primordial He from the mantle plus the solar-wind ^3He can approximately balance the ^3He losses from the atmosphere.

All groundwaters contain dissolved He which entered solution because of equilibrium with the atmosphere at recharge. Andrews and Lee (1979) and Heaton and Vogel (1981) have demonstrated that additional radiogenic He accumulates in an amount which increases with groundwater age, as a consequence of radioelement decay in the aquifer, and diffusion from underlying formations. The $^3He/^4He$ ratio of He in groundwaters decreases from that characteristic of atmospheric He for sources close to recharge to much lower values as radiogenic He is progressively dissolved during groundwater migration. When the amount of radiogenic He dissolved greatly exceeds the amount of atmospheric He acquired at recharge, the $^3He/^4He$ ratio approaches a limiting value which is characteristic of the source rock for the radiogenic He.

Characteristic $^3He/^4He$ ratios for various rock types

The $^3He/^4He$ ratio of the radiogenic He produced within a rock matrix has been shown to be independent of the U content of the rock, and almost independent of its Th content, but to be strongly influenced by the Li-content of the rock (Andrews, 1985). The variation of this ratio with Li-content in rocks which otherwise have average geochemical compositions as defined by Parker (1967), is shown in Figure 3. This shows that ultramafic and basaltic rocks, which do not contain more than a few ppm of Li, cannot contain

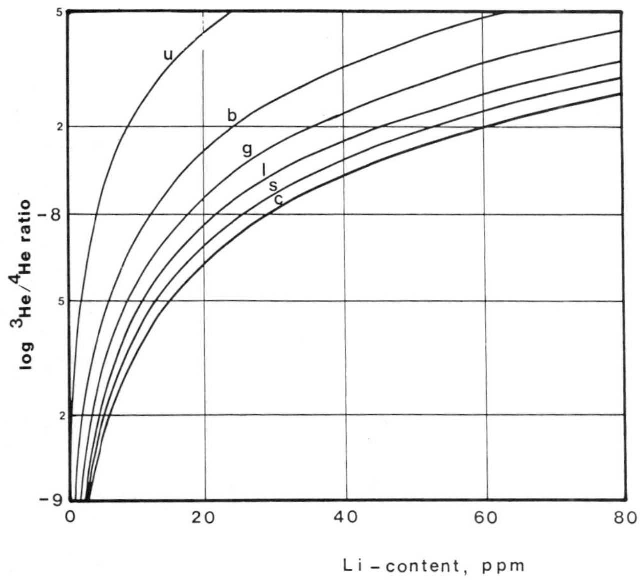

Figure 3. *The characteristic $^3He/^4He$ ratio of radiogenic He as a function of rock Li-content. (u = ultrabasic; b = basalt; g = granite; l = limestone; s = sandstone; c = clays and shales).*

radiogenic He with a $^3He/^4He$ isotopic ratio much greater than 10^{-9}. Granitic rocks, which commonly have Li contents in the range 40-80 ppm, can contain radiogenic He with an isotopic ratio in the range 1-5 $\times$ 10^{-8}. For similar Li-contents, the isotopic composition of radiogenic He in sedimentary rocks is generally somewhat less than would be expected for granites. Provided that sufficient data on the Li-contents of the various stratigraphic horizons are available, measurements of the limiting $^3He/^4He$ ratio of He in migrating groundwaters may be used in conjunction with Figure 3 to identify the source of radiogenic He.

Variations in the $^3He/^4He$ ratio have been applied to earthquake prediction (Mamyrin et al., 1979; Nagao et al., 1980). Such variations reflect a change in the source rock from which the He is diffusing or a possible contribution from mantle He in deeply faulted or thin crustal regions. Mantle He has been identified in geothermal fluids associated with volcanic activity in thin crustal regions (Hooker, et al., 1985).

Radiogenic He accumulation in groundwaters

Considerable amounts of He in excess of that due to equilibration with the atmosphere at recharge have been observed in groundwaters from fractured rocks. The source of this excess He is the generation of ^4He by radioactive decay of U, Th and their alpha-emitting daughters in the rock matrix. Marine (1979) demonstrated a linear relationship between excess ^4He and groundwater conductivity (or salinity) in the crystalline matamorphic basement beneath the Savannah River Plant in South Carolina and concluded that the samples with the highest He-content and salinity best represented the groundwater of the basement. He calculated the age of the groundwater according to the equation:

$$t = [He].\rho.\phi^{-1} \{1.19 \times 10^{-13}[U] + 2.88 \times 10^{-14}[Th]\}^{-1} \text{ years} \qquad (9)$$

where [He] is the groundwater He content, cm^3STP/cm^3H_2O, ϕ is the rock porosity and ρ is the rock density (g/cm^3), [U] and [Th] are the U and Th contents (ppm) of the rock matrix. This gave an age of 840 ka for the groundwater and since equation (9) assumes that all of the He generated leaves the mineral phases of the rock and is dissolved in the fracture/interstitial fluids, it represents a minimum groundwater age provided that no diffusive movement of He has occured. Zaikowski et al. (1984) have determined the ^4He and ^{40}Ar contents of deep brines in the Texas Panhandle and have calculated brine ages from radiometric accumulation rates. The estimated brine ages are, however, more than an order of magnitude greater than hydrological estimates.

Increasing He contents with ^{14}C-age of groundwaters in intergranular media aquifers have been demonstrated (Andrews and Lee, 1979; Bath, et al., 1979; Heaton and Vogel, 1979) but it was shown that equation (9) gave groundwater ages which were much greater than ^{14}C-derived ages. It was suggested that this discrepancy could be due to diffusive movement of ^4He into the aquifer from other geological horizons (Andrews and Lee, 1979). Such diffusion into intergranular aquifers has since been identified (Andrews et al., 1985; Torgersen and Clarke, 1985) and groundwater ages have been estimated from the solution of the diffusive flux of He by the groundwater. If the rate of groundwater movement is greater than the rate of He diffusion in the retrograde direction, the groundwater must dissolve the diffusive flux of He from underlying formations during its migration. Groundwater age may then be estimated by dividing its accumulated He content by the He flux.

In the case of fractured rocks, diffusion processes are generally more difficult to quantify because of the absence of any well-defined groundwater flow path. Processes which may contribute to ^{4}He accumulation in fracture fluids are:

(i) diffusive loss to the fracture fluid of radiogenic ^{4}He stored in the mineral phases of the rock
(ii) diffusive movement of ^{4}He within fracture fluids from greater depths
(iii) general diffusion of ^{4}He from depth through both the crystal lattices of the mineral phases and along fluid-filled fracture paths

The first of these processes is dependent upon lattice diffusion coefficients, grain boundary diffusion, the fracture spacing and the crystallisation age of the rock. The diffusion coefficient for He in water is about 1000 times greater than that in crystal lattices so that the second process can permit He diffusion relatively rapidly from depth. At all depths the He diffusion into the fracture fluids is controlled by mechanism (i). The overall He diffusion from depth, process (iii), is a combination of crystal lattice diffusion and diffusion along fracture fluids. The He concentration gradient with depth may additionally be modified by He transport in mobile fracture fluids. It is extremely difficult to model accurately the ^{4}He concentration/depth profile because of the number of parameters involved. Also, the frequency of conductive fractures may vary considerably within particular rock masses and the fracture permeability would generally be greater at shallow depths, especially in the weathering zone.

SAMPLING TECHNIQUES FOR NOBLE GAS ANALYSIS

Careful attention to the conditions of sampling is required to ensure that the gas analysis is truly representative of the in-situ gas contents of a groundwater. Many groundwaters, especially those which have circulated to great depths, evolve gases if they flow under artesian pressure to the surface. The groundwater becomes supersaturated with gases during its circulation and these gases are generally held in solution by the increasing hydrostatic pressure with depth. In karstic systems, supersaturation with respect to the HCO_3^-/CO_2 system commonly occurs and "bubbling springs" in such systems are well known. Additionally, the entrainment of excess air at recharge can lead to supersaturation as such air dissolves at depth. The amount of such excess air which is present in most groundwaters is small. It may be represented by a contamination index which is defined as the ratio of the groundwater Ne-content to the Ne-solubility at the recharge temperature for the groundwaters. Values of this index are generally less than two and the amounts of such gas could be held in solution by a hydrostatic pressure of less than 10 m. The composition of dissolved gas containing excess air is indicated in Table II. Groundwaters may also dissolve biogenic and/or thermogenic methane and nitrogen during circulation and the amount of such gas may be such that very high hydrostatic pressures are needed to retain all of the gases in solution. For example, if a groundwater contained CH_4 and N_2 in the ratio 10:1, a hydrostatic pressure of 9 bars (90 m) would be necessary to retain all the gases in solution at 10°C. (Andrews and Wilson, 1986). Brines in the Canadian Shield (Fritz et al., 1986) contain large amounts of dissolved methane and gases are continuously exsolved from wells which are de-pressurised to atmospheric pressure.

To interpret noble gas data in terms of recharge temperature and paleoclimate, it is necessary to ensure that all gases are held in solution. This generally requires the samples to be taken under pressure and also that the well has previously been pressurised and flushed under pressure so that any zone of degassed water is removed. This requirement is often the most difficult sampling condition to satisfy. Subsequently, samples may be collected at pressures up to 30 bars in soft annealed copper tubing which is closed by swage-clamps. It is possible to reconstruct the in-situ gas contents of a groundwater which has undergone de-gassing if the flow rates and compositions of the water and gas phases can be determined (Andrews and Wilson, 1986). This reconstruction must assume that the collected gas and water phases are in equilibrium with respect to gas dissolution. In practice, it is often difficult to ensure that such equilibrium has been attained and to measure the flow rates of both phases. There is an urgent need to develop sampling techniques for gas-rich brines which will permit the gas contents in the undisturbed formation fluids to be estimated. Such in-situ gas contents are required before geochemical relationships between different gas-rich brines can be fully investigated.

APPLICATIONS OF NOBLE GAS MEASUREMENTS

Determinations of groundwater recharge temperatures

Noble gas derived recharge temperatures may be used to identify groundwaters which have been infiltrated under different climatic regimes. The existence of waters due to pre-glacial recharge in predominantly intergranular-flow confined aquifers has been well established (Bath et al., 1979; Andrews et al., 1984; Bath, 1984; Andrews et al., 1985). The identification of such Pleistocene recharge in fractured rock systems has proved more difficult than for intergranular flow systems. This is partly due to the complex flow patterns which can lead to mixing of groundwaters with different residence times in such systems, thus obscuring changes in both recharge temperature and isotopic composition. The large amounts of excess air which are generally entrained during recharge in fractured systems also contribute to the difficulty of estimating recharge temperatures.

At Altnabreac (Scotland), although both ^{14}C and ^{4}He measurements indicate that some recharge probably occurred during the last interstadial, the oldest groundwaters show only a marginal decrease in recharge temperature (Andrews and Kay, 1985). Some flow zones in the deeper boreholes have waters which are isotopically slightly more negative in δ^{18}O (Kay and Bath, 1982) but the overall conclusion was that most groundwater up to depths of 300 m in this system was infiltrated during the Holocene.

Noble gas measurements on groundwaters which circulate to 300 m depth in the Carnmenellis granite also show that only Holocene recharge has been involved (Edmunds et al., 1984).

In the Stripa (Sweden) granite, estimated recharge temperatures for shallow groundwaters at depths of <100 m are similar to present-day mean annual air temperatures. Deeper groundwaters in this system trend towards lower recharge temperatures (Fig. 4) but discrete fracture zones such as V2/4 and V2/1 may contain groundwaters with recharge temperatures which are different from the general depth trend. Some zones in this borehole yield groundwaters with greater ^{14}C contents than are present at intermediate depths in the mine, again suggesting discrete zones of more rapid infiltration (Nordstrom et al., 1985). The recharge temperatures of the minewaters are distinctly lower than those for groundwaters from shallow private wells. The recharge temperature decreases as the isotopic composition of the groundwater becomes lighter (Fig. 5). This shows that groundwaters which are presently encountered at depths below 350 m in the mine were recharged under cooler climatic conditions than presently exist in southern Sweden.

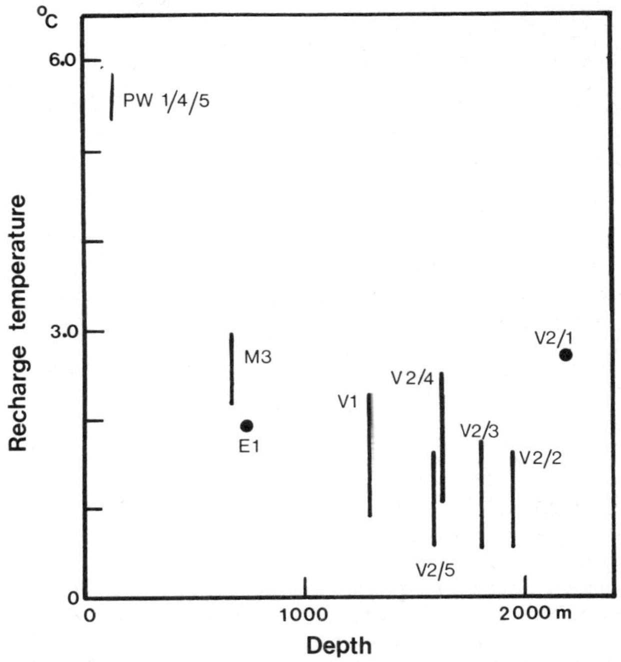

Figure 4. *The noble-gas derived recharge temperature of Stripa groundwaters plotted against their depth of origin.*

Radiogenic helium

The dependence of ^4He content on the nature of the fracture flow system is well illustrated by a study of the Altnabreac granite (Kay et al., 1984). Figure 6 is a schematic flow profile for this system in which localised recharge and discharge zones are controlled by topographic variation. In borehole C which is in a recharge zone, recent water penetrates rapidly to depth and acquires little radiogenic He in so doing. At all depths in this borehole the water has high ^3H and ^{14}C contents. However, close to the bottom of borehole B, which is situated well away from the recharge, the fracture fluids have acquired large excess He contents due to their long flow path at depth. The shallow boreholes sample relatively localised flow systems and their waters are characterised by low ^4He, high ^{14}C contents and often by high ^3H contents.

Very large amounts of dissolved ^4He, up to $3 \times 10^{-3} cm^3$ STP/$cm^3 H_2O$, have been reported in groundwaters from the fractured granites at Stripa, Sweden (Andrews et al., 1982; Nordstrom et al., 1985). The variation of He content with groundwater depth in the granite is illustrated in Figure 7. The groundwater He-contents may be compared with the

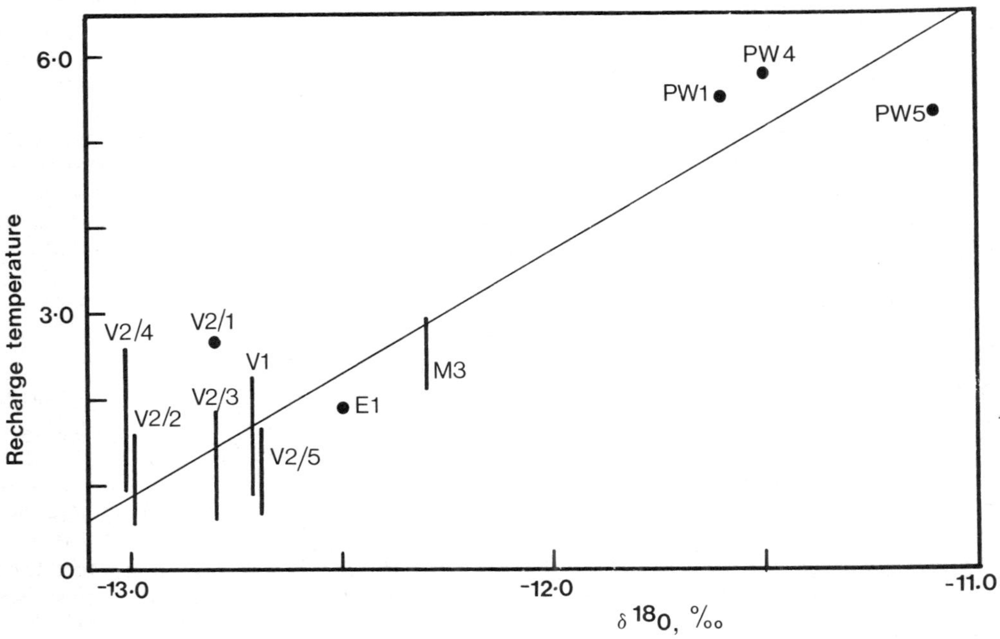

Figure 5. *The relationship between the recharge temperatures of Stripa groundwaters and their isotopic composition ($\delta^{18}O$).*

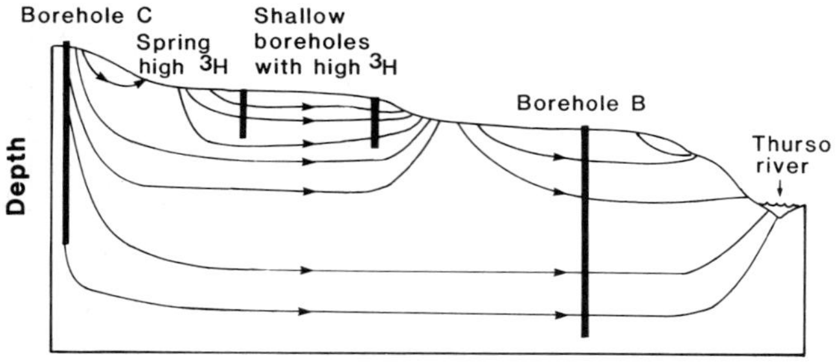

Figure 6. *A schematic flow profile for groundwaters in the Altnabreac granite. The hydraulic conductivity ranges from 10^{-6} ms^{-1} in the shallow part of the system to 10^{-9} ms^{-1} in the deepest part of the system.*

He-concentration/depth profiles (A and B), calculated for diffusive loss of He from the 1640 Ma old granite. Profile A was calculated for a diffusion coefficient of $3.16 \times 10^{-4} m^2a^{-1}$ which is an order of magnitude greater than the crystalline lattice diffusion coefficient and may be expected to apply to a crystalline rock matrix. Profile B was calculated for the diffusion coefficient of He in water ($3.16 \times 10^{-2} m^2a^{-1}$). Groundwater which circulated in relatively small fractures should readily equilibrate with the rock-matrix He-content, so that the groundwater He contents are a measure of the matrix He-content. The groundwater He-contents are generally rather less than those expected for diffusive loss from the rock matrix with a diffusion coefficient similar to that in water. The Stripa intrusion appears to have lost He by diffusion processes which involve an aqueous phase. The very high ^{222}Rn-contents which have been observed for the Stripa groundwaters (Andrews et al., 1982) also suggest that gaseous species very effectively diffuse within the fracture system. The granite has an abundance of both continuous and discontinuous fractures on a microscopic scale (Nordstrom et al., 1985) which could considerably aid diffusion of He.

The He-contents of the groundwaters from depths greater than 800m are close to or slightly less than the diffusion profile for water as the diffusive medium. The He-contents of the shallower groundwaters are generally much less than those suggested by the diffusion profile. The cause of such differences is likely to be additional He-loss by transport in circulating groundwaters. It is likely that the conductive fracture system is much more extensive at shallow depths than for the deeper groundwaters. If the shallow fracture system is flushed more rapidly than diffusive equilibrium can be established, then He can be swept from the system more rapidly than it can be replenished. This can explain the greater departures from the diffusion profile at depths up to 350 m in the granite.

$^3He/^4He$ isotope ratio measurements for He in groundwaters from the Stripa granite show that the radiogenic He has the composition expected for the average Li-content of the rock (Nordstrom et al., 1985). The isotopic composition of dissolved He in groundwaters from the Austrian Molasse have shown that the He can only have been generated within the granitic basement and hence that it has migrated by diffusion processes (Andrews et al., 1985). The $^3He/^4He$ ratio for these groundwaters decreased with increasing dissolved He content (Fig. 8) and was attributed to mixing of atmospheric He with radiogenic He. The limiting $^3He/^4He$ ratio was equal to the calculated ratio for radiogenic He from the granitic basement which has a Li content of 60 ppm. The ages of groundwaters which circulate in the Miocene sediments were estimated on the basis of solution of the diffusive He-flux from the basement.

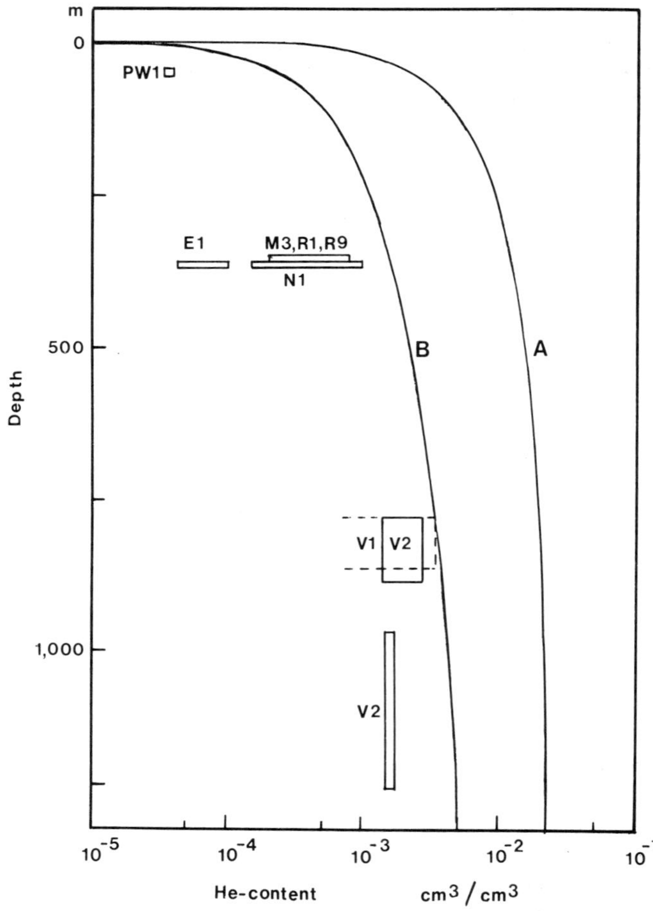

Figure 7. *The ^4He-content of Stripa groundwaters in relation to their depth of origin. ^4He concentration/depth profiles are shown for the diffusion coefficients of a rock matrix (A) and of water (B), for a granite with the average U and Th contents and of the same age as the Stripa granite.*

Radiogenic argon

The biotites and feldspars of crystalline rocks generally retain most of the ^{40}Ar which is generated by ^{40}K decay within the mineral. Diffusive loss of radiogenic ^{40}Ar at temperatures up to 200°C is negligibly small (Dalrymple and Lanphere, 1969). The ^{40}Ar content of a mineral which contains 5% K and is 650 Ma old is 4.0×10^{-4} cm^3 STP/cm^3. This is almost equal to the Ar-content of groundwaters which have been equilibrated with the atmosphere during recharge at 10°C. No concentration gradient which could cause Ar diffusion from mineral to fracture fluid can exist for such conditions. Such diffusion is only possible for mineral ages in excess of 650 Ma and/or K-contents > 5%.

The presence of excess ^{40}Ar in groundwaters can be detected by measurement of the $^{40}Ar/^{36}Ar$ isotopic ratio. Ratios which are greater than that for atmospheric Ar (295.5) indicate that radiogenic ^{40}Ar has been added. The presence of radiogenic Ar in groundwaters from the Carnmenellis granite (Edmunds et al., 1984) and from the Stripa granite (Nordstrom et al., 1985) can most readily be explained as a consequence of mineral alteration processes. Such alteration would readily release ^{40}Ar from the lattice without involving diffusion processes.

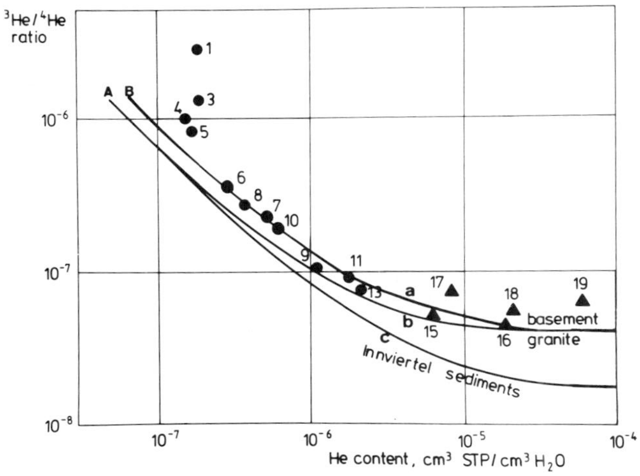

Figure 8. *The $^3He/^4He$ ratio as a function of the He-content of Miocene (circles) and deeper (triangles) groundwaters from the Molasse of Upper Austria. The point A corresponds to air-saturated water and point B to additional excess air. Line 'a' is the dilution line for recharge containing excess air, line 'b' is the air-saturation dilution line (both for dilution with radiogenic He from the basement) and line 'c' is the air-saturation dilution line for radiogenic He from the Miocene sediments. Samples 1-5 indicate the presence of ^3He due to thermonuclear ^3H decay.*

CONCLUSIONS

The amounts of the noble gases, other than He, which are dissolved in groundwaters are principally determined by air-equilibration during recharge. At sufficient depth in flow systems, the concentrations of these noble gases generally behave conservatively even when large amounts of biogenic and thermogenic gases such as N_2 and CH_4 are present. The contents of radiogenic ^4He in groundwaters in crystalline rocks can be used qualitatively to establish the pattern of groundwater flow and to identify those groundwaters with the longest residence times. Departures from the expected ^4He concentration/depth profile for diffusive loss from a fractured system can indicate the most conductive part of the flow system.

The presence of radiogenic ^{40}Ar in groundwaters from crystalline rocks is more likely to arise because of mineral alteration processes than because of diffusive loss from potassic minerals.

Noble gas contents may be used to estimate groundwater recharge temperatures which can be used to place groundwater age in its palaeoclimatic context. The noble gas derived recharge temperatures can usually be correlated with the stable isotope composition of the groundwater. Care must be taken during groundwater sampling to avoid gas losses as the hydrostatic pressure in flowing wells is reduced.

ACKNOWLEDGEMENTS

I would like to thank the two anonymous referees for their constructive criticisms and suggestions which have led to improvements in this paper.

REFERENCES

Andrews, J.N., 1985. The isotopic composition of radiogenic helium and its use to study groundwater movement in confined aquifers: Chemical Geology, v. 49, p. 339-351.

Andrews, J.N. and Wilson, G.B., 1986. The composition of dissolved gases in deep groundwaters and groundwater degassing. (Accompanying paper in this volume).

Andrews, J.N., Goldbrunner, J.E., Darling, G., Hooker, P., Wilson, G.B., Youngman, M.J., Eichinger, L., Rauert, W. and Stichler, W., 1985. A radiochemical, hydrochemical and dissolved gas study of groundwaters in the Molasse basin of Upper Austria: Earth and Planetary Science Letters, v. 73, p. 317-332.

Andrews, J.N. and Kay, R.L.F., 1985. Dissolved helium, inert gases, radium and radon in groundwaters from the Altnabreac research site: British Geological Survey, Report FPU-85-2.

Andrews, J.N., Balderer, W., Bath, A.H., Clausen, H.B., Evans, G.V., Florkowski, T., Goldbrunner, J.E., Ivanovich, M., Loosli, H. and Zojer, H., 1984. Environmental isotope studies in two aquifer systems: a comparison of groundwater dating method: In: Isotope Hydrology 1983, p. 535-576, IAEA Vienna, 1984.

Andrews, J.N., Giles, I.S., Kay, R.L.F., Lee, D.J., Osmond, J.K., Cowart, J.B., Fritz, P., Barker, J.F. and Gale, J., 1982. Radioelements, radiogenic helium and age relationships for groundwaters from the granites at Stripa, Sweden: Geochimica et Cosmochimica Acta, v. 46, p. 1533-1543.

Andrews, J.N. and Lee, D.J., 1979. Inert gases in groundwater from the Bunter sandstone of England as indicators of age and palaeoclimatic trends: Journal of Hydrology, v. 41, p. 233-252.

Banwell, G.M. and Parizek, R.R., 1985. Relationship between ^4He, ^{222}Rn and lineaments: EOS, v. 66, p. 1115.

Bath, A.H., 1984. Stable isotopic evidence for palaeo-recharge conditions of groundwater: Palaeoclimates and Palaeowaters; A Collection of Environmental Isotope Studies: In: Isotope Hydrology 1983, p. 169-186, IAEA Vienna, 1984.

Bath, A.H., Edmunds, W.M. and Andrews, J.N., 1979. Palaeoclimatic trends deduced from the hydrochemistry of a Triassic Sandstone aquifer, U.K.: In: Isotope Hydrology 1978, Vol. II, p. 545-568, IAEA Vienna, 1979.

Benson, B.B. and Krause, D., 1976. Empirical laws for dilute aqueous solutions of non-polar gases: Journal of Chemical Physics v. 64, p. 639-709.

Bulashevich, Yu.P. and Bashorin, V.N., 1973. On the detection of faults along the Sverdlorsk DSS profile from high concentrations of He in underground water: Physics of the Solid Earth, v. 9, p. 185-189.

Clarke, W.B., Begg, M.A. and Craig, H., 1969. Excess ^3He in the sea: evidence for terrestrial primordial helium: Earth and Planetary Science Letters, v. 6, p. 213-220.

Coon, J.H., 1949. ^3He isotopic abundance: Physical Review, v. 75, p. 1355-1357.

Craig, H. and Lupton, J.E., 1976. Primordial neon, helium and hydrogen in oceanic basalts: Earth and Planetary Science Letters, v. 31, p. 369-385.

Craig, H., Clarke, W.B. and Begg, M.A., 1975. Excess ^3He in deep water on the East Pacific Rise: Earth and Planetary Science Letters, v. 26, p. 125-132.

Dalrymple, G.B. and Lanphere, M.A., 1969. Potassium-Argon Dating: San Francisco, W.H. Freeman, 258 p.

Edmunds, W.M., Andrews, J.N., Burgess, W.G., Kay, R.L.F. and Lee, D.J., 1984. The evolution of saline and thermal groundwaters in the Carnmenellis granite: Mineralogical Magazine, v. 48, p. 407-24.

Fritz, P., Frape, S.K. and Miles, M., 1986. Methane in the crystalline rocks of the Canadian Shield. (Accompanying paper in this volume).

Glueckauf, E., 1946. A micro-analysis of the He and Ne contents of air: Proceedings of the Royal Society, v. A185, p. 98-119.

Glueckauf, E. and Kitt, G.P., 1956. The Kr and Xe contents of atmospheric air: Proceedings of the Royal Society, v. A234, p. 557-565.

Hays, J.D., Imbrie, J. and Shackleton, N.J., 1976. Variations in the Earth's orbit: pacemaker of the ice ages: Science, v. 194, p. 1121-1132.

Heaton, T.H.E. and Vogel, J.C., 1981. "Excess air" in groundwater: Journal of Hydrology, v. 50, p. 201-216.

Heaton, T.H.E. and Vogel, J.C., 1979. Gas concentration and ages of groundwaters in Beaufort Group sediments, South Africa: Water SA, v. 5, p. 160-170.

Hooker, P.J., Bertrami, R., Lombardi, S., O'Nions, R.K. and Oxburgh, E.R. Helium-3 anomalies and crust-mantle interaction in Italy: Geochimica et Cosmochimica Acta, v. 49, p. 2505-2514.

Jenkins, W.J., Rona, P.A. and Edmond, J.M., 1980. Excess ^3He in deep water over the Mid-Atlantic Ridge at 26°N: evidence for hydrothermal activity: Earth and Planetary Science Letters, v. 49, p. 39-44.

Johnson, H.E. and Axford, W.I., 1969. Production and loss of ^3He in the Earth's atmosphere: Journal of Geophysical Research, v. 74, p. 2433-2438.

Kay, R.L.F., Andrews, J.N., Bath, A.H. and Ivanovich, M., 1984. Groundwater flow profile and residence times in crystalline rocks at Altnabreac, Caithness, UK.: In: Isotope Hydrology 1983, p. 231-247, I.A.E.A., Vienna, 1984.

Kay, R.L.F. and Bath, A.H., 1982. Groundwater geochemical studies at the Altnabreac research site: British Geological Survey Report, ENPU-82-12.

Libby, W.F., 1946. Atmospheric helium three and radiocarbon from cosmic radiation: Physical Review, v. 69, p. 671-673.

Mamyrin, B.A. and Tolstikhin, I.N., 1984. Helium isotopes in nature: Amsterdam, Elsevier, 273 p.

Mamyrin, B.A., Tolstikhin, I.N. and Khabavin, L.V., 1979. $^3He/^4He$ ratios in earthquake forecasting: Geochemistry International, v. 16, p. 42-44.

Marine, I.W., 1979. The use of naturally occurring helium to estimate groundwater velocities for studies of geologic storage of radioactive waste: Water Resources Research, v. 15, p. 1130-1136.

Mazor, E., 1972. Palaeotemperatures and other hydrological parameters deduced from noble gases dissolved in groundwaters, Jordan Rift Valley, Israel: Geochimica et Cosmochimica Acta, v. 36, p. 1321-1336.

Morrison, T.J. and Johnstone, N.B., 1954. Solubilities of inert gases in water: Journal of the, Chemical Society, v. 36, p. 3441.

Nagao, K., Takaoka, N. and Matsubayashi, O., 1981. Rare gas isotopic compositions in natural gases of Japan: Earth and Planetary Science Letters, v. 53, p. 175-188.

Nagao, K., Takaoka, N., Wahita, H., Matsuo, S. and Fujii, N., 1980. Isotopic compositions of rare gases in the Matsushiro earthquake fault region: Geochemical Journal, v. 14, p. 63-69.

Nordstrom, D.K., Andrews, J.N., Carlsson, L., Fontes, J.-C., Fritz, P., Moser, H. and Olsson, T., 1985. Hydrogeological and hydrogeochemical investigations in boreholes – geochemical and isotope characterization of the Stripa groundwaters: SKB Stripa Project Report 85-06, 250 p.

Ozima, M. and Podosek, F.A., 1983. Noble gas geochemistry: Cambridge, Cambridge University Press, 367 p.

Parker, R.L., 1967. Data of geochemistry: composition of the earth's crust: U.S. Geological Survey, Professional Paper 440-D.

Rankama, K., 1963. Progress in Isotope Geology: New York, Wiley, 705 p.

Rankama, K., 1954. Isotope Geology: London, Pergamon, 535 p.

Rudolph, J., Rath, H.K. and Sonntag, C., 1984. Noble gases and stable isotopes in ^{14}C-dated palaeowaters from central Europe and the Sahara: In: Isotope Hydrology 1983, p. 467-477, IAEA, Vienna, 1984.

Scholz, C.H., Sykes, L.R. and Aggarwal, Y.P., 1973. Earthquake prediction: a physical basis: Nature, v. 181, p. 803-810.

Sugisaki, R., 1981. Deep-seated gas emissions induced by the earth tide; a basic observation for geochemical earthquake prediction: Science, v. 212, p. 1264-1266.

Sugisaki, R., 1978. Changing He/Ar and N_2/Ar ratios of fault air may be earthquake precursors: Nature, v. 275, p. 209-211.

Teegarden, B.J., 1978. Cosmic ray production of deuterium and tritium in the earth's atmosphere: Journal of Geophysical Research, v. 72, p. 4863-4868.

Tolstikhin, I.N., 1978. Some recent advances in isotope geochemistry of light rare gases: In: E.C. Alexander, Jr. and M. Ozima (Editors), Advances in Earth and Planetary Sciences, v. 3, Tokyo, Japan Scientific Societies Press, p. 33-62.

Top, Z. and Clarke, W.B., 1981. Dissolved He isotopes and tritium in lakes; Further results for He prospecting in central Labrador: Economic Geology, v. 76, p. 2018-2031.

Torgersen, T. and Clarke, W.B., 1985. Helium accumulation in groundwater, I: An evaluation of sources and the continental flux of crustal ^4He in the Great Artesian Basin, Australia: Geochimica et Cosmochimica Acta, v. 49, p. 1211-1218.

Zaikowski, A., Kosanke, B.J. and Hubbard, N., 1984. Progress on radiometric dating of Wolfcamp brines using ^4He and ^{40}Ar: Materials Research Society, Annual Meeting Symposium Proceedings, v. 26, p. 943-949.

The Composition of Dissolved Gases in Deep Groundwaters and Groundwater Degassing

J.N. Andrews and G.B. Wilson
School of Chemistry, University of Bath, Bath BA2 7AY, U.K.

Andrews, J.N. and Wilson, G.B.; The composition of dissolved gases in deep groundwaters and groundwater degassing; in Saline water and gases in crystalline rocks, Editors: Fritz, P., and Frape, S.K.; Geological Association of Canada Special Paper 33, 1987. p. 245–252.

Abstract

Groundwaters dissolve gases by air equilibration in the unsaturated zone at recharge. The amount of gas in solution is determined by the partial pressures of gases in the atmosphere and by the recharge temperature. In addition "excess air" is entrained by the groundwater in its downward movement through the unsaturated zone and enters solution as the hydrostatic pressure increases in the aquifer.

Deeply circulating groundwaters may gain additional dissolved gases by geochemical, biogenic and thermogenic processes. Thus the bicarbonate evolution leads to an equivalent partial pressure of dissolved CO_2, nitrate reduction may lead to enhanced N_2 and thermo-degradation of organic materials may lead to further N_2 and CH_4 in solution. Geothermal heating of the groundwater during its circulation, together with these increases in its gas content, may cause the groundwater to be supersaturated with gases. Drill holes which sample such groundwaters under artesian pressures, will permit the exsolution of dissolved gases unless a sufficiently high pressure is applied at the well-head.

The composition of exsolved gas and the degassed solution has been modelled for various possible dissolved gas compositions in formation waters and the use of exsolved gas and solution compositions to estimate the in-situ composition of dissolved gas in formation waters is discussed.

Résumé

Les eaux souterraines dissolvent des gaz par l'équilibrage avec l'atmosphère dans la zone de recharge non-saturée. La teneur en différents gaz en solution est déterminée et par la pression partielle d'un gaz dans l'atmosphère, et par la température de recharge. En plus, le mouvement de l'eau qui descend à travers la zone non-saturée entraîne un "excès" d'air, qui se dissout au fur et au mesur que la pression hydrostatique s'élève dans l'aquifère.

Les eaux souterraines qui circulent à grand profondeur peuvent acquérir des gaz dissous supplémentaires qui résultent de différents processus géochimiques, biogènes ou thermogènes. Ainsi l'évolution de bicarbonates mène á une pression partielle équivalente de CO_2 dissous, la réduction de nitrates peut donner un niveau plus élevé de N_2, et la thermodégradation de matières organiques peut créer à son tour du N_2 et du CH_4 en solution. Le chauffage géothermique de l'eau souterraine pendant sa circulation, pris ensemble avec ces croissances dans le contenu gazeux, peut causer une supersaturation de gaz dans l'eau souterraine. Tout forage qui prend des échantillons de telles eaux sous pression artésienne permet la fuite des gaz dissous, sauf l'application d'une pression suffisament élevée à la tête du puits.

La composition du gaz émis et de la solution dègazée est postulée pour de différents combinaisons possibles de gaz dissous dans les eaux de formation. Nous traitons aussi de l'utilisation de l'analyse de la composition des gaz évadés comme des gaz encore en solution pour estimer la composition des gaz dissous dans les eaux souterraines sur place.

INTRODUCTION

There have been several recent studies of dissolved gases in groundwaters which have related the amounts of gas in solution to geochemical and hydrological parameters. Noble gas contents and their solubility/temperature relationships have been used to derive the climatic conditions which prevailed during groundwater recharge (Mazor 1972; Andrews and Lee, 1979). Correlations between such derived "recharge temperatures" and the isotopic composition of the groundwater have been demonstrated (Rudolph et al., 1984; Andrews et al., 1985). These relationships also reflect the "oceanic" and "continental" influences on climate (Andrews et al., 1985). The accumulation of radiogenic gases, especially ^4He, has been suggested as a basis for the estimation of groundwater residence time (Andrews and Lee, 1979; Marine, 1979). The isotopic ratio $^3He/^4He$ has been shown to change systematically as atmospherically-derived He is diluted by radiogenic He produced in an aquifer matrix (Andrews, 1985). The $^3He/^4He$ ratio may also be used to determine the extent of diffusion of crustal He into a migrating groundwater (Andrews, et al., 1985). Groundwater residence times may then be estimated from their He contents and the calculated crustal He flux from underlying formations, if the diffusing He is dissolved by the groundwater. Mantle-derived He cannot diffuse through regions of stable continental crust (Andrews, 1985). Most dissolved N_2 and Ar is of atmospheric origin and the N_2/Ar ratio ($\simeq 38$) of an atmosphere-equilibrated groundwater is characteristic of the equilibration temperature and is different from the N_2/Ar ratio in air ($\simeq 80$).

A groundwater which has been equilibrated with the atmosphere in the unsaturated zone, frequently entrains further air by capillary processes as it moves into the saturated zone. Such "excess air" (Heaton and Vogel, 1981; Herzberg and Mazor, 1979) contains atmospheric gases in relative amounts which are characteristic of the atmosphere. The ratio of gas abundances in excess air is therefore different from the ratios in solution in an atmosphere-equilibrated water. The derivation of "recharge temperatures" from dissolved gas analyses requires the subtraction of any excess air present.

Denitrification processes in groundwaters can cause an enhancement of the N_2/Ar ratio of the dissolved gas (Vogel et al., 1981). Larger enhancements of the N_2/Ar ratio may occur where thermodegradation of organic matter has occurred. The Ar content of groundwaters is generally almost totally derived from air-equilibration at recharge with very little leakage of stored radiogenic ^{40}Ar from minerals of the aquifer matrix. For fracture flow in crystalline rocks, however, significant leakage of ^{40}Ar due to mineral alteration processes may occur (Edmunds et al., 1984; Nordstrom et al., 1985).

The presence of excess air, the solution of radiogenic gases, the development of HCO_3^-/CO_3^{2-} equilibria and most significantly the generation of nitrogen and hydrocarbon gases due to biogenic and thermogenic processes can cause the development of supersaturation in a mature groundwater. Groundwaters which circulate to sufficient depths to become heated as a consequence of the geothermal gradient, can also become significantly supersaturated with respect to their atmospheric gases. At depth, the gases are generally held in solution because of the increased hydrostatic pressure. On depressurisation of the groundwater to 1 bar at the wellhead, for example, exsolution of gas usually occurs. This paper examines the process of equilibration between a groundwater and such exsolved gas. The composition of gases dissolved in such re-equilibrated water and in the gas phase are estimated for various possible geochemical processes which cause gas generation in the groundwater. The requirements for estimating the groundwater gas contents before depressurisation from analyses of the exsolved gases and solution, are discussed.

EVOLUTION OF GASES FROM GROUNDWATERS

The geochemically significant quantity which must be determined to establish the origin of gases in groundwaters is the gas content whilst the groundwater is still under hydrostatic pressure in the aquifer and no degassing has occurred. We shall refer to such gas contents as the "in-situ" gas content of the groundwater. The evolution of gases from groundwaters may be the result of geochemical changes in gas compositions during groundwater maturation, temperature changes during circulation and depressurisation during groundwater abstraction. The factors which influence the in-situ gas content of a groundwater and subsequent degassing are:

1. recharge temperature and hence the initial solution of atmospheric gases.
2. the incorporation of excess air during recharge.
3. radiogenic He and Ar accumulation.
4. N_2 production by denitrification and/or thermodegradation of organic matter.
5. CH_4 production by thermodegradation of organic matter.
6. the evolution of HCO_3^-/CO_3^{2-} system and the pH on degassing.
7. the development of increased salinity during groundwater maturation.
8. the temperature and pressure of the groundwater at the abstraction point.

The in-situ dissolved gas composition may be determined directly if sufficient pressure can be applied to the well during abstraction. This is, however, not always possible and it is of interest to be able to predict the applied pressure necessary to retain gases in solution and also to be able to re-construct the in-situ gas compositions in cases where degassing has occurred and analyses are available for the exsolved gases and the gases retained by the degassed solution.

Mass Balance between Gases Exsolved and Retained in Solution

A system can only exsolve gas if the total gas pressure in any gas bubble created equals the applied hydrostatic pressure. For equilibrium between the solution and the gas phase formed, a series of mass balance equations may be written:

$$S_i = S_i' + v_i' \qquad (1)$$

where S_i is the in-situ volume of gas i in solution, S_i' is the volume of gas i remaining in solution after degassing and v_i' is the volume of gas i in the gas phase. All gas volumes are normalised to STP and are relative to 1 litre of groundwater. The equilibrium between the gas phase and dissolved gas is determined by Henry's Law, thus:

$$S_i' = P_i H_i \qquad (2)$$

where P_i is the partial pressure of gas i in the gas phase and H_i is its Henry's Law constant expressed in volumetric terms (Bunsen coefficient) at the equilibration temperature. The volume of gas i in the gas phase is related to the total gas volume, V, evolved at the equilibrium pressure and temperature, T, by:

$$v_i = P_i V \quad (3)$$

from which, on converting to STP, we obtain:

$$v_i' = P_i V.273/T \quad (4)$$

Substituting for S_i' and v_i' in equation 1 from equations 2 and 4 and solving for P_i gives:

$$P_i = S_i/(H_i + 273.V/T) \quad (5)$$

Calculation of the composition of exsolved gases

For stability of the gas phase formed, the total gas pressure must equal the pressure, P, at which degassing is allowed to occur, and:

$$P = \Sigma P_i \quad (6)$$

where the partial pressures of all the gaseous species present are summed. A stable gas bubble cannot exist if $\Sigma P_i < P$. The solution of equation 5 requires that the in-situ volume of gas i and the total exsolved volume of gas, V, is known. Ar is unlikely to be significantly effected by geochemical changes after its initial solution at recharge. Other initial parameters may be selected to allow for excess air addition, radiogenic gas addition and the addition of biogenic or thermogenic gases. The partial pressure of any constituent gas in the gas phase may then be estimated provided an initial estimate of the total gas exsolved can be made. The total gas pressure, ΣP_i, is then estimated and compared with the degassing pressure, P. Any mismatch between these pressures may be used to correct the first estimate of the volume of gas evolved, V, and a new calculation of the partial pressure summation made.

A computer programme, "OUTGAS", has been written to make iteritive calculations in this manner until a satisfactory match has been obtained. The solubility data of Benson & Krause (1976), and Weiss (1970, 1971) were used in these calculations. The first estimate of the volume of gas evolved need only be an approximation. The gases which are likely to have the greater partial pressures in the gas phase are N_2, H_2O, and CH_4 or higher hydrocarbons. If the ratio, R, of hydrocarbons/nitrogen in the exsolved gases is known, then:

$$R = P_{HC}/P_N \quad (7)$$

where P_{HC} and P_N are the partial pressures due to hydrocarbons and nitrogen respectively, in the gas phase. The sum of the gas phase partial pressures is approximately equal to the degassing pressure, thus:

$$P = P_N + P_{H_2O} + P_{HC} \quad (8)$$

$$= P_N + P_{H_2O} + R.P_N \quad (9)$$

from which $$P_N = (P - P_{H_2O})/(1 + R) \quad (10)$$

Mass balance for nitrogen between the in-situ solution, the outgassed solution and gas evolved, requires that:

$$P_N V = S_N - H_N.P_N \quad (11)$$

from which $$V = S_N/P_N - H_N \quad (12)$$

The first approximation for the volume of gas exsolved obtained from equation (12) is used to calculate the partial pressures of each exsolved gas by equation 5. If the partial pressure summation is not equal to the degassing pressure (according to equation 6) within declared limits, the volume of gas evolved is modified by the factor, $\Sigma P_i/P$, and the calculation repeated. This iteration process is continued until ΣP_i equals P within the set limits.

EFFECT OF VARIOUS PARAMETERS ON GROUNDWATER DEGASSING

The "OUTGAS" programme has been used to investigate the influence of various parameters on the exsolution of gases from groundwaters. The compositions of both the exsolved gas phase and the dissolved gases which remain in solution, have been calculated. The in-situ gas contents which were present in the groundwater prior to degassing have been assumed to be due to air-saturation of the groundwater at 5°C in the unsaturated zone with the addition of some excess air and radiogenic ^4He. The additional excess air was equivalent to a 10% excess Ne content and the radiogenic ^4He content was 10^{-5} cm^3 STP/cm^3H_2O. Although this amount of ^4He is several hundred times the air saturation value, it is negligibly small compared with the total dissolved gas volume and has little effect on the degassing process. Oxygen which is dissolved in the unsaturated zone, is generally removed by redox processes in the confined zone of an aquifer and it has therefore been assumed that it will not be present when outgassing takes place.

Groundwater degassing due to temperature and salinity changes

Increases in temperature and salinity during groundwater migration can cause supersaturation at a pressure of 1 bar so that degassing will occur if the groundwater is sampled under atmospheric pressure. Figure 1 shows that if a groundwater with an in-situ N_2/Ar ratio close to that for air-saturated water (38) undergoes degassing, the N_2/Ar ratio in the gas phase is somewhat less than for air (83.5) for a small extent of degassing (temperature < 40°C). As temperature and hence the extent of degassing increases, the N_2/Ar ratio for the exsolved gas decreases and that for the gas remaining in solution must increase. Similar trends are observed for other gas ratios, for example Ne/Ar and Kr/Xe (Fig. 2). Salinity changes can also markedly effect the extent of degassing.

Effect of bicarbonate-carbonate equilibria on groundwater degassing

In solution, equilibria exist between the species CO_2aq; H_2CO_3; HCO_3^- and CO_3^{2-} and these are discussed by Stumm and Morgan (1970). At low pH, CO_2aq is the dominant species in solution whilst at the pH of natural waters, HCO_3^-, or CO_3^{2-} at pH > 10 are the dominant species in solution (Fig. 3). Evolved groundwaters with high HCO_3^-/CO_3^{2-}

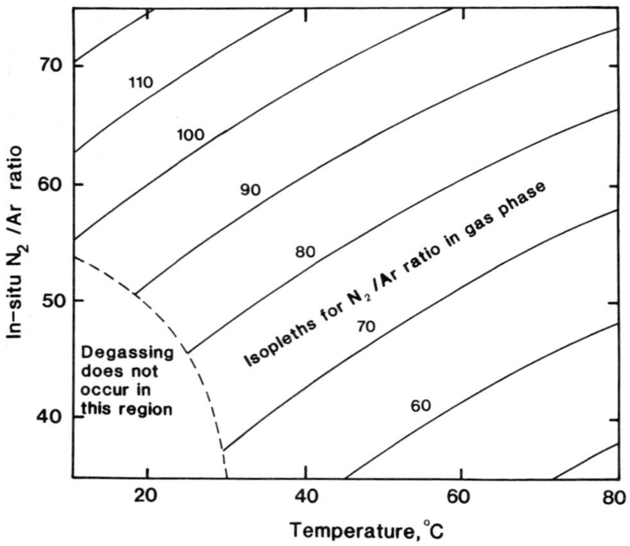

Figure 1. *The N_2/Ar ratios in the gas phase and in solution for groundwater degassing at bar 1 as a function of temperature. The in-situ gas content of the groundwater was as defined for solution E of Table I. Oxygen which was dissolved at recharge is assumed to have been removed by oxidation processes during groundwater migration. The groundwater HCO_3^- content was 100 mg l^{-1}. In-situ N_2/Ar ratios greater than that for solution E may occur due to dentrification processes and due to thermodegradation of organic matter.*

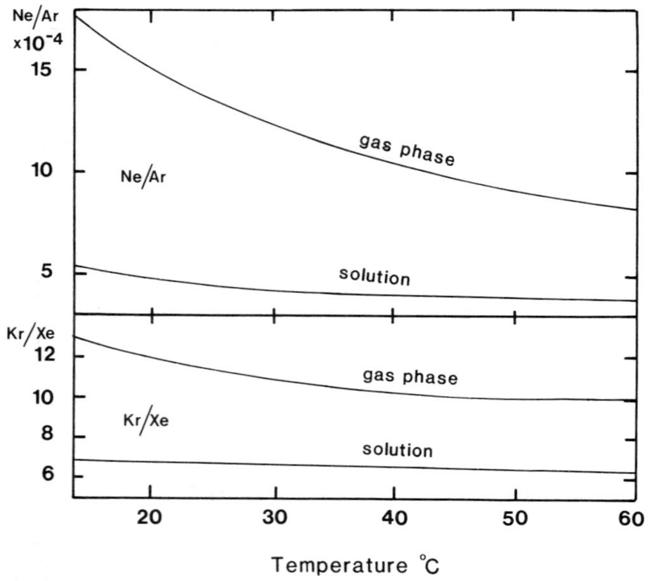

Figure 2. *The Ne/Ar and Kr/Xe ratios in the gas phase and in solution for partial degassing of groundwater at atmospheric pressure as a function of temperature. The in-situ gas content of the groundwater is the same as for figure 1.*

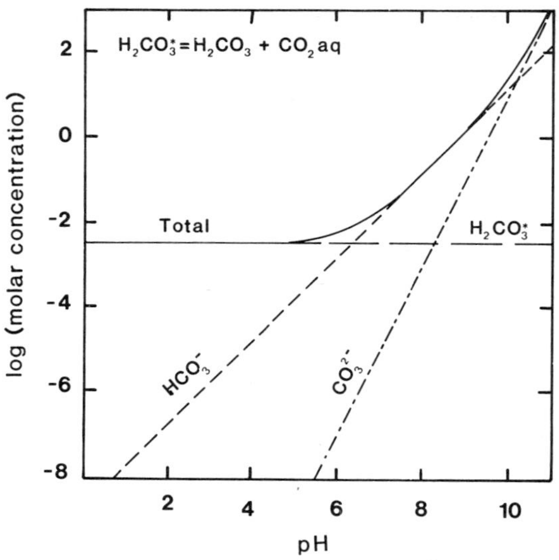

Figure 3. *Dissolved $CO_2/HCO_3^-/CO_3^{2-}$ equilibria as a function of pH. This figure corresponds to a groundwater HCO_3^- content of 300 mg l^{-1} at pH 6.5. (This is equivalent to P_{CO_2} = 0.1 atm. at 25°C.). At pH < 5, the dominant species in solution are H_2CO_3 and CO_2.*

contents may therefore contain significant volumes of dissolved CO_2 which will be released on re-equilibration with air at atmospheric pressure. Figure 4 illustrates the extent to which CO_2 is present in the evolved gas when groundwater degassing occurs. The CO_2 content of the exsolved gas is greatest for groundwaters of low pH and high HCO_3^- content. The fraction of the total dissolved N_2 which is exsolved is also influenced by the pH and HCO_3^- content of the groundwater (Fig. 5). A large fraction of the in-situ N_2 is exsolved when a low pH/high alkalinity groundwater is geothermally heated. Figure 6 shows the influence of temperature, pH and HCO_3^- content on the minimum pressures necessary to retain all dissolved gases in solution. It may be seen that geothermal heating of the groundwater would have a much greater influence on its potential for degassing than any increase in its alkalinity.

Effect of biogenic and thermogenic nitrogen on groundwater degassing

Groundwaters generally do not contain more than a few mg l^{-1} of NO_3^- from natural sources. Anthropogenic NO_3^- inputs are usually less than 50 mg l^{-1} although in a few cases considerably higher values have been reported (Towler, 1982). Such nitrate is converted to nitrogen by denitrifying bacteria, usually soon after the groundwater enters the confined zone of the aquifer. The presence of such denitrification can markedly change the in-situ N_2/Ar ratio for dissolved gases in the groundwater, as the Ar content is not generally changed by geochemical process after recharge. Figure 7 shows the effect of denitrification on the in-situ N_2/Ar ratio. Such changes in aquifer systems have been observed (Towler, 1982; Vogel et al., 1981) and the highest anthropogenic inputs of nitrate can cause in-situ N_2/Ar ratios of up to about 60. Higher ratios must be the

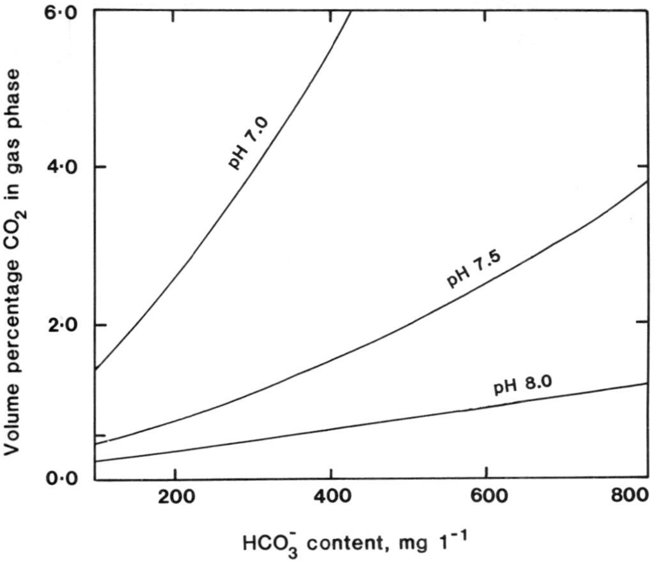

Figure 4. *The CO_2 content of the gas phase for degassing at 1 bar and 40°C of groundwaters with various HCO_3^- contents. The in-situ gas contents of the groundwater are otherwise similar to those for figure 1. This figure illustrates how a reduction in pH can cause enhanced exsolution of CO_2 and other dissolved gases, mainly N_2.*

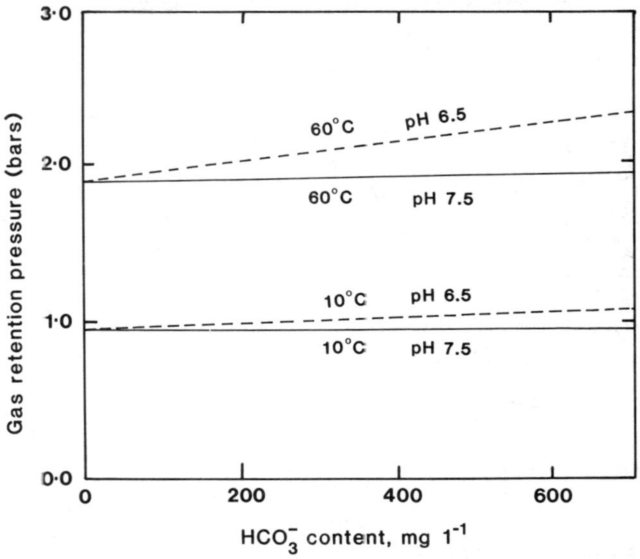

Figure 6. *The hydrostatic pressure necessary to prevent groundwater degassing as a function of HCO_3^- content for in-situ gas contents otherwise similar to those for figure 1. The gas retention pressure is more strongly influenced by temperature than by alkalinity.*

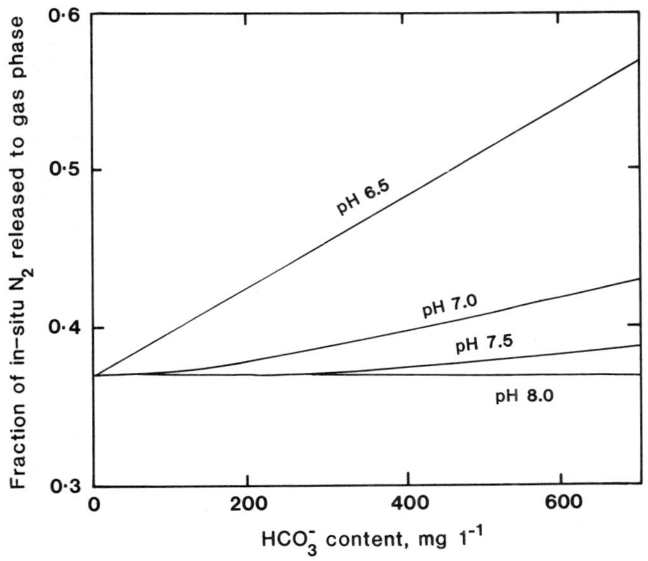

Figure 5. *The fraction of the total dissolved N_2 content of the groundwater, which is released into the gas phase for groundwater degassing at 1 bar and 40°C for in-situ gas contents otherwise similar to those for figure 1. This figure illustrates how a reduction in pH can cause enhanced exsolution of N_2. Degassing at pH > 8 is due entirely to the increased temperature of 40°C.*

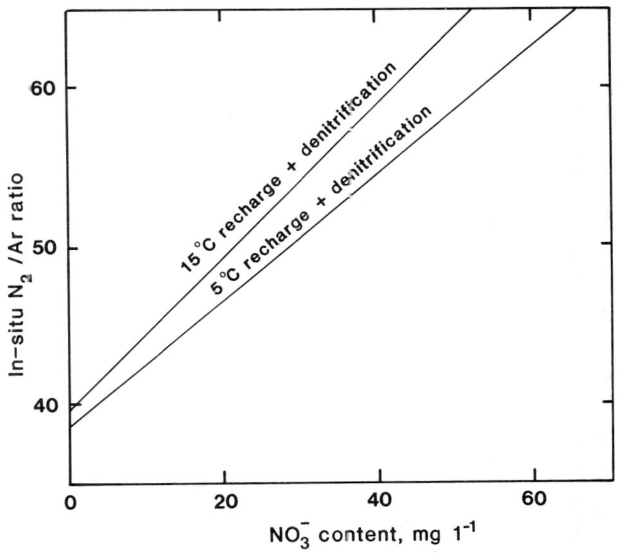

Figure 7. *The effect of denitrification on the in-situ N_2/Ar ratio of dissolved gases in groundwaters. The initial in-situ gas contents are as for figure 1 with additional N_2 due to denitrification.*

result of thermodegradation of organic nitrogen compounds. Figure 8 indicates the minimum pressures needed for retention of in-situ gases as a function of the in-situ N_2/Ar ratio and temperature. The N_2/Ar ratios in the gas exsolved at 1 atmospheric pressure and in the residual water phase are shown in figure 9 as a function of the degassing temperature. Figure 9 also shows the same ratios in the case of a high salinity brine. The N_2/Ar ratios for the brine are much lower than for low-salinity groundwaters. This is because of the low gas solubilities in brines and the larger amounts of degassing that consequently occur.

Influence of thermogenic methane on groundwater degassing

Thermodegradation of organic compounds will produce methane as well as nitrogen. Since nitrogen is a minor component relative to carbon in most organic molecules, the CH_4/N_2 ratio for the thermogenic gas is likely to be large. Figures 10 to 12 illustrate the effects of thermogenic nitrogen and methane on groundwater degassing. The minimum pressures needed for gas retention at 10°C and 60°C are plotted in Figures 10 and 11 against the in-situ CH_4/N_2 and in-situ N_2/Ar ratios. The presence of methane in excess of dissolved nitrogen is likely to cause considerable outgassing of all in-situ gases if a groundwater is depressurised to 1 bar.

Figure 12 shows the CH_4/N_2 ratio in the exsolved gas for various in-situ CH_4/N_2 ratios for a groundwater which contains no thermogenic N_2. Generally, N_2 is exsolved in preference to CH_4 so that the CH_4/N_2 ratios in the exsolved gas are less than the in-situ CH_4/N_2 ratios. The degassing temperature has only a slight influence on this effect.

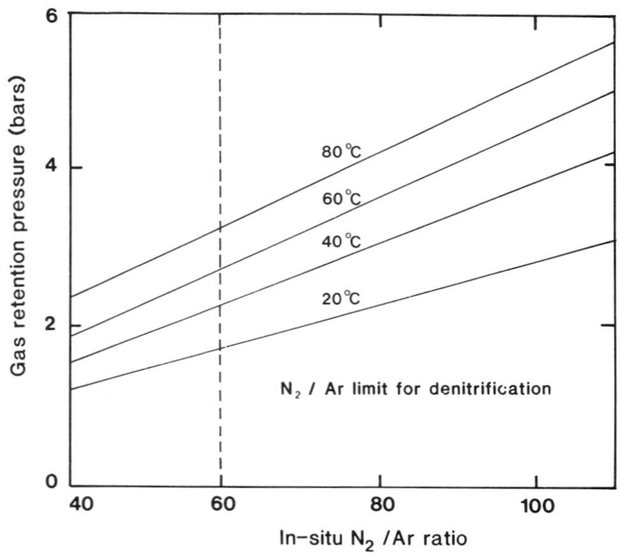

Figure 8. *The hydrostatic pressure necessary to prevent groundwater degassing at various temperatures as a function of the in-situ N_2/Ar ratio for the dissolved gas. The initial in-situ gas contents are as for figure 1 with additional N_2.*

ESTIMATION OF IN-SITU GAS CONTENTS

For a groundwater which has been depressurised to 1 bar, the N_2/Ar and CH_4/N_2 ratios may be determined on samples of the exsolved gases. The in-situ N_2/Ar and CH_4/N_2 ratios may then be estimated using Figure 1 and Figure 12.

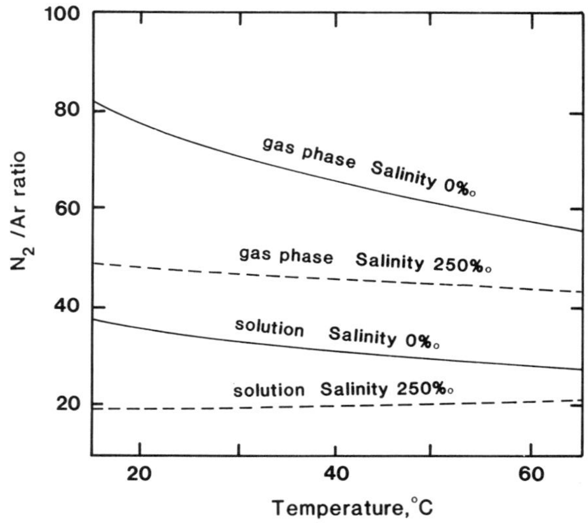

Figure 9. *The N_2/Ar ratios in the exsolved gas and in the degassed groundwater as a function of the groundwater salinity and degassing temperature. The initial in-situ gas contents are as for figure 1 and degassing has occurred because of depressurisation to atmospheric pressure.*

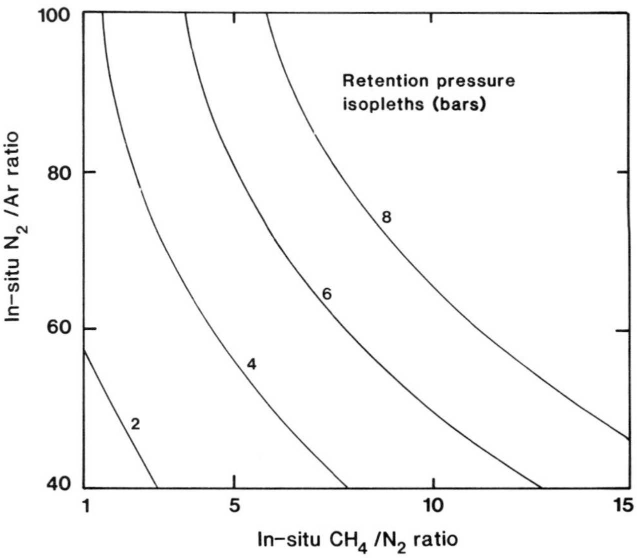

Figure 10. *The hydrostatic pressure necessary to retain gases in solution at 10°C for in-situ N_2/Ar and CH_4/N_2 ratios due to thermogenic production of N_2 and CH_4.*

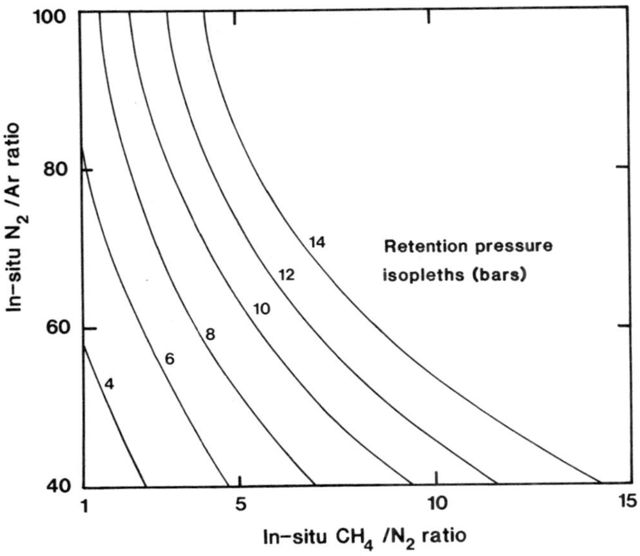

Figure 11. *The hydrostatic pressure necessary to retain gases in solution at 60°C for in-situ N_2/Ar and CH_4/N_2 ratios due to thermogenic production of N_2 and CH_4.*

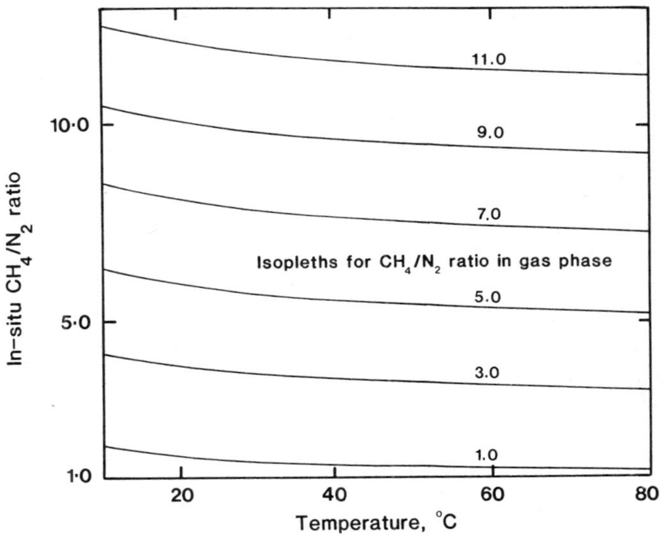

Figure 12. *The CH_4/N_2 ratio in the exsolved gas as a function of the in-situ CH_4/N_2 ratio and temperature for groundwater degassing at atmospheric pressure and for an in-situ N_2/Ar ratio of 38.*

This procedure is sufficiently accurate to determine the extent of denitrification processes or thermogenic N_2 and CH_4 production, provided that these processes do not occur simultaneously.

The estimation of recharge temperatures from dissolved noble gas data requires the determination of the in-situ contents of these gases in the groundwater. This can be most readily achieved by collection of the sample under sufficient pressure to avoid degassing. Figures 11 and 12 show that pressures up to 20 bars may be necessary to achieve this for samples with significant amounts of thermogenic N_2 and CH_4 present, especially if the groundwater temperature is high. Such samples would commence evolution of gas at depths up to 200 m if the borehole is at atmospheric pressure at the surface.

In cases where it is not possible to collect pressurised samples, the in-situ noble gas contents of the groundwater may be reconstructed from analyses of the exsolved gas, the degassed solution and the flow rates of the gas and water phases. It may readily be shown that for depressurisation to atmospheric pressure:

$$V_i = v_i' + v_i = v_i'' f'/f + v_i$$

where

V_i = in-situ contents (cm^3STP/litre H_2O) of gas i in the groundwater.
v_i' = volume (cm^3STP) of gas i released from 1 litre of water into the gas phase.
v_i = volume (cm^3STP) of gas i retained in solution in 1 litre of degassed groundwater.
v_i'' = volume (cm^3STP) of gas i in 1 litre of gas phase at atmospheric pressure.
f′ = flow rate of the gas phase, l s^{-1}.
f = flow rate of the water phase, l s^{-1}

Table 1

Volumes (cm^3STP/cm^3H_2O) of dissolved gases in groundwaters

Solution	He $\times 10^{-8}$	Ne $\times 10^{-7}$	Ar $\times 10^{-4}$	Kr $\times 10^{-8}$	Xe $\times 10^{-8}$	N_2 $\times 10^{-2}$	O_2 $\times 10^{-2}$	N_2/Ar
A	4.86	2.17	4.38	10.73	1.59	1.69	0.90	39.3
B	4.75	2.06	3.89	9.24	1.32	1.47	0.80	37.7
C	4.67	1.98	3.50	8.07	1.12	1.33	0.72	38.0
D	5.24	2.18	3.60	8.19	1.13	1.41	0.74	39.3
E	1005.24	2.18	3.60	8.19	1.13	1.41	0.74	39.3

Key to solutions:

A is air-equilibrated water at 5°C.
B is air-equilibrated water at 10°C.
C is air-equilibrated water at 20°C.
D is the same as solution A with the addition of "excess air" such that the amount of Ne in solution is 10% in excess of that due to air-equilibration.
E is the same as solution D with the addition of $10^{-5}cm^3$ of radiogenic He.

The samples of the gas and water phases may be obtained from a closed system gas/water separator which must be designed to avoid air contamination. The gas and water phase flow rates may also be determined using such equipment. In cases where large amounts of thermogenic methane and nitrogen are present, a large proportion of the dissolved noble gases are released from solution at 1 bar. The measurement of the relative flow rates of the gas and water phases then becomes critical for estimating the in-situ gas contents of the groundwater.

CONCLUSIONS

Most groundwaters which are sampled at atmospheric pressures will spontaneously evolve gases. Supersaturation of the groundwaters occurs because of geothermal heating, evolution of the bicarbonate/carbonate system, formation of N_2 in denitrification processes and biogenic or thermogenic production of N_2 and CH_4.

Groundwaters which evolve gas at atmospheric pressure containing up to 95% CH_4 can be prevented from degassing by pressurisation to <20 bars for temperatures up to 60°C. Noble gas analyses are most readily performed on such pressurised samples. In cases where de-pressurisation has occurred, the in-situ groundwater gas contents can be reconstructed from analyses of the exsolved gas, the degassed water and the relative flow rates of the gas and water phases.

REFERENCES

Andrews, J.N., Goldbrunner, J.E., Darling, G., Hooker, P., Wilson, G.B., Youndman, M.J., Eichinger, L., Raubert, W. and Stichler, W. 1985. A radiochemical, hydrochemical and dissolved gas study of groundwaters in the Molasse basin of Upper Austria: Earth and Planetary Science Letters, v. 73, p. 317-332.

Andrews, J.N., 1985. The isotopic composition of radiogenic helium and its use to study groundwater movement in confined aquifers: Chemical Geology, v. 49, p. 339-351.

Andrews, J.N. and Lee, D.J., 1979. Inert gases in groundwater from the Bunter sandstone of England as indicators of age and palaeoclimatic trends: Journal of Hydrology, v. 41, p. 233-252.

Benson, B.B. and Krause, D., 1976. Empirical laws for dilute aqueous solutions of non-polar gases: Journal of Chemical Physics, v. 64, p. 639-709.

Edmunds, W.M., Andrews, J.N., Burgess, W.G., Kay, R.L.F. and Lee, D.J., 1984. The evolution of saline and thermal groundwaters in the Carnmenellis granite: Mineralogical Magazine, v. 48, p. 407-24.

Heaton, T.H.E. and Vogel, J.C., 1981. "Excess air" in groundwater: Journal of Hydrology, v. 50, p. 201-216.

Hertzberg, O. and Mazor, E., 1979. Hydrological applications of noble gases and temperature measurements in underground water systems: examples from Israel: Journal of Hydrology, v. 41, p. 217-231.

Marine, I.W., 1979. The use of naturally occurring helium to estimate groundwater velocities for studies of geologic storage of radioactive waste: Water Resources Research, v. 15, p. 1130-1136.

Mazor, E., 1972. Palaeotemperatures and other hydrological parameters deduced from noble gases dissolved in groundwaters, Jordan Rift Valley, Israel: Geochimica et Cosmochimica Acta, v. 36, p. 1321-1336.

Nordstrom, D.K., Andrews, J.N., Fontes, J.C., Fritz, P., Moser, H. and Olsson, T., 1985. Hydrogeological and Hydrogeochemical investigations in boreholes – geochemical and isotope characterization of the Stripa groundwaters: SKB Stripa Project Report 85-06, Stockholm, 250 p.

Rudolph, J., Rath, H.K. and Sonntag, C., 1984. Noble gases and stable istopes in ^{14}C-dated palaeowaters from central Europe and the Sahara: In: Isotope Hydrology 1983, Vienna, IAEA, p. 467-477.

Stumm, W. and Morgan, J.J., 1981. Aquatic Chemistry: an introduction emphasising equilibria in natural waters: New York, Wiley, 2nd edition, 780 p.

Towler, P.A., 1982. The geochemistry of nitrogen species in groundwaters: Unpublished PhD thesis, University of Bath, Bath, U.K.

Vogel, J.C., Talma, A.S. and Heaton, T.H.E., 1981. Gaseous nitrogen as evidence for denitrification in groundwater: Journal of Hydrology, v. 50, p. 191-200.

Weiss, R.F., 1971. Solubility of helium and neon in water and seawater: Journal of Chemical Engineering Data, v. 16, p. 235-241.

Weiss, R.F., 1970. The solubility of nitrogen, oxygen and argon in water and seawater: Deep Sea Research, v. 17, p. 721-735.

Index

Notes

Notes

Notes

Notes

Notes

Notes

Notes